Evolutionary Ecology

Som B. Ale
University of Illinois Chicago (Illinois)

Tania L. S. Vincent
Anchorage (Alaska)

Joel S. Brown
Moffitt Cancer Center (Florida)

CRC Press
Taylor & Francis Group
Boca Raton London New York

CRC Press is an imprint of the
Taylor & Francis Group, an **informa** business
A SCIENCE PUBLISHERS BOOK

Cover credit: Image reproduced by kind courtesy of Tania L. S. Vincent.

First edition published 2026
by CRC Press
2385 NW Executive Center Drive, Suite 320, Boca Raton FL 33431

and by CRC Press
4 Park Square, Milton Park, Abingdon, Oxon, OX14 4RN

CRC Press is an imprint of Taylor & Francis Group, LLC

Library of Congress Cataloging-in-Publication Data (applied for)

ISBN: 978-0-367-43763-3 (hbk)
ISBN: 978-1-041-29444-3 (pbk)
ISBN: 978-1-003-00853-8 (ebk)

DOI: 10.1201/9781003008538

Typeset in Times New Roman
by Prime Publishing Services

Foreword

Most educated folk, when asked what comes to mind upon hearing the name Charles Darwin, likely first think of "evolution" or maybe "evolution by natural selection." Yet Darwin was very much an ecologist, using the phrase "the economy of nature" multiple times in *On the Origin of Species* (the word "ecology" had not yet been coined). He can be viewed as one of the founding fathers of ecology (Harper 1967) and indeed as an early evolutionary ecologist (Egerton 2011). Darwin puzzled over issues akin to those worked on by evolutionary ecologists today, such as why some plant seeds are brightly colored, and consumed by birds, yet non-nutritious. It turns out that these seeds are mimics of other, nutritious ones (see https://www.darwinproject.ac.uk/commentary/life-sciences/beauty-and-seed for a discussion of this fascinating side eddy in Darwin's life). He was also an accomplished geologist (LaPorte 1996), and developed his theory of evolution in the broader context of what we now call Earth science.

This splendid introductory textbook by Ale, Vincent and Brown carries forward the Darwinian paradigm of the interplay of ecology and evolution, framed within the grand forces shaping Earth, such as plate tectonics and climate drivers. The authors take great care to lay out the historical origins of the ideas they present, and they draw on a wide range of compelling natural history examples to illustrate core concepts. Charming illustrations by one of the authors, Tania Vincent, effectively illuminate many of these examples and theoretical concepts.

Like most ecology textbooks, the authors cover bread-and-butter issues such as population dynamics, interspecific dynamics, food webs, succession, and nutrient cycling. But they also crisply describe important issues in the realms of foraging ecology and life histories, themes which sometimes get neglected in ecological training these days. Almost unique among ecology texts, they introduce evolutionary game theory as a key conceptual 'hook' for catching numerous insights about the interplay of ecology and evolution, illustrated deftly with examples from plant ecology (e.g., what determines how high trees grow?). The text is deliberately light on mathematical models, translating equations into more readily interpretable and intuitive graphical formulations. The book ends with a set of chapters on topics of great practical importance: the maintenance of biodiversity, the impacts of climate change, how evolutionary ecology informs our understanding of human origins and traits, and how it offers a unique (and still developing) perspective on the scourge of cancer.

The book is firmly grounded in the reticulate details of the natural history of organisms, and in how the Earth operates as a physical machine setting the context for both ecology and evolution to play out. I warmly recommend this book not just to students, but also to professionals wanting a fresh take on familiar themes.

References

Egerton, F. N. 2011. History of ecological sciences, part 40: Darwin's evolutionary ecology. Bulletin of the Ecological Society of America 92: 351–374.

Harper, J. L. 1967. A Darwinian approach to plant ecology. Journal of Ecology 55: 247–270.

Laporte, L. F. 1996. Rock stars: Darwin the geologist. GSA Today 6: 8–10

Dr. Robert D. Holt
University of Florida, USA

Preface

The intersection of ecology and evolution—what Gordon Orians (1962) termed "evolutionary ecology"—has been closely intertwined since Darwin (1859), who was as much a pioneer in ecology as in evolution. From the 1920s to the 1950s, the scientific focus shifted decisively toward populations as fundamental biological units. This period saw two parallel developments: population dynamics, advanced by Lotka, Volterra, Gause, and Elton, who revealed how species interact and change over time; and population genetics, elucidated by Fisher, Wright, Haldane, Ford, and Dobzhansky—along with their successors—who demonstrated how variation and natural selection shape populations.

During this formative era, ecology emerged as the discipline that illustrates how environmental factors drive evolutionary adaptation, what Hutchinson (1965) aptly called "the ecological theater" for "the evolutionary play." It was also the period when individual selection (rather than group selection) became central to evolutionary thinking, influencing models of life-history evolution, ecologically relevant behaviors, and other adaptive traits (Williams 1964). Inspired by Hutchinson, Robert MacArthur and his collaborators—including Richard Levins, E. O. Wilson, Eric Pianka, Michael Rosenzweig, Martin Cody, Robert Ricklefs, Robert May, Jared Diamond, Jonathan Roughgarden, and Robert D. Holt—investigated species interactions (e.g., competition and predation) and their consequences for ecological community structure. Their work helped establish a theoretical framework grounded in approximate equilibrium, applicable to both ecological dynamics and evolutionary adaptation.

This book builds on the contributions of these intellectual pioneers. Its academic lineage traces back to the naturalist Joseph Grinnell, who emphasized natural history—the "mud-and-boots" fieldwork of biology—and draws inspiration from Hutchinson's theatrical metaphor. As Theodosius Dobzhansky (1973) famously stated, "Nothing in biology makes sense except in the light of evolution." We extend this axiom: nothing in evolution makes sense except in the light of ecology. Rather than presenting a mere catalog of facts, we offer readers conceptual frameworks for interpreting ecological patterns through the lens of evolutionary processes. Our approach integrates ecological and evolutionary principles into a cohesive narrative, revealing the interconnectedness of patterns and processes in nature. Throughout, we explore ecological relationships by examining the fit between form and function—how natural selection shapes adaptations to environmental challenges. While adaptations are context-specific, the principles governing foraging behavior, population dynamics, and evolutionary processes are universal. Often, higher levels of ecological organization emerge from interactions among these dynamics, revealing nature's underlying patterns, as Darwin famously called them "the entangled bank."

This book originated around the turn of the millennium in the undergraduate Ecology Lab course at the University of Illinois Chicago (UIC). It synthesizes lecture materials developed over 15 years by Dr. Joel S. Brown and later expanded by his former student, Dr. Som B. Ale. Dr. Brown (now at the Moffitt Cancer Center in Florida) studied under evolutionary game theorist Thomas L. Vincent and evolutionary ecologist Michael L. Rosenzweig, whose academic roots trace back to Robert MacArthur—a towering figure in modern ecology—and ultimately to the legendary G. Evelyn Hutchinson. Dr. Tania L. S. Vincent's perspective on evolutionary ecology was shaped by mentors including T. L. Vincent, J. S. Brown, M. L. Rosenzweig, J. R. McAuliffe, P. A. Abrams,

D. Tilman, and (for scientific illustrations) D. B. Sayner. Like a lineage in the tree of life, we hope to pass these inherited insights on to you.

We gratefully acknowledge Dr. Sandro Lovari and Dr. Miquel Gonzalez-Meler for their valuable edits on earlier drafts regarding climate change, Dr. Mary Ashley for her suggestions on biodiversity, and Dr. Emily Minor for her ideas on population dynamics and food webs. Stafford Garbutt improved the English of the preliminary draft, and Dr. Robert D. Holt helped us with the final spit and polish.

Last but not least, we extend our deepest gratitude to Dr. Chris Whelan for his thorough review of the entire manuscript and meticulous fact-checking.

References

Darwin, C. 1859. On the Origin of Species by Means of Natural Selection, or Preservation of Favoured races in the Struggle for Life. John Murray, London, UK.

Dobzhansky, T. 1973. Nothing in biology makes sense except in the light of evolution. American Biology Teacher 35: 125–129.

Hutchinson, G. E. 1965. The Ecological Theater and the Evolutionary Play. Yale University Press, New Haven, CT.

Orians, G. H. 1962. Natural selection and ecological theory. American Naturalist 96: 257–263.

Williams, G. C. 1964. Adaptation and Natural Selection. Princeton University Press. Princeton, New Jersey.

Drs. S. B. Ale, T. L. S. Vincent, and J. S. Brown

Contents

Chapter 1

Relationships Between Organisms and Their Environment

Nature is not an art gallery, and species are not art!

From space, Earth, our marvelous blue planet, appears serene and still. But as you plunge into the biosphere, you experience rushing air, pouring water, and the extraordinary hustle and bustle of life; ten million or so different species of living organisms. All have unique genetic codes and diverse life cycles. Why are there so many kinds of life? How does each organism make a living and succeed amid all that fantastic complexity? As a species, we *Homo sapiens* are undeniably not alone on this planet. Even as individuals, we are never truly alone, even when we feel so. Life murmurs its presence wherever you go: a country lane, the beach, a city parking lot, and even within your own body. Ecologists are those who want to understand the murmur and unravel the awesome mystery of this seemingly complex nature using a science called "ecology."

That leads to the question: What do we mean by "ecology?"

Ecology is the study of "how many," "where," "who does what to whom," and ultimately "why" there is life where it is, doing what it does. From time immemorial, humans have been curious about their place and role in the grand scheme of nature. Some of our curiosity is driven by necessity, as in the case of subsistence hunters, or by economic practicality, as in the case of farmers seeking to enhance nature's bounty while staving off pathogens and pests. Some of this curiosity has been driven by the benefits we gain from healthy ecosystems, which are serviced by diverse suites of species. Some are aesthetic and entail finding joy and refreshment from nature and its charismatic species. Others are spiritual, providing a way to find meaning, solace, connection, and sanctity among the other denizens of our planet. To satisfy this curiosity, we must understand the relationships between organisms and their environment. These relationships include interactions with the physical and chemical worlds, as well as with members of the same and other species; they form the crux of all ecological questions.

This book is titled *Evolutionary Ecology*, and throughout, we will show how **evolution** (the change in species traits over time) is intimately tied to ecology. The Greeks used the word *oikos* to describe a home, a place where one lives and is familiar with the local surroundings. From this word, over a century and a half ago, German zoologist Ernst Haeckel (1870) coined the term "ecology." While defining it succinctly, he also linked it with the evolution of adaptive traits as follows: "Ecology is the study of all those complex interactions referred to by Darwin[1] as the conditions of the struggle for existence" (Watts 2019). In other words, to understand an organism and its evolutionary adaptations, we must study the complex interplay between the organism and its "home."

[1] Charles Darwin (1809–1882) is considered the father of evolutionary biology and is best known for writing the seminal work *On the Origin of Species by Means of Natural Selection, or the Preservation of Favoured Races in the Struggle for Life*.

DOI: 10.1201/9781003008538-1

Although Haeckel coined the term "ecology" in 1866, ecology as a science broke into the forefront with Danish botanist Eugen Warming's seminal 1895 work on the sand dune plants of Denmark (Goodland 1975), which addressed the following question: "Why do plant species unite into certain communities and why have they possessed certain characteristics?" From his work, Warming recognized that growth forms of plants are adaptations to various ecological factors such as water, light, temperature, and substrate.

The "home" where an organism lives, something we call its environment, may include not just the sun, water, and atmospheric chemicals such as oxygen and carbon dioxide, but also other organisms, which may help or hinder the survival and even the reproduction of the organism of interest. The external physical and chemical parts of an organism's environment are the **abiotic**, or nonliving, components, whereas all organisms, tiny or gargantuan, form the **biotic**, or living, components of the environment. Both biotic and abiotic environmental components influence the survival, growth, development, and reproduction of each organism.

Many of the traits that distinguish organisms from one another are related to their morphologies, physiologies, and behaviors, which pertain to survival and reproduction. These traits allow an animal to find and capture food, a plant to obtain carbon, nitrogen, and other essential soil nutrients, and microbes to uptake a wide range of molecules, organic material, and even other microbes. In their quest for food, organisms must also have means to protect themselves from becoming food for other organisms; if successful, they will pass on these traits to their offspring. This is the vaunted "struggle for existence" in the sense that Charles Darwin meant in *On the Origin of Species* (1859).

It may seem like a trivial point, but no organism exists separate from its environment. Resonating with William Shakespeare's "All the world's a stage," the British ecologist G. Evelyn Hutchinson (1965) referred to the place—the physical location in time and space—in which each organism carries out the "struggle for existence" as the "ecological theater." Important interactions—who eats whom, how individuals and species compete, and who breeds with whom—all influence how an individual's traits get passed along to successive generations in what Hutchinson calls "the evolutionary play." We cannot understand an organism without understanding how it is integrated with its environment. This includes how populations of organisms both modify and are modified by their environment, how individuals respond behaviorally or physiologically to their circumstances, how individuals of the same or different species interact, and ultimately how these myriad interactions and environmental factors influence how natural selection and other evolutionary forces shape the adaptations and characteristics of individuals and populations. Together, organisms and their environments form a thin layer on, above, and below the Earth's surface, which we collectively refer to as the **biosphere**. This biosphere currently makes Earth different from all other planets in the solar system.

Levels of Ecology

All organisms live within Earth's biosphere, which is that portion of the Earth upon which life depends and exists. The biosphere is approximately three billion years old and comprises ecosystems that include land, water, and air. Although the biosphere extends from about 11 kilometers beneath sea level in the deepest oceanic trenches to the top of the troposphere, about 9 kilometers high, most of Earth's organisms exist only about half a kilometer beneath the ocean surface to about six kilometers above sea level. This integrated system of living things seems unfathomably complex, yet it reveals an order that can be defined and understood. To reveal this order, ecological systems can be studied hierarchically, from individuals and populations to communities, ecosystems, landscapes, biomes, and finally, the entire biosphere.

At the level of individuals, ecologists study **morphology** (the size and shape of the body), **physiology** (the functions and parts of the body), and **behavior** to answer the question, "How and why does an organism live in some environments but not in others?" The general answer is that certain traits make an organism well-suited to its local environment; we call these traits **adaptations**, and there often appears to be a fit between the form and function of adaptations to the environment. However, traits that are considered adaptive in one kind of environment may be maladaptive in another. The cryptic coloration of many animals is a good example of this. Concealment via camouflage is a great adaptation, but camouflage is highly dependent on context. For example, moths that are perfectly disguised as lichen-covered oak bark survive as long as lichens are plentiful, but if the lichens decline, the moths will be picked off by ever-vigilant birds in the new "plain bark" environment (Walton and Stevens 2018). Thus, adaptive traits may often be a double-edged sword, enhancing survival in one locale or habitat, but restricting the organism to that very environment.

At the level of populations, ecologists examine the factors that influence birth and death rates, as well as immigration and emigration rates, to answer the question: "How and why do populations change in size and location over time?" Changes in the number of individuals in a local population reflect the balance of births, deaths, immigration, and emigration—collectively known as **demographic rates**. Interactions with other individuals (such as competition or predation) and environmental conditions can separately or together influence these demographic rates. For example, the survival rate of juvenile Chinook salmon (*Oncorhynchus tshawytscha*) is strongly related to ocean conditions that influence the availability of high-quality prey (Hertz et al. 2016).

Ecologists studying life at the level of communities aim to answer the question, "Why does a certain number of species live in a particular place, and why are some species more abundant than others?" The focus is on interactions between species that can either promote or limit their coexistence. For example, one might ask how the presence of semi-feral horses, which graze on grasses, affects the abundance of other species, such as bison, which also consume grass (Jung et al. 2015). All organisms within an ecological community interact directly and indirectly, thereby impacting population distributions and abundances in time and space. These interactions also form the basis for many ecosystem properties and processes, such as nutrient cycling and food webs.

Ecologists studying the environment of an ecological community or "ecosystem" (Tansley 1935) focus on the storage and transfer of energy and matter between the biotic and abiotic components. This includes the various chemical elements essential to life, such as oxygen, carbon, nitrogen, and phosphorus. Ecosystem ecologists ask the question, "How do the activities of organisms influence the movements of energy and matter through the physical and chemical environment?" For example, the beavers (*Castor canadensis*) of Voyageurs National Park, Minnesota, USA, build dams that create wetlands and selectively harvest woody plants, transforming their physical habitat into a mosaic of different vegetation and forest types. Beaver ponds and their associated meadows have been shown to trap nutrients and soil from their surrounding landscape, thereby improving soil quality through increased soil nitrogen, phosphorus, and potassium (Naiman et al. 1994). These changes enhanced plant productivity, resulting in more diverse plant communities and greatly increasing the overall number of different species, including plants, animals, and even microbes.

Clearly, we cannot understand the nature of life from a snapshot. Nature is not an art gallery, and species are not art! Nature is a dynamic, interactive system that changes both temporally and spatially at all ecological levels. In this ever-changing world of nature, many relationships between plants, animals, and their environment can appear baffling to us in their complexity. In most cases, the relationships are so intricate that interfering with the natural environment results in unexpected outcomes. This is because ecological relationships create a system that is more than the sum of its component parts.

The Idea of Fixity to the Idea of Evolution

For millennia, scholars of Western philosophy strove to identify, describe, and understand the diversity of life by scrutinizing the behavior and anatomical forms of organisms in search of relationships among them (Egerton 2012). Greek philosophers such as Plato (427–347 BCE), a student of Socrates, and Plato's pupil Aristotle (384–322 BCE) believed that the links—whether among animate or inanimate objects—revealed the pattern of divine creation (Armstrong 1993).

Plato, deeply influenced by the study of geometry—where a triangle is always a triangle—promoted beliefs in essentialism (Mayr 1982). Essentialism is the view that every entity has a set of attributes that are necessary to its identity and function. His theory of form posited that everything, whether living or nonliving, has a form which is its essence. He believed that each natural type of object in the observed world is an imperfect manifestation of an abstract "ideal" form, which exists completely outside space and time and therefore can only be perceived through the mind, not through sensual experience. Plato argued that because they are changeless, ideal forms possess a higher degree of reality than their earthly manifestations.

Aristotle, often considered the first biologist, apparently spent his youth studying marine organisms. He later wrote books on the diversity of life he observed, explicitly contradicting his mentor, Plato, by stating that observations from one's senses have primacy over reason (Mayr 1982). Aristotle also rejected Plato's theory of ideal forms—arguing that forms did not exist independently of things—while still believing that forms remained unchanging and were essential to identity and function. He believed that the features of living organisms suited their function. Aristotle also saw that organisms vary in complexity and can be arranged in order of increasing complexity as a *scala naturae* ("ladder of nature"; Singer 1931). The *scala naturae*, described in *Historia Animalium*, one of Aristotle's four books on biology, classified organisms in relation to a hierarchical but static scale that portrayed inanimate objects as the simplest beings, which were relegated to the lowest rung. Progressively higher rungs were occupied by plants, simple animals, and then more complex animals, with humans placed on the rung above these.

Despite the fading of ancient Greek philosophy during the downfall of the Roman Empire and the subsequent rise of Christianity in Europe, starting in the late 4th and early 5th centuries (CE), the idea of species adhering to an essential type still remained. The Bible's first chapter of Genesis states that "each living thing was created after his kind," which, taken literally, meant that species were immutable and had the same identity from the very beginning. Reflecting this view of the time, the philosopher Augustine (426) asserted that species are maintained by the direct action of God (Wilkins 2009).

A thousand years later, in the 13th century, with Christianity firmly entrenched, St. Tomas Aquinas popularized "natural theology," which is the study of God through His natural works. Although forms of natural theology can be traced back to antiquity (including Plato and Aristotle), it was used by Aquinas and others to understand the nature of the Biblical God and was used to prove the existence of divine reason (Davies and Stump 2011). Adherents to this philosophy believed that the perfections (adaptations) observed in all beings must necessarily be caused by God.

The idea of species fixity reigned strongly until the Middle Ages (Mayr 1982). However, the world was turned upside down by the Black Death, also known as the bubonic plague, a multi-continent pandemic that first raged through Asia and then Europe from 1347 to 1353. An estimated 25 million people succumbed in Europe, a third of its population. People were aghast, and their mental outlook changed dramatically. They prayed, but it did not ameliorate their suffering. "God is deaf nowadays and will not hear us…" wrote a 14th-century English cleric (Cantor 2001). In the wake of this catastrophe, some began to question the long-held belief that the world was fixed and divinely ordained.

In many ways, recovery from the bubonic plague and the Turkish conquest of Byzantium, causing Greek scholars to flee to Italy, fueled the Renaissance, a cultural and intellectual movement blossoming in the 15th and 16th centuries, which sparked a heightened interest in Greco-Roman learning and culture, much of which had gone into eclipse in Western Europe (Egerton 2003). The Renaissance, which flowered first in Italy and then through much of Europe, gave the world scholars such as the philosopher Niccolò Leoniceno and artists who studied nature like Leonardo da Vinci and Albrecht Dürer (Egerton 2003). While it inspired the fields of medicine and anatomy, natural theologians such as Charles de Bouelles (1512) still embraced Plato's fixed forms and *Aristotle's scala naturae*, interpreting them as reflections of God's perfect design, although this latter idea had evolved into the "Ladder of Life," also known as the "Great Chain of Being," which now included spiritual beings, with God positioned at the uppermost rung of the ladder (Ragan 2009).

The Renaissance led to the European Enlightenment in the 17th and 18th centuries, centered in France and also referred to as the Age of Reason. The Enlightenment gave birth to intellectual giants such as Issac Newton, Francis Bacon, René Descartes, and Voltaire. They, unlike Renaissance philosophers, no longer sought validation in the texts of the Greco-Roman philosophers but were grounded in rationalism and empiricism, a concept derived from the Arabs (Mayr 1982). However, if there was a common belief among Enlightenment scholars about the divine, it was deism—belief in the existence of a supreme deity who does not interact with humankind. The influence of the church was so powerful that scholars, even if skeptical, could not openly question the idea of the fixity of species, which, of course, was enshrined in doctrine. The "Great Chain of Being" was a popular metaphor that supported the orderliness of creation fixed by divine design. Extinction—the sudden disappearance of a species—was absurd in such a divine scheme, because it meant that the chain would lose a link.

One of the most famous naturalists during the Age of Reason, Carolus Linnaeus (1707–1778), the architect of **binomial nomenclature** (a two-part system for naming species and the classification scheme that we use today), insisted firmly on the immutability of species. But species were not whimsical creations. Reflecting the rationalism of his time, Linnaeus was struck by the apparent kinships among living things; he compared the anatomies of species and sought clues to the relationships among them (Linnaeus 1758). He used his perceptions of similarity to place organisms into their respective taxonomic groupings. While he saw, as did others of his time, common design rules, these rules belonged to a Creator, not to evolution.

Meanwhile, more evidence-based perspectives on nature emerged in the British Isles during the "Midlands Enlightenment" in the second half of the 18th century, forming a pivotal link between the earlier Enlightenment and the later Industrial Revolution, facilitating the exchange of ideas between experimental science, polite culture, and practical technology (Jones 2017). The Renaissance emphasized the centrality of humans, whereas the Midland's Enlightenment sought general principles and laws of nature independent of humans. Erasmus Darwin, Charles Darwin's grandfather, was part of this intellectual group.

This period of Enlightenment also saw the rise of the new science of geology. Ironically, the work of one of the first paleontologists, Georges Cuvier (1769–1832), who upheld religious convictions regarding the fixity of species and saw fossils as evidence of the Biblical deluge, inspired contemporary geologists to resolve the fossil conundrum by invoking natural causation (Wool 2001). Fossils presented a paradox to natural philosophers—different rock layers reveal different groups of fossilized organisms, many of them now extinct—how could species created by God become extinct and why were species on the bottom layers more different from modern species than those on the top? The conflict between ideology and observation widened.

During this formative period in the science of geology, James Hutton (1726–1797) proposed that geological land formations could be explained by processes that are currently in operation, such as erosion and sedimentation. This idea, known as uniformitarianism, was in sharp contrast to the Biblical catastrophism of the day. Fellow geologist Charles Lyell (1797–1875) later popularized Hutton's ideas in his book *Principles of Geology* (1833). Uniformitarianism states that the natural processes observed today, such as erosion, rain, and wind, are the same processes that have occurred throughout millions of years. These gradual processes, which have sculpted the landscape for eons and continue to operate at approximately the same rate today, explain the forms, types, and layering of rocks. Hutton and Lyell demonstrated that the landforms seen on modern Earth were subject to slow but continuous cycles of rock formation and erosion, requiring vast amounts of time that contradicts the notion of a 6,000-year-old-Earth shaped by sudden catastrophes (Wool 2001). Uniformitarianism also explains how erosion produces dirt and rock debris that are washed into rivers, transported to oceans, and then deposited in thick layers, which are converted over time into sedimentary rock. These layers of sedimentary rock (Figure 1), often containing fossils, are then lifted from below sea level to form land during geological upheavals. Keen-eyed observers could

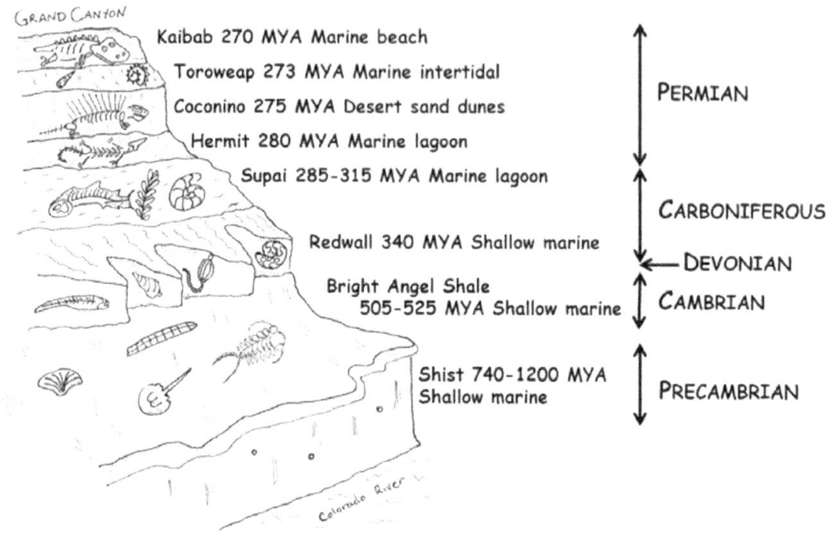

Figure 1. Sedimentary rock layers such as those in the Grand Canyon often contain fossils of organisms that lived and died when the layer was formed. Layers near the bottom are older than those on top. (Imaginary sketch based on Timmons [2009]).

not help but notice the change in fossil organisms among the layers of rock, observing a progression from simpler to more complex organisms preserved from deeper to shallower layers.

In the late 18th century, Jean-Baptiste de Lamarck (1744–1829) was the first to suggest a viable theory of species change. Lamarck's evolutionary driver was the inheritance of acquired characteristics: anatomical features employed by an organism in response to its environment would become exaggerated by use and then "inherited" by its offspring. For instance, the son of a blacksmith would inherit the strong muscles his father built up working at the forge. Although Lamarck used many examples in his writings, *Philosophie Zoologique* (Lamarck 1809), the development of the giraffe's long neck is most frequently cited. The long necks of giraffes, according to Lamarck, developed over time because their necks grew as they stretched to reach food in tall trees (Figure 2), and this longer neck was then passed on to their offspring. Lamarck's notion of the inheritance of acquired characteristics quickly ran into difficulties. In the 1880s, the German biologist August Weismann (1834–1914) disproved Lamarck's theory of the inheritance of acquired characteristics. In a series of experiments, he removed the tails of mice for many generations and reported that the final generation grew tails no shorter than those of the first (Weismann 1888), thereby discrediting the concept of inheritance of acquired traits.

Furthermore, without a mechanism of inheritance, explanations like Lamarck's became mere "just-so stories" about adaptation, although he was not the only evolutionary biologist to run up against this problem. Yet Lamarck's broader thinking about adaptation was correct, and he did correctly ascertain that biological characteristics are inherited (Mayr 1982). Modern geneticists acknowledge the possibility that the action of the environment may bring about certain transmissible changes (e.g., via DNA methylation; see Chapter 13 *Evolutionary Ecology*), but the Lamarckian concept of inheritance was ultimately disproven with the discovery of genes and the laws of heredity.

These intellectual upheavals regarding the fit of form and function, and the possible fluidity of species, did not reach a watershed until 1858, when the theory of evolution by natural selection was jointly proposed by Charles Darwin (1809–1882) and Alfred Russel Wallace (1823–1913) to the Linnean Society of London. However, reaching this point was a long and tortuous process, especially for Darwin.

When Charles Darwin was young, he was assigned as a naturalist to the *HMS Beagle*, setting off in 1831 on a five-year voyage around the world to study continental coastlines and observe the flora and fauna along the way. What Darwin observed and collected on this voyage inspired his thinking

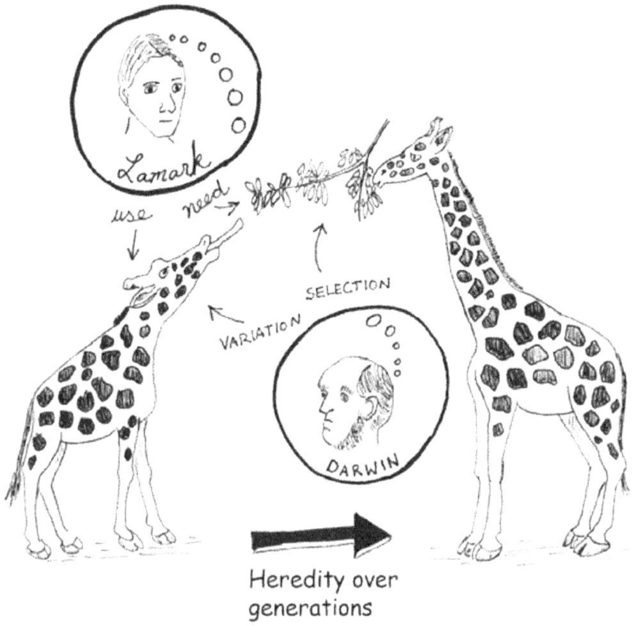

Figure 2. Lamarckian versus Darwinian explanations of the evolution of the giraffe's long neck.

about nature, species, and their origins. He was particularly intrigued, if not utterly astonished, by what he observed on the Galapagos Islands in the Pacific Ocean about 650 miles west of Ecuador. What most attracted Darwin's attention was the variety of finches on the islands. The birds were divided into at least 13 varieties, distinguished from one another mainly by differences in the sizes and shapes of their bills. These varieties, which turned out to be entirely different species, did not exist anywhere else in the world; however, they resembled a single species that was apparently a close relative on the South American mainland. Darwin thought that all of them might have descended from the same mainland species of finch, and then differentiated into separate species while isolated on the islands.

Darwin's ideas about evolution by natural selection were strongly influenced by Thomas Malthus's (1766–1834) conjecture concerning human population growth, namely that human populations seem to expand exponentially, whereas their food supply grows arithmetically, or much more slowly (Malthus 1798). This imbalance creates constant shortages, leading to competition for resources. Darwin extended the Malthusian Principle beyond the boundaries of the human population and political economy to all of nature, noting how organisms can produce far more offspring than can possibly survive. In doing so, Darwin proposed that factors in the environment set the stage for competitive survival by selecting for traits that enable plants and animals to survive long enough to reproduce. The notion of a struggle for existence was easy to appreciate; even those on a small islet off Northern Ireland were known to say of their sheep, "Ten to fatten, eleven to maintain, and twelve to starve."

The time was clearly ripe for the notion that evolution can occur via survival of the fittest. Like Darwin, Alfred Russel Wallace also spent years on biological collecting expeditions (to Brazil and the Malay Archipelago). However, unlike Darwin, who spent 20 years after the *Beagle* voyage carefully assembling an argument for evolution using natural selection, Wallace conceived the same idea in a fevered dream brought on by a tropical illness in 1857. He promptly wrote the idea down in an essay (Oosterzee 1997). Wallace, knowing of Darwin's interest in the subject of evolution, sent the essay to him (Van Wyhe and Rookmaaker 2013). Despite his careful work, Darwin realized he must now publish his idea. To avoid the inevitable conflict of priorities, he presented a paper jointly with Wallace and then promptly set to work on *On the Origin of Species by Means of Natural Selection* (1859), which was published 15 months later.

How is Nature Dynamic?

If nature is not fixed and change is the norm, what kinds of dynamics can we see in nature? It is a basic fact that organisms are born and eventually die. Birth and death give rise to changes in population size, which ecologists refer to as **population dynamics**. The ebb and flow of population size over time is one example of dynamics we see in nature.

Organisms require food for energy to grow, maintain homeostasis, and reproduce. Their acquisition of energy to survive and reproduce reveals a second kind of dynamics in nature—foraging dynamics. **Foraging dynamics** are adjustments in behavior with regard to changing information in the environment, and they result from allocation decisions made by organisms regarding what, where, when, and how to forage to acquire necessary nutrients, resources, and food. They also arise when organisms face trade-offs between foraging opportunities and the risk of becoming food themselves or trade-offs between foraging and other life activities such as mating, caring for young, etc. Foraging dynamics can affect the distribution and abundance of species over time and space.

Over longer periods, the heritable characteristics of a population or species can change, and even the diversity of species can evolve via speciation and extinction events. Changes in heritable trait values and their diversity represent **evolutionary dynamics**. This is a third type of dynamics recognized in nature. Evolutionary dynamics result in the adaptations and diversity of species.

We can also describe dynamics at different levels in the ecological hierarchy, including individuals, populations, communities, and ecosystems. At the individual level, dynamics include individual growth and dispersal, and foraging. Individuals also move around for other reasons, such as finding shelter or a mate. In addition to size, populations change in distribution and range. At the community level, there are **successional dynamics**, as one group of species is replaced by another. At the largest scale, we can consider climate change. Often, dynamics at higher levels of ecological organization are manifestations of foraging, population, and evolutionary dynamics. In any case, populations, communities, and ecosystems are always in a constant state of flux. Individuals acclimate and adjust, populations evolve, and species change.

Due to these dynamics, the diversity we witness today differs from what existed in the past, and this is evident from the differing arrays of fossils apparent between layers of rocks. Consider the trilobites, a fossil group of extinct marine arthropods. They appeared in the Cambrian period over 500 million years ago (mya), at a time when the Earth was warming in a "greenhouse" episode following a period known as "Snowball Earth." Trilobites dominated the ocean world for over 250 million years, with more than 20,000 known species, exhibiting a varied range of lifestyles—predators, mutualists, scavengers, filter feeders, and others—across many geographic regions (NHM 2021). Despite their success for over 200 million years, trilobites went extinct during one of Earth's largest mass extinction events. Much later in Earth's history, a different group of fossil organisms, called dinosaurs, diversified and dominated the Earth for nearly 100 million years, until they too perished 65 mya in another mass extinction event at the end of the Cretaceous Period in the Mesozoic Era. *Tyrannosaurus rex* and other creatures of titanic size gave way to the smaller, furry creatures that diversified in the age of mammals. In fact, Wake and Vredenburg (2008) assert that five major extinction episodes impacted the history of life on Earth, resulting in profound losses of biodiversity and the subsequent radiation of new species (see Chapter 11 *Biodiversity*). All these events indicate that the Earth relentlessly changes, and that life, in turn, evolves over time.

The Four Big Quests in Evolutionary Ecology

Ecology seeks to understand the complex interactions of the myriad organisms and their environment. Evolutionary ecology, in turn, aims to elucidate these interactions and ecological relationships through the adaptations of organisms to their environments. This scientific approach represents a systematic quest to explain the following four fundamental characteristics observed in nature.

Diversity of Life

Why are there so many species? How do they coexist? These were perhaps the first questions asked by naturalists following on the heels of the age of exploration, from the 15th to 17th centuries, and continued through the Victorian era in the 19th century. During this period, advancements in sailing technology and the expansion of colonial enterprises prompted Europeans to explore the world. Naturalists such as the pirate William Dampier (1651–1715), the scholar Georg Wilhelm Steller (1709–1746), the well-heeled gentleman Joseph Banks (1743–1820), and both Darwin and Wallace, took advantage of the era to observe the Earth's varied forms, catalog them, and collect them for posterity. As numerous museum trays filled with bewildering arrays of different life forms—from delicate mosses and flowering plants to bizarre monotremes and fantastically feathered birds—it became increasingly apparent that life is vastly diverse. Larsen et al. (2017) estimate there are between one and six billion species on Earth, although only about 1.5 million have been formally described.

Most living things can be readily recognized and categorized as distinct from others using the term "species." The quest to understand what generates so many species is a difficult one. If species are shaped and evolved by natural selection (a necessarily competitive, Malthusian process that weeds out variations), what are the generative mechanisms that create species diversity? Does it require isolation of large populations? Does it require shifting Earth landscapes? Does it depend on invasions or interbreeding? And after this diversity is generated, what allows species utilizing the same habitat to coexist without going extinct? Do they avoid competition? Do some species reduce the effects of competition between others? Do some species benefit from the presence of others?

Answering these questions requires not only an understanding of interactions at the level of individuals, but also an understanding at the level of populations and communities.

Distribution and Abundance

Understanding the biotic and abiotic factors affecting the abundance and distribution of arrays of diverse organisms is the second big quest in evolutionary ecology. The distribution of a species is defined in space by the presence of individuals of that species in a particular area. While distribution defines a population's spatial extent, abundance describes the number of individuals or the "size" of the population. Abundance is a function of population density, understood as the number of individuals per unit area over which the population is distributed.

Life is ubiquitous throughout Earth. Organisms can be found in places as extreme as the alkaline soda lakes of California, the hot acid pools of Yellowstone, and the pressurized depths of the oceanic trenches. There is almost no environment on Earth without life (except, for example, the Atacama Desert of Chile; Navarro-Gonzáles et al. 2003). But no single species can tolerate the full range of the planet's environments. All species live between certain ranges of environmental and resource conditions that are suitable for their survival, growth, and reproduction (Willi and Van Buskirk 2019). Certain species find some environments too warm, cold, saline, or otherwise unsuitable. The vagaries of the physical environment limit their geographic distribution. For instance, no reptiles occur in the western Arctic and no polar bears live in areas without sea ice (Barry et al. 2013). However, differences in the physical (abiotic) environment only partially explain the distributions of species. Interactions with other organisms (biotic environment) constitute an important part of an organism's overall environment. Such biotic interactions can also expand or restrict the range of a species. For example, the distribution of the yucca moth (*Tegeticula yuccasella*) is highly dependent on the presence of yucca plants, which are found in warm, dry areas throughout North America.

Distribution can be local, regional, or global. People have long known that if you want to see a kangaroo in its native land, you need to travel to Australia, but if you wish to observe a polar bear, you must visit the Arctic. Why are kangaroos and other marsupials abundant and widely distributed in Australia and New Guinea, but, in the case of kangaroos, originally found nowhere else on Earth? Does this depend on Earth's geological history, on dispersal events, or abiotic and biotic interactions with the environment? Organisms are not spread randomly across space and time, and their distributions reveal certain patterns. In the case of kangaroos, about 250 million years ago, all of Earth's landmasses were joined together in the supercontinent Pangaea, which later split into Gondwana and Laurasia. They, in turn, split into the modern-day continents. Because Australia separated from Gondwana so early, mammal diversity remained low, and only monotremes and marsupials had evolved. For tens of millions of years, no terrestrial placental mammals existed in Australia, and kangaroos evolved to graze Australian grass instead of placentals such as cattle, horses, and sheep that evolved elsewhere. Therefore, to understand the distribution of species and factors that limit their distribution, we need to study the history of Earth and how landforms have influenced and restricted dispersal events.

The history of continents separating and coming together has strongly influenced the distributions of present-day animals and plants. However, once organisms are dispersed to a geographical region, their distribution and abundance are influenced by interactions with their biotic and abiotic environments. Some organisms seem ubiquitous and excessive in numbers. For example, the red maple (*Acer rubrum*) is geographically widespread in eastern North America. Species of roaches and rats occupy most geographical parts of the globe. In contrast, a species with a restricted distribution is referred to as endemic. Endemic species, owing to their specialized habitat requirements or small geographic ranges, have low population sizes, though within their range they may occur at high densities. The Amani sunbird (*Hedydipna pallidigaster*) of East Africa is a notable example. The entire species occurs in just several small patches of habitat in Kenya and Tanzania. Kenya's entire population inhabits just 8,000 hectares of the Arabuko-Sokoke Forest, though in this forest, it is one of the most abundant bird species. In addition to the Amani sunbird, this small area is home to numerous species of endemic plants, several endemic butterflies, and an endemic mammal, the golden-rumped elephant shrew (*Rhynchocyon chrysopygus*; Oyugi et al. 2012).

Neither the distribution nor the abundance of a species remains static. Consider the distribution and abundance of the puma or mountain lion (*Puma concolor*) in the United States. In the recent past, the cat's population may have been over 100,000, but the population has diminished, with the current estimate at 30,000, mostly confined to the western United States. Why has its distribution and abundance changed? Evidence points to persecution by humans via bounty hunting of eastern puma populations (Anderson Jr. et al. 2010). The puma is more abundant and widely distributed in Central and South America than in North America, but its historical range in both North and South America matches that of deer and deer-like camelid species such as guanacos and vicuñas. More recently, perhaps because of high deer populations and less human persecution, the range of the puma has been recovering and advancing eastward again, colonizing South Dakota, Nebraska, and Missouri.

Distribution and abundance are population-level descriptors that are the consequence of both ecological and evolutionary interactions through time. Answering questions about the distribution and abundance of species requires not only an understanding of how differential survival and reproduction of individuals change population size in ecological time, but also an understanding of the historical forces that have shaped both the biotic and abiotic environments in which individuals live and how those forces have provided or prevented opportunities for dispersal into new areas.

Fit of Form and Function

Consider any organism and examine how its physiological, morphological, and behavioral features make it adept at performing certain functions essential to its survival. The traits of organisms appear amazingly well-suited to where they live, what they eat, and how they find shelter and avoid predators. Fish have fins to swim. Birds are equipped with wings to fly. Whether it is a bushy-tailed squirrel, with its superb climbing ability, or a cheetah with its non-retractable claws for traction and long limbs for speed, the structures and forms of all organisms have evolved to perform specific functions. The high-crowned molars of zebras and rhinos act as mulching mowers for grinding grass, but none of these animals possesses the sharp canines that carnivores such as the tiger have. Arrays of organisms appear remarkably suited and well-engineered for their particular environment and biological role.

The third big quest of evolutionary ecology is to deconstruct adaptive design while also seeking to discover how biotic and abiotic interactions and constraints engineered these remarkable features. For example, the finches of the Galápagos Islands show a remarkable array of beak adaptations, and their feeding ecology closely matches these adaptations. The array of beak shapes has been likened to a set of pliers: linesman, diagonal, long chain nose, gripping, curved, and needlenose (Grant and Grant 2008). There are robust beaks specialized for feeding on arthropods and mollusks, beaks for crushing seeds, and even a long beak useful for sucking the blood of seabirds and drinking nectar.

What are the forces that shaped these adaptations? Was it competition for food? Was it selection for novel food resources? Did interbreeding and climate extremes help drive the evolution? Answering these questions requires understanding interactions at the level of individuals, the selective forces on both survival and reproduction at the population level, and an understanding of heredity, the concept of evolutionary "fitness," and what it means to "win" the game of life in evolutionary time.

The Common Ancestry of Life (Procession of Life)

Despite the diversity of life and the "fit of form and function," all organisms on Earth are connected by common ancestry and common design rules. Despite or perhaps because of this, there is almost always a trend toward increasing levels of complexity in the progression of life. Notions such as the tree of life as a regular, yet increasing sophistication of organisms in terms of size, behavior, and the number and specialization of traits, fit this view. Linnaeus, as he worked to organize the biological collections brought back by world-traveling naturalists into his binomial nomenclature (still used today), which was based on systematic physical similarities, formalized the recognition of design rules. And Darwin's only illustration published in On the Origin of Species was a tree of life, depicting the notion of common ancestry.

Two examples clearly illustrate the common origin and design of all life-forms, whether a whale, bacterium, or pine tree:

The genetic code is universal to all living organisms. This code provides the instructions for synthesizing proteins, regulating the expression of other genes and metabolic pathways, as well as choreographing the processes and molecules needed for cell division. This molecular language consists of just four letters—A, C, T, and G—the initials of four chemicals that form the string of nucleic acids known as DNA. Combinations of three of these letters specify each amino acid used by the cell to build proteins. Today, we know that 98.8% of human DNA is similar to that of chimpanzees and 70% to that of zebrafish. In 1987, when British researchers inserted a human gene into the cells of yeast, remarkably, the gene—which controls cell division—functioned effectively (Lee and Nurse 1987). Genes have been "conserved" throughout evolution; genes that perform certain functions in simpler animals have been preserved in the human genome, though they are sometimes modified to carry out more complex functions.

A second example of conservation of design comes from the cytochrome complex (Cyt c), which is a protein composed of 104 amino acids common to bacteria and universal in the mitochondria of all eukaryotes (a group that includes most things that are not bacteria). Cyt c performs a key step in cellular energy production and metabolism, and perhaps because it is so critical to life, it has changed little in millions of years of evolution. Systematic biologists have used Cyt c to create an evolutionary tree that spans all of life, including groups of species that had previously been impossible to compare: humans, dogs, horses, rabbits, kangaroos, pigeons, ducks, turtles, snakes, moths, bread mold, and yeast (Yamanaka 1967). Although Cyt c has been conserved over the course of Earth's history, there have been small changes, and these can be used to organize organisms into more closely related and less closely related groups. The Cyt c protein of the duck is much like that of pigeons and other birds, but it differs slightly from the Cyt c proteins of mammals. These proteins are all very similar within mammals, but the Cyt c proteins within primates, including humans, are even more similar.

The fourth quest of evolutionary biology is to understand the ancestral relationships between various types of organisms and how ancestry may have shaped organismic traits and population distributions. This requires teasing apart what has been conserved and what has evolved.

The Past and Legacy of Evolutionary Ecology

All organisms, in a struggle for existence, interact with their biotic and abiotic environments in varied ways. Such interactions, in turn, define their distribution and abundance and set the stage for natural selection. They exhibit, in their characteristics spawned by diversity and sculpted by natural selection, superb engineering as the fit of form and function. Their engineering shows a commonality and connectedness of design across all life, from simple to complex.

Evolutionary ecology was founded out of Western natural philosophy by pre-Darwinian scholars who recognized the engineering component to nature and, drawing a logical connection, concluded: "Tools exist because humans design and fashion them with purpose and forethought of intent; a watch is proof of a watchmaker!" Evolutionary ecology experienced its infancy in the 19th century when scientific rigor came into fashion. Lyell's geology showed how ongoing forces and non-personified natural processes, such as erosion, rain, or wind, could explain the forms, types, and layering of rocks, and, within many of these non-living rocks, the distinct remains of previous life. These natural processes explain such features as the amazing landscape of the Grand Canyon, sculpted by the Colorado River and eroded over eons. Could similarly non-personified natural forces explain the origins and changes of life over time? Just as raindrops carve landscapes by eroding softer and harder stones at different rates, the ecological raindrops of births and deaths, striking the softer and harder rocks of heritable characters, mold myriad forms of life.

It is easy to see how the original ideas of a "Ladder of Life" influenced early evolutionary theories. However, the concept of a static ladder of life has changed. When naturalists brought back samples of life's diversity, and people across the world read Darwin's *On the Origin of Species*, evolutionary ecology grew in the 19th century and flourished in the 20th and 21st centuries through diverse contributions. We now understand that the ladder's rungs—that is, different species in the web of life—interact in myriad ways, influencing each other's distribution and abundance. These interactions, in turn, affect how species relate in real and evolutionary time. Indeed, the web of life is composed of two distinct threads: one that links the dazzling varieties of organisms at any given moment in time through the flow of energy and cycling of materials in the business of eating and being eaten, and another linking all life through deep time via genetic information and shared common ancestry. The first represents ecology, and the second evolution. Ecology and evolution as a unified theme—evolutionary ecology—provides a robust scientific platform for understanding the big questions of life and appreciating the nature of Nature. That same evolutionary perspective that allows us to understand Earth's biodiversity is equally essential for being good ecological stewards and meeting the global opportunities and challenges of preventing extinctions, conserving biodiversity, promoting human health and healthy ecosystems for humans, and mitigating the consequences of global climate change.

References

Anderson, Jr., C. R., Lindzey, F., Knopff, K. H., Jalkotzy, M. G. and Boyce, M. S. 2010. Cougar management in North America. pp. 41–54. *In*: Hornocker, M. and Negri, S. [eds.]. Cougar Ecology and Conservation, University of Chicago Press, Chicago.

Armstrong, K. 1993. A History of God: The 4,000-year Quest of Judaism, Christianity and Islam. A. A. Knopf, New York.

Barry, T., Berteaux, D., Bültmann, H., Christiansen, J. S., Cook, J. A., Dahlberg, A. et al. 2013. Arctic Biodiversity Assessment 2013. Conservation of Arctic Flora and Fauna (CAFF).

Cantor, N. F. 2001. In the Wake of the Plague: The Black Death and the World It Made. The Free Press, New York.

Darwin, C. 1859. On The Origin of Species by Means of Natural Selection. John Murray, London, UK.

Davies, B. and Stump, E. [eds.]. 2011. The Oxford Handbook of Aquinas. Oxford University Press.

Egerton, F. N. 2003. A history of the ecological sciences, part 10: Botany during the Italian Renaissance and beginnings of the scientific revolution. Bulletin of the Ecological Society of America 84: 130–137.
Egerton, F. N. 2012. Roots of Ecology: Antiquity to Haeckel. University of California Press.
Goodland, R. J. 1975. The tropical origin of ecology: Eugen Warming's jubilee. Oikos 26: 240–245.
Grant, P. R. and Grant, B. R. 2008. How and Why Species Multiply: The Radiation of Darwin's Finches. Princeton University Press, Princeton, NJ.
Haeckel, E. 1870. Ober Entwicklungsgang und Aufgabe der Zoologie. Jena. Z. Med. Naturwiss 5: 353–370.
Hertz, E., Trudel, M., Tucker, S., Beacham, T. D., Parken, C., Mackas, D. et al. 2016. Influences of ocean conditions and feeding ecology on the survival of juvenile Chinook salmon (*Oncorhynchus tshawytsha*). Fisheries Oceanography 25: 407-419.
Hutchinson, G. E. 1965. The Ecological Theater and the Evolutionary Play. Yale University Press, New Haven, CT.
Jones, P. M. 2017. Industrial Enlightenment: Science, Technology and Culture in Birmingham and the West Midlands: 1760–1820. Manchester University Press.
Jung, T. S., Stotyn, S. A. and Czetwertynski, S. M. 2015. Dietary overlap and potential competition in a dynamic ungulate community in northwestern Canada. The Journal of Wildlife Management 79: 1277–1285.
Lamarck, J-B. 1809. Philosophie Zoologique ou exposition des considérations relatives à l'histoire naturelle des animaux. National Museum of Natural History, France.
Larsen, B. B., Miller, E. C., Rhodes, M. K. and Weins, J. J. 2017. Inordinate fondness multiplied and redistributed: The number of species on earth and the new pie of life. The Quarterly Review of Biology 92: 229–265.
Lee, M. G. and Nurse, P. 1987. Complementation used to clone a human homologue of the fission yeast cell cycle control gene cdc2. Nature 327: 31–35.
Linnaeus, C. 1758. Systema Naturae per regna tria naturae, secundum classis, ordines, genera, species cum characteribus, differentiis, synonymis, locis. Tenth Edition, Vol. 1, Laurentii Salvii, Stockholm.
Lyell, C. 1833. The Principles of Geology. John Murray, London, UK.
Malthus, T. 1798. An Essay on the Principle of Population. J. Johnson, London, UK.
Mayr, E. 1982. The Growth of Biological Thought. Harvard University Press, Cambridge, MA.
Naiman, R. J., Pinay, G., Johnston, C. A. and Pastor, J. 1994. Beaver influences on the long-term biogeochemical characteristics of boreal forest drainage networks. Ecology 75: 905–921.
Natural History Museum (NMH). 2021. Trilobite Research. The Natural History Museum. London. https://www.nhm.ac.uk/our-science/our-work/origins-evolution-and-futures/trilobite-research.html (accessed Dec16, 2021).
Navarro-González, R., Rainey, F. A., Molina, P., Bagaley, D. R., Hollen, B. J., de la Rosa, J. et al. 2003. Mars-like soils in the Atacama Desert, Chile, and the dry limit of microbial life. Science 302: 1018–1021.
Oosterzee, P. 1997. Where Worlds Collide: The Wallace Line. Cornell University Press, Ithaca, New York.
Oyugi, J. O., Brown, J. S. and Whelan, C. J. 2012. Foraging behavior and coexistence of two sunbird species in a Kenyan woodland. Biotropica 44: 262–269.
Ragan, M. A. 2009. Trees and networks before and after Darwin. Biology Direct 4: 43.
Singer, C. 1931. A Short History of Biology. Clarendon Press, Oxford, UK.
Tansley, A. G. 1935. The use and abuse of vegetational concepts and terms. Ecology 16: 284–307.
Timmons, S. 2003. Grand Canyon geology training manual: learning to read the pages of a book. Oregon State University, Corvallis, OR. http://npshistory.com/publications/grca/geology-trn-man.pdf.
Timmons, S. 2009. Grand Canyon Geology Training Manual: Learning to Read the Pages of a Book (Doctoral dissertation, Oregon State University).
Van Wyhe, J. and Rookmaaker, K. [eds.]. 2013. Alfred Russel Wallace: Letters from the Malay Archipelago. Oxford University Press, Oxford.
Wake, D. B. and Vredenburg, V. T. 2008. Are we in the midst of the sixth mass extinction? A view from the world of amphibians. PNAS 105: 11466–11473.
Walton, O. C. and Stevens, M. 2018. Avian vision models and field experiments determine the survival value of peppered moth camouflage. Communications Biology 1: 118.
Watts, E., Uwe, H. and Levit, G. S. 2019. Ecology and evolution: Haeckel's Darwinian paradigm. Trends in Ecology & Evolution 34: 681–683.
Weismann, A. 1888. The supposed transmission of mutilations. Chapter 8. *In*: Poulton, E. B., S. Schönland and A. E. Shipley [eds.]. Essays upon Heredity and Kindred Biological Problems. Oxford: 1889. https://www.gutenberg.org/files/48132/48132-h/48132-h.htm (accessed 1/24/2022).
Wilkins, J. S. 2009. Species: A History of the Idea (Vol. 1). University of California Press.
Willi, Y. and Van Buskirk, J. 2019. A practical guide to the study of distribution limits. The American Naturalist 193: 773–785.
Wool, D. 2001. Charles Lyell–"the father of geology"–as a forerunner of modern ecology. Oikos 94: 385–391.
Yamanaka, T. 1967. Cytochrome c and evolution. Nature 213: 1183–1186.

CHAPTER 2

Population Dynamics

If "nothing in biology makes sense except in light of evolution" (Dobzhansky 1973), then nothing in evolution makes sense except in light of ecology. And nothing in ecology makes sense except in light of population dynamics.

Between 2003 and 2005, swarms of desert locusts (*Schistocerca gregaria*) spread throughout a huge swath of territory, decimating vegetation over thousands of hectares in northern Africa, reaching from Mauritania as far north as southern Portugal in Europe and as far east as Israel in the Middle East (Sánchez-Zapata et al. 2007). It was indeed a plague of biblical proportions, and the sheer number of locusts had not been seen since the last plague occurred between 1987 and 1988. How did this population increase so rapidly, and what kept it in check for 15 years?

While not as extreme as the desert locust, most populations are **dynamic**—that is, they change in size and other properties over time. Births and deaths, universal properties of life, affect these dynamics. A population grows when births surpass deaths, and it decreases when the opposite happens. Births and deaths are part of the dynamic processes that govern the distribution and abundance of all organisms. With that in mind, the study of population dynamics focuses on the factors influencing the expansion, decline, and maintenance of populations.

As Pianka (1983) puts it, a **population** is an abstract entity that is nonetheless real. While it can be useful to define a population as a group of interbreeding individuals from the same species occupying a defined area at a given time, in practice it might be difficult to draw boundaries between populations (see Chapter 11 *Biodiversity*). For example, European Starling (*Sturnus vulgaris*) seen on the lawns of Chicago (Illinois, USA) this past spring are definitely a different population than the same species flitting through Parisian trees in France 30 years ago. However, it might be hard to decide whether contemporary starlings in Chicago and New York form two populations or not.

In the broadest sense, a population may be equivalent to a species, but in a narrower sense, a species may comprise many populations, each separated from the others by distance, time, and other barriers. Even if they are the same species, different populations might be genetically differentiated into distinct ecotypes or clades. The geographical extent of a population is delimited by the species in question and the questions being asked. A far-from-exhaustive list of studied populations includes all moose on Michigan's Isle Royale in the United States (Murie 1934), elephants in East Africa (Laws 1970), grizzly bears in Yellowstone National Park (USA; Craighead et al. 1974), beech trees in the Central Valley of Central Chile (Vergara et al. 2014), mites on four oranges in a tray (Huffaker 1958), and plasmodium gametocytes (the sexually reproducing life history state of the protozoan causing malaria) in a single mosquito (Vaughan et al. 1992).

Some species, such as clovers, dandelions, squirrels, and starlings, are stubbornly common, whereas others are acutely rare. For example, Venus flytraps (*Dionaea muscipula*) occur only in marshy coastal areas of North and South Carolina in the United States, and giant pandas (*Ailuropoda melanoleuca*), numbering fewer than 2,000, inhabit temperate forests in a few mountain ranges of south-central China's Sichuan, Shaanxi, and Gansu provinces. In many cases, we have observed

DOI: 10.1201/9781003008538-2

throughout recorded history that common species generally remain common, just as rare ones remain rare. What factors determine this difference?

Likewise, some insect species appear and disappear from one year to the next, while others are consistently present year after year. Again, what causes this difference? Fisheries scientists are concerned that some salmon populations are declining, but why are they declining? Are they declining because of overfishing, low spawning rates due to changes in stream temperatures, or because of poorer survival due to changing ocean conditions? Or are they declining due to a combination of factors? By studying the factors that affect population dynamics, we can better understand and prevent a species' decline or extinction, manage harvested populations, or control noxious pest species such as parasites and pathogens.

Hug your dog, pet a horse, and observe an oriole through binoculars. One can see, hold, and touch an individual, making it feel "real." In contrast, when discussing a population, it is often done in abstract terms. This is because a population usually contains a large number of individuals and spans a substantial area that cannot be observed in its entirety at once. Therefore, ecologists use various statistics and relationships to describe a population, including its size, growth rate, limitations on growth, and the nature of its dynamics.

Population Size

There are three primary methods for quantifying population size: total population, biomass, and density.

Total Population

This is a count of the total number of individuals. In the United States, compare California's human population of almost 40 million (as of 2023) to that of Wyoming's, which is just a little over half a million (2023). The most populous country in the world is India (2023), with a population of 1.42 billion people. The least populated is Vatican City at around 764 individuals (2023).

We might assume that the larger the total population, the less likely it is that a particular population will become rare, but that is not necessarily the case. Over 1,500 golden toads (*Bufo periglenes*) were counted in the Monteverde rainforest of Costa Rica in 1987 (Crump et al. 1992). Unfortunately, only ten were found a year later, and the year after that, there was only one. No one has seen a golden toad since; this species has gone extinct (Ochoa-Ochoa et al. 2013).

Biomass

Plant, fish, and invertebrate (especially arthropod) populations generally have too many individuals to count, or they have many life history stages and sizes. In such cases, it is more practical to measure their biomass instead. Common direct measurements include weighing wet weight (generally "unprocessed" body weights, e.g., 100 kg of fresh-caught halibut), dry weight (individuals are dried and measured, e.g., 50 g of grama grass), or ash-free dry weight (dried material is oxidized in a furnace and weighed, allowing measurement of just the organic content, which can be compared across taxa). However, small body size and high abundance often hamper direct quantification of biomass in many organisms. Indirect estimation methods have been proposed, such as using body surface area or volumes as proxies for individual mass based on linear measurements.

Density

The third way of quantifying population size is by calculating its density, which is the number of individuals per unit area (e.g., square kilometer, hectare, square meter) or per unit volume (e.g., liter, cubic meter). Populations of the same species in different-sized areas can be compared using density. For example, a small state like New Jersey, with over 498 individuals per square kilometer (in 2024), has a higher human population density than a large one, such as Alaska, which has about one individual per square kilometer. A contrasting but equally informative density comparison can also be made between species. Good habitat for tigers may support six individuals per 100 square kilometers, whereas good habitat for deer can support 1,000 individuals per 100 square kilometers.

How do you count individuals of certain grass species in your lawn or a grassland? An alternative measure of density is to estimate the percentage of such cover (referred to as "percent cover"). For instance, a square meter of lawn may contain 15% clover, 20% dandelion, and so on. By this metric, the dandelions cover one-fifth of the available space when viewed two-dimensionally from above.

It must be remembered that density, measured as the number of individuals per unit area, is viewed as "crude" density. This is because individuals do not occupy all the available space; not all habitats within a particular area may be suitable, and different portions of the space may be used at different times or seasons. Thus, crude density can vary widely from location to location and from one time to the next. Measurements of crude density may also be sensitive to how we define areas in space and time. However, the reason for using density rather than size (in terms of numbers or biomass) is that often the frequency of interactions among individuals depends on the density of the population rather than the number of individuals in the population. These interactions, in turn, exert effects on individual organisms.

When populations are quantified in terms of density, the spatial position of individuals relative to each other can also be described. Individuals may be distributed randomly, uniformly, or clumped. Random distribution occurs when each individual's position is independent of others. When uniform, individuals are more or less evenly spaced. This pattern of distribution can arise when there is some form of negative interaction, such as competition among individuals, that creates a minimum distance between neighbors. A uniform distribution is common in animal populations where individuals maintain territoriality, or in plant populations where severe competition exists for water, nutrients, or both. A clumped distribution may occur for a variety of reasons: resources are often distributed in patches, leading to clumps of consumers (e.g., birds at feeders or impala around a water hole). Additionally, some species form social groups, such as fish that move in schools or caribou that move in herds. Across the globe, the distribution of humans is clumped because of factors such as social behavior, economics, and geography. Within countries, humans also exhibit clumping in urban centers, where resources and economic opportunities are often concentrated.

Population size, however we measure it, is not static and can grow because of births and immigration, or decline due to deaths and emigration. The change in population size can be expressed in terms of an absolute number per unit time (e.g., the population grew by 2,000 individuals in 1 year) or as the per capita rate of change (e.g., the population grew by 15% in 1 year). Similarly, birth rates or death rates can be computed as the total number of births or deaths per unit time, or as per capita birth and death rates when these rates are scaled by the number of individuals. Changes in birth and death rates are often affected by environmental factors and interactions with individuals of the same or different species. Understanding the factors that affect birth and death rates allows us to predict whether a population will grow, decline, or remain the same.

Population Growth

Early in the 19th century, Thomas Malthus (1798), in his book, *An Essay on the Principle of Population*, wrote: "Through the animal and vegetable kingdoms nature has scattered the seeds of life abroad with the most profuse and liberal hand … population when unchecked increases in a geometrical ratio; it's a mathematical inevitability." According to Malthus, populations exhibit exponential growth. Charles Darwin (1859) returned to this point in his *On the Origin of Species*: "There is no exception to the rule that every organic being naturally increases at so high a rate that, if not destroyed, the earth would soon be covered by the progeny of a single pair…"

Any population has the potential to grow exponentially under ideal conditions. This is the **first law of population ecology**. However, as Malthus recognized in his essay, populations cannot grow exponentially forever because resources are finite. In other words, there are limits to growth. This is the **second law of population ecology**.

Exponential Growth

When populations are sparse and uncrowded, or at low densities, they may grow very rapidly. If this population is linked to other populations, immigrants may be the cause of this increase as they move from areas of high density to areas with fewer individuals and more resources available. When immigrants and emigrants contribute to changing population size (in addition to births and deaths within the population), this is called an **open population**. If there is no significant immigration or emigration, that population is referred to as a **closed population**. It is easier to see the power of exponential growth in closed populations, where the change in the number of individuals over time depends solely on the balance between births and deaths.

Birth and death rates are commonly expressed per capita rather than as absolute numbers, allowing for the comparison of growth between populations, whether one has one million individuals or just ten. Per capita births can be expressed as the number of births divided by the population size. However, because we are interested in how the population is changing, we need to express this as a rate. For example, if we start with a population size (N) of 10,000 at the beginning of the year and 1,300 births occur over the course of that year, then the per capita birth rate (b) would be 1,300/10,000 = 0.13 per individual (capita) per year. Likewise, the per capita death rate (d) can be obtained in the same way. If 700 die in one year, the death rate (d) would be 700/10,000 = 0.07 per capita per year. Any unit of time can be used. For example, 0.13 births per individual per year is the same, on average, as 0.01 births per capita per month or 0.0004 births per capita per day. But what if we wanted to know how fast a population is growing at any instant in time, not just on average between two points?

Calculus proves to be a handy tool for predicting instantaneous rates of change. To predict how a population changes over time depending on constant per capita birth and death rates, we can use the following equation:

$$\frac{dN}{dt} = (b - d)N \qquad \text{Equation 1}$$

The calculus term "*d*" on the left-hand side of Equation 1 is not to be confused with the death rate. It is shorthand for "the change in," and it expresses a rate of change in N (the population size) at a singular point in t (time). This is known as the "instantaneous rate of change." For example, if $b = 0.13$ and $d = 0.07$, then the instantaneous rate of change of the population is 0.06 multiplied by the current population size (N). If we start with a population of 10,000 and $(b - d) = 0.06$, the instantaneous change in population size is 600, meaning the population is growing by 600 individuals at that point in time.

However, it is much easier to write the difference between the per capita birth and death rates $(b - d)$ collectively as r, so we can express Equation 1 as follows:

$$\frac{dN}{dt} = rN \qquad \text{Equation 2}$$

Dividing a population's growth rate by the number (or density) of individuals in the population (N) gives the following equation:

$$\frac{dN}{dt}\frac{1}{N} = r \qquad \text{Equation 3}$$

In this form, we see that r is the **per capita rate of increase**, although ecologists use the terms per capita rate of increase and **instantaneous rate of increase** interchangeably. Since r is the difference between *b* and *d*, a population grows if *r* is positive and declines if *r* is negative. A stable population has an *r* of zero.

Many of us are accustomed to thinking of growth as a linear process. A quantity grows linearly when it increases by a *constant amount* over a certain period. For example, a child who grows one inch taller each year is growing linearly. Similarly, if a person saves $100 annually by hiding it under their mattress, their savings also increase at a linear rate. In both cases, the annual increase is not affected by the size of the child in the former case or the amount of money already under the mattress in the latter. This type of growth is called **arithmetic growth**.

A quantity exhibits **exponential growth** when it increases by a *constant percentage* over time. A colony of yeast cells in which each cell divides into two cells every ten minutes is growing exponentially. After ten minutes, where there was one cell, there will be two cells, an increase of 100% (doubling). After another ten minutes, there will be four cells, then eight, then 16, and so forth. Similarly, if a miser takes $100 from their mattress and invests it at an annual interest rate of 5%, the total amount accumulated increases by 5% each year. In this case, the invested money will grow faster and non-linearly compared to a constant amount of money hidden under the mattress every year. The amount added each year to a bank account, or every ten minutes to a yeast colony, is not constant; in fact, it increases over time due to compounding. The growth rate of the total population (or investment) increases with the size of the population (or investment). Such exponential growth is a common process in biological, financial, and many other systems.

Equation 2 depicts one of the simplest models of population growth, where the number of individuals increases at a constant rate over time. Models are simplified extractions of the real world. They are useful to ecologists because they facilitate the formulation of ideas about important ecological processes and testing our understanding of them. And often, they can reveal properties of a system that we hadn't thought of before. We can also observe what happens when we change our simplifying assumptions. So far, we have assumed that there is no immigration or emigration, that per capita birth and death rates (and therefore *r*) are constant, that there are no demographic complexities, and that there are no time lags.

The exponential population growth model, Equation 2, is a continuous-time model. Ecologists use continuous-time growth models (that is, a differential equation) for populations such as bacteria, yeast, hydra, and humans, where births and deaths are constantly occurring. In many populations, however, births occur nearly simultaneously at discrete times. For instance, births are seasonal in plants and many animals. Consider Rocky Mountain bighorn sheep in North America—they rut in November and give birth in March and April. Thus, the population will exhibit a changing pattern of seasonal abundance, with a large increase in March and April due to births, and a decline in abundance for the rest of the year due to deaths. Therefore, to measure this population's growth without seasonal interference, one should count its individuals at regular intervals (perhaps every fall). For such species, the continuous growth model in Equation 2 does not reflect **finite** (discrete-time) rates of birth and death; it would underestimate growth early in the year and overestimate growth late in the year. A discrete-time (difference equation) model can be used as an alternative. For example, if we were to measure the ratio of population sizes at the end of one year (say 10,600) versus the beginning of the year (10,000), this would be the finite rate of increase of the population, which in this example would be 6%. Put another way, the finite rate of increase (called λ) is the ratio of the population size after the end of a period (t+1) over the population size at the beginning of the period (t), written as follows:

$$\frac{N_{t+1}}{N_t} = \lambda \qquad \text{Equation 4}$$

Which can be rearranged to

$$N_{t+1} = N_t\lambda \qquad \text{Equation 5}$$

If you were to multiply the finite rate of growth by the population size every year, you would realize that it is possible to predict the population at any finite time t in the future from a starting population size of N_0 as follows:

$$N_t = N_0\lambda^t \qquad \text{Equation 6}$$

As we've seen, population growth rates can be expressed as either an instantaneous rate (r; Equation 3) or as a finite rate (λ; Equation 4). For instantaneous rates, the time element is very short—that is, an instant, whereas for finite rates, the time element is a measurable period, like a day, week, or year—that is, a finite time interval. Although similar, r and λ are not identical. The former is a per capita rate of change at a particular instant in time. The latter refers to the change in population over a specified period. However, it is possible to find their relationship.

Equation 2 expresses the instantaneous *rate of change* in a population at time t, but what if we wanted to know the population *size* at time t? Differentiating N with respect to t gives us a rate, so if we "undifferentiate" the differential equation, we should get N at time t. In calculus, this is referred to as taking the integral. If we take the integral of Equation 2, the resulting equation would look like this:

$$N_t = N_0e^{rt} \qquad \text{Equation 7}$$

Does this look familiar? It turns out $\lambda = e^r$!

Although the meaning of λ is easy to grasp (it is just a ratio of the new population size to the old one), e^r is not easy to grasp in a finite growth context. However, λ has the undesirable property of taking on values between one and infinity when the population is increasing, and between zero and one when the population is declining (Caughley 1994). Therefore, biologists prefer to use the per capita instantaneous rate of increase, r, which, as we mentioned earlier, is positive when the population is growing, negative when declining, and zero when it does not change. Because $\lambda = e^r$, we can take the natural log to get $r = \ln \lambda$. When a population doubles each year ($\lambda = 2$), r is about 0.69 (ln 2); when it halves each year ($\lambda = 0.5$), r is −0.69. Note that if we have population size data over several years, r can be estimated as the slope of a linear regression of ln λ on time t.

Whether we are considering continuous or finite growth, when r is positive, we soon have a world "covered by the progeny of a single pair," as Darwin said, whether they are slow growing elephants or fast-breeding mice; it just takes a little longer for the elephants (Figure 1). Under optimal conditions—meaning birth rates per individual are as high as possible and death rates per individual are as low as possible—r reaches a maximum value called the intrinsic rate of increase (r_{max}), and the population attains its maximum rate of growth (which can be modeled by substituting r_{max} for r into the growth equations above). The intrinsic rate of increase, r_{max}, is biologically important because it can be heritable and is a function of the life history traits of the organism (see Chapter 6 *Life Histories and Life History Tradeoffs*). For example, a mouse population can increase more quickly than an elephant population because female mice can have a litter of six to eight pups five to ten times a year, whereas female elephants can only have one, or rarely two, calves every four to six years. Likewise, in species such as dandelions or fruit flies, which grow fast and produce many offspring each year, r_{max} is high; whereas in species like palm trees and pandas, which take years to mature and produce only a few offspring, r_{max} is low. In other words, a species' r_{max} is a function of its life history traits. But even if r_{max} is low, if a population is growing exponentially without limit, it does not take long for the population to increase to staggering numbers. Fortunately for us, few organisms ever attain their maximum rate of growth, and if they do, it is usually not sustained for long.

The r_{max} value varies among animals by several orders of magnitude. Small, short-lived organisms, such as the common human intestinal bacterium *Escherichia coli*, have high r values, whereas large, long-lived vertebrates have low r values. *E. coli*, which doubles every 15 minutes

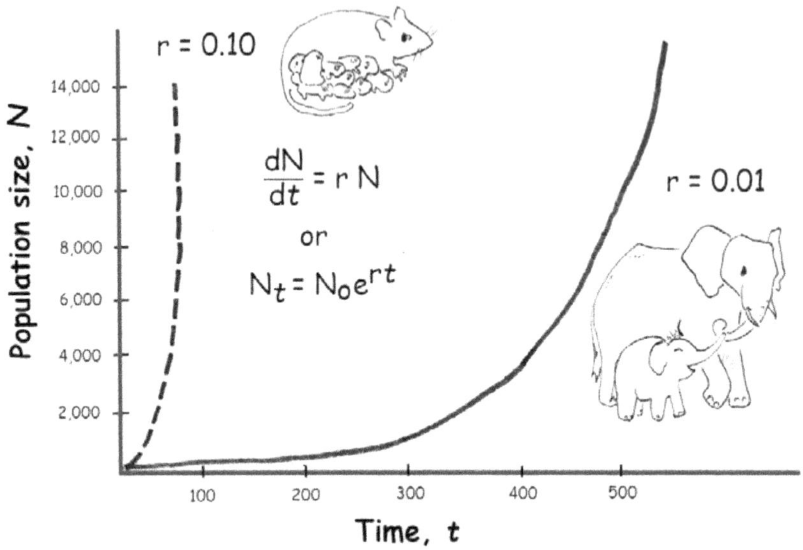

Figure 1. Exponential population growth for two different kinds of animals. The solid curve (less steep curve) might be elephants (with a low intrinsic growth rate), and the dashed curve (steep curve) might be mice (with a high intrinsic growth rate).

or so, has an r_{max} of around 60 (per individual per day). For comparison, the r_{max} of the protozoan *Paramecium aurelia* is 1.24. For the insect *Tribolium confusum*, it is 0.12. For the mammal *Rattus norvegicus*, it is 0.015. Finally, for humans, r_{max} is 0.0003 per capita per day (Pianka 2000).

For finite periods and under the right conditions, populations can grow exponentially (the first law of ecology). Good examples of this occur when humans either intentionally or unintentionally introduce a nonnative species to a new habitat, such as when New Zealand mud snails (*Potamopyrgus antipodarum*) first arrived in the Yellowstone ecosystem, reindeer were introduced to St. Paul Island in Alaska, English sparrows were transported to New York City, and rabbits were brought to Australia. Exponential growth can also occur naturally under two circumstances: when a few individuals establish a new population in a suitable habitat, and after a population has been devastated by a storm, fire, or other disturbance and begins to recover, starting with a few surviving individuals within an otherwise favorable habitat.

In 2016, NOAA Fisheries (USA) estimated the size of the eastern North Pacific gray whale (*Eschrichtius robustus*) population to be 27,000. Although the western Pacific (Asian population) is seriously endangered, the eastern Pacific (California) population has recovered from its earlier decimation by whaling, beginning in 1845. By the early twentieth century, the density of gray whales had become so low that it was no longer profitable to hunt them, and whaling was largely abandoned. In 1946, an international agreement granted protection to the California gray whales, allowing their numbers to recover. By 1995, the population had reached its pre-whaling level of approximately 20,000, a recovery so dramatic that the U.S. National Marine Fisheries Service recommended that the species be removed from the list under the US Endangered Species Act. Between 1967 and 1980, the gray whale population experienced an annual increase of 2.5% increase, which translated to an annual per capita growth rate of $r = 0.25$ (Reilly et al. 1983). This growth pattern fits the exponential model (Equation 3). In other words, the population grew at an exponential rate.

However, populations cannot grow exponentially indefinitely (not even humans) because there are limits to the resources a habitat can support for an ever-increasing number of individuals. When a population stabilizes at the maximum number that the available resources can support, it has reached the habitat's carrying capacity.

Logistic Growth

The second law of population ecology recognizes limits to growth. This can be best defined by the continuous-time logistic model (Figure 2), which produces a "sigmoid" (or S-shaped) growth curve. The key to logistic growth is the introduction of carrying capacity (K) into a continuous growth model. Carrying capacity refers to the maximum number of individuals that a habitat can sustainably support.

The logistic equation uses exponential growth as its base, but then adds a "braking force" as the population size approaches the carrying capacity. The equation is as follows:

$$\frac{dN}{dt} = rN\left(\frac{K-N}{K}\right) \qquad \text{Equation 8}$$

The portion rN is the exponential growth equation. The new term, $(K-N)/K$, discounts exponential growth by the difference between the carrying capacity and the actual population density. For a small population, the new term is near 1.0, meaning growth is essentially exponential. As the population nears its carrying capacity, growth slows and can even be negative for $N > K$.

The logistic model of population growth is derived following this simple assumption: as population size N increases, per capita population growth $\frac{dN}{dt}\frac{1}{N}$ declines linearly with N. We can visualize this negative straight-line relationship by imagining population size (N) on the x-axis and per capita population growth ($dN/dt \cdot 1/N$) on the y-axis (Figure 3). We know that when N is very small, the population will be approaching exponential growth (r), so we can put the y-intercept of the line at r. Likewise, when the population nears carrying capacity (K), the per capita population growth will approach zero; therefore, we can set the x-intercept at K.

By plugging in N, r, and K into the equation for the straight line, $Y = b + mX$, we derive the logistic growth model where $Y = dN/dt \cdot 1/N$, $X = N$, $b = r$, and $m = -r/K$ as follows:

$$\frac{dN}{dt}\frac{1}{N} = r + \left(-\frac{r}{K}\right)N$$

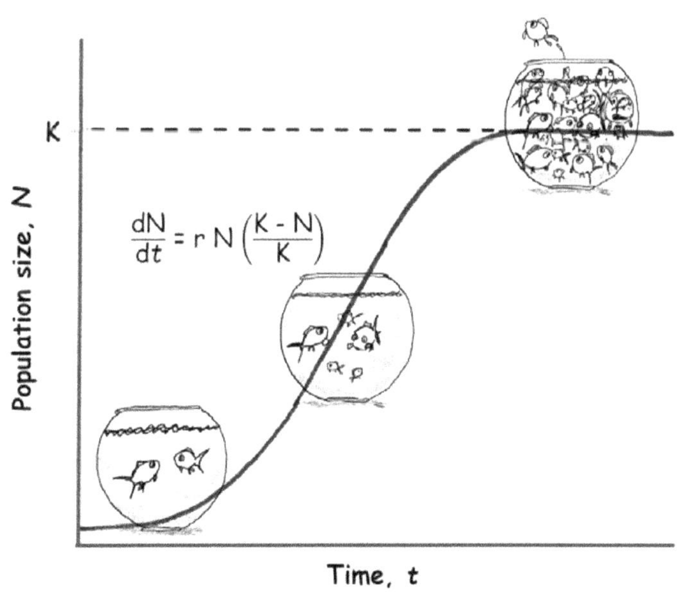

Figure 2. Logistic population growth. The plot shows population size (N) as a function of time (t). The population grows at first exponentially. At a population size of $K/2$, the growth rate begins to decline and eventually reaches an asymptote at the carrying capacity, K. The term $(K-N)/K$ acts as a "brake" on growth.

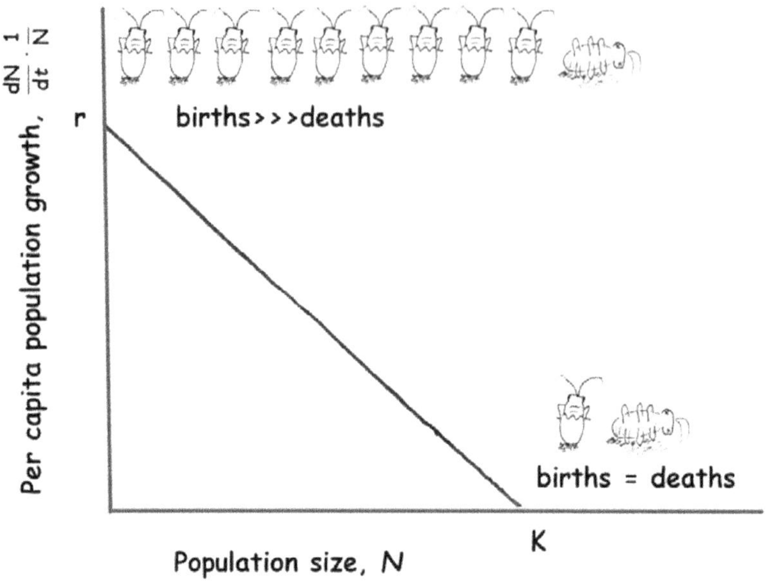

Figure 3. In the logistic model, per capita population growth decreases as population size increases.

or

$$\frac{dN}{dt}\frac{1}{N} = r + \left(\frac{K-N}{K}\right)$$

or

$$\frac{dN}{dt} = rN\left(\frac{K-N}{K}\right)$$

The discrete-time version of the continuous-time model can be written as

$$N_{t+1} = N_t e^{r\left(\frac{K-N}{K}\right)} \qquad \text{Equation 9}$$

A peculiar tension exists regarding population growth: while populations tend to expand, this growth is ultimately opposed by environmental factors that limit expansion. This dynamic is represented by *J*-shaped and *S*-shaped (or sigmoid) curves, which represent exponential and logistic growth models, respectively (Figure 4). These models correspond to the first and the second laws of population ecology.

The potential for exponential population growth and limits to growth generate the Malthusian struggle for existence (Gause 1934). In this struggle, individuals that are better adapted will survive and outcompete those that are less well adapted. Heritable variation among individuals within a population influences this struggle, driving evolution by natural selection. The link between Malthus and Darwin (see Chapter 13 *Evolutionary Ecology*) can be expressed in terms of logistic growth. As Mallet (2012) points out, the original equation for logistic growth was

$$\frac{dN}{dt}\frac{1}{N} = r - \alpha N \qquad \text{Equation 10}$$

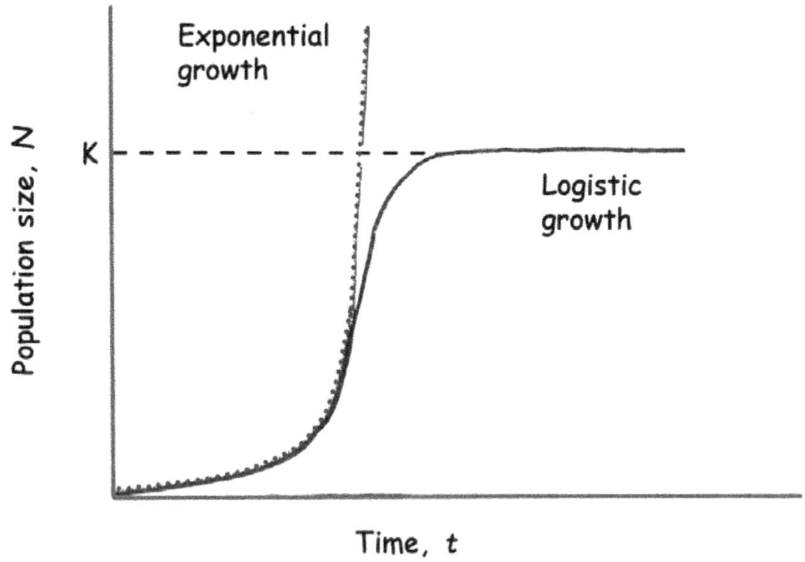

Figure 4. The *J*-shaped and *S*-shaped curves represent exponential and logistic growth models, respectively, which correspond to the first and the second laws of population ecology.

Here, α represents the density-dependent crowding effect (Chapter 3 *Species Interaction: Competition*). Figure 3 shows that the population reaches equilibrium size (i.e., stops growing) at K. Setting the left-hand side of Equation 10 to zero yields an equilibrium population size of r/α. These two formulations are equivalent. However, the latter shows more clearly that the logistic equation models a balance between density-independent exponential growth (r) and density-dependent regulation (α, or the struggle for existence), rather than a fixed (density-independent) resource limit (Mallet 2012).

What Factors Limit Population Size?

Both density-independent and density-dependent factors limit population growth. Density-independent factors alter birth rates and death rates regardless of the number of individuals in a population. They usually involve changes in the abiotic environment, such as variations in weather patterns, or natural disasters including fires, earthquakes, floods, volcanic eruptions, or droughts. For example, a hurricane may severely damage birds' nests in a forest, killing many individuals regardless of the density of the bird population. Similarly, a fire in a grassland may destroy populations of annual plants. The effects of oil spills killing fish and invertebrates are density-independent.

Density-dependent factors, which are often biotic, change birth rates and death rates as a function of population size. Intraspecific and interspecific interactions can influence density-dependent factors. Within a population, competition for space, nutrients, sunlight, food, mates, and breeding sites are all density-dependent factors. They are responsible for the decline in population growth seen in logistic growth models. When many individuals share limited resources, fewer may be able to reproduce, and population growth slows or even crashes. For example, a population of weedy plants thrives at high density when the plants are seedlings, but as the population becomes crowded, the seedlings compete for access to water and light, causing their individual growth and survivorship to decline.

Models of density-dependent growth do not necessarily include a *K* term. For plants, when weight is used to measure population size rather than numbers, Yoda's Law (a so-called self-thinning relationship) is used where $W = CN^{-3/2}$. W is the total dry weight of surviving plants, N is their density, and C is a growth constant that is unique to each plant species (Shinozaki et al. 1964; Weller 1987). This function behaves like an exponentially decaying curve where the total weight of a particular plant population drops dramatically as the density of the plant population increases.

Other biotic factors, such as predation and pathogens, often act in a density-dependent manner. When populations are crowded, they are more vulnerable to infectious diseases and parasites, exposure to toxic waste buildup, and increased predation rates. The alarming transmission of the COVID-19 virus in cities, as opposed to rural areas, provides a vivid example. Ecologists are divided in their opinions as to the relative importance of density-dependent and density-independent factors in natural populations (Andrewartha and Birch 1954; Lack 1954; Nicholson 1957). Both are important. Both play significant roles, with their impacts varying from one population to another as the size of the population changes.

r- and *K*-Selected Species

In situations of abundant resources, natural selection favors organisms that produce large numbers of small offspring that mature rapidly, usually without parental care. However, near the carrying capacity of the environment, when per capita resources become scarce, natural selection favors those that produce fewer but larger, more competitive offspring, often with extensive parental care. The former is categorized as an *r*-selected species, whereas the latter describes *K*-selected species (MacArthur and Wilson 1967). These terms were borrowed from growth-curve terminology, where *r* represents the rate of increase and *K* is the maximum asymptotic size in the logistic equation. *r*- and *K*-selection have also been referred to as birth and death selection, respectively (Hairston et al. 1970). By channeling proportionately more energy into reproduction, the increased reproductive output of birth-selected species can fill unoccupied space or harvest abundant, unexploited resources. In contrast, death-selected species channel energy and resources towards reducing the probability of offspring mortality.

The *r*-selected species, such as bacteria, insects, mice, and small birds at temperate latitudes, reproduce rapidly, mature early, and usually have a short lifespan. In contrast, *K*-selected species, such as large mammals, take a long time to mature, produce fewer offspring, and have long lives. Not all *r*-selected species have a short life. For instance, intestinal parasites like the tapeworms, which are common in humans, can live up to 25 years and produce up to 100,000 eggs per worm, provided there is a steady supply of nutrients. Nevertheless, most *r*-selected species are opportunists, taking advantage of temporarily favorable conditions for rapid growth. For instance, some mosquito species thrive in puddles after rainfall, wood-boring beetles favor trees damaged by fire, and bacteria and flesh flies seek carcasses. Due to their capacity for high rates of increase, *r*-selected species can rapidly colonize new habitats and may become destructive pests.

As introduced at the beginning of this chapter, migrating locusts in parts of Africa and Asia, sometimes numbering in the billions, can travel thousands of kilometers, devouring crops and wild plants. They grow exponentially when conditions are favorable and resources are abundant, but decline rapidly once conditions are harsh and resources dwindle. Species like the locust allocate so much of their metabolism (energy) into reproduction that their lifespans become short, and future opportunities for reproduction are limited. This is known as a life history tradeoff, where investing heavily in reproduction comes at the cost of lifespan (see Chapter 6 *Life Histories and Life History Tradeoffs*). For example, lizards in the tropics have a long growing season but are mainly short-lived, *r*-selected species that rapidly reach sexual maturity. They are seldom iteroparous (multiple litters of young per adult), rarely show parental care, and have high juvenile and adult mortality (Tinkle 1969). By allocating much energy to reproduction in the short run, they produce more and larger clutches, but this comes at the cost of their survival.

In contrast, *K*-selected species are adapted to tighter ranges of more permanent resources and maintain their overall numbers close to the carrying capacity of their habitat. They reproduce slowly and generally produce a single, large offspring per reproductive bout, which requires parental care and protection. Unlike *r*-selected species, *K*-selected species thrive in more predictable environments. This predictability allows long-lived *K*-selected species to exhibit later sexual maturity. They allocate little energy and risk to reproduction in the short term but invest a great deal of energy over time in their survival and parental care. These species are often viviparous (live-bearing) to minimize the risk of reproductive failure. All large mammals, such as elephants, deer, cattle, sheep, and goats, as well as whales and dolphins, exhibit *K*-selected traits. They produce single offspring that require parental care, typically provided by the mother. However, in these particular cases, the young are born relatively mature and instantly mobile for a variety of ecological and evolutionary reasons (Sibly and Brown 2009).

Within a genus or even a species, different populations can exhibit different life history strategies that can impact population growth. For the lizard *Lacerta (Zootoca) vivipara*, the Austrian mountain population appears to exhibit *K*-selected traits, characterized by low mortality, high maximum age, and delayed sexual maturity at the fourth year. The lowland population in Brittany (France) has more *r*-selected traits, with 50% of females reproducing at one year of age. In the Austrian population, most females live much longer, albeit with lower fecundity, compared to the lowland populations in the Netherlands and France, where they breed more and live shorter adult lives (Strijbosch and Creemers 1988). The growth rates of the *Lacerta vivipara* populations from the cold Austrian mountains and the warmer lowland France—differing in both abiotic and biotic elements—align with expected patterns for *K*-selected and *r*-selected populations. This example demonstrates that populations of the same species living under different climatic conditions may exhibit varying growth rates due to location-specific adaptations. Such locally evolved populations are known as ecotypes.

Indeed, when the notion of categorizing species into *r*- and *K*-selected species was conceived, the argument was that in temperate and arctic regions, populations face harsh weather conditions and individuals die independent of their population size. Frequent population crashes are followed by longer periods of population increase. During such conditions, adaptations that increase rapid intrinsic population growth are selected (*r*-strategies). In contrast, in stable tropical environments where populations fluctuate minimally, adaptations that improve the competitive efficiency of resource utilization are favored (*K*-strategies). No organism is completely r- or *K*-selected; rather, an organism may be an ***r*-strategist** or a ***K*-strategist** *relative* to another species. This makes an *r*-selection and *K*-selection dichotomy a continuum, with each organism positioned along it depending on its environment at a given instant in time (Pianka 1972).

Although the idea of *r* and *K* selection helped biologists think more systematically about species characteristics and life history variation, this dichotomy does not do well in generalizing a great deal of known variation in species and their life histories (Reznick et al. 2002). For instance, among warm-blooded organisms, a species with a small body size is generally categorized as an *r*-strategist, while those with a large body size are categorized as *K*-strategists. However, warthogs and pigs, which can easily weigh more than 100 kilograms, are *r*-strategists because females produce multiple offspring that mature early. Similarly, while the longevity and competitiveness of trees might characterize them as *K*-strategists, many tree species display an *r*-strategy, typically producing thousands of offspring in each reproductive bout. Reptiles such as sea turtles also display traits that are both *r*- and *K*-selected. Although they are large organisms with long lifespans (*K*-strategy), they produce large numbers of offspring once they reach adulthood (*r*-strategy).

How Many Kinds of Population Dynamics Does Nature Reveal?

The logistic equation is a model of population growth—a simplified representation of the world that helps us understand basic processes. The growth rates of simple, lab-grown organisms, such as yeast and paramecium, fit nicely with the logistic equation; however, the *S*-shaped curve provides a poor approximation to the growth rates of actual populations (Slobodkin 2001). One reason for this discrepancy is that certain assumptions of the model may not accurately reflect what is observed in nature (Allee et al. 1949). For instance, not all individuals in a population may be identical with respect to reproductive output or resource consumption. Likewise, for most species, r and K are not constants because these parameters vary spatially, temporally, and, in the case of r, even with respect to density among various populations. The logistic model also assumes no time lags; however, we know that there is an inevitable lag in feedback between population density and the actual instantaneous rate of increase. Because life is messier (and more interesting) than the simple logistic equation suggests, we must remember that the basic logistic growth model is one of the simplest tools in our kit for understanding population dynamics.

The simple logistic growth equation is especially poor at modeling species with a complex life history, such as those of many insect species, as well as populations with age or stage structure, like Pacific salmon, because the effects of birth rates may be delayed. Such a delay may cause the population to overshoot its carrying capacity, leading to a subsequent decline. The population may then stabilize at the carrying capacity, fall below it, and then grow again, or continue oscillating. In some real populations, such oscillations may stabilize, while in others, they may increase, leading to population crashes. Population behavior depends on a species' ecology, its interactions with the abiotic environment (such as seasonal variation or climate change), as well as those with the biotic environment (such as predators and competitors), so complexity in population growth is inevitable.

That leads to the question: How many kinds of population dynamics in natural populations do we find other than exponential and logistic growth?

Dynamics of Real Populations

Natural populations have revealed four complex kinds of population dynamics: irregular (stochastic or chaotic), cyclic, flush-crash, and episodic.

Irregular Population Dynamics

Some populations, such as those of small birds in higher latitudes, or some types of insects, are subject to environmental stochasticity. Their populations are characterized by random fluctuations due to changes in environmental factors. For example, spruce bark beetle populations are sensitive to the availability of storm-felled trees (Figure 5).

Irregularity may not necessarily be stochastic (random) but can arise from chaotic population dynamics. Chaotic systems are highly sensitive to initial conditions, meaning very small changes in environmental factors can produce significantly different population outcomes. Rogers et al. (2022) searched for chaos in 172 population time series and discovered it was most prevalent in plankton and insect populations and least prevalent in bird and mammal populations.

Cyclic Population Dynamics

A cyclic population exhibits regular fluctuations in population size—or boom-and-bust cycles—over regular periods. Examples include the collared lemming (*Dicrostonyx groenlandicus*) in northern Greenland, whose population dynamics are driven by predators such as the stoat (*Mustela erminea*), which specializes in hunting lemmings (Gilg et al. 2003). Forest grouse populations in Finland also exhibit cyclical population dynamics in response to predation, fluctuating every two years as predation by goshawks increases or decreases (Figure 6; Tornberg et al. 2005).

Other studies have implicated predators as the driving force behind cycles of field voles in Scandinavia (Korpimäki and Norrdahl 1998). But when predators were removed in a large-scale field experiment in England, field vole cycles were not affected (Graham and Lambin 2002). Studies like these lead us to question whether population cycles are driven more by food availability than predation.

Canada lynx (*Lynx canadensis*) and snowshoe hares (*Lepus americanus*) exhibit population dynamics with nine to 11-year cycles (Figure 7). To explain the hare-lynx cycle, ecologists distinguish between two hypotheses: the bottom-up (food) hypothesis and the top-down (predation) hypothesis.

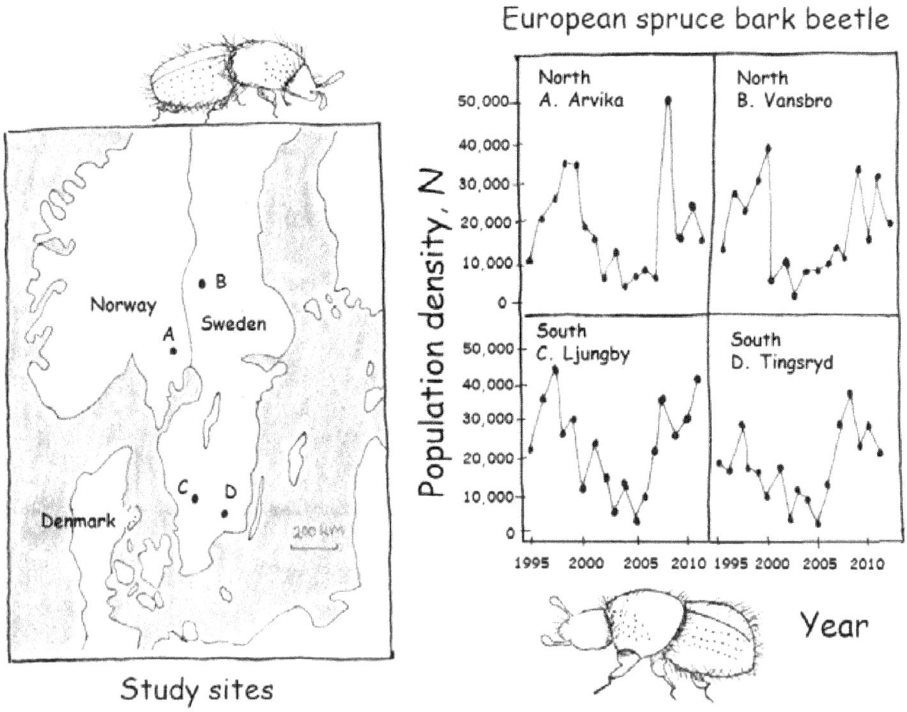

Figure 5. Inter-annual fluctuations in population density of the European spruce bark beetle, *Ips typographus*, across four regions between 1995 and 2011 (interpreted from Figure 1 in Marini et al. 2013). The population dynamics were driven by the availability of breeding substrates in the form of storm-felled trees (an abiotic environmental factor), whereas the subsequent intra-specific competition for host trees was the main regulator of population density (a biotic environmental factor).

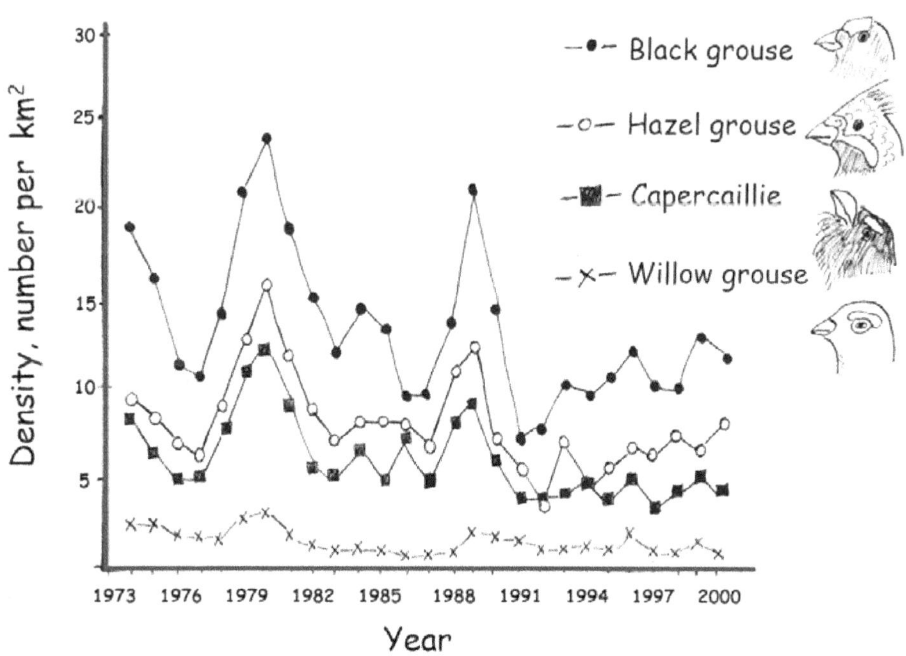

Figure 6. Density (individuals per km^2) changes of four forest grouse species in the province of Vaasa, Finland, during 1974–2000 (interpreted from Figure 2 in Tornberg et al. 2005). Goshawk predation contributes to the generation of multi-year cycles in forest grouse. The occupancy rate of goshawk territories tracked grouse fluctuations with a two-year lag.

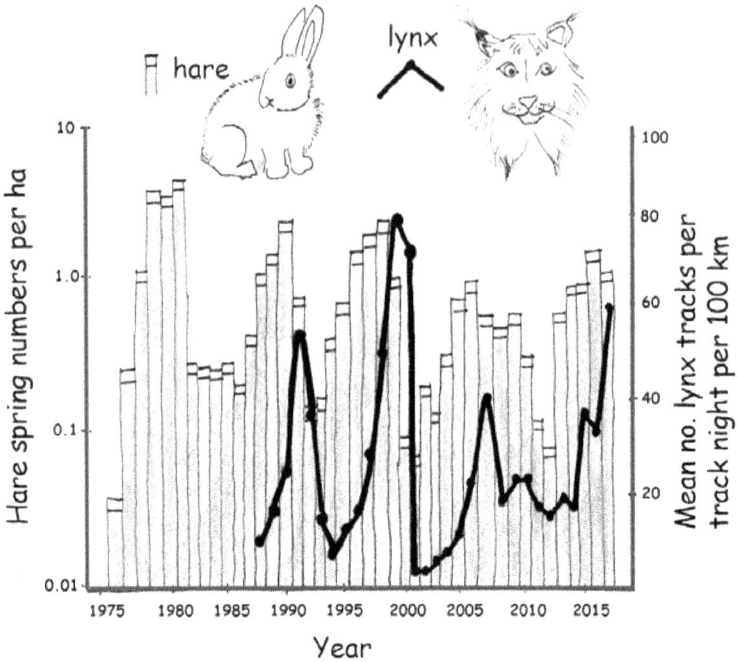

Figure 7. Population trend for snowshoe hares (bars) and lynx track index (line) in the Kluane Lake region of the Yukon Territory (interpreted from Krebs et al. 2018).

The bottom-up hypothesis posits that when hares reach high densities, they consume all their food and subsequently starve, causing the lynx population to decline due to the decline in the hare population. The top-down hypothesis, in contrast, predicts that lynx populations reach high density in response to increases in hares, but with a time lag that drives the cycle. Increasing numbers of predators eventually **overkill** their prey, both through direct mortality and indirectly through stress or behavioral effects, and the prey population begins to decline. However, due to time-lag effects, the predators continue to increase for a period, driving the prey to an even lower density. Finally, at low prey densities, many predators starve, allowing the prey to recover, and the cycle repeats itself. Is the hare population controlling the lynx population (intraspecific density-dependent factor affecting hares), or is the lynx population controlling the hare population (interspecific density-dependent factor affecting lynx), or is it a combination of both? The work of Charles Krebs and colleagues has demonstrated that predators drive the snowshoe hare cycles, both via direct mortality and through indirect stress-related or behavioral effects on hares (Lavergne et al. 2021).

Flush-Crash or Epidemic Dynamics

Flush-crash or epidemic dynamics are observed in pathogens, parasites, and bacteria—organisms characterized by high reproductive rates ("*r*"-strategists). The population crash of ash trees infested by ash borer beetles across several U.S. states, costing millions in timber revenue, exemplifies this type of dynamic in nature. It is common for populations to crash to low levels following a disease outbreak. The surviving individuals repopulate following the decimation until another outbreak of parasites occurs. Defoliating parasitic insects (e.g., spruce budworm) show this dynamic, affecting a staggering area of North American forests. Spruce budworm in the 1970s affected over 50 million hectares of fir and spruce forests at its peak in Eastern Canada and the Northeastern United States, making the species one of the most economically and ecologically important forest insects on the continent (Figure 8).

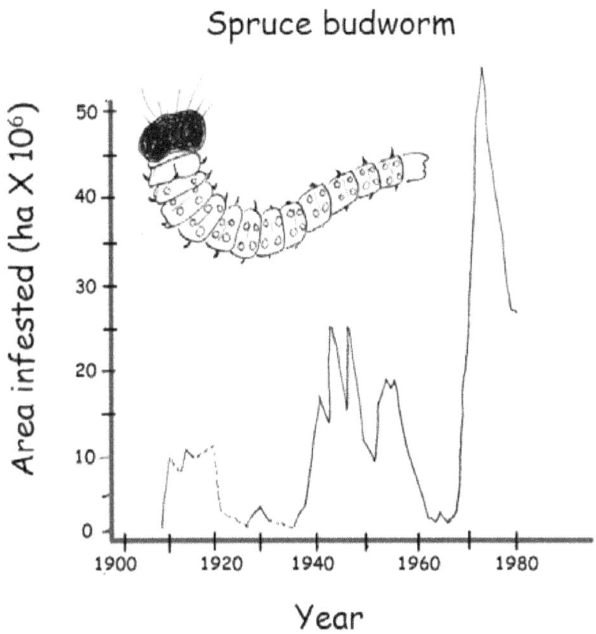

Figure 8. The total area of budworm infestation east of Ontario, including Maine (interpreted from the Figure in Kettela 1983). Changes in the size of areas of infestation were likely correlated with population density.

Another example includes rinderpest (cattle plague), which has recurred throughout history. It hit Europe particularly hard in the 18th century and, then in the 1890s, killed 80 to 90% of all cattle and wildebeest in southern Africa (Sinclair 1977).

Episodic Dynamics

Episodic population dynamics are irregular, but governed by regular weather events, rather than environmental stochasticity. As a result, changes in population size are more predictable than those in populations with truly irregular population dynamics. For example, a population of mosquitoes irrupts when the temperature is favorable but plummets once the temperature drops. Similarly, fluctuations and crashes in small bird populations, such as Darwin's medium ground finches, may occur due to severe weather events like drought (Gibbs and Grant 1987). In Southern Australia, the red kangaroo population reached 2,175,000 in 1981 but plummeted precipitously to 745,000 by 1984 due to the El Niño–caused droughts of 1982 and 1983 (see Chapter 12 *Climate Change*). Incredibly, red kangaroos rebounded in 1985, only to be decimated again in 1987 by another El Niño event. The population then increased again in 1988 and 1989 in response to abundant rains (Cairns and Grigg 1993).

Population as Unit of Evolution

Individual animals, plants, and microorganisms grow, develop, reproduce, and eventually die. The characteristics expressed at the level of the individual organism collectively determine the dynamics of the population. A clear example of this is *r*- and *K*-selected traits: some individuals exhibit *r*-strategies, such as producing large clutch size, while others follow *K*-strategies, such as providing long-term parental care. These individual traits, in turn, influence population-level dynamics. Population dynamics give rise to evolutionary dynamics. Individuals do not evolve; populations and species do. Individuals retain the same genetic constitution throughout their lives. However, the genetic constitution of a population varies to a greater or lesser extent, and therefore, the characteristics of populations change over time as population size changes from one generation to the next. The population is the smallest unit where evolutionary change is possible.

Evolutionary ecologists have documented a dramatic shift in a physical characteristic of finch populations in the Galápagos Islands (Grant 1999; Grant and Grant 2007; Grant and Grant 2014). Mean beak size in the population of medium ground finch (*Geospiza fortis*) in Daphne Major, one of the Galápagos Islands off the coast of Ecuador, changed over generations. Beak size is a trait that influences birds' feeding behavior. Individuals with large beaks feed on a wide range of seeds, from small to large, whereas individuals with smaller beaks feed only on smaller seeds. The 1977 drought caused substantial mortality in the medium ground finch population; however, this mortality did not fall equally on all segments of the population. As seeds were depleted during the drought, the birds ate the smallest and softest seeds first, leaving the largest and toughest seeds. Because birds with smaller beaks could not crack the remaining larger and tougher seeds, mortality fell most heavily on smaller birds. By the end of the drought, the surviving population consisted mainly of larger individuals with larger, stronger beaks that could feed on hard seeds. The characteristic (beak size) of medium ground finches changed, within three generations, as their population size fluctuated from 1976 to 1998. This illustrates how population dynamics (a dramatic drop in population size) led to evolutionary dynamics (a change in beak size).

References

Allee, W. C., Emerson, A. E., Oark, O., Park, T., and Schmidt, K. P. 1949. Principles of Animal Ecology. Saunders, Philadelphia.

Andrewartha, H. G. and Birch, L. C. 1954. The Distribution and Abundance of Animals. University of Chicago Press, Chicago, IL.

Cairns, S. C. and Grigg, G. C. 1993. Population dynamics of red kangaroos (*Macropus rufus*) in relation to rainfall in the south Australian pastoral zone. Journal of Applied Ecology 30: 444–58.

Caughley, G. 1994. Directions in conservation biology. Journal of Animal Ecology 63: 215–244.

Craighead, J., Varney, J. R. and Craighead Jr, F. C. 1974. A population analysis of the Yellowstone grizzly bears. University of Montana Bulletin: Forestry Series, 1949–1982.

Crump, M. L. Hensley, F. R. and Clark, K. L. 1992. Apparent decline of the golden toad: underground of or extinct? Copeia 1992: 413–420.

Darwin, C. 1859. On The Origin of Species by Means of Natural Selection. John Murray. London, UK.

Dobzhansky, T. 1973. Nothing in biology makes sense except in the light of evolution. American Biology Teacher 35: 125–129.

Gause, G. F. 1934. The Struggle for Existence. Williams and Wilkins, Baltimore, Maryland.

Gibbs, H. L. and Grant, P. R. 1987. Ecological consequences of an exceptionally strong El Niño event on Darwin's finches. Ecology 68: 1735–46.

Gilg, O., Hanski, I. and Sittler, B. 2003. Cyclic dynamics in a simple vertebrate predator-prey community. Science 302: 866–868.

Graham, I. M. and Lambin, X. 2002. The impact of weasel predation on cyclic field-vole survival: the specialist predator hypothesis contradicted. Journal of Animal Ecology 71: 946–956.

Grant, P. 1999. Ecology and Evolution of Darwin's Finches. Princeton University Press. Princeton, NJ.

Grant, P. R. and Grant, B. R. 2007. How and why species multiply: the radiation of Darwin's finches. Princeton University Press.

Grant, P. R. and Grant, B. R. 2014. 40 years of evolution: Darwin's finches on Daphne Major Island. Princeton University Press.

Hairston, N. G., Tinkle D. W., and Wilbur, H. M. 1970. Natural selection and the parameters of population growth. The Journal of Wildlife Management 34: 681–690.

Huffaker, C. B. 1958. Experimental studies on predation: dispersion factors and predator-prey oscillations. Hillgardia 27: 343–383.

Kettela, E. G. 1983. A cartographic history of spruce budworm defoliation from 1967 to 1981 in eastern North America. Canadian Forest Service Information Report DPC-X14. Environment Canada, Ottawa, Canada.

Korpimäki, E. and Norrdahl, K. 1998. Experimental reduction of predators reverses the crash phase of small-rodent cycles. Ecology 79: 2448–2455.

Krebs, C. J., Boonstra, R., Kenney, A. J. and Gilbert, B. S. 2018. Hares and small rodent cycles: a 45-year perspective on predator-prey dynamics in the Yukon boreal forest. Australian Zoologist 39: 724–732.

Lack, D. 1954. The natural regulation of animal numbers. Oxford University Press, New York.

Lavergne, S. G., Krebs, C. J., Kenney, A. J., Boutin, S., Murray, D., Palme, R. and Boonstra, R. 2021. The impact of variable predation risk on stress in snowshoe hares over the cycle in North America's boreal forest: adjusting to change. Oecologia 197: 71–88.

Laws, R.M. 1970. Elephants as agents of habitat and landscape change in East Africa. Oikos 21: 1–15.

MacArthur, R. H. and Wilson, E. O. 1967. The Theory of Island Biogeography. Princeton University Press, Princeton, NJ.

Mallet, J. 2012. The struggle for existence. How the notion of carrying capacity, K, obscures the links between demography, Darwinian evolution and speciation. Evolutionary Ecology Research 14: 627–665.

Malthus, T. 1798. An Essay on the Principle of Population. J. Johnson, London, UK.

Marini, L., Lindelöw, Å, Jönsson, A. M., Wulff, S. and Schroeder, L. M. 2013. Population dynamics of the spruce bark beetle: a long-term study. Oikos 122: 1768–1776.

Nicholson, A. J. 1957. The self-adjustment of populations to change. Cold Spring Harbor Symp. Quant. Biol. 22: 153–173.

Murie, A. 1934. The Moose of Isle Royale. University of Michigan Press, Ann Arbor.

Ochoa-Ochoa, L. M., Whittaker, R. J. and Ladle, R. J. 2013. The demise of the golden toad and the creation of a climate change icon species. Conservation and Society 11: 291–319.

Pianka, E. R. 1972. r and K Selection or b and d Selection? The American Naturalist 106: 581–588.

Pianka, E. R. 1983. Evolutionary Ecology, 3rd edition. Harper and Row.

Pianka, E. R. 2000. Evolutionary Ecology. Sixth Edition. Benjamin-Cummings, Addison-Wesley-Longman. San Francisco.

Reilly, S. B., Rice, D. W., and Wolman A. A. 1983. Population assessment of the gray whale, *Eschrichtius robustus*, from California shore censuses, 1967–80. Fisheries Bulletin 81: 267–280.

Reznick, D., Bryant, M. J. and Bashey, F. 2002. r-and K-selection revisited: the role of population regulation in life-history evolution. Ecology 83: 1509–1520.

Rogers, T. L., Johnson, B. J. and Munch, S. B. 2022. Chaos is not rare in natural ecosystems. Nature Ecology and Evolution 6: 1105–1111.

Sánchez-Zapata, J. A., Donázar, J. A., Delgado, A., Forero, M. G., Ceballos, O. and Hiraldo, F. 2007. Desert locust outbreaks in the Sahel: resource competition, predation and ecological effects of pest control. Journal of Applied Ecology 44: 323–329.

Shinozaki, K., Yoda, K., Hozumi, K., and Kira, T. 1964. A quantitative analysis of plant form-the pipe model theory: I. Basic analyses. Japanese Journal of Ecology 14: 97–105.

Sibly, R. M. and Brown, J. H. 2009. Mammal reproductive strategies driven by offspring mortality-size relationships. The American Naturalist 173: E185–E199.

Sinclair A. R. E. 1977. The African Buffalo: A Study in Resource Limitations of Populations. Chicago University Press, Chicago.

Slobodkin, L. B. 2001. The good, the bad and the reified. Evolutionary Ecology Research 3: 1–13.

Strijbosch, H. and Creemers, R. C. M. 1988. Comparative demography of sympatric populations of *Lacerta vivipara* and *Lacerta agilis*. Oecologia 76: 20–26.

Tinkle, D. W. 1969. The concept of reproductive effort and its relation to the evolution of life histories of lizards. The American Naturalist 103: 501–516.

Tornberg, R., Korpimäki, E., Jungell, S., and Reif, V. 2005. Delayed numerical response of goshawks to population fluctuations of forest grouse. Oikos 111: 408–415.

Vaughan, J. A., Noden, B. H. and Beier, J. C. 1992. Population dynamics of *Plasmodium falciparum* sporogony in laboratory-infected *Anopheles gambiae*. The Journal of Parasitology 78: 716–724.

Vergara, R., Gitzendanner, M. A., Soltis, D. E. and Soltis, P. S. 2014. Population genetic structure, genetic diversity, and natural history of the South American species of *Nothofagus* subgenus Lophozonia (Nothofagaceae) inferred from nuclear microsatellite data. Ecology and Evolution 4: 2450–2471.

Weller, D. E. 1987. Self-thinning exponent correlated with allometric measures of plant geometry. Ecology 68: 813–821.

Chapter 3

Species Interaction

Competition

"[T]he evidence is strong that the observed competition with [the barnacle] Balanus [is] the principal factor determining the local distribution of [the barnacle] Chthamalus." (Connell 1961)

No species lives in isolation. Although members of a species interact with their physical abiotic environment in numerous complex ways, they also live with other species together in communities, embedded in webs of biotic interactions. Species interactions begin at the level of the individual organisms that make up a population. These interactions among individuals can have significant impacts on their growth, survival, and reproduction, ultimately affecting population-level dynamics for the interacting species.

The way species interact directly with each other is mostly through feeding behavior, whether they share a resource or one is a resource for the other. Species in a community are also indirectly interconnected through trophic (feeding) relationships. We will discuss indirect interactions through trophic relationships in Chapter 8 *Food Webs*. In this chapter, we will mostly focus on direct interactions between pairs of species. To better understand the roles species play within their communities and how they can positively or negatively affect other species, we can categorize direct pairwise interactions between participating species (Table 1; Odum 1971).

When two species live together in the same area but neither population affects the other, the interaction is called **neutralism** (that is, 0/0 effects). Neutralism occurs when two species share no common resource or when they have partitioned the resource in some way. For instance, a study of the European roe deer (*Capreolus capreolus*) and the brown hare (*Lepus europaeus*), both herbivores found in Central Italy, showed significant differences in their dietary composition. The roe deer's staple foods comprised deciduous and evergreen woody plants (about 70% of the total diet), whereas cereals, spontaneous herbs, and graminoids were the staple of hares (about 90% of the total diet). The roe deer and hares had neutral effects on one another due to the abundance of cultivated cereals and spontaneous herbs for hares, as well as high-quality woody browse for the deer. This reduced interspecific competition between these herbivores because they largely foraged in different habitats within Mediterranean agricultural areas (Sangiuliano et al. 2016). In addition to dividing a resource in space, neutralism can occur when resources are divided in time (e.g., when insectivorous birds feed on diurnal insects, while insectivorous bats feed on nocturnal insects). Neutralism can also occur between two species that are on different trophic levels and have no feeding relationship between them.

Amensalism occurs when one species is inhibited or destroyed while the other remains unaffected. One classic example is the bread mold *Penicillium*, which produces an enzyme that inhibits the growth of gram-positive bacteria, while the mold remains unaffected. Certain plants also secrete substances that inhibit the growth of other plants; for example, red pine (*Pinus densiflora*)

DOI: 10.1201/9781003008538-3

Table 1. Kinds of direct (pairwise) interactions between species (modified from Odum 1971).

Interaction	Effect on Species 1	Effect on Species 2	Description
Neutralism	0	0	Neither species affects the other.
Amensalism	–	0	One species is disadvantaged, the other not unaffected.
Commensalism	+	0	One species is advantaged, the other is not affected
Mutualism	+	+	Both species benefit (it can be obligatory or facultative)
Competition	–	–	Each species disadvantages the other.
Predation	+	–	Carnivore advantaged, prey disadvantaged (often lethal)
Herbivory/ Granivory	+	–	Herbivore/granivore advantaged, prey disadvantaged (rarely lethal to the adult plant)
Parasitism	+	–	Parasite advantaged, host disadvantaged (may be lethal but often not)
Parasitoid	+	–	Similar to parasitism, but always lethal

forests often have no understory because the secretions by red pine destroy the plants underneath them (Kato-Noguchi et al. 2017). Amensalism may be difficult to find among co-occurring species because a species that inhibits another species with zero effect on itself will exert a selective pressure on the inhibited species, causing the disadvantaged species to be eliminated (hence the lack of understory growth above) unless it evolves adaptive traits that neutralize the effect (in which case, the relationship will be neutral).

Commensalism is an interaction in which one species benefits while the other remains unaffected (+/0). For example, most humans have tiny mites inhabiting their hair follicles. The mites benefit by getting nourishment, but their presence is so minimal that it does not affect us. Likewise, the reproductive success of a tree is neither helped nor harmed by the moss or lichens that benefit by growing on its trunk and branches. The endemic pipit, *Anthus steindachneri*, on the Antipodes Islands of New Zealand, feeds on flies that grow and breed in penguin droppings (Slack 2010). The pipit benefits from the presence of penguins without affecting the penguins at all.

Interactions that benefit both populations (+/+) are classified as **mutualisms**. Flowering plants and their coevolved pollinators are a good example of mutualism. Many of the bacteria in our intestines are also mutualistic. Mutualistic interactions may be obligatory, meaning neither population can exist without the other, or facultative, where the mutualism benefits both species but is not essential for their survival.

Predation (+/–) occurs when individuals of one species consume individuals from another species' population to the detriment of the prey and benefit of the predator. In **competition**, both species are disadvantaged by the overlapping use of a limiting resource, and the relationship is characterized by (–/–).

Competition, predation, and mutualism are seemingly more prevalent in nature than the other categories of interactions. In this chapter, we focus on competition. Predation will be explored in the next chapter.

Intraspecific and Interspecific Competition

Competition occurs when individuals restrict each other's access to limited resources by either depressing resource abundance or using antagonistic behaviors to prevent access to those resources. Competition is an inevitable consequence of increasing population size and limited resources. In *"On the Origin of Species,"* Darwin (1859) emphasized the struggle for existence, arguing that not all individuals can be supported by the existing resources essential for life. For Darwin, the competition between individuals of the same species—**intraspecific competition**—was the mechanism behind adaptive traits. Over many generations, natural selection, driven by intraspecific competition (the struggle for existence), could change a population's characteristics or even give rise to new species. Ecologists have also long recognized that competition for scarce resources among species—**interspecific competition**—influences the distribution and abundance of interacting species and can also impact their evolution.

Competition need not necessarily imply a literal battle between individuals—in nature, we rarely observe organisms fighting to the death over resources—but we often see antagonistic (threatening) behaviors. For instance, hummingbirds are notorious for their outsized fierceness when it comes to defending a nectar source against others, and hyenas may threaten lions over a carcass. However, competition is not necessarily an open conflict either. It may also result from common use of a resource—for example, two crustaceans consuming algae, thereby reducing its availability in a freshwater lake, or introduced gray squirrels consuming acorns, a prominent food source for the smaller red squirrels native to England. As with individuals in the same population, when individuals of different species require the same limiting resources, the presence of each species inhibits the population growth of the other; that is, they compete. Food is often a limiting resource, whether it is the soil nutrients used by bacteria or the grass consumed by deer. In arid ecosystems, water may be a limiting resource. Plants, especially trees, compete for limited light to produce glucose through photosynthesis. Plants, algae, and animals also consume space—which can be a resource like food, water, and light—by defense or occupation, thereby making it unavailable to their own or other species for use, ultimately limiting population growth.

Interspecific competition and intraspecific competition are similar in that, in both cases, the interactions between individuals cause competition to occur. However, one major difference between intraspecific and interspecific competition is the population consequences of those individual interactions. Intraspecific competition leads to stable population dynamics that center around a carrying capacity. In crowded populations, intraspecific competition is so intense that it results in a process known as **self-thinning**, leading to increased mortality among less competitive individuals and relative success by more competitive individuals. For instance, as a local population of plants in an open patch of forest develops from seeds to seedlings to saplings to adults, individual plants take up increasing quantities of nutrients, water, and space, with some individuals competing more successfully than others. Those that cannot acquire enough resources die, resulting in a thinning or decrease in population density. Over time, the population is composed of a few large individuals, and the population size reaches a steady state.

Interspecific competition can also be intense, as the acquisition of a resource by individuals of one species reduces its availability to individuals of another species, eventually affecting the abundance and distribution of both species, which, in turn, contributes to the structural and functional organization of ecosystems (Cody and Diamond 1975). Unlike intraspecific competition, interspecific competition can have population consequences that result in the exclusion of one species, coexistence for both, or the evolution of traits that reduce competition, which can then influence the distribution and dynamics of the evolved species.

Exploitative vs. Interference Competition

Two categories of competition are recognized by ecologists, defined by the types of interactions that occur between individuals: exploitative competition, which involves indirect interactions, and interference competition, which involves direct interactions.

Most competitive interactions are **exploitative**. In exploitative competition, individuals contribute to the depletion of a resource without affecting each other's ability to obtain that resource (Elton and Miller 1954). For example, harvester ants in the Sonoran Desert (USA) do not directly interact with seed-eating rodents, but they use the same resource. When Brown and Davidson

(1977) experimentally excluded rodents, the number of ant colonies increased by 70%, and when ants were excluded, rodent biomass increased by 24%, indicating that the presence of one species decreased the population size of the other. Under exploitative competition, a common limiting resource becomes less available, impacting population growth; however, the competitors do not directly interact with one another.

In **interference** competition, the ability of an individual to use a resource is affected by the presence of competing individuals (Elton and Miller 1954). In this type of competition, while limited resources may be available, they are much less accessible to individuals or species that are competitively inferior. This kind of competition can occur over space to grow. For example, the barnacle *Chthamalus stellatus* is competitively inferior to the barnacle *Semibalanus balanoides* in the struggle for space on intertidal rocks because *Semibalanus* lifts, crushes, or smothers *Chthamalus* to seize the space for itself (Connell 1961). Bryozoans and corals also exhibit this type of competition via direct contact and overgrowth, thereby inhibiting competitors' access to light and food (Connell 1979).

Interference competition also happens when one species interferes with another species at feeding sites. In Italy, the introduction of fallow deer (*Dama dama*), recorded over a 4-year study, exhibited significant behavioral interference with the foraging behavior of roe deer (Ferretti et al. 2011). The fallow deer engaged in direct aggression in 90% of interactions with the roe deer, leading to an increase in the density of fallow deer and a decrease in the density of roe deer.

Although interference competition often occurs over space or territory, it can also occur over the resource itself. For example, the hummingbird *Amazilia tzactl* chases skipper butterflies (Hesperiidae) away from the upper and middle portions of flowering hedges—where most inflorescences are present—forcing the butterflies to use the sparser lower inflorescences (Primack and Howe 1975). Under interference competition, direct interactions between species can limit access to resources for one another; however, this comes at a cost to the superior competitor, which is paid in terms of time and energy spent usurping or defending those resources, ultimately impacting its population growth as well.

Interference competition has not been as extensively documented as exploitative competition, which occurs anytime species consume a common limiting resource. It has been argued that interference competition should not evolve unless the energetically costly "squabbling" potentially leads to increased access to a limiting resource. In other words, a costly interactive behavior is unlikely to evolve if there is no payoff (Roughgarden 1983). However, we see in many cases of interference competition, the "squabble" may not be as energetically costly to the competitor as the benefit gained from increased resource access. For example, it costs hummingbirds relatively little to chase butterflies compared to the increased amount of nectar that becomes available by chasing them away. Similarly, in a forest, the cost of investing more carbon in a taller trunk pays large dividends when the tallest trees get all the light.

Both interference competition and exploitative competition can appear lopsided or asymmetrical (in that one species appears to be a much better competitor than another), but assuming that neither species obtains all of the resource, the effects of competition are negative for both. In a competitive interaction between two species, each population affects the fitness (per capita growth rate) of the other adversely. Negative interactions manifest themselves as reduced abundance, lower fitness, or a decrease in one or more fitness components such as body size, growth rate, fecundity, or survivorship.

Competitive Exclusion, Niche, and Coexistence

When individuals of one species compete with another species, one may wonder: Why do competing species coexist? Why do the two species not intrude upon each other until one is excluded? Would not the one that has any advantage at all outdo the other and cause its exclusion?

Russian microbiologist Georgy Gause (1934) tested the idea that interspecific competition can result in coexistence. Gause, under laboratory conditions, grew two different species of Paramecium, *P. caudatum* and *P. aurelia*, by feeding them bacteria—both separately and together. In nature, he argued, there were too many variables to draw conclusions about how the mechanisms of competition work; however, in a controlled experimental setup, stronger inferences could be obtained than from observing natural systems. When cultured separately, Gause found that each population achieved stable densities (remember the logistic equation?). But when he reared both species in a single tube containing a fixed amount of bacterial food, *P. caudatum* died out while *P. aurelia* continued to exist, demonstrating the exclusion of the former species by the latter. However, in another experiment, Gause reared *P. caudatum* with *P. bursaria*, and both species coexisted because *P. caudatum* fed on bacteria suspended in solution, whereas *P. bursaria* confined its feeding to bacteria at the bottom of the tubes. Each species partitioned the resource in space, allowing them to coexist because their feeding site (ecological requirements) was not identical.

The results of Gause's study and findings from other lab experiments with fruit flies, flour beetles, and annual plants led to the concept known as the **competitive exclusion principle**, which states that no two species possessing identical ecological requirements can long coexist. The idea of ecological difference can be more easily grasped by the concept of the **niche**.

Originally, "niche" referred to a recessed place in a church wall, often used for displaying objects. Ecologists have co-opted the term as a way to envision all the environmental factors that influence a species' growth, survival, and reproduction. Because this single word encompasses so much, the term "niche" has been defined in different ways by different ecologists. Joseph Grinnell (1917) defined the niche of an organism as the habitat or environment it occupies, whereas Elton (1927) saw it as the role a species plays in the community—essentially everything it does to survive and reproduce. A species' niche, in Elton's view, is its profession. Later workers, following Grinnell and Elton, recognized that a species' niche could be used to describe both physical and biological factors, lumped together as ecological requirements, needed for growth, survival, and reproduction (e.g., Chase and Leibold 2003).

Each species, or more precisely, each local population within a species, has a specific range of environmental conditions, such as temperature and humidity, in which the population lives and reproduces successfully. The population also requires certain types of food (or soil minerals in the case of plants) on which it must subsist. For our imaginary population, it is possible to draw a three-dimensional graph with a range of temperatures on one axis, humidity on another, and concentration of nutrients on the third, showing the blob of niche space that the population occupies. We can continue to add other ecological requirements to the "niche," such as soil pH in the case of plants or salinity in the case of kelp. Using such metrics, we can analyze a niche, component by component, until a complete picture of the ecological requirements is constructed. This was what G. Evelyn Hutchinson (1957) envisioned when he crystallized the niche concept as "an n-dimensional hypervolume," where "n" equals the number of all ecological requirements a species needs to survive, grow, and reproduce. Hutchinson also distinguished between an organism's fundamental niche and its realized niche. The fundamental niche, according to Hutchinson, defines the physical conditions under which a species might live in the absence of interactions with other species, whereas the realized niche is the actual niche of a species whose distribution is limited by biotic interactions such as competition, predation, disease, and parasitism.

In a classic study—mentioned briefly before in the context of interference competition—Connell (1961) investigated competitive interactions between two species of barnacles on intertidal rocks along the coast of Scotland. *Chthamalus stellatus* (a smaller barnacle) inhabits shallower water, where tidal action often exposes it to air, while *Semibalanus balanoides* (a bigger barnacle) lives lower down (deeper intertidal zone), where it is rarely exposed. In the deeper zone, *Semibalanus* outcompeted *Chthamalus* by crowding it off the rocks. When Connell removed *Semibalanus* from the area, however, *Chthamalus* occupied the deeper zone, indicating that no physiological constraints prevented it from inhabiting the zone. In contrast, *Semibalanus* cannot survive in the shallow-water habitats where *Chthamalus* normally occurs; it evidently lacks the special adaptations to thrive in that zone. Thus, the fundamental niche of the barnacle *Chthamalus* included both shallow and deeper zones, but its realized niche was much narrower because of the presence of the superior competitor *Semibalanus.* Conversely, *Semibalanus*' realized and fundamental niches were both identical.

The principle of competitive exclusion, then, states that unless the niches of two species differ, the species cannot coexist indefinitely. Because competitive interactions have negative effects on population fitness factors, natural selection is expected to favor individuals that minimize competitive overlap. This leads to species that are distinct in their ecological requirements, suggesting that the populations we see today are composed of descendants of individuals that survived competition because they were able to avoid it. The principle of competitive exclusion predicts that speciation is partly a process of avoiding competition by promoting differences.

What we see most often in nature is not competitive exclusion, but the coexistence of species over time. Nature abounds with coexisting species: we see a coral reef with over a hundred species of fish, a hectare of tropical rainforest supporting over 300 species of trees, and a temperate lake containing 30 to 40 species of planktonic algae. Hutchinson (1961), in a paper entitled "The paradox of the plankton," asked: "Why are there so many species of phytoplankton in the same habitat of open water of a lake with little habitat structure?" Most algae compete for the same array of resources, including carbon dioxide, nitrogen, phosphorus, sulfur, trace elements, and vitamins. Competition theory suggested that the best competitor should exclude weaker competitors; however, this did not seem to be happening in the lake ecosystem. Was it the apparent failure of the competitive exclusion principle? Hutchinson suggested that competition between algae might be mitigated by the effect of gradually changing physical conditions within the lake and that this environmental variation might provide a resolution to the paradox. Today, ecologists are aware of a wide variety of factors—such as variations in temperature, spatial and temporal variations in resource availability, and resource partitioning—that provide different competitive advantages to different species occupying the same habitat (e.g., Klęparski et al. 2022). When species occupy different niches, this allows for coexistence in the same place.

General Descriptive Model of Competition

The concept of niche, by itself, provides only a crude and tautological understanding of competitive exclusion—a species occupies a niche because it is not identical to the niche of any other species. However, there is much more to the saga of competition that we need to know than just that two competing species are not identical in their ecological requirements. To understand competition more completely, we must return to the elemental study of species' population dynamics and growth rates. Because competitive interactions happen at the individual level, the effect of competition increases with density. In population dynamic terms, the competitive exclusion principle dictates that competitive exclusion occurs only when one species produces enough individuals to cause population decline in another. The converse of competitive exclusion is coexistence. Again, in population terms, this will happen only if the two species' niches are sufficiently different that neither population suppresses the other to exclusion. Each species will increase logistically until it reaches its equilibrium population size (as observed by Gause with *P. caudatum* and *P. bursaria*). If, at equilibrium, neither population size is big enough to cause population decline in the other, the two species will coexist in the same geographical range.

Working independently, Lotka (1925) and Volterra (1926) developed a theoretical basis for interspecific competition by modifying the logistic equation, which was initially proposed by Pearl and Reed (1920). Recall that for one species, population growth can be modelled by

$$\frac{dN}{dt} = rN\left(\frac{K-N}{K}\right) \qquad \text{Equation 1}$$

where N is the population size, r is the intrinsic rate of increase (the species' maximum possible growth rate achieved only under ideal conditions), and K is the carrying capacity at which the population stops increasing in size (when $N = K$, by definition). In the logistic equation, the first term (rN) on the right side of the equation describes the exponential population growth that occurs when no (intraspecific) competitors are present. The second term denotes intraspecific competition and increases its effect on the growth term as a function of population size, taking on a value of 1 when there are no intraspecific competitors and a value of 0 when the population is at carrying capacity (K). The addition of this term puts a break on potential exponential growth, slowing it to a maximum ($rN \times 0 = 0$) when the population reaches the carrying capacity. Both Lotka (1925) and Volterra (1926) realized that if *interspecific* competition is also occurring, the competition term can be modified to reflect the additional competition. And because there are now two species to consider, there needs to be two interconnected equations to represent interspecific competition between the two species.

Like intraspecific competition, the effect of interspecific competition depends on both population size and the competitor. Because the effect of competition may be scaled differently between one species and its competitor, a new variable is introduced called the "coefficient of competition." Thus, the Lotka-Volterra equations for interspecific competition take the form of the following coupled equations for population growth of species 1 and species 2, respectively, using subscripts to denote values for species 1 and species 2:

$$\frac{dN_1}{dt} = r_1N_1\left(\frac{K_1 - N_1 - \alpha_{12}N_2}{K_1}\right) \qquad \text{Equation 2a}$$

$$\frac{dN_2}{dt} = r_2N_2\left(\frac{K_2 - N_2 - \alpha_{21}N_1}{K_2}\right) \qquad \text{Equation 2b}$$

In Equations 2a and 2b, all the terms are directly analogous to those in the single species logistic equation, except for the new terms, α_{12} and α_{21}, which are the coefficients of competition. α_{12} is a characteristic of species 2 that measures its per capita competitive value with respect to an individual of species 1 (Equation 2a). If individuals of species 1 and 2 were identical in their competitive effects on each other (e.g., both were capable of catching and consuming identical resources with equal ability), then α_{12} would equal 1, and every individual of species 2 would reduce the carrying capacity of species 1 in the same way that every individual of species 1 reduces the carrying capacity of species 1. If α_{12} is less than one, then the competitive effect of every individual of species 2 would have a slightly less effect on the carrying capacity of species 1 than an individual of species 1 would. If $\alpha_{12} = 3$, each individual of species 2 would decrease the growth of species 1 by the same amount as three additional individuals of species 1. Similarly, α_{21} measures the per capita impact of species 1 on the carrying capacity of species 2 (Equation 2b). Note that the first number of the subscript always refers to the species being affected.

To summarize, when α_{21} is < 1, the effect of species 2 on species 1 (interspecific competition) is less than the effect of species 1 on its own members (intraspecific competition). In contrast, when α_{12} is > 1, the effect of species 2 on species 1 is greater than the effect of species 1 on its own members. The α_{21} term in Equation 2b is interpreted in the same way. When α_{12} or α_{21} equals one, intra- and interspecific competition are directly equivalent.

There are essentially three possible outcomes of two-species competition: the exclusion of species 1 by species 2, the exclusion of species 2 by species 1, or coexistence, although slight variations on these outcomes are also possible. The differing outcomes of two-species competition are directly the result of the relationship between the competition coefficients, and the key to predicting the outcome is whether a species inhibits its growth more than it inhibits the growth of the other species.

In other words, whether intraspecific competition is stronger than interspecific competition (Adler et al. 2018). If intraspecific competition is stronger for both species, then the population of one species cannot become abundant enough to drive the population of the other species to extinction. However, if intraspecific competition is stronger only for species 1 but not species 2, the population of species 1 is likely to increase and reach its carrying capacity while the population of species 2 is likely to decrease to zero, so species 1 "wins" the competitive interaction and "excludes" species 2. In general, the Lotka-Volterra model predicts the coexistence of two species only when, for both species, interspecific competition is weaker than intraspecific competition. Otherwise, one species is predicted to eventually exclude the other.

This concept is easily demonstrated by programming a computer to solve Equations 2a and 2b under different scenarios where the competitive coefficients are set at values greater than or less than one, but a graphical approach makes it much easier to examine the conditions under which each species' population would be expected to increase or decrease. We begin by determining when the populations of species 1 and 2 cease to grow. Since $\frac{dN}{dt}$ represents population growth in the Lotka-Volterra model, a population is neither increasing nor decreasing when $\frac{dN}{dt}$ equals zero (indicating an equilibrium population size).

To find the conditions of zero population growth for species 1 we start with Equation 2a,

$$\frac{dN_1}{dt} = r_1 N_1 \left(\frac{K_1 - N_1 - \alpha_{12} N_2}{K_1} \right)$$

set population growth to zero,

$$0 = r_1 N_1 \left(\frac{K_1 - N_1 - \alpha_{12} N_2}{K_1} \right)$$

and simplify,

$$0 = K_1 - N_1 - \alpha_{12} N_2$$

Rearranging the equation, we get

$$N_2 - \frac{K_1}{\alpha_{12}} \quad \frac{1}{\alpha_{12}} N_1 \qquad \text{Equation 3a}$$

and find that Equation 3a is the equation for straight line. We can graph it thus with N_1 representing the *x*-axis and N_2 representing the *y*-axis (Figure 1a):

Note that Figure 1a depicts a "state-space" where each point in the space represents a combination of the abundances of the two species (represented by N_1 and N_2). The straight line derived from Equation 3a is an isocline. "Iso" means "same." Because we derived the line for all points where species 1's population growth is zero, this line is called the **zero growth isocline** for species 1, and it represents the combined values of the population sizes for species 1 (N_1) and species 2 (N_2) at which the population growth rate of species 1 is zero ($\frac{dN_1}{dt} = 0$). This occurs when the combined population sizes (in terms of species 1) are equal to the carrying capacity of species 1. We see from Figure 1a that when $N_2 = 0$, the population size (N_1) of species 1 is at its carrying capacity K_1. However, as N_2 increases along the isocline, to maintain zero growth, N_1 decreases. When N_2 reaches K_1/α_{12}, the population of species 1 ceases to exist ($N_1 = 0$).

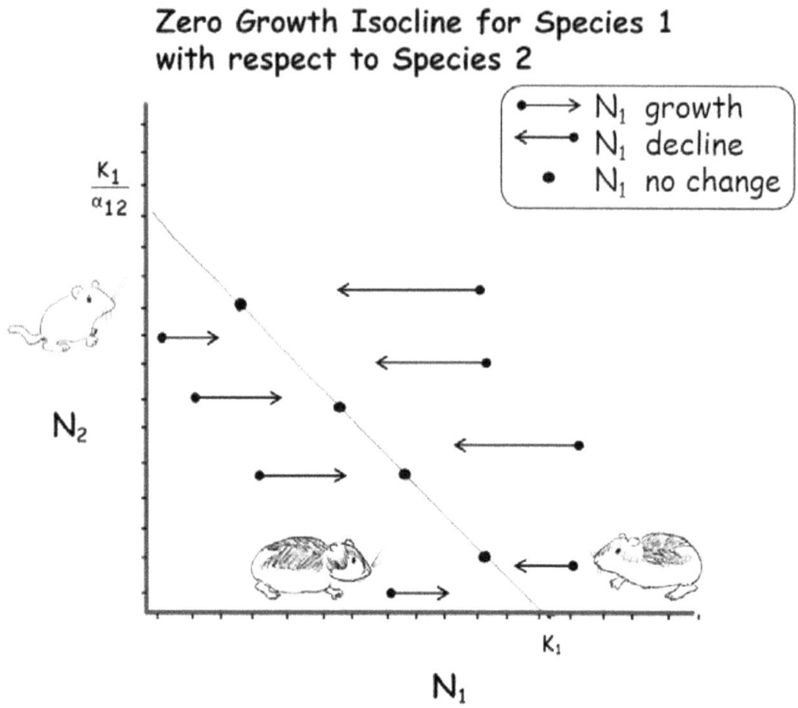

Figure 1a. Species 1's zero growth isocline in the population size state space for the Lotka-Volterra competition equations.

Above the zero growth isocline, the population size N_1 must decrease because there are too many individuals of species 1 for the environment to support population growth in species 1. Below the zero growth isocline, N_1 is increasing because the environment can still support more individuals of species 1. The arrows in Figure 1a depict the areas of the state space where N_1 must decline or increase. Note that while we are focused on what is happening to species 1, the values of N_1 and N_2 must be considered jointly because an increase in N_2 means that something is being taken away from species 1, resulting in a decrease in its growth rate.

Applying the same logic for species 2, we can start with Equation 2b:

$$\frac{dN_2}{dt} = r_2 N_2 \left(\frac{K_2 - N_2 - \alpha_{21} N_1}{K_2} \right)$$

set population growth to zero,

$$0 = r_2 N_2 \left(\frac{K_2 - N_2 - \alpha_{21} N_1}{K_2} \right)$$

and simplify,

$$0 = K_2 - N_2 - \alpha_{21} N_1$$

Rearranging the equation, we get

$$N_1 = \frac{K_2}{\alpha_{21}} - \frac{1}{\alpha_{21}} N_2 \qquad \text{Equation 3b}$$

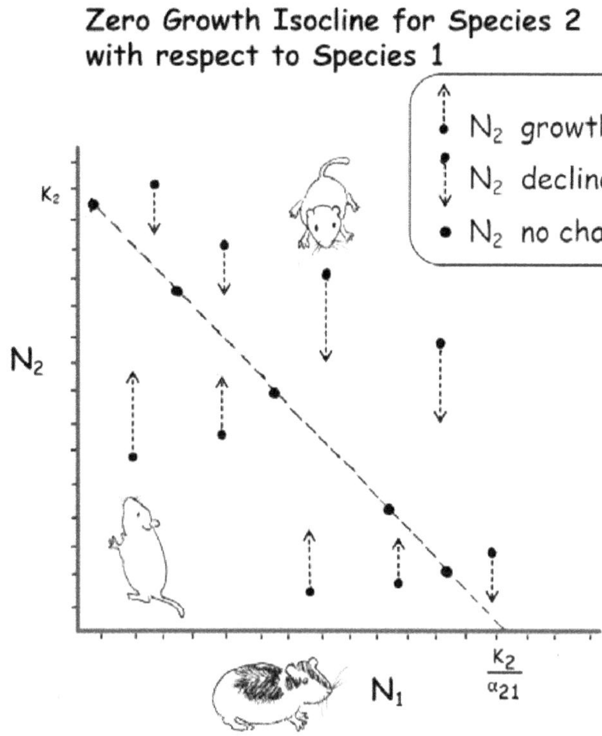

Figure 1b. Species 2's zero growth isocline in the population size state space for the Lotka-Volterra competition equations.

Equation 3b is another equation for a straight line. We can graph it thus (Figure 1b). Note that in Figure 1b, the arrows now reflect what is happening to N_2 in various parts of the state-space.

Now, suppose we want to consider the population growth of both species simultaneously. Plotting the zero growth isoclines for species 1 and species 2 in the same state-space produces four possible outcomes of competition between the two species (Figures 2a–2d). These outcomes depend on where each species' isocline lies in relation to the other. In each graph, a solid line represents species 1's isocline and the dashed line represents species 2's isocline. The thick black arrows represent the joint trajectory of two populations, and the thinner arrows indicate trajectories of the individual populations.

Scenario 1: When the isocline for species 1 is entirely above the isocline for species 2, species 1 "wins" in competition, and species 2 is excluded (Figure 2a). In the lower region, each species exhibits positive growth, whereas in the upper region, both species exhibit negative growth. In the middle region, between the two isoclines, the population growth of species 2 is negative, while that of species 1 is positive. The joint movement of two populations (thick black arrows) is down and to the right, causing species 2 to go extinct and species 1 to increase until it reaches the carrying capacity (K_1). The "bullseye" at this point represents a stable equilibrium. Species 1 always outcompetes species 2, and this is referred to as the competitive exclusion of species 2 by species 1. Looking at the intercepts in Figure 2a, we can write:

$$\frac{K_1}{\alpha_{12}} > K_2 \text{ and } K_1 > \frac{K_2}{\alpha_{21}}$$

Rearranging the equations gives

$$K_1 > K_2\,\alpha_{12} \text{ and } K_1\alpha_{21} > K_2$$

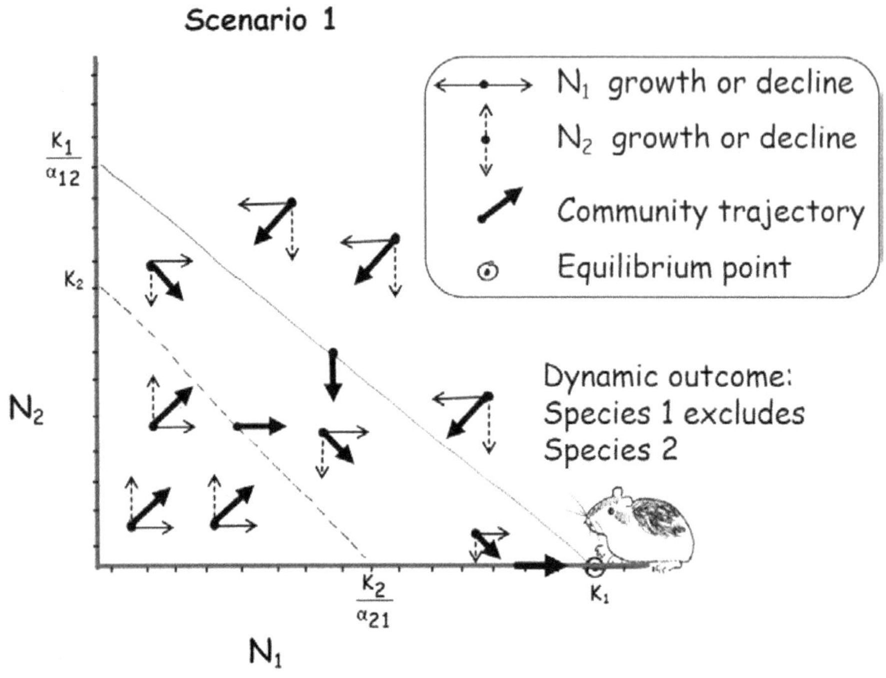

Figure 2a. Species 1 and 2 zero isoclines drawn together (scenario 1).

The first inequality ($K_1 > K_2\,\alpha_{12}$) means the **intraspecific effects** of species 1 on itself, denoted by K_1, are greater than the **interspecific effects** of species 2 on species 1 (represented by $K_2\alpha_{12}$). This suggests that species 2 is an inferior interspecific competitor. The second inequality ($K_1\alpha_{21} > K_2$) points out that species 1 can exert more of an effect on species 2 than species 2 can exert on itself. Species 1 is a superior interspecific competitor. And the verdict? Species 1 excludes the inferior interspecific competitor species 2 and reaches its carrying capacity.

Scenario 2: By the same logic, when the isocline for species 2 is entirely above the isocline for species 1, species 2 "wins" and excludes species 1 (Figure 2b). The situation is reversed in Figure 2b with respect to the inequalities, such that $K_2 > K_1\,\alpha_{21}$ and $K_2\,\alpha_{12} > K_1$. Thus, species 2 excludes the inferior interspecific competitor species 1 and reaches its carrying capacity.

In Figures 2a and 2b, neither isocline crossed the other. However, with graphical depictions of population dynamics, any time two zero-growth isoclines cross, an equilibrium point is formed. That means if the population sizes of species 1 and species 2 lie exactly on that point, neither population will grow nor decline; each will be at **equilibrium**. However, this point can be either **stable**, meaning the populations will converge toward that point if they are in a nearby area of the state space, or **unstable**, meaning the populations will move away from that point in a nearby area of the state space. Given the variability of the environment, it becomes very unlikely that N_1 and N_2 can remain at equilibrium on an unstable equilibrium point, which leads us to the following scenario.

Scenario 3: If the isoclines are oriented as in Figure 2c, the population trajectories point away from the equilibrium within the boundaries of the two isoclines. Thus, population densities in these quadrants will converge to fixation at either K_1 or K_2 (that is, either species 1 or species 2 will win), depending on which side of the equilibrium point the population densities are in the state-space. Thus, the outcome of competition in this case depends on the starting densities of species 1 and species 2.

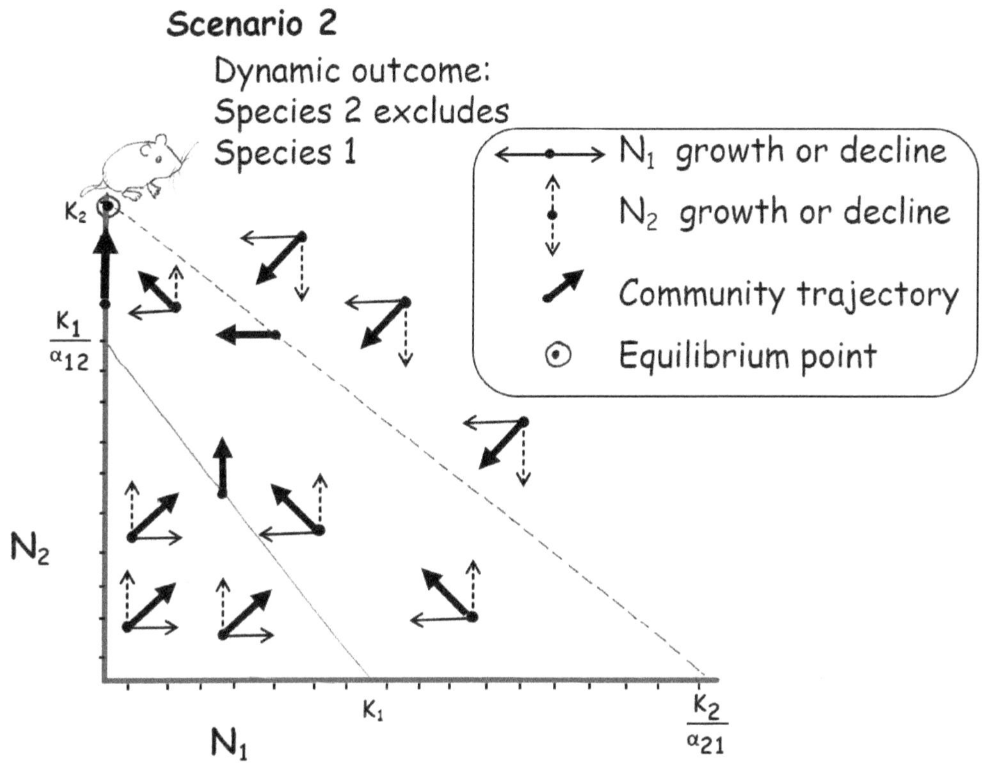

Figure 2b. Species 1 and 2 zero isoclines drawn together (scenario 2).

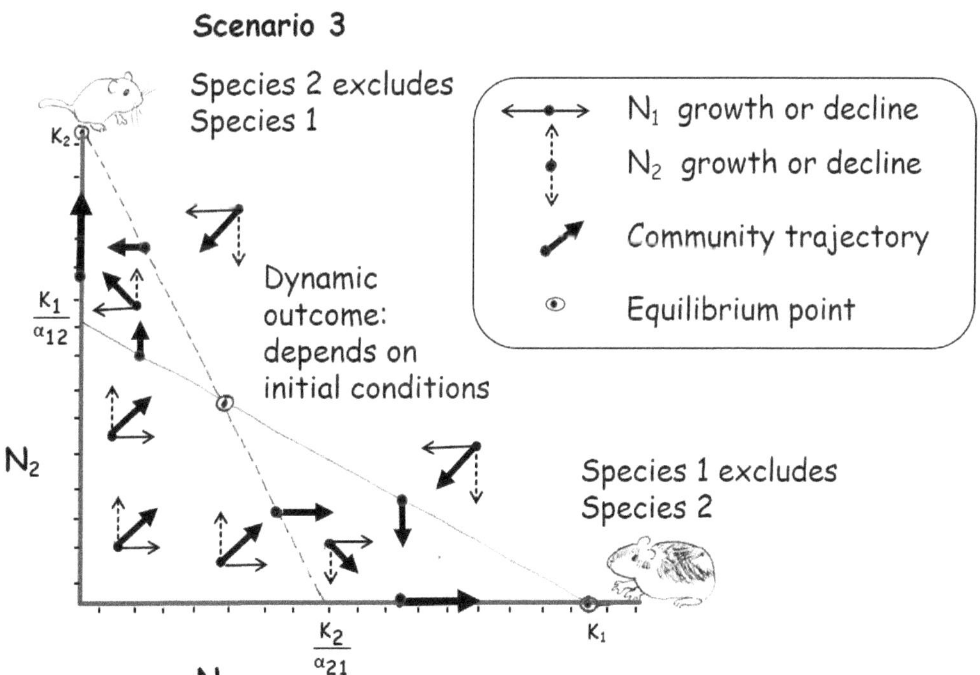

Figure 2c. Species 1 and 2 zero isoclines drawn together (scenario 3).

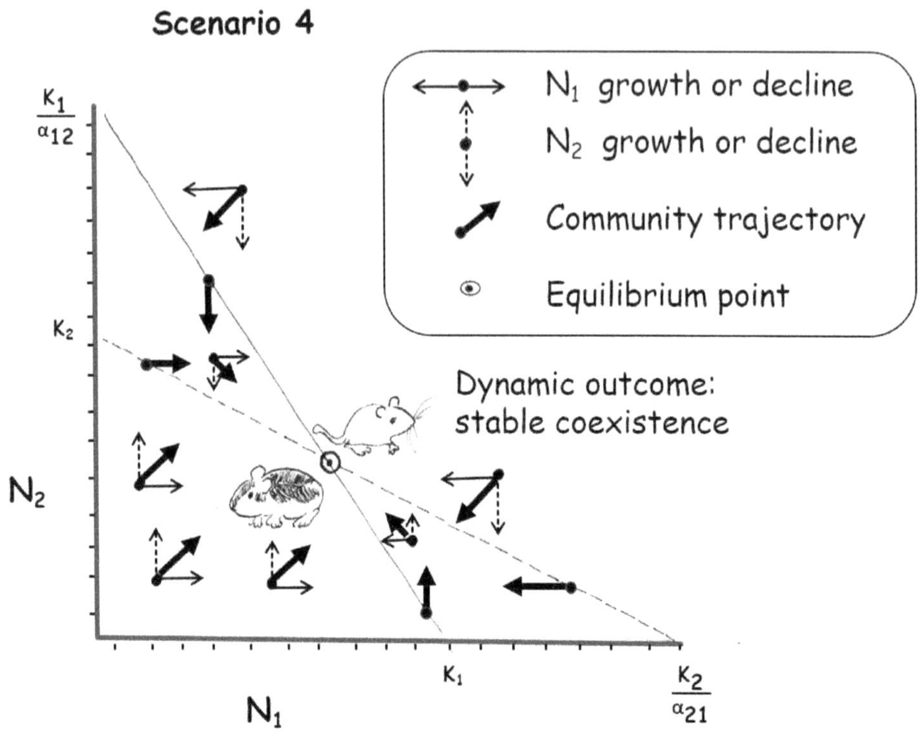

Figure 2d. Species 1 and 2 zero isoclines drawn together (scenario 4).

The intercepts in Figure 2c, $\frac{K_1}{\alpha_{12}} < K_2$ and $K_1 > \frac{K_2}{\alpha_{21}}$, give us the following inequalities: $K_2\alpha_{12} > K_1$ and $K_1\alpha_{21} > K_2$.

Individuals of both species have a greater interspecific competitive effect on individuals of the other species than the intraspecific competitive effect each species has on itself. In the lower-right side of the graph, species 1 reaches its carrying capacity, excluding species 2, and in the upper-left side of the graph, species 2 reaches its carrying capacity, excluding species 1. In other words, both species are strong interspecific competitors. Under the right conditions, each species is capable of driving the other's population to zero, but which does so cannot be predicted unless it is known which species started with a higher population density. Whichever species has the upper hand can capitalize on that and outcompete the other.

Scenario 4: When the isoclines are oriented as in Figure 2d, the two species coexist because the trajectories for both populations always point toward the equilibrium point, and the densities of both species will become positive and stable when they reach it. In this case, both species' carrying capacities are lower than the other's carrying capacity divided by the competition coefficient. Rather than outcompeting one another, the two species coexist at the stable equilibrium point (bullseye). This is the outcome regardless of either species' initial abundance.

When examining the intercepts in Figure 2d and simplifying as above, we see that $K_1 > K_2\alpha_{12}$ and $K_2 > K_1\alpha_{21}$. In this case, both species experience less interspecific competition than intraspecific competition, indicating they are weak interspecific competitors. In such a scenario, due to the relative strength of intraspecific competition, neither population can grow to such an extent that it completely displaces the members of the other population. The outcome of this type of competition is the stable coexistence of the competitors in the same habitat.

Lotka–Volterra Competition Model in Theory and Practice

The Lotka–Volterra competition model predicts that, in most cases, competitive exclusion should occur, and in only one special case, when intraspecific competition is stronger than interspecific competition within both species, will coexistence occur. Our world is not a monoculture; therefore, is this special condition true in most instances of competition between species, or is competition just not a common interaction? Regarding the latter, 90% of the 164 ecological studies of competition in Schoener's (1983) analysis, and 83% of the 54 studies surveyed by Connell (1983) demonstrated interspecific competition, and the observation that competition is a strong force in nature was recently corroborated by Adler et al. (2018). Interspecific competition is pervasive. That leaves us with the special condition. We might imagine a case where this special condition is met if members of species 1 squabble over resources but ignore those of species 2, and vice versa, but this seems like a restrictive rather than universal case. So why the seemingly mismatch between theory and nature? Why do species coexist?

The Lotka–Volterra competition model is, of course, a vast simplification of how competition might occur in nature among real species. Simplifying assumptions include the following:

1. Environmental conditions are constant or homogenous across space.
2. Environmental conditions do not change over time.
3. The system is closed; that is, there are no immigrations and emigrations within and among populations.
4. Species exhibit consistent behavior.

Because the Lotka–Volterra model prioritizes competitive exclusion, in contrast to the coexistence we observe in nature, what it suggests is that the key to coexistence—other than the special condition that intraspecific competition be greater than interspecific competition—may lie in the violation of these assumptions.

Environmental Heterogeneity in Space

The assumption that environmental conditions are constant across space is rarely met in practice. In nature, we see a heterogeneous environment that, in many cases, promotes species coexistence. Even a seemingly homogenous beech–maple temperate forest reveals substantial variations in the availability of nitrogen and potassium (Lechowicz and Bell 1991), and even the variation in light penetrating through a forest may allow different species to coexist. For example, beech thrives in shade, whereas maple is favored in gaps (Poulson and Platt 1996). This is an example of habitat partitioning.

Habitat partitioning occurs when species consume the same food, but they do it in distinct places; different species possess different spatial niches, so they do not engage in exploitative competition. We recall that Gause (1934), inspired by the Lotka–Volterra competition equations, set up a series of carefully controlled experiments with three species of *Paramecium*. In the first case, he grew *P. aurelia* and *P. caudatum* together, and *P. caudatum* was always excluded. In the second case, when *P. caudatum* was grown with *P. bursaria*, he found coexistence. However, upon closer inspection, he discovered that his careful experiments were violating a Lotka–Volterra assumption: both species coexisted because they used distinct microhabitats for feeding. *P. caudatum* foraged on bacteria in the open water column, while *P. bursaria* fed on bacteria at the bottom of the test tubes. Gause had found an excellent example of spatial habitat partitioning.

Other examples of habitat partitioning exist. MacArthur (1958) found five species of warblers living together in the same spruce forests of northeastern North America. All five warbler species—Cape May (*Dendroica tigrina*), Yellow-rumped (*D. coronata*), Black-throated Green (*D. virens*), Blackburnian (*D. fusca*), and Bay-breasted (*D. castanea*)—are similar morphologically, and they overlap extensively in their insect diets. How could several warbler species with apparently similar ecological requirements live together in the same forest? MacArthur showed that the warblers seemed to partition each tree into different zones and confined their foraging within a particular one. Although the Blackburnian warbler's feeding zone overlapped broadly with the Cape May warbler's, it tended to feed farther down the tree from the top zone. The Black-throated green warbler fed within the tree's interior, the Bay-breasted warbler fed in the tree's middle zone, and the Yellow-rumped warbler fed in the lower parts of the tree and on the ground. This habitat partitioning allowed the species to coexist.

Connell's (1961) study of barnacles found that one species, which could inhabit both low and high intertidal habitats, was excluded from the lower intertidal and confined to the upper intertidal habitat due to competition with another species, which specialized in the lower habitat. Coexistence occurred, but only because one species had a refuge from competition. In some cases, coexistence may occur because the preferred habitats do not overlap at all. For example, guilds of related species with similar morphology and diets, such as rodent herbivores in northern latitudes, would normally compete with one another for limited resources (Morris 2005). However, on Herschel Island in Canada's Yukon Territory, three species of rodents can coexist at low densities along a wet-dry habitat continuum. Two of the species exhibit distinct habitat preferences, which are caused by habitat specialization rather than competitive exclusion. The tundra vole (*Microtus oeconomus*) specializes in the extreme wet microhabitat, while the collared lemming (*Dicrostonyx groenlandicus*) favors the extreme dry microhabitat. A third species, the Siberian brown lemming (*Lemmus sibiricus*), is a habitat generalist on Herschel Island, using moderately wet and moderately dry microhabitats, although it exhibits a strong preference for wetter habitats. Competition from *Lemmus* appears to shift the collared lemming into more extreme dry microhabitats, thereby reducing competition between the two. Consequently, *Lemmus* avoids competition with *Microtus* by utilizing more moderate wet microhabitats than those occupied by *Microtus* (Ale et al. 2011).

Environmental Heterogeneity in Time

Another simplifying assumption of the Lotka–Volterra model is that the environment remains constant over time. However, environmental conditions often vary over time, providing competitive advantages to species at different times. Coexistence can occur when no individual species reaches sufficient density to displace its competitors, thereby allowing both competitor species to coexist. An excellent example of this is the annual variations in the relative abundance of grass species in a savanna community in southwest Zimbabwe. Between 1971 and 1981, the dominant grass species shifted from *Urochloa mosambicensis* to *Heteropogon contortus* in response to changes in precipitation patterns. *Urochloa mosambicensis* competed successfully under the drier conditions prevailing from 1971 to 1973, but with increasing rainfall beginning in 1973, *Heteropogon contortus* started to dominate the site (Dye and Spear 1982).

Species can also avoid competition and coexist when resources fluctuate in availability over time, and the consumer species have trade-offs in their relative ability to exploit low and high resource densities. For example, a tree species that thrives in high-light environments may coexist with one that shades out its competitors if a patchwork of light gaps is generated at different times. With such temporal and spatial variation, the outcome of competition between the two species cannot be predicted by a simple Lotka–Volterra model. Although one species is the inferior competitor under a constant environment, it may turn out to be more successful at colonizing open gaps created by disturbances such as fire, storms, or the grazing of an herbivore, and thus coexist with the superior competitor in the same habitat.

Like spatial habitat partitioning, coexistence can also occur due to habitat partitioning in time. In Europe, two mosquito species, *Culex pipiens* (the main vector for West Nile viruses) and *Aedes albopictus* (which transmits not only West Nile but also dengue and Zika viruses) have co-occurred since the last century when *A. albopictus* was introduced from Asia. Since its arrival, *A. albopictus* has rapidly spread and occupied habitats already exploited by *C. pipiens*. There is competition between the two species, and according to Marini et al. (2017), this competition amplifies the seasonal separation in time, with the *C. pipiens* population arising early in the season and declining (quicker) than otherwise with the rise in numbers of *A. albopictus*. Marini and colleagues also demonstrated that competition induces significant reductions in the total abundance of *C. pipiens*. Another example of temporal partitioning of the environment is the relationship between swallows, a type of bird which are diurnal, and bats, which are nocturnal. Both are insectivores, but they coexist in the same habitat because swallows feed on diurnal insects, while bats feed on a different set of nocturnal insects. Similarly, herbivorous insects that live and grow during the spring season feed on different plants from those that live and grow in the fall. In each of these examples, the partitioning of time allows the species to coexist in the same habitat.

Open Systems

The third assumption of the Lotka–Volterra model is that there are no immigrations and emigrations of individuals to or from either species' populations. However, these kinds of population dynamics can lead to coexistence. The Lotka–Volterra model assumes a closed system with no dispersal to replace individuals lost to the effects of competition in a population. However, when a small fraction of the individuals from one habitat patch disperses to other habitat patches, this can prevent competitive exclusion from happening. Such a system exemplifies Levin's (1974) source-sink dynamics. Species coexist when individuals from the competing populations emigrate from "source" patches where they are doing well to "sink" patches where their fellows have declined, although this kind of coexistence may be unstable and, over a longer period of time, may end up with the exclusion of one or the other species.

Behavioral Plasticity

In the central Himalaya, two closely related species of large cats, the snow leopard (*Panthera uncia*) and the common leopard (*Panthera pardus*), living in sympatry (the same area), have coexisted through habitat separation and exploitation of different prey species. The snow leopard uses open habitats (grassland and shrubland), whereas the common leopard uses forest (closed habitat). Despite habitat separation, both cats prey upon the Himalayan tahr (*Hemitragus jemlahicus*), musk deer (*Moschus moschiferus*), and domestic animals, utilizing both open and closed habitats as potential prey. Dietary overlap between the snow leopard and the common leopard was found to be 69% annually, with 76% during the colder months and 60% in the warmer months. Rather than exploiting different resources, these cats notably avoid interspecific aggression. This behavioral adjustment may facilitate the coexistence of potentially competing large predators (Lovari et al. 2013).

Another example of behavioral adjustment can be found in the black rhinoceros (*Diceros bicornis*) in the Addo Elephant National Park (South Africa). Landman et al. (2013) found that in the presence of African elephants (*Loxodonta africana*), black rhinoceros switched their preference to grasses they normally avoided in the absence of elephants. Rhinos increased their diet breadth and shifted their diet along the grass-browse continuum (elephant present: 20.8% grass, 79.2% browse; elephant absent: 6.1% grass, 93.9% browse). Since black rhinoceros are strict browsers, their lack of specialized grazing adaptations may increase foraging costs by reducing their efficiency in harvesting and handling grasses. This insight may also have management implications, as managing elephants for high densities may compromise the foraging opportunities of coexisting browsers such as rhinos.

Predictions from the Lotka–Volterra competition model apply only under equilibrium conditions; however, real ecological systems may often be unsaturated (not yet at equilibrium), or they may violate the model assumptions—as we saw above. Despite these shortcomings, simple models of competition, such as the Lotka–Volterra equations, have played a prominent role in making ecological theory quantitative and predictive (Levins 1968; MacArthur 1972). Although the Lotka–Volterra equations have been extensively used to model and better understand competition, they remain phenomenological, as they overlook the actual mechanisms of competition and fail to explicitly include resources. Using this critique of the 1920s-era Lotka–Volterra model as a stepping stone, more mechanistic models of exploitative competition that include multiple resources have been developed over the years, including those by MacArthur (1970, 1972), May (1971), Tilman (1982), and McPeek (2022). All these models have broadened our understanding of the mechanisms of competition that lead to coexistence or exclusion.

The Competitive Exclusion Principle and Its Consequences

In the Lotka–Volterra interspecific competition model, competition is abstractly depicted as the negative effect of individuals of one species on the population growth of a second species. Coexistence occurs only when, for both species, intraspecific competition is stronger than interspecific competition. Although this may seem like an unlikely condition, when considered alongside the (also abstract) idea of the niche, it begins to make sense. One expects the strongest competition will be between individuals of the same species because they, by definition, occupy the same niche. Therefore, coexistence can occur only if two species occupy different enough niches not to have an undue effect on the other's population growth. Alternatively, when two species have overlapping niches, this will lead to competitive exclusion. This concept is encapsulated in the **Competitive Exclusion Principle,** which states that "complete" competitors—species with complete niche overlap—cannot coexist (Hardin 1960). Thus, we expect to see adaptations that place similar species into different niches.

Competition may also be avoided in evolutionary time through the action of natural selection and evolution. When two species interact negatively, selection occurs in each population for physiological, morphological, and behavioral adaptations that allow individuals to exploit common resources in a manner that reduces the adverse effects of competition. Thus, we can expect species to have evolved characteristics that enable them to compete less or not at all with members of other similar species. This implies that observations of co-occurring species should reveal differences in behavior, physiology, or morphology that lessen competitive interactions—such as the spatial, temporal, and behavioral resource partitioning discussed earlier.

Connell (1980) aptly referred to this line of thinking as the **Ghost of Competition Past**. Although we cannot directly observe the role of past competition on the evolution of adaptive traits, we can still examine differences in the traits of subpopulations of a species that face different competitive environments to infer that natural selection has taken its course. Grant (1986), for instance, in his studies of Darwin's finches on the Galapagos Islands, reported that the medium ground finch (*Geospiza fortis*) and the small ground finch (*Geospiza fuliginosa*) both feed on an overlapping array of seed sizes. When the two species occur alone on different islands—*G. fortis* on Daphne Major and *G. fuliginosa* on Los Hermanos—they show substantial overlap in beak size and, consequently, in seed size preference. However, when the two species live (and compete) on the same island (Santa Cruz), there is almost no overlap in beak size and little overlap in resource use. This divergence in beak size is presumed to be the result of evolutionary forces on beak size due to competition. These competing species on the island of Santa Cruz exhibit **character displacement**, a phenomenon in which phenotypic traits (in this case, beak size) diverge as a result of the exploitation of shared, limited resources.

In general, it is expected that when two species of animals overlap geographically, the differences between them will be accentuated in the zone of sympatry and weakened or lost entirely in the parts of their ranges outside this zone (Lack 1947; Brown and Wilson 1956). Although the competitive exclusion principle can be nearly impossible to disprove (Hardin 1960), it remains a powerful idea in understanding the evolutionary consequences of competition for limited resources.

References

Adler, P. B., Smull, D., Beard, K. H., Choi, R. T., Furniss, T., Kulmatiski, A. et al. 2018. Competition and coexistence in plant communities: Intraspecific competition is stronger than interspecific competition. Ecology Letters 21: 1319–1329.

Ale, S. B., Morris, D. W., Dupuch, A., and Moore, D. E. 2011. Habitat selection and the scale of ghostly coexistence among Arctic rodents. Oikos 120: 1191–1200.

Brown, J. H. and Davidson, D. W. 1977. Competition between seed-eating rodents and ants in desert ecosystems. Science 196: 880–882.

Brown, W. L., Jr. and Wilson, E. O. 1956. Character displacement. Systematic Zoology 5: 49–64.

Chase, J. M. and Leibold, M. A. 2003. Ecological Niches: Linking Classical and Contemporary Approaches. University of Chicago Press.

Cody, M. L. and Diamond, J. M. 1975. Ecology and evolution of communities. Belknap Press of Harvard University, Cambridge, MA.

Connell, J. H. 1961. The influence of interspecific competition and other factors on the distribution of the barnacle *Chthamalus stellatus*. Ecology 42: 710–723.

Connell, J. H. 1979. Tropical rain forest and coral reefs as open nonequilibrium systems. pp. 141–163. *In*: Anderson, R. M., Turner, B. D. and Taylor, L. R. [eds.]. Population Dynamics. Blackwell Scientific Publications, Oxford.

Connell, J. H. 1980. Diversity and the coevolution of competitors, or the ghost of competition past. Oikos 35: 131–138.

Connell, J. H. 1983. On the prevalence and relative importance of interspecific competition: evidence from field experiments. The American Naturalist 122: 661–696.

Darwin, C. 1859. On the Origin of Species by Means of Natural Selection, or Preservation of Favoured races in the Struggle for Life. John Murray, London, UK.

Dye, P. J. and Spear, P. T. 1982. The effects of bush clearing and rainfall variability on grass yield and composition in south-west Zimbabwe. Zimbabwe Journal of Agricultural Research 20: 103–118.

Elton, C. S. 1927. Animal Ecology. University of Chicago Press, Chicago.

Elton, C. S. and Miller, R. S. 1954. The ecological survey of animal communities: with a practical system of classifying habitats by structural characters. The Journal of Ecology 42: 460–496.

Ferretti, F., Lovari, S., Minder, I. and Pellizzi, B. 2014. Recovery of the snow leopard in Sagarmatha (Mt. Everest) National Park: effects on main prey. Eur. J. Wildl. Res. 60: 559–562.

Ferretti, F., Sforzi, A. and Lovari, S. 2011. Behavioural interference between ungulate species: roe are not on velvet with fallow deer. Behavioral Ecology and Sociobiology 65: 875–887.

Gause, G. F. 1934. The Struggle for Existence. Williams and Wilkins, Baltimore.

Grant, P. R. 1986. Ecology and Evolution of Darwin's finches. Princeton University Press, Princeton, NJ.

Grinnell, J. 1917. Field test of theories concerning distributional control. The American Naturalist 51: 115–128.

Hardin, G. 1960. The competitive exclusion principle. Science 131: 1292–1297.

Hutchinson, G. E. 1957. Concluding remarks. Cold Spring Harbor Symposium on Quantitative Biology 22: 415–427.

Hutchinson, G. E. 1961. The paradox of the plankton. The American Naturalist 95: 137–145.

Kato-Noguchi, H., Kimura, F., Ohno, O. and Suenaga, K. 2017. Involvement of allelopathy in inhibition of understory growth in red pine forests. Journal of Plant Physiology 218: 66–73.

Kléparski, L., Beaugrand, G. and Kirby, R. R. 2022. How do plankton species coexist in an apparently unstructured environment? Biology Letters 18: 20220207.

Lack, D. 1947. Darwin's Finches. Cambridge University Press, Cambridge.

Landman, M., Schoeman, D. S. and Kerley, G. I. H. 2013. Shift in black rhinoceros diet in the presence of elephant: evidence for competition? PLoS ONE 8: e69771. doi: 10.1371/journal.pone.0069771.

Lechowicz, M. J. and Bell, B. 1991. The ecology and genetics of fitness in forest plants. II. Microspatial heterogeneity of the edaphic environment. The Journal of Ecology: 687–696.

Levin, S. A. 1974. Disturbance, patch formation, and community structure. PNAS 71: 2744–2747

Levins, R. 1968. Evolution in Changing Environments. Princeton University Press, Princeton, NJ.

Lotka, A. J. 1925. Elements of Physical Biology. Williams and Wilkins, Baltimore.

Lovari, S., Ventimiglia, M. and Minder, I. 2013. Food habits of two leopard species, competition, climate change and upper treeline: a way to the decrease of an endangered species? Ethology Ecology and Evolution 25: 305–318.

MacArthur, R. H. 1970. Species packing and competitive equilibrium for many species. Theoretical Population Biology 1: 1–11.

MacArthur, R. H. 1958. Population ecology of some warblers of Northeastern coniferous forests. Ecology 39: 599–619.
MacArthur, R. H. 1972. Geographical Ecology. Harper & Row, New York.
Marini, G., Guzzetta, G., Baldacchino, F., Arnoldi, D., Montarsi, F., Capelli, G., et al. 2017. The effect of interspecific competition on the temporal dynamics of *Aedes albopictus* and *Culex pipiens*. Parasites and Vectors 10: 102. doi 10.1186/s13071-017-2041-8.
May, R. M. 1971. Stability in multispecies community models. Mathematical Biosciences 12: 59–79.
McPeek, M. A. 2022. Coexistence in ecology: a mechanistic perspective (Vol. 66). Princeton University Press.
Morris, D. W. 2005. On the roles of time, space, and habitat in a boreal small-mammal assemblage: predictably stochastic assembly. Oikos 109: 223–238.
Odum, E. P. 1971. Fundamentals of Ecology. Third Edition, W. B. Saunders Co., Philadelphia.
Pearl, R. and Reed, L. J. 1920. On the rate of growth of the population of the USA since 1790 and its mathematical representation. PNAS 6: 275–288.
Poulson, T. L. and Platt, W. J. 1996. Replacement patterns of beech and sugar maple in Warren Woods, Michigan. Ecology 77(4): 1234–1253.
Primack, R. B. and Howe, H. F. 1975. Interference competition between a hummingbird (*Amazilia tzactl*) and skipper butterflies (Hesperiidae). Biotropica 7: 55–58.
Roughgarden J. 1983. Competition and theory in community ecology. The American Naturalist 122: 583–601.
Sangiuliano, A., Lovari, S., and Ferretti, F. 2016. Dietary partitioning between European roe deer and European brown hare. Eur. J. Wildl. Res. 62: 527–535.
Schoener, T. W. 1983. Field experiments on interspecific competition. The American Naturalist 122: 240–285.
Slack, N. G. 2010. G. Evelyn Hutchinson and the invention of modern ecology. Yale University Press, New Haven, Connecticut.
Tilman, D. 1982. Resource competition and community structure. Princeton University Press, Princeton, NJ.
Volterra, V. 1926. Variations and fluctuations in the numbers of individuals in animal species living together. (Reprinted in 1931). pp. 409–448. *In*: Chapman, R. N. [ed.]. Animal Ecology. McGraw-Hill, New York. http://dx.doi.org/10.1086/284409.

CHAPTER 4

Species Interaction

Predation

In the game of predator-prey, it's better to eat lunch than be lunch.

When we talk of predation, most of us visualize lions pouncing on zebras or wolves chasing moose. Predation, however, can be recognized in many different forms, including herbivory, parasitism, and parasitoidism. In general, predation occurs when an individual of one species (the predator) consumes all or part of an individual of another species (the prey), benefiting the predator while harming the prey. Classic or "true" predation—where the predator consumes the prey, such as lions killing a zebra—involves lethal interactions. However, other forms of predation are not necessarily lethal to the prey. Herbivory, for example, is a type of predation in which the predators (herbivores) are rarely lethal. Whether they are zebras browsing leaves of acacia trees, or nudibranchs grazing bryozoans, only parts of the prey are exploited. Another form of predation that is not necessarily lethal is parasitism, where the prey (a host) is usually not killed outright, but parasites (the "predators") consume its tissues without necessarily killing it. Examples include hookworms, roundworms, and tapeworms that infest the intestines, blood, or tissues of their hosts. Pathogens, such as bacteria, viruses, and fungi, can be parasites also.

In contrast to parasites, parasitoids are always lethal to their hosts, laying eggs on living hosts that succumb to hatched parasitoid larvae. If the eggs are laid within the host, the larvae eat the host from the inside out; if the eggs are laid on the outside of the host, the larvae feed by sucking out the host's body fluids. Fortunately for us, this lethal form of predation is confined to insects. Parasitoid behavior is common among wasps, although some types of flies, beetles, and moths have also been identified as parasitoids. Thus, predation comes in many forms, including true predators, grazers, parasites, or parasitoids, yet all predation is fundamentally similar. In all cases, interactions between predators and prey reduce either the fecundity or the chance of survival of depredated individuals, while increasing the survival of the predators, thereby impacting the population abundances of both predator and prey.

Origin and Evolution of Predation

The phrase "nature, red in tooth and claw" evokes an image of nature as a ruthless predator striking fear into its quivering prey, but nature was not always filled with predators. Earth's earliest life forms were probably bacteria and archaea that could convert carbon dioxide, hydrogen, and sulfur into energy–they were **chemoautotrophs** (Sleep and Bird 2007). Then came the photosynthesizing **autotrophs** such as the microbes that create stromatolites. But at what point did organisms start consuming other organisms? The first predators remain hidden in the obscurity of time due to a lack of early fossil evidence. However, there is ample later fossil evidence showing that predation originated multiple times across many different groups of organisms.

DOI: 10.1201/9781003008538-4

We visualize prehistoric predators such as *Tyrannosaurus rex* using jaws and sharp teeth to rip chunks of flesh from their prey. Indeed, the origin of modern vertebrate jaws dates back approximately 420 million years (Barford 2013), and teeth date back approximately 500 million years (Murdock et al. 2013). But the fossil record indicates that predators existed before fangs and teeth evolved, dating back to the earliest periods of abundant animal life. At the beginning of the last century, Evans (1912) proposed that the sudden appearance of skeletal tissues in the Cambrian era (about 550 million years ago) was due to the evolution of predators. Fossils from a whole group of animal life collectively known as the *anomalocarids* reveal a shrimplike marine predator that grew up to two meters long. These organisms had well-developed eyes, a small brain, and a pair of dexterous graspers near the mouth. These giants were undoubtedly terrors of the archaic seas. In fact, there is evidence that predators existed even before the *anomalocarids*. Following the Gaskiers glaciation 580 million years ago (Hoffman et al. 1998), unknown animals evolved hard shells and exoskeletons composed of common minerals, most often calcium carbonate or silica. Some of the fossils of these earliest exoskeletons, paleontologists noticed, were punctured with circular holes about 0.4 mm across. It is believed such evidence indicates punctures left by predators that bored through the tough armor to feast on the soft tissue hidden within (Bengtson and Zhao 1992). Undoubtedly, predatory behavior evolved as a feeding strategy in these ancient ecosystems. Since predators must have found it much easier to eat soft-bodied organisms than to waste energy breaking through a tough mineralized exoskeleton, there must have been strong selection against vulnerable prey with weaker exoskeletons, leaving those with tougher exoskeletons to survive. Like the quip says, "I don't have to outrun the bear, I only have to outrun you!" In short order, selection for slightly tougher exoskeletons than your neighbor would have resulted in an "evolutionary race" to make a tougher shell or die. That quantum leap in self-protection ultimately resulted in the plethora of amazingly diverse organisms found in the sedimentary layers, known as the Cambrian Explosion.

Its history and origin aside, whenever and however predation evolved, we can safely assume that it became a strong evolutionary force. This is because predation, in the strict sense, involves mortality of the prey, directly shaping the selection of "anti-predator" traits within prey populations. Contemporary evolutionary ecologists have acknowledged species interactions as major evolutionary forces, with predation exerting particularly strong selective pressure. Because predation directly reduces chances of survival and reproduction—both critical components of "evolutionary fitness"—prey that are better at eluding predation will have a greater likelihood of passing their genes to the next generation. Whether through more cryptic (camouflaged) coloration, better sensory equipment, longer legs, harder shells, or other adaptations, individuals with superior defenses gain a selective advantage. Over time, any advantage is strongly selected, as the traits of survivors will propagate throughout the population, while the traits of those who fall victim to predation will not.

Of course, once prey become more adept at avoiding predation, there will be selective pressure on the predators to evolve as well. Naturally, predators that are better able to capture prey by whatever means, including seeking, capturing, and efficiently consuming it, should do better in terms of surviving and producing offspring than those less able at capturing prey. Those traits that foster predation success will spread throughout the population via natural selection. Thus, the two selective forces, one acting on predators and the other on prey, oppose each other, leading to improved offenses being met by improved defenses, and vice versa. Is an "arms race" inevitable? It seems logical there might be an asymmetry in the evolutionary responses of predator and prey, aptly depicted as the "life–dinner principle" by Dawkins and Krebs (1979): the prey is running for its life, but the predator is only running for its meal, implying greater evolutionary pressure on the prey than the predator. Yet McLean et al. (2024) note that this represents a simplistic, single-encounter scenario that may not necessarily reflect the broader evolutionary dynamics between species. Many factors can generate significant evolutionary pressure for adaptations in both predator and prey species. Whether or not a coevolutionary "arms race" occurs may be contingent upon several complexities, such as whether predators are greatly outnumbered by their prey or whether predators pursue only one type of prey or they can switch prey species when a primary one becomes scarce (Abrams 1986).

Adaptation related to predation can be among the most remarkable we see in nature. Witness, for instance, the overpowering speed of the cheetah, one of the world's fastest land animals, in pursuit of the swift and graceful escaping gazelle. Other such adaptations include the complex social hunting behaviors of lions and wolves, the utilization of long, sticky tongues and accurate aim of some fish and reptiles, as well as spiders and their webs. Prey have equally elaborate predator escape mechanisms, including the posting of sentinels, predator alarm calls, camouflage, toxins, and thorns.

Ecological Impacts of Predation

In ecological, rather than evolutionary time, prey–predator interactions can also affect the distribution and abundance of species, which in turn affect communities and ecosystems. Predators and herbivores can change the outcome of competition for prey and plants. Inferior competitors may increase in abundance when they are in the presence of a predator that decreases the abundance or performance of a more dominant competitor. For example, when a predatory sea star (*Pisaster*) was removed from an intertidal community by Paine (1974), it resulted in the local extinction of all large invertebrates except mussels. In the absence of the sea star, mussels—being dominant competitors—outcompeted and eliminated the other large invertebrates from the area (Paine 1974).

In another example, arctic foxes (*Vulpes lagopus*) were introduced to some, but not all, of the Aleutian Islands off the coast of Alaska in the late 19th and early 20th centuries. This allowed for an opportunistic study of the effects predators can have on plant communities by consuming herbivore species (Croll et al. 2005). On islands without foxes, diverse seabirds thrive at high densities, excreting guano rich in ocean-produced nitrogen and phosphorus onto the otherwise nutrient-poor land. But when foxes were introduced, the density of breeding seabird populations decreased nearly 100% because of fox predation. The decrease in seabird numbers, in turn, reduced the input of guano, shifting the plant community from one dominated by grasses to one with a more equitable distribution of grasses, shrubs, and forbs. Because grasses outcompete forbs and shrubs at high nutrient levels, the introduction of arctic foxes transformed nutrient-rich islands dominated by grasses to ones where nutrient limitation allowed more shrubs and forbs to exist together with grasses (Croll et al. 2005).

Predation as Population "Regulator"

The ancient Greeks, such as Herodotus, believed that natural law prevented predators from eating up all their prey, thus maintaining the balance of nature. Based on hearsay, Herodotus argued that lions—which roamed much of Europe and the Mediterranean during that time—had low reproductive rates, allowing them to maintain their numbers well below the available prey. This, he suggested, served a divine purpose to maintain balance in the universe (Egerton 1973).

This idea of a balance between the numbers of predators and prey dominated much thinking until the birth of modern ecology and into the late 1960s, when debates about what regulates (balances) population numbers raged. Many ecologists were searching for a single general mechanism that could regulate the density of all populations, whether predator or prey. Andrewartha and Birch (1954) emphasized the role of the fluctuating abiotic environment. Lack (1954) and Milne (1957) advocated the importance of food supply driven by intraspecific and interspecific competition, whereas Errington (1946) and Varley (1947) argued that predation itself plays a role as a population regulator.

Although it seems obvious that predators are harmful to individual prey, their impact on prey populations is not always straightforward. For instance, when food is scarce and competition is intense, predation may relieve competitive pressures among prey species, allowing at least some prey to survive under better resource conditions. This, in turn, can promote the survival of youngsters, ensuring the continuation of the population. In this case, predation is a compensatory action that reduces intraspecific competition among survivors. Likewise, predation may have less impact on a prey population than it otherwise would if predators kill only the old, weakest, and most vulnerable (naïve and young), including the sickest individuals. In northern Tanzania's Serengeti, cheetahs and wild dogs kill a disproportionate number of younger groups of Thomson's gazelles (FitzGibbon and Fanshawe 1989). Lions have also been observed to disproportionately kill the aged, young, and sick (Schaller 1962).

Does this compensatory effect mean that predators do not regulate prey populations? According to Schaller (1962), lions kill large numbers of wildebeests and zebras, yet the number represents only a small fraction of each prey population—too few to regulate them. On the other hand, experimental studies on small mammal communities indicate that predators can affect not only individual prey but also the whole population. In a series of four one-hectare enclosures in Finland, Klemola et al. (2000) protected voles from predation by weasels, stoats, and avian predators using fenced enclosures with net roofs. The abundance of voles increased over 20 times in the enclosures compared to the control grids. This experiment showed that predators can indeed reduce prey populations and influence their population dynamics. Predation can also have an effect when the predator returns to an area from which it was previously extirpated. In Mt. Everest National Park (Nepal), the snow leopard (*Panthera uncia*) reappeared after an absence of over forty years, leading to a significant decline in its primary prey, the Himalayan tahr (*Hemitragus jemlahicus*). The tahr decreased by more than 66% from 2003 to 2010, mainly due to predation on young individuals. Subsequently, in turn, the density of snow leopards itself decreased by 60% from 2007 to 2010 (Ferretti et al. 2014).

Based on both a large number of both experimental and observational studies, it is clear that there is no universal population regulator. Instead, population fluctuations may track seasonal or annual changes in the availability of resources, or changes in population size may appear stochastic (random). But some population dynamics show strong evidence of being driven by predation. These population dynamics fluctuate so regularly that they can be called cyclic. Among the best-known examples of cyclic oscillations are those from a study by Elton (1924), who used historical records kept by the Hudson Bay Company to trace the fluctuations of fur-bearing animal populations of North American boreal forests. He observed a cyclic relationship between the numbers of lynx and hare pelts, which suggested that these predators regulate their prey population. Heralded as one of the best-studied cases of animal population cycles, the lynx (predator) and snowshoe hare (prey) populations were shown to fluctuate in nine- to ten-year cycles over about 200 years. This record demonstrates a remarkable match in the cycles of the two populations, with lynx population peaks slightly lagging behind those of the hare population. By the 1950s, several hypotheses had been proposed to explain these and other cycles among northern populations of mammals and birds. One of them, the predation hypothesis, suggests that predators (e.g., lynx) drive prey (e.g., hare) populations by increasing in number in response to increasing prey availability. Eventually, when the predator population becomes large enough, it reduces the prey numbers; the predator then declines, allowing the prey to rebound, and the cycle starts again.

Mathematical models buttress the idea that predators drive prey populations. Vito Volterra, the eminent mathematician of the Lotka–Volterra competition models discussed in Chapter 3, became interested in ecological problems in the 1920s when his daughter's fiancé, ecologist Umberto D'Ancona, asked him to analyze data from Adriatic Sea fisheries. Umberto D'Ancona had observed that the virtual suspension of fishing around the time of the First World War led to a marked increase in the numbers of predatory fish species and oscillations in the abundance of both predators and prey (Boyce et al. 2023). Volterra took up this challenge, perhaps partly to impress his future son-in-law, but it ultimately became a lifetime pursuit. When Volterra (1926) eventually developed a model of predation in a two-species system, apart from the equally impressive model of two-species competition, his primary focus was not so much on energy transfer but on the population dynamics caused by species interactions. This work coincidentally had been anticipated by Alfred Lotka (1925), an American mathematician and demographer. The two-species model of predation is now known as the "Lotka–Volterra equations" and is the starting point for most modern discussions of predation (Kingsland 1995).

The Lotka–Volterra Equations

In the Lotka–Volterra predator–prey model, the prey population is assumed to grow exponentially in the absence of predators. Recall that we can model this using $dN/dt = rN$, where N is the prey population size and r is its per capita population growth rate.

But while the prey population is growing, predators will also be removing prey at the same time. We can assume that any particular predator will search for and encounter individual prey at a rate that scales (increases or decreases) with the size of the prey population N. We can also assume that only a portion of the time a predator spends foraging is actually spent encountering, attacking, and successfully subduing a prey item. We will refer to this as the attack rate a (note that, per unit of foraging time, the attack rate varies between 0 and 1). For now, we will assume that a successful attack results in the prey being instantly consumed. If we multiply the attack rate by the number of prey in the population (N), we obtain aN, which represents the consumption rate of a single predator (i.e., number of prey consumed per predator per unit time). If we multiply aN by the number of predators, we get aNP, which is the total number of prey consumed per unit time by the predator population. Thus, the net rate of change in the prey population can be written as follows:

$$\frac{dN}{dt} = rN - aNP \qquad \text{Equation 1}$$

The growth of the predator population depends on the number of prey consumed per unit time, which is equal to aNP, and on the efficiency b with which the predators convert consumed prey (food) into newborn predators. Thus, the birth rate of the predator population is $baNP$. However, in the absence of any prey, the predators will starve to death at a particular rate, d. We can now define the net rate of change in the predator population as follows:

$$\frac{dP}{dt} = baNP - dP \qquad \text{Equation 2}$$

Equations 1 and 2 together constitute the Lotka–Volterra equations. These equations link the two populations; therefore, the model assumes mutual regulation of the predator and prey populations. A single term, aNP, describes the link between the predator population and the prey population. For prey, this term represents the regulation of population growth through density-dependent mortality, while for predators, it represents the regulation of population growth through density-dependent birth rate.

We can use the same kind of graphical approach we used for competition between two species to understand how the predator and prey populations will interact. In the absence of predators or at low predator density, the prey population will grow exponentially: $\frac{dN}{dt} = rN$. As the predator population increases, prey mortality, aNP, increases from near zero until it equals the prey population growth rate, rN, at which point the *net* prey population growth rate becomes zero: $\frac{dN}{dt} = 0$. At this state, the prey population has reached equilibrium. We can find the predator population size, P, at which this occurs by starting with the prey population growth model:

$$\frac{dN}{dt} = rN - aNP$$

setting the growth rate equal to zero

$$0 = rN - aNP$$

and simplifying

$$aNP = rN$$

$$P = \frac{r}{a}$$

This means the prey population growth rate is zero when the predator population is equal to the per capita prey growth rate, r, divided by the per capita predator attack rate, a. This value defines the **prey zero growth isocline** (Figure 1a). Recall that a zero growth isocline defines the conditions in which a population is neither growing nor declining.

In Figure 1a, the two axes of the graph represent the two interacting populations. The x-axis represents the size of the prey population, and the y-axis represents the size of the predator population. Note that the prey zero growth isocline is independent of the prey population size, N, and is represented by a line parallel to the x-axis at a point along the y-axis, represented by the value $P = r/a$. With respect to the zero-growth isocline, when predator densities are less than r/a, the prey population will increase (arrows to the right in Figure 1a), and when predator densities are greater, the prey population will decrease (arrows to the left in Figure 1a). A specific number of predators ($P = r/a$) will maintain the prey population at equilibrium (zero growth). P can be large (prey support a large predator population at equilibrium) if the prey either has a very high growth rate (r), or a very low attack rate (a) by the predator, or both.

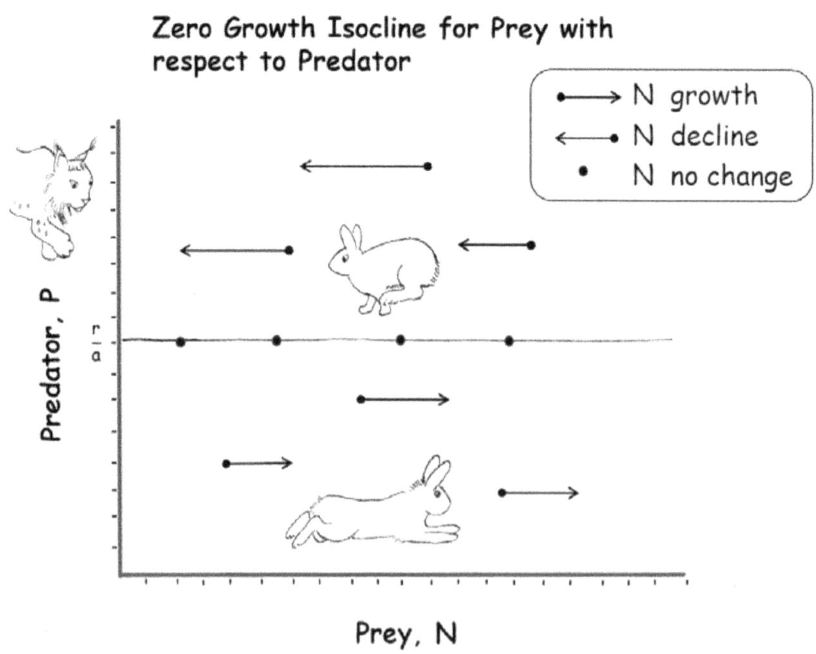

Figure 1a. Prey zero growth isocline (r/a) in the population size state space for the Lotka–Volterra predation equations. When the number of predators is lower than r/a, the prey population increases; when greater than r/a, it decreases.

Likewise, to find the **predator zero growth isocline**, if we start with:

$$\frac{dP}{dt} = baNP - dP$$

Set predator population growth to zero:

$$0 = baNP - dP$$

Simplify to:

$$baNP = dP$$

The predator population will be at equilibrium when:

$$N = \frac{d}{ba}$$

As with the prey population, the value $N = d/ba$ defines the predator zero growth isocline (Figure 1b), and it is independent of the predator population size, P. This means that the growth rate of the predator population is zero when the prey population equals the per capita mortality rate of the predator, d, divided by the product of per capita predator attack rate, a, and the efficiency at converting consumed prey into new offspring, b. Note that these two factors, a and b, determine the per capita predator birth rate. Therefore, more prey will exist at equilibrium when either the natural death rate of predators is high or the predator birth rate is low, or both. When prey densities are less than d/ba, the predator population will decrease (down arrows in Figure 1b), and when prey

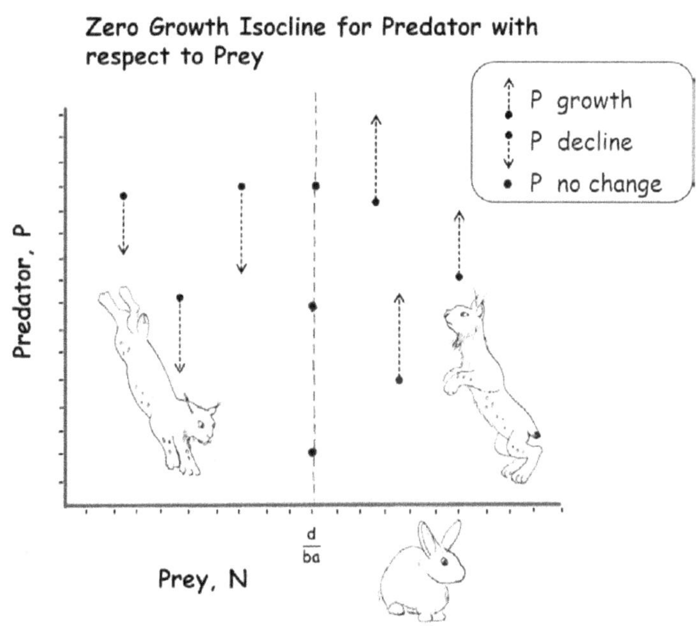

Figure 1b. Predator zero growth isocline (*d/ba*) in the population size state space for the Lotka–Volterra predation equations. When the number of prey is lower than *d/ba*, the predator population decreases, but when it is greater than *d/ba*, it increases.

densities are greater than *d/ba*, the predator population will increase (up arrows in Figure 1b). When the number of prey is equal to *d/ba*, the predator population is maintained at equilibrium.

In the Lotka–Volterra predator–prey model, the predator population determines whether the prey population is at equilibrium $\left(\frac{dN}{dt}=0\right)$. Similarly, the prey population determines whether the predator population is at equilibrium $\left(\frac{dP}{dt}=0\right)$. Because *d, b, a*, and *r* are all constants in the Lotka–Volterra model, the zero growth isoclines for both the predator and prey are straight lines. Plotting these two growth isoclines on the same graph (Figure 1c), we see that they intersect at a single point, an equilibrium point, at which both predator and prey populations remain constant. Away from the intersection of the predator and prey isoclines, the population sizes of the predators and prey change over time. For instance, both prey and predators become abundant in the upper right-hand region of Figure 1c. The predator population increases in this region (up arrows in Figure 1b) because the prey population is above that needed to support zero growth in predators (*r/a*). In this same region, however, the prey population is declining (downward arrows in Figure 1a) because the predator population is above that required to hold the prey population at zero growth (*d/ba*). Consequently, the joint trajectory of the two populations, illustrated by the heavy solid arrow (Figure 1c), will move the system over time into the upper left-hand region. In this region, both predator and prey abundances decrease, and the combined trajectory drives the system to the lower left-hand region, and so on. If we follow the changes of the predator and prey populations in all quadrants, the population sizes will track counterclockwise in state space.

If we graph the predator and prey populations as a function of time, the two populations rise and fall in oscillations, with the predators lagging behind the prey. As the prey population increases, the predators also increase, and they consume more prey until the prey begin to decline. With the decreasing prey population, predators now face a food shortage, starve, and fail to reproduce. Once

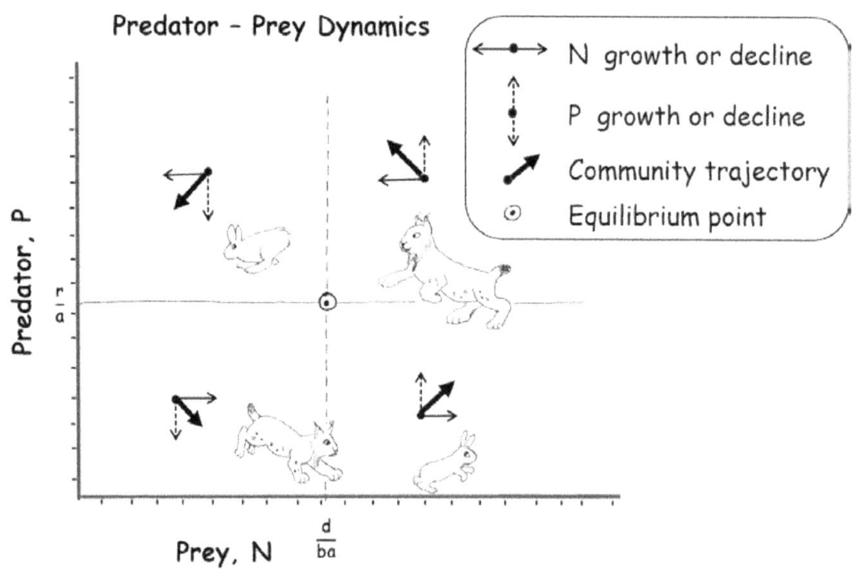

Figure 1c. Predator and prey zero growth isoclines are plotted on the same graph. The point at which the two isoclines intersect defines an equilibrium point (bullseye) when both populations remain constant. Away from the intersection, both populations change over time. In the upper right-hand region, predators increase and prey decrease. The joint trajectory of the two populations (solid arrow) moves the system into the upper left-hand region, where both predator and prey abundance decrease. The combined trajectory drives the system to the lower left-hand region, and so on. The population sizes track counterclockwise in state space, and although the trajectory is closed and circular, such that both populations forever oscillate, it is not necessarily a circle and depends on the constants in the model.

the predator population declines, the reproduction of prey more than compensates for its losses through predation. The prey population then increases, followed by a corresponding rise in the number of predators. As long as the system remains undisturbed, the cycle will continue for eternity. The prey is never completely wiped out, nor does the predator die out completely.

The Lotka–Volterra model shows an innate tendency for predator-prey systems to oscillate. However, does it explain the lynx and hare cycles seen by Elton (1924), and do we expect to see this in predator–prey systems in general? In a series of lab experiments, the Russian biologist G. F. Gause (1934) empirically tested the predictions of the predator–prey model in a lab setting. Gause raised ciliated protozoans *Paramecium caudatum* (prey) and *Didinium nasutum* (voracious paramecium predator) together in a medium of oats. In the first set of experiments, *Didinium* always exterminated the *Paramecium* and then self-annihilated due to starvation. In a second set of experiments, Gause added sediment to the oat medium, which functioned as a refuge for the prey. The paramecium avoided predation, but its success was so great that it led to the extinction of the predator. After the predator went extinct, the prey species emerged from hiding and increased in abundance. In a third set of experiments, Gause introduced one new predator individual and one new prey individual to the populations every third day—and this time... Eureka!—the two populations oscillated as predicted by the model. Gause concluded from this experiment that predators characteristically overexploit their prey, and that in nature, immigrants must repopulate local environments where prey have been annihilated.

The predator–prey populations also cycled in another lab experiment conducted by Carl Huffaker (1958), using the six-spotted mite (*Eotetranychus sexmaculatus*), which fed on oranges, and another mite, *Typhlodromus occidentalis*, which fed on the six-spotted mite and the orange. When two predatory mites were introduced to a single orange infested by the six-spotted mite, they eliminated the prey population and then died out from starvation. However, when Huffaker made the system more complex by using a rectangular tray of oranges, adding barriers, and partially covering oranges, all of which functioned as refuges for the prey population, the predator–prey populations cycled as expected from theory.

This clearly indicates that to persist in the face of exploitation by predators, prey need refuges. This unambiguous message from the lab indicates that predator–prey cycles can result from the direct link between predator and prey populations, as suggested by the Lotka–Volterra model, but only in the presence of environmental heterogeneity that allows refuge from predation. Ironically, the environmental aspect is not explicitly incorporated into the Lotka–Volterra model.

Elton's (1924) Hudson Bay Company pelts clearly show 10-year cycles in Canada lynx (*Lynx canadensis*) and snowshoe hare (*Lepus americanus*) populations (Figure 2). The snowshoe hare, the dominant herbivore of the region, lives in dense growths of understory shrubs within the boreal forest, which provide the hare with cover and winter food. Lynx are the dominant predators of snowshoe hare, upon which they largely specialize (Roth et al. 2007). Elton's (1924) work "inspired a tidal wave of literature" (Krebs et al. 2018), and various ecologists put forth a number of possible mechanisms to explain the lynx–hare cycles, including predation by specialist predators (Roth et al. 2007), as predicted by the Lotka–Volterra model, but also food shortage (Keith 1983), the production of plant defensive chemicals (Fox and Bryant 1984), interactions between food and predation (Keith et al. 1984), and stress (Boonstra et al. 1998). To explicitly test these competing hypotheses, Krebs et al. (2018) conducted a series of large-scale field experiments. They found that the immediate cause of death for most snowshoe hares was predation by a variety of predators, including the lynx, but also that the driver of decline in the hare cycles was due to chronic stress caused by predator chases that resulted in low reproduction. While this was not a "smoking gun" for lynx, the study nevertheless demonstrated predation as a regulating force on the prey population.

Population cycles have been documented for a wide variety of animals living in northern latitudes, including southern pine beetles, voles, lemmings, snowshoe hares, and moose. Much evidence indicates that predators contribute strongly to the cycles (Gilg et al. 2003; Turchin 2003). However, this cyclic pattern appears to disappear in more southern latitudes. For instance, several

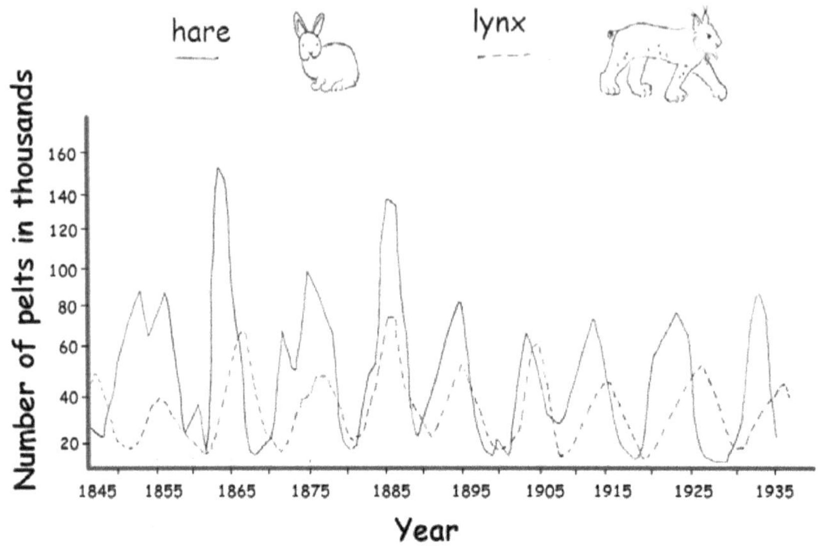

Figure 2. Changes in population sizes of lynx (*Lynx canadensis*), dashed line, and snowshoe hare (*Lepus americanus*), solid line, as documented in records of the numbers of pelts brought into the Hudson Bay Company (interpreted from Figure 5 in Elton 1924).

rodent species show cyclic populations in the northern parts of their range but not in the southern parts (Krebs 2013). The question of why cycles are not shown by more predators and their prey in nature has kept ecologists deeply interested. One explanation may be that detecting these cycles requires several decades, if not centuries, of carefully collected data, such as that of the Hudson Bay Company, which is a fairly rare situation. Another explanation is that multiple predator species attack most prey species, and most predator species attack more than one species of prey. This lack of specialization may hinder population cycles. Roth et al. (2007) found that the celebrated lynx–hare cycle from the higher latitudes attenuates or is lost in more southern reaches of the boreal forest, where the diversity of prey is higher, and the diet of Canada lynx becomes less specialized on hare. This supports the idea that the availability of alternative prey can decrease population cycling. Likewise, rodent species showing cyclic populations in the northern parts of their range are subject to fewer predator species than in the southern parts of their range (Krebs 2013).

It is impressive that despite minimal assumptions, the simple Lotka–Volterra predator-prey model made valuable contributions to our understanding of predator and prey systems. Specifically, it shows that predator-prey interactions, in theory, can produce population cycles without any external influences. However, the Lotka–Volterra model behaves in a very "unrealistic" way—it is neutrally stable, meaning its predator and prey populations will cycle indefinitely, with the same amplitude and period. If a disturbance perturbs the system and changes predator or prey numbers, the system will cycle with a similar period but a new amplitude determined by the size of the perturbation. These oscillations, which depend on initial conditions, do not represent the real world, where perturbations are common.

Perhaps the model is too simplistic. One glaring, unrealistic assumption of the Lotka–Volterra model is that neither the prey nor the predator species is self-limited (i.e., neither population is subject to a carrying capacity), but we know that in nature, no population grows without bounds because eventually space or resources become limiting (the second law of ecology). This complexity can be easily added to the predator model (as we shall see below).

Another simplistic assumption of the Lotka–Volterra model is that as the prey population increases, the predator's per capita prey consumption rate increases linearly with prey abundance. The relationship between the per capita rate of consumption and the number of prey is referred

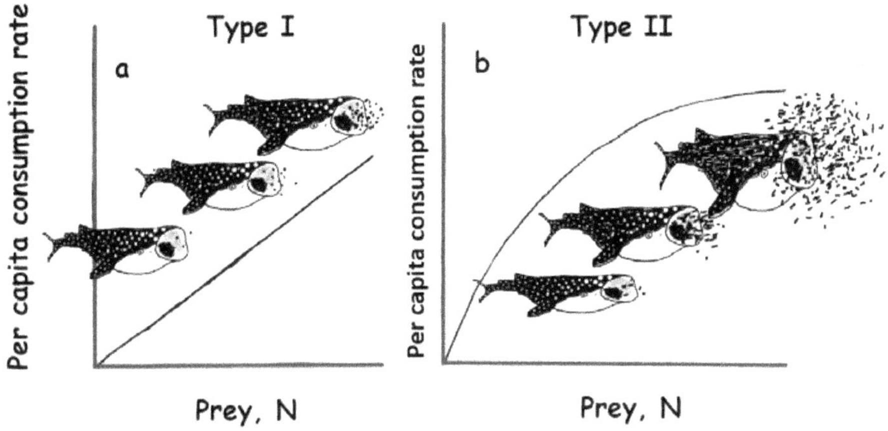

Figure 3. Illustrations of Type I (a) and Type II (b) functional response of the predator's per capita rate of consumption with respect to the number of prey.

to as the predator's **functional response**. The Lotka–Volterra model assumes a Type I functional response (Figure 3a), which may be a reasonable approximation of consumption rate at lower prey population densities, but at higher prey numbers, we would expect a predator to become satiated, and the functional response would be expected to level off as in what is termed a Type II functional response (Figure 3b; Holling 1959). This idea can also be incorporated into a predator–prey population dynamic model.

Rosenzweig–MacArthur Model

Rosenzweig and MacArthur's (1963) predator-prey model incorporates both a Type II functional response and self-limitation in the prey (using the logistic equation) into the original Lotka–Volterra model. For self-limitation in the prey, we use the logistic equation: $dN/dt = rN(K-N)/K$. For predator consumption rate, we introduce the Type II functional response (derived in Holling 1959) with the new variables, T (total time available to forage) and h (the time it takes to catch and eat a single prey item once encountered), along with the old variable a (attack rate), such that the per predator consumption rate equals $aN/(1+ahN)$.
We can make these modifications to the Lotka–Volterra model as follows:

$$\frac{dN}{dt} = rN\left(\frac{K-N}{K}\right) - \frac{aNP}{1+ahN} \qquad \text{Equation 3}$$

$$\frac{dP}{dt} = \frac{baNP}{1+ahN} - dP \qquad \text{Equation 4}$$

Now, when we plot the isoclines of the predator population and the prey population together on a graph, we see a humped prey isocline and a vertical predator isocline (Figure 4). In this more complex model, increases in both predator density and prey density regulate the growth of the prey population. At low prey density, prey population growth is mostly limited by predator density, whereas at high prey density, prey population growth is primarily limited by prey density, not predator density. While prey density *does* regulate population growth of the predators, the inclusion of the Type II functional response, $\frac{aN}{1+ahN}$, means the predator per capita rate of consumption decelerates as prey abundance increases and predators become satiated, reaching a maximum rate at some high prey population size (Figure 3b). Even in the absence of satiation, predators will be limited by handling time, which is the time needed to chase, capture, and consume each prey item. This somewhat puts the brake on predator population growth, but those with sharp eyes will note that P can be factored out of the birth and death rates on the right side of Equation 4. So, unlike the prey population, the predator population has no density-dependent self-regulating mechanism.

As with the Lotka–Volterra model, the predator–prey isoclines meet at an equilibrium point. But in this case, the dynamics near the equilibrium point will vary depending on the position of the predator's isocline relative to the hump in the prey curve. If the predator isocline intersects the prey isocline to the left of the hump (a very efficient predator that is less limited by satiation [bigger a] and/or handling [smaller h]; Figure 4a), the equilibrium point will be unstable, and the population trajectories will swing away from the equilibrium, heading toward the axes, leading to the extinction of either the predator or prey (and then the predator). If the predator isocline intersects the prey isocline at the hump, the predator and prey populations will enter a limit cycle, in which predator and prey densities cycle endlessly (as in the original L-V predation model). Intersections to the right of the hump (the predator is highly limited by satiation [smaller a] and/or handling [bigger h]) result in a stable equilibrium point, to which predator and prey densities return if perturbed away from it (Figure 4b). You will note that this model explains what the original Lotka–Volterra model could not—that is, the lab experiments of Gause (1934) and Huffaker (1958) demonstrate that lowering the attack rate of the predator makes their lab systems more stable. In Figure 4b, with a relatively inefficient predator, the prey population exists at a high density relative to its carrying capacity and is self-limited. Intersections to the left of the hump (Figure 4a) describe an efficient predator that exploits its prey at densities far lower than the prey's carrying capacity. Stability in any predator-prey system, therefore, is more likely when the prey population is close to its carrying capacity (and thus self-limited) and when the predator is relatively inefficient, requiring high densities of prey to survive, which makes it less likely to overexploit the prey and drive it to low densities.

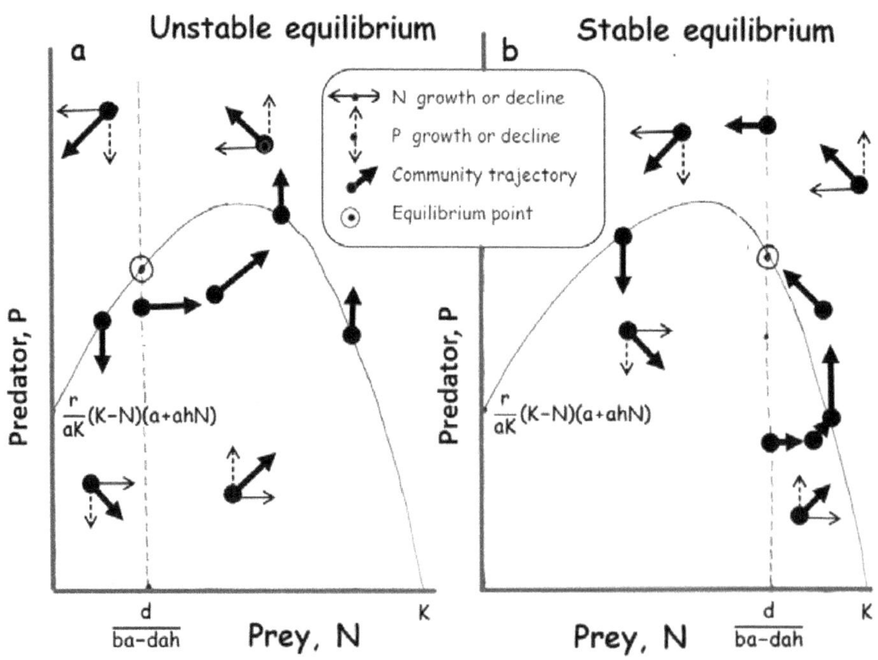

Figure 4. Zero growth isoclines in Rosenzweig–MacArthur model. The prey zero growth isocline is humped, not the straight line in the original Lotka–Volterra models. (a) Near the equilibrium, the dynamics lead the system away from the equilibrium point, making the system unstable. (b) Near the equilibrium, the dynamics lead the system toward the equilibrium point, making the system stable.

Thus, the Rosenzweig–MacArthur model exhibits a more realistic range of dynamics than the original Lotka–Volterra model, and the dynamic behavior can vary widely (Case 2000; Turchin 2003). Predator and prey populations can coexist at a stable equilibrium point (with unchanging population numbers), can exhibit a stable limit cycle (like the lynx–hare cycle), or one or both populations can go extinct.

Although we will not consider it here because of its complexity, cannibalism requires a short mention as a special form of predation, where one animal is eaten by another of the same species. Although not a common form of predation, this can occur frequently in piscivorous fish, such as northern pike, when alternative prey species are not available (Pereira et al. 2017). It is interesting to consider the population dynamic consequences when the predator does not go extinct but switches its diet to its species after crashing the prey population.

Paradox of Enrichment

Intuition suggests that an increase in the prey's food supply, allowing for an increase in its carrying capacity, *K*, would help the prey population better withstand the negative impacts of predation. However, the Rosenzweig–MacArthur model predicts otherwise: increasing the prey's carrying capacity may have precisely the opposite result. That is, increasing carrying capacity causes the predator isocline to intersect the prey isocline to the left of the hump, thereby destabilizing the equilibrium and potentially leading to extinction of one or both species. This is what Rosenzweig (1971) termed the "paradox of enrichment" (Figure 5). It may even explain the loss of species in ecosystems experiencing increased nutrient input, like eutrophication in aquatic ecosystems.

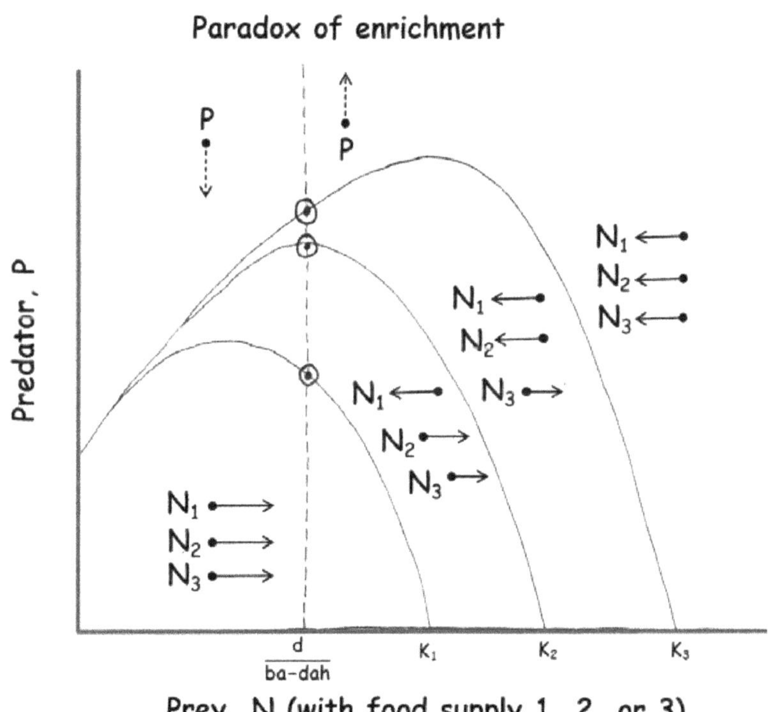

Figure 5. Paradox of enrichment. Intersection of the vertical predator isocline with prey isoclines for three increasing values of prey carrying capacity: K_1, K_2, K_3. Increasing the prey's carrying capacity shifts the intersection of the predator isocline from the right to the left of the prey isocline's hump. Intersections to the right of the hump are stable, while intersections to the left are unstable.

Ecology of Fear

In the Rosenzweig–MacArthur predator-prey model, the predator isocline is a straight, vertical line. This means the growth of the predator population is dependent on its prey density, not on its own density. However, in nature, this need not be the case. Predators interfere with one another and impact their growth, meaning their population growth can be density-dependent. A density-dependent predator isocline can be visualized as bending toward the right (Figure 6). Predator interference occurs through defending a territory, chasing away other predators, and fighting with another predator over food. As predator density increases, each predator will expend more energy interacting with other predators. This means each predator is required to kill a higher number of prey to obtain enough energy to replace itself, resulting in the rightward bend.

This rightward bend of the predator isocline can also occur when higher densities of predators are more likely to scare the prey. In effect, what happens is that fearful prey will spend more time in a refuge and less time out and about (foraging or seeking mates), thereby effectively reducing prey abundance for the predator by lowering the encounter rate (Abrams 1982). This is just one small aspect of the ecology of fear (*sensu*, Brown et al. 1999), which can have ramifications on not only the population dynamics of predator and prey but also on the foraging behavior of both prey and predators, the food choice of the prey (via trophic cascades), community interactions among prey and predator species (mechanisms of coexistence), and even coadaptations between behaviors and morphologies (coevolution; Brown and Kotler 2007; Kotler and Brown 2017).

For example, lions and other predators influence their prey's ecology more for the fear they instill than the mortality they cause directly (Sinclair and Arcese 1995). In the classic predator-prey models (Lotka–Volterra or Rosenzweig–MacArthur), the predators merely require a fixed standing crop of prey, regardless of the number of predators present. However, when fear-induced vigilance decreases prey catchability merely in response to the risk of predation, each predator may require an increased population size of prey to maintain a stable predator population. An important outcome of including direct density dependence in the predator population is that it tends to stabilize the predator–prey interaction: the predator isocline is more likely to intersect the prey isocline to the right of the hump, and the curve in the predator isocline will tend to create a stronger drive toward the equilibrium point (Brown and Kotler 2007; Arditi and Ginzburg 2012).

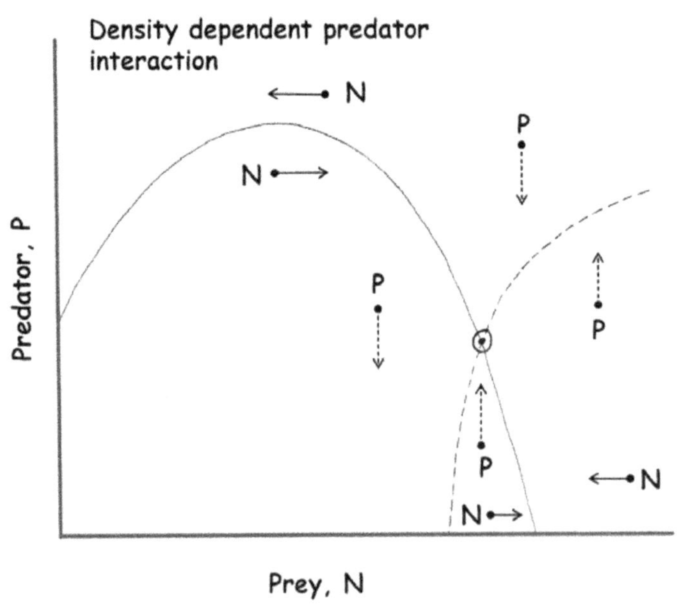

Figure 6. The predator isocline bends under density-dependent population growth due to interference competition or the fear it instills in its prey (the prey is more vigilant; ecology of fear *sensu*, Brown et al. 1999).

Once predators evolved on Earth, their exploitation of other species undoubtedly produced complex, multifold population dynamics as well as adaptive and counter-adaptive evolutionary dynamics. While theoretically fascinating on the one hand, our ability to understand these complex ramifications is also essential for managing sustainable fisheries and wildlife in ways that maximize their benefit, and for responding to the global spread of pests and pathogens in ways that minimize their harm.

References

Abrams, P. A. 1982. Functional responses of optimal foragers. The American Naturalist 120: 382–390.

Abrams, P. A. 1986. Adaptive responses of predators to prey and prey to predators: the failure of the arms-race analogy. Evolution 40: 1229–1247.

Abrams, P. A. 2000. The evolution of predator-prey interactions: theory and evidence. Annual Review of Ecology and Systematics 31: 79–105.

Andrewartha, H. G. and Birch, L C. 1954. The Distribution and Abundance of Animals. University of Chicago Press, Chicago, IL.

Arditi, R. and Ginzburg, L. R. 2012. How Species Interact. Oxford University Press. New York, NY.

Barford, E. 2013. Ancient fish face shows roots of modern jaw. Nature. https://doi.org/10.1038/nature.2013.13823.

Bengtson, S. and Zhao, Y. 1992. Predatorial borings in late Precambrian mineralized exoskeletons. Science 257: 367–369.

Boonstra, R., Hik, D., Singleton, G.R. and Tinnikov, A. 1998. The impact of predator-induced stress on the snowshoe hare cycle. Ecological Monographs 79: 371–394.

Boyce, M. S., Carpentier, C. A. and Linnell, J. D. 2025. Coexisting with large carnivores based on the Volterra principle. Conservation Biology: e14448.

Brown, J. S. and Kotler, B. P. 2007. Foraging and the ecology of fear. pp. 437–482. *In*: Foraging: behavior and ecology. D. Stephens, R. Ydenberg and J. S. Brown [eds.]. University of Chicago Press, Chicago, IL.

Brown, J. S., Laundre, J. W. and Gurung, M. 1999. The ecology of fear: Optimal foraging, game theory, and trophic interactions. Journal of Mammalogy 80: 385–399.

Case, T. J. 2000. An Illustrated Guide to Theoretical Ecology. Oxford University Press, Oxford.

Croll, D. A., Maron, J. L., Estes, J. A., Danner, E. M. and Byrd, G. V. 2005. Introduced predators transform subarctic islands from grassland to tundra. Science 307: 1959–1961.

Dawkins, R. and Krebs, R. J. 1979. Arms races between and within species. Proceedings of the Royal Society B 205: 489–511.

Egerton, F. N. 1973. Changing concepts of the balance of nature. Quarterly Review of Biology 48: 322–350.

Elton, C. S. 1924. Periodic fluctuations in the numbers of animals: their causes and effects. British Journal of Experimental Biology 2: 119–163.

Errington, P. L. 1946. Predation and vertebrate populations. Quarterly Review of Biology 21: 144–177.

Evans, J. W. 1912. The sudden appearance of the Cambrian fauna. Report of the Session of the International Geological Congress, Stockholm 1910: 543–546.

Ferretti, F., Lovari, S., Minder, I. and Pellizzi, B. 2014. Recovery of the snow leopard in *Sagarmatha* (Mt. Everest) National Park: effects on main prey. Eur. J. Wildl. Res. 60: 559–562.

FitzGibbon, C. D. and Fanshawe, J. 1989. The condition and age of Thomson's gazelles killed by cheetahs and wild dogs. Journal of Zoology 218: 99–107.

Fox, J. F. and Bryant, J. P. 1984. Instability of the snowshoe hare and woody plant interaction. Oecologia 63: 128–135.

Gause, G. F. 1934. Experimental analysis of Vito Volterra's mathematical theory of the struggle for existence. Science 79: 16–17.

Gilg, O., Hanski, I. and Sittler, B. 2003. Cyclic dynamics in a simple vertebrate predator-prey community. Science 302: 866–868.

Hoffman, P. F., Kaufman, A. J., Halverson, G. P. and Schrag, D. P. 1998. A Neoproterozoic snowball Earth. Science 281: 1342–1346.

Holling, C. S. 1959. The components of predation as revealed by a study of small-mammal predation of the European pine sawfly. Canadian Entomologist 91: 293–320.

Huffaker, C. B. 1958. Experimental studies on predation: dispersion factors and predator-prey oscillations. Hilgardia 27: 343–383.

Keith, L. B., Cary, J. R., Rongstad, O. J. and Brittingham, M. C. 1984. Demography and ecology of a declining snowshoe hare population. Wildlife Monographs 90: 3–43.

Keith, L.B. 1983. Role of food in hare population cycles. Oikos 40: 385–395.

Kingsland, S. E. 1995. Modeling Nature: Episodes in the History of Population Ecology. University of Chicago Press, Chicago.
Klemola, T., Koivula, M., Korpimaki, E. and Norrdahl, K. 2000. Experimental tests of predation and food hypotheses for population cycles of voles. Proceedings of the Royal Society B 267: 351–356.
Kotler, B. P. and Brown, J. S. 2017. Fear and loathing on the landscape: What can foraging theory tell us about vigilance and fear? Animal Sentience 15. doi: 10.51291/2377-7478.1280.
Krebs, C. J. 2013. Population Fluctuations in Rodents. University of Chicago Press, Chicago, IL.
Krebs, C. J., Boonstra, R. and Boutin, S. 2018. Using experimentation to understand the 10-year snowshoe hare cycle in the boreal forest of North America. Journal of Animal Ecology 87: 87–100. https://doi.org/10.1111/1365-2656.12720.
Lack, D. 1954. The natural regulation of animal numbers. Oxford University Press, London, UK.
Lotka, A. J. 1925. Elements of Physical Biology. Williams and Wilkins, Baltimore.
McLean, D. J., Herberstein, M. E. and Kokko, H. 2024. Asymmetric arms races between predators and prey: a tug of war between the life–dinner principle and the rare-enemy principle. Proceedings of the Royal Society B 291: 20241052.
Milne, A. 1957. The natural control of insect populations. Canadian Entomologist 89: 193–213.
Murdock, D. J. E., Dong, X.-P., Repetski, J. E., Marone, F., Stampanoni, M. and Donoghue, P. C. 2013. The origin of conodonts and of vertebrate mineralized skeletons. Nature 502: 546–549. https://doi.org/10.1038/nature12645.
Paine, R. T. 1974. Intertidal community structure: experimental studies on the relationship between a dominant competitor and its principal predator. Oecologia 15: 93–120.
Pereira, L. S., Agostinho, A. A. and Winemiller, K. O. 2017. Revisiting cannibalism in fishes. Reviews in Fish Biology and Fisheries 27: 499–513.
Rosenzweig, M. L. 1971. Paradox of enrichment: destabilization of exploitation ecosytems in ecological time. Science 171: 385–387.
Rosenzweig, M. L. and MacArthur, R. H. 1963. Graphical representation and stability conditions of predator-prey interactions. The American Naturalist 97: 209–223.
Roth, J.D., Marshall, J. D., Murray, D. L., Nickerson, D. M. and Steury, T. D. 2007. Geographical gradients in diet affect population dynamics of Canada lynx. Ecology. 88: 2736–2743.
Schaller, G. 1962. The Serengeti Lion. University of Chicago Press, Chicago, IL.
Sinclair, A. R. E. and Arcese, P. 1995. Population consequences of predation-sensitive foraging: the Serengeti wildebeest. Ecology 76: 882–891.
Sleep, N. H. and Bird. D. K. 2007. Niches of the pre-photosynthetic biosphere and geologic preservation of Earth's earliest ecology. Geobiology 5: 101–117.
Turchin, P. 2003. Complex Population Dynamics: A Theoretical/Empirical Synthesis. Princeton University Press, Princeton, NJ.
Varley, G. C. 1947. Ecological aspects of population regulation. Trans. IXth Int. Cong. Ent. 2: 210–214.
Volterra, V. 1926. Variation and fluctuations of the numbers of individuals in animal species living together. Reprinted. pp. 409–448. *In*: Chapman, R. M. [ed.]. Animal Ecology. 1931. McGraw-Hill, New York.

CHAPTER 5

Foraging Dynamics

"There is no love sincerer than the love of food." (Bernard Shaw, 1903)
"Feed me!" (Audrey 2, Little Shop of Horrors)

All living organisms must forage for energy and matter, but the ways they go about it can be vastly different. Plants harvest sunlight with their chloroplasts, take up carbon dioxide (CO_2) through their leaf stomata, and absorb molecules containing nitrogen, phosphorus, and trace nutrients through their roots. Bacteria absorb food molecules across their membranes either through passive diffusion or active transport. A paramecium can harvest bacteria or particulates into its gullet, which are then transported within its body as a food vacuole. Mosquitoes, octopuses, barnacles, frogs, parrots, elephants, blue whales, cheetahs, and other animals each have their unique means of finding, handling, ingesting, and digesting food. But despite all these differences, the need for energy and matter is a universal property of life. And the ecological and evolutionary implications of this need are profound. Individual feeding dynamics—how resources enter the body—give rise to population dynamics through the processes of survival and reproduction. Heritable feeding strategies that increase survival and reproduction will result in the evolution of adaptive feeding behaviors within a population; thus, what you eat ultimately determines what you are.

The energy and matter that organisms forage are called **resources**. When foraging for resources, three aspects contribute to an adaptive feeding strategy: efficiency, speed, and safety. As the well-known adage says, "it takes money to make money," the same is also true of resources. All organisms must expend resources to acquire more resources. Witness the spectacular chase of a cheetah running down a gazelle. Such a chase often ends with the gazelle escaping. The cheetah is not only expending energy but also putting wear and tear on its muscles and bones that require rest and repair. More costly still, the cheetah may suffer injury from punctures caused by the gazelle's swift kicks or sharp horns. The key to the cheetah's survival, then, is to be sure to gain more than it loses. Therefore, being able to acquire resources inexpensively, in less time, and more safely provides a competitive edge over other foragers.

Efficiency is about minimizing the energetic cost of obtaining a particular amount of a resource (foraging inexpensively). For example, the giant Pacific octopus (*Enteroctopus dofleini*) employs two strategies to penetrate the protective shells of bivalve prey and reach the tasty morsel inside. They can either pull the valves apart with their arms or drill through the valve with a specialized radula to inject a paralytic toxin, which relaxes the muscles holding the valves. The former action is much less costly to perform in terms of time and resources than the latter. But there is a limit to an octopus's pulling strength. Octopuses can easily pull apart thin-shelled bivalves like mussels but have a harder time with thick-shelled clams, resorting instead to drilling. When presented with a choice, giant Pacific octopuses prefer to eat mussels (Anderson and Mather 2007). However, when already opened mussels and clams are presented, the octopuses find clams a tastier choice (Anderson and Mather 2007). This shows how efficiency can impact foraging choices.

DOI: 10.1201/9781003008538-5

Speed is also important. Foragers that can respond swiftly to a sudden appearance of resources can gain a competitive edge over others, especially if the resource can be monopolized. The giant saguaro cactus (*Carnegiea gigantea*) in the Sonoran Desert of the southwestern United States and northwestern Mexico, or its close relative, the cardon cactus (*Pachycereus pringlei*) from Argentina, can swiftly take up water immediately following infrequent rains. Their trunks are pleated, allowing them to expand in girth like an accordion to store water for later. A shallow spreading root system allows these cacti to take up water that might otherwise evaporate from the soil, and perhaps even more importantly, to preempt neighboring plants with deeper root systems, such as the palo verde (*Parkinsonia microphyllum*), from getting the water first. McAuliffe (1984) cites this interception of water as the reason he found an increased probability of death in palo verdes when closely associated with saguaros (Figure 1).

Putting **safety** first can also be adaptive because it lowers the "mortality cost" of foraging. While seeking food, most organisms risk becoming food themselves. A North American squirrel foraging for acorns may fall prey to a hawk, just as an octopus foraging for crabs on the Pacific coast may fall prey to a sea otter. Likewise, an Antarctic penguin hunting fish and krill may fall prey to a leopard seal, and so on. The list of examples is nearly as long as the list of foragers. Therefore, there is strong selection for "safety" adaptations, which can come in numerous forms.

Sessile foragers expend resources on growing physical defenses. For example, whistling thorn acacia trees (*Acacia drepanolobium*) in Africa respond to damage from large herbivores by growing more and longer thorns (Young et al. 2003). Similarly, the calcium carbonate fortress built by a delicate coral polyp allows it to retreat quickly from predators. Even organisms that can move may evolve physical defenses. Some organisms grow specialized spines (e.g., porcupines, Hystricidae and Erethizontidae) or hard plate armor (e.g., tortoises, Testudinidae) as a defense against predation. Some use toxins—either self-produced or gained as part of a specialized diet—as a deterrent or lethal defense against predators, such as the skin of poison dart frogs (Dendrobatidae), the seeds of the apple (*Malus domestica*), the bodies of monarch butterflies (*Danaus plexippus*), the internal

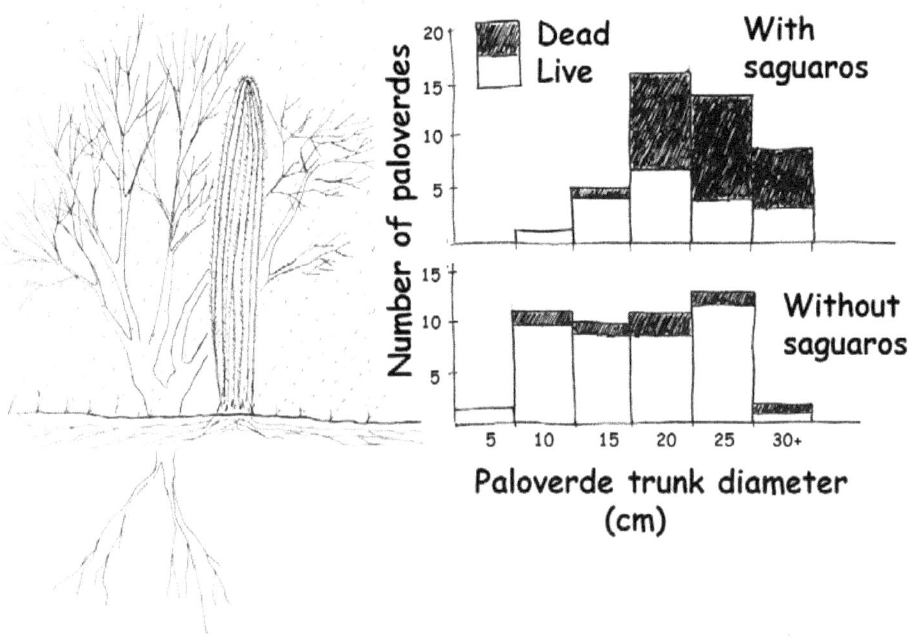

Figure 1. Size distribution of live and dead palo verde trees with respect to the presence or absence of saguaros (interpreted from Figure 1 in McAuliffe 1984).

organs of the pufferfish (Tetraodontidae), and the skin and feathers of the Hooded Pitohui (Pitohui dichrous).

Like a flounder on a sandy ocean floor, some foragers use their specialized skin and coats as camouflage, enhancing their survival when living in certain environments, while other foragers, such as octopuses and chameleons, can change color, pattern, and texture to match the background they are in. Safety adaptations include not just physical defenses, but also traits that give them the ability to fight back, such as the electric fields of moray eels and the tusks of elephants. Some organisms exploit others for safety. For example, many *Acacia* plant species use biting ants as a means of defense against herbivores, and many animals use "the herd" as protection from predators, leading to the African spectacle of hundreds, if not thousands, of flamingos taking to the air at the approach of a goshawk. Finally, many organisms have habitat-provided "panic rooms" that are either conveniently placed or carefully modified, such as the burrows of sand crabs or the rock crevices of lizards, where they quickly flee from approaching predators. "Better safe than sorry" is the strategy used by most organisms that, for the sake of safety, would rather lose a meal than become a meal. This is the mortality cost of foraging: dying while feeding does not enhance fitness; surviving does.

Foraging theory (Stephens et al. 2007) helps us understand how the astounding diversity in how organisms acquire resources came about, whether they are plants, animals, or microbes. It examines how feeding success is influenced by the distribution and abundance of a forager's resources. It considers how the tradeoffs or conflicting demands of an organism's morphology, physiology, and behavior affect how efficiently, swiftly, and safely it can acquire resources. It also evaluates the evolutionary "fitness" of foraging traits and behaviors in individuals foraging for limited energy and resources.

Evolutionary ecologists define fitness as the per capita growth rate; however, instead of attempting to measure per capita growth rate with respect to a foraging trait, most ecologists use an economic approach where gains are balanced against losses to measure a net gain in fitness. This is because natural selection should favor individuals that maximize the net gain of their foraging activity, given the constraints they face. For instance, a foraging European starling may not be able to forage for food at its maximum rate (which depends on how adept it is at discovering the location of insects or berries) because it is trying to avoid a peregrine falcon. It must balance foraging gains (benefit) against loss of life (cost). However, since it cannot hide forever (a constraint), the starling must forage in a way that maximizes its intake while minimizing the chance of being spotted by a peregrine falcon. Perhaps our starling minimizes its search time in a risky habitat, or it may forage for longer periods in a less risky habitat. Foraging theory can be used to explain these behaviors as adaptations.

Contexts of Foraging

An animal searching for food makes many decisions. We use the term "decision" here, but it does not necessarily imply that the animal makes a conscious decision. Plants and bacteria can make foraging "decisions" as well, so a decision simply means an organism is using one of a given set of behavioral options in its repertoire in response to opportunities or hazards in its environment. Feeding decisions can be understood in three contexts: what should I eat, where should I look for this food, and how thoroughly should I deplete an area of this resource once I find it? These questions fall into **diet choice**, **habitat selection**, and **patch use**, respectively.

Diet Choice

Are you a picky eater, or do you eat just about anything? For humans, a wide variety of foods may seem appealing, but even in that context, there are things we do not eat: oyster shells, pine needles, earthworms (most of us anyway), and so on. But these may be consumed by other organisms. Ground oyster shells can provide calcium for laying hens. Pine needles decompose slowly but steadily via microorganisms, and some birds, particularly the North American robin and European blackbird, feast on earthworms. Whether a plant, animal, or microbe includes a particular food item or prey in its diet largely depends on resource availability, the probability of a successful encounter, the handling time required to process the food, the energetic or nutritional reward, and the cost of foraging.

Resource availability refers to the abundance in terms of density or concentration of a food item or nutrient. This could be the number of acorns per square meter under an oak tree for a squirrel, the concentration of soluble nitrogen molecules in the soil for a plant, or the dissolved CO_2 in pond water available to aquatic algae. We will refer to this as ***R*** for **resource availability**.

Encounter probability refers to the likelihood that a certain resource item or molecule of a nutrient will be successfully detected and acquired within a foraging period; it is a way of defining "hunting success." A squirrel may use its senses of sight and smell to detect an acorn, at which point collecting it is relatively easy. Similarly, a plant must have root hairs that have grown close enough to water and soil nutrients for successful uptake to occur. When prey are camouflaged, encounters by a visually orienting predator may not lead to success if the prey are not seen. Even predators that see their prey may not be able to catch them successfully. Flyswatters do not always hit the fly, and for predators like the cheetah, fewer than half of their hunts may successfully end in a kill (e.g., Hilborn et al. 2012). Thus, an "encounter" occurs when both predator and prey can potentially engage, but a "successful encounter" requires both successful detection and capture by the predator.

We will refer to the probability of successfully detecting and capturing a given food item or molecule of nutrient within a foraging period as ***a*** for (successful) **encounter probability**. Note that this is the probability of a successful encounter per unit of foraging time. For example, suppose a predator such as a cheetah has, on average, over its foraging day, a successful encounter (detects, stalks, catches, and subdues a gazelle) in one out of every 500 gazelles encountered ($a = 1/500/\text{day}$). How does R (resource availability) relate to this? By multiplying the encounter probability by the density of resources, we can calculate the encounter rate.

Suppose there are $R = 200$ gazelles within a cheetah's daily foraging area (50 km^2) on any particular foraging day. On average, the cheetah will have a (successful) **encounter rate** of $\boldsymbol{aR} = (1/500/\text{day}) \times (200\ \text{gazelle}/50\ \text{km}^2) = 0.008$ gazelles/day/km^2, meaning that on average, the cheetah catches two gazelles in its foraging area of 50 km^2 every 5 days (Figure 2).

Handling time is the foraging cost of a successful encounter. Once successfully detected, time is required to acquire, ingest, and digest the food item to the point at which its energy and nutrients can be absorbed into the body of the consumer. For our cheetah, once detected, it takes time to stalk, pursue, capture, kill, consume, and digest a gazelle. As we have seen for the giant Pacific octopus, it may take considerably more time to break through the shell of a clam than a mussel. For some organisms, handling time may be negligible, such as the time it takes for a plant to capture a photon of light in its chlorophyll. On the other hand, sometimes handling time can be extremely long. The anaconda, a very large snake native to South America, upon capturing a deer, must spend time constricting the deer until it is killed before swallowing it whole. It may take weeks to digest before the snake can forage again for another prey.

We will refer to the time required to take and consume a food item or molecule of nutrient as **handling time**, ***h***. In general, the handling time refers to the idea that while an organism is consuming its prey, it cannot be simultaneously searching for other prey. A predator's total foraging time thus contains a fundamental tradeoff between search time and handling time.

Handling, consuming, and digesting a food item not only takes time but also involves a metabolic (energetic) **cost** (***C***). However, after consuming and digesting a food item, there is usually a reward in terms of energy and/or nutrients. We will refer to the **reward** from a consumed food item as ***E***.

Now that we have described foraging in terms of R, a, h, E, and C, we have a logical basis for discussing diet choice.

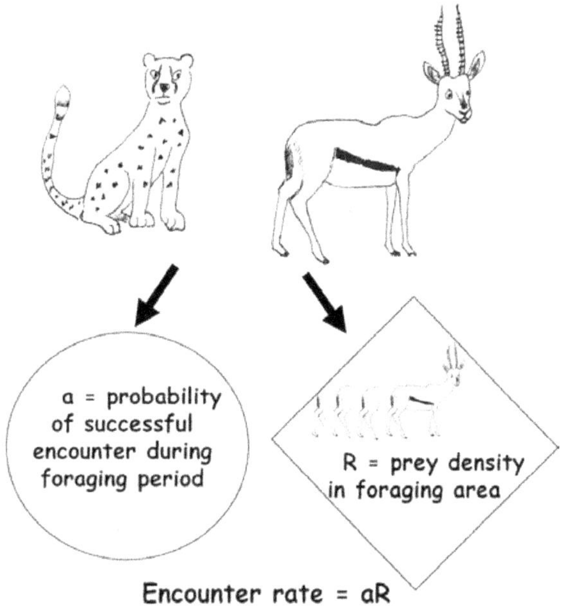

Figure 2. Encounter rate of the prey by the predator is determined by the probability of successfully detecting, catching, and subduing a prey item during a foraging period (a) multiplied by the density of prey items in the foraging area (R).

Most simply, something should be considered a potential part of the diet only if the reward E compensates for the metabolic cost C. Thus, there will be a **net energetic gain (*e*)** if:

$$e = E - C > 0 \qquad \text{Equation 1}$$

But if $e < 0$, then there will be a net loss from consuming and utilizing a food item. For instance, if we were to eat wood, the amount of energy gained would be less than the energy involved in chewing, swallowing, and digesting what little can be digested. In fact, only a subset of plant parts (usually tubers such as potatoes, fruits, sprouts, and, most importantly, seeds) is foods that provide a net caloric gain for humans. Interestingly, Carmody and Wrangham (2009) summarized a number of studies showing that our "cooking" behavior is an adaptive way to decrease C in starchy foods by increasing their digestibility. Furthermore, Boback et al. (2007) used pythons fed diets of cooked versus raw beef and showed that cooking significantly decreases the energy expended on digestion, absorption, and assimilation of meat, reducing its metabolic cost.

There has to be some net energetic reward (e) for an organism to include an item in its diet, but many animals seem to prefer some energetically rewarding foods ($e > 0$) over others. What drives preference in diet, and when should an animal include a less preferred item? This was the question asked by MacArthur and Pianka (1966) in what is considered the first paper on optimal foraging theory (Begon et al. 1996). Their work produced two insights. Their first insight was that foods can be ranked in terms of **profitability**. Charnov (1976a) formalized this idea in terms of energy per handling time. Food 1 is more profitable than Food 2 when

$$\frac{e^1}{h^1} > \frac{e^2}{h^2} \qquad \text{Equation 2}$$

This explains why a chocolate bar seems so much tastier than a banana; it may take the same amount of time to peel the wrappers ($h^1 = h^2$), but a chocolate bar has 200 Cal compared to a banana with 100 Cal ($\frac{e^1}{h^1} > \frac{e^2}{h^2}$). Wild rabbits behave the same way when offered similarly sized pellets (same handling time) of foods but differing energy content. They prefer augmented alfalfa over regular alfalfa and timothy grass, ranking them in order of preference according to e/h (Somers et al. 2008).

When calories are easily consumed (low h), our preference for high-energy foods can work against us, sometimes resulting in obesity. Yet even obesity can be adaptive. Under conditions of starvation or calorie deprivation, favoring a candy bar or other energy-dense food items makes evolutionary sense and explains why brown bears in Katmai National Park in Alaska spend their late summer and fall fishing for oil-rich salmon returning to the Brooks River—so they can pack on the pounds before winter hibernation. Preferring fatty salmon over grasses, flowers, and roots, which bears will also eat, makes sense energetically, but not all preferred foods may be the most energetically rewarding. We also need to consider handling time. This explains why octopuses preferred mussels over clams when they were unopened, and clams over mussels when they were already opened (Anderson and Mather 2007). Thus, it is profitability, measured in terms of e/h rather than just e, that drives preference.

Once encountered, a forager should always spend time handling its most profitable food. But when should it accept a less profitable food item? MacArthur and Pianka's (1966) second insight was that a forager should harvest a less preferred food only if it is unable to gain more by searching for and handling its preferred food than by simply handling the less preferred food.

To understand this, let us suppose that you are the forager, foraging for blueberries in the Alaskan alpine tundra. Your preferred berry is the blueberry because it is sweet (lots of calories), but while foraging for blueberries, you also encounter crowberries, which occur in the same habitat. However, crowberries taste bland because they have less sugar and fewer calories. Both of these berries are a similar size, and the time it takes to reach your hand out, pluck a berry from the bush, and plop it into your bucket is equivalent. So, because the handling time is similar and the blueberry is sweeter, it is easy to rank blueberries above crowberries in terms of profitability, which explains why you are foraging for blueberries rather than crowberries (Figure 3).

Even though crowberries are rather bland, is there ever a time when you should harvest them? Put another way, if you are sitting in front of a crowberry, under what circumstances should you grab that crowberry rather than searching for another blueberry? We gain insight by stating it mathematically.

First, we need to know what our (the forager's) rate of gain (H^1) would be if we were to search for and collect only blueberries (rank 1) and ignore crowberries (rank 2). H^1 is the reward from

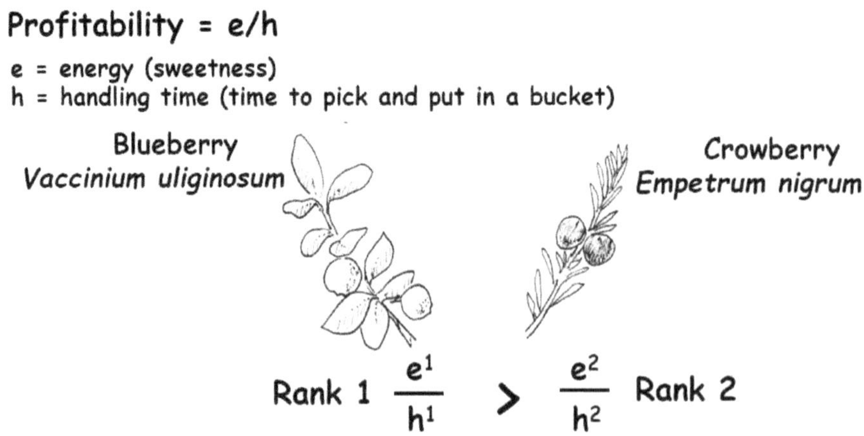

Figure 3. Foraging theory predicts that foragers will rank food items with respect to profitability; this ratio of reward to handling time can influence diet choice.

consuming one food item divided by the time required to search for and handle that item; this rate might be stated in terms of calories (as the reward) per unit time. Stated mathematically:

$$H^1 = \frac{e^1}{S^1 + h^1} \qquad \text{Equation 3}$$

The reward is e^1 and the handling time is h^1, but it is useful to consider search time S^1 in terms of the encounter rate aR of the preferred food item. Remember that R reflects the density of the resource, and a is the probability of a successful encounter per unit foraging time. Therefore, there should be an inverse relationship between encounter rate and search time (if encounter rate is high, search time is low). If we restrict a to just the probability of successfully detecting a food item per unit search time and do not worry about the probability of successfully handling the food item once captured, then within the foraging area, the number of items encountered per search time (aR) can be inverted ($1/aR$) to give the search time per item encountered, S.

In our blueberry example, let us say the foraging area is the size of a plant stem, so the density of berries is the number per stem, maybe $R = 4$ berries/stem. But we must also consider a in a unit of search time. Leaves sometimes hide berries, so perhaps you can detect a blueberry about half the time; therefore, the probability of a successful encounter is $a = 0.50$/second. That is an encounter rate of $aR = 0.50$/second × 4 berries/stem = 2 berries/second within a stem-sized search area. Therefore, the search time it takes to find a single blueberry is $S^1 = 1/a^1R^1 = 0.50$ seconds/berry (Figure 4).

Returning to our question of when a forager should take a less profitable food item: if the rate of gain when searching for and handling the more profitable item (blueberry) is greater than the profit in simply taking the less preferred item (crowberry), then the forager should spend time searching for another preferred item (blueberry). Stated mathematically, if the rate of gain of the preferred item (blueberry) is:

$$H^1 = \frac{e^1}{\frac{1}{a^1 R^1} + h^1} \qquad \text{Equation 4}$$

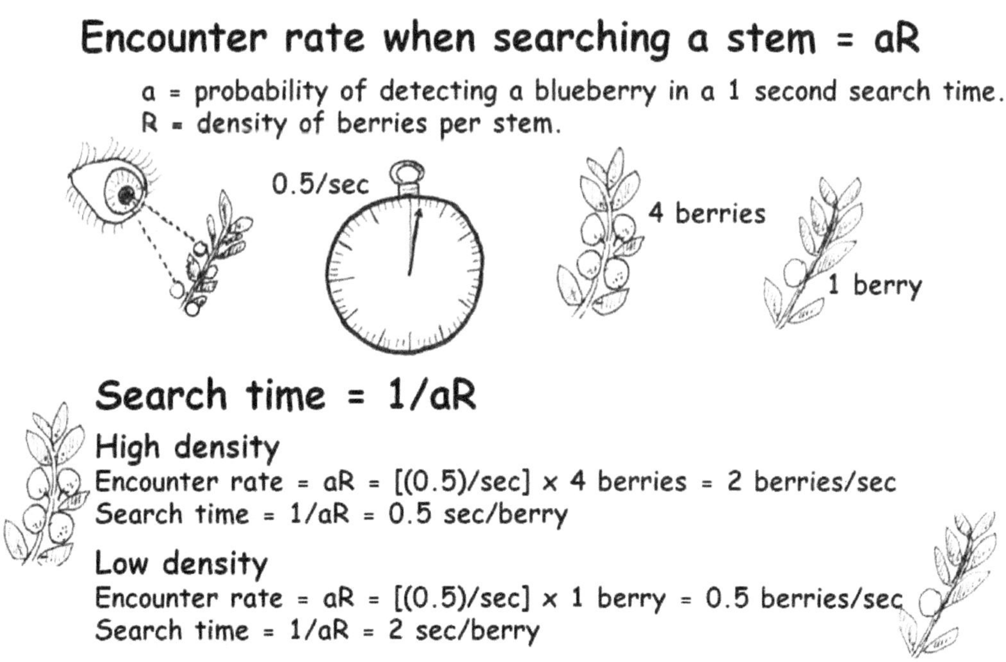

Figure 4. Illustrated example of the inverse relationship between encounter rate and search time.

which can be simplified by multiplying the top and bottom by a^1R^1 to:

$$H^1 = \frac{a^1R^1e^1}{1 + a^1R^1h^1}$$

Then a forager should reject the less preferred food (crowberry) when

$$\frac{a^1R^1e^1}{1 + a^1R^1h^1} > \frac{e^2}{h^2} \qquad \text{Equation 5}$$

Remember, we already know the reward ratio is greater for the "preferred" item by definition (Figure 3); what we are comparing in Equation 5 is whether the time spent foraging for another preferred item (rate of gain for a blueberry) will be worth it compared to just picking up the less preferred food item sitting right there in front of the forager's nose. From this expression (Equation 5), we can derive *two key predictions* for understanding diet choice. First, the decision to be picky or not depends only on the density of the preferred food and not the density of the less preferred food (R^1 is in inequality (Equation 5), but R^2 is not). Second, a forager is more likely to be picky when the density of the preferred food is high and/or the difference in reward is large, though the reward has a much greater impact. This is because increasing R^1 increases H^1 in a saturating fashion, whereas increasing the reward e^1 with respect to the reward e^2 directly makes the inequality larger. However, e^1 and e^2 are unlikely to change over time or across different areas, whereas the presence of other foragers or foraging in different habitats can change the density R^1, prompting a behavioral change in foraging. In fact, there should be some critical density of the preferred food at which the forager switches from being picky to not.

In our blueberry example, if a single blueberry contains 1 Cal and its handling time is 2 seconds, then using Equation 3 and our calculated search time of 0.50 sec/berry, the rate of gain H^1 = [1 Cal/berry] / [0.50 sec/berry + 2 sec/berry] = 0.40 Cal/sec. Because a crowberry also takes 2 seconds to handle, you should therefore include that crowberry in your bucket if it is greater than or equal to 0.80 Cal. However, if crowberries are only half as sweet (e^2 = 0.5 Cal), we would say "nah" and search for the next blueberry. Now suppose the patch of blueberries is sparse. Your probability of successfully detecting a blueberry from amongst the leaves has not changed (a = 0.50/sec), but now blueberries are so sparse that there is only one berry per stem. Thus, encounter rate a^1R^1 = 0.50/second × 1 berry/handful = 0.90 berries/sec. Now the rate of gain for blueberries is H^1 = 0.25 Cal/sec. Yes, that crowberry, even at 0.5 Cal (e^2/h^2 = 0.25 Cal/sec), would be worth putting in your bucket (Figure 5).

In our example above, we mathematically demonstrated how density can influence diet choice. This is also evident in the real world. In a classic study (Krebs et al. 1977), caged great tits (*Parus major*) were fed mealworms of two sizes. When large and small worms were present in low density, the birds were not selective. However, when the density of the large prey was increased sufficiently, the birds ignored the small prey and became highly selective. Even when the density of large prey was kept at this level and the density of small prey was increased to twice that of the large prey, the birds remained highly selective and continued to ignore the small prey. If you recall, Equation 5 explains why the density of the less preferred prey item does not matter in terms of diet choice—only the density of the more preferred prey item.

Although we held handling time equal in our blueberry-crowberry example, in many cases, handling time differs, and this can have a big impact on diet choice. In a study of pied wagtails (*Motacilla alba*) in a pasture near Oxford, England (Davies 1977), the birds were allowed to feed on various dung flies and beetles of various sizes. The wagtails revealed a preference for medium-sized prey, ignoring smaller and larger prey. The size of the preferred prey corresponded to the prey the birds could handle most profitably. Although easy to handle (low h), small prey did not return sufficient energy (e), and large prey required too much time and effort to handle (high h) relative to the energy gained.

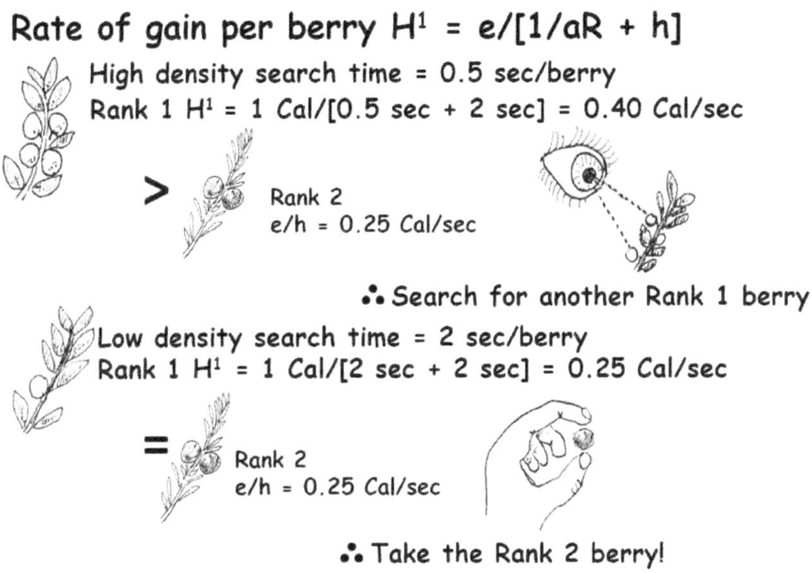

Figure 5. Illustrated example of how a difference in search time influences diet choice.

Patch Use

Most of us have had the experience of scraping the last bits of peanut butter or jam from a jar or trying to get the last drops of soda or water from a can or bottle. We never quite get everything—not because it is impossible, but because the effort becomes so high and the reward so low that we stop. Tom Witham (1977) observed this same behavior in both bees and hummingbirds—they left flowers before all the nectar had been drained. Where did they go? To seek out a better patch. This is the essence of patch use.

Recall from our Chapter 2 *Population Dynamics* that when discussing population density, we considered that individuals within a species may be randomly, uniformly, or patchily distributed in the environment. This must also be considered when discussing the density of resources sought by foragers. Most often, in fact, resources are neither randomly nor uniformly distributed in space and time. Rather, nutrients and resources usually occur in patches, with areas of both high and low availability. In other words, the density of food items (R) may not be the same in every spot, and their occurrence may vary over time. For example, those blueberries and crowberries we were foraging for are only present and mature during a short period at the end of August and into September, before the high-latitude alpine snows make them inaccessible. Furthermore, foraging in a patch reduces R. Because of this, when a forager is in a patch, its harvest rate declines. A squirrel foraging under an oak tree may deplete the acorns to a point at which it is better off leaving the current tree and looking for another tree to hunt for acorns. Because of the patchiness of the environment, a forager must choose when to forage and which patch to forage, and how thoroughly to exploit any given patch before moving on to another.

The theory of patch use addresses the question of when a forager should leave a food patch. Like diet choice, this is an economic decision—how long should a forager remain in a patch before the benefits of foraging no longer exceed the costs? One of the first people to put foraging in these terms was Eric Charnov (1976b). To answer this question, he first assumed that a forager knows its habitat. This assumption is reasonable for foragers that can accumulate experience about their habitat over days, weeks, or even years. Given this knowledge, foragers should not only be aware of the average rate of food capture within the current patch but also across the entire habitat. If a forager possesses this information, it should continue exploiting a depletable resource patch until its instantaneous harvest rate in the patch (called the marginal value) falls to the environmental average. This principle is called the **Marginal Value Theorem** (Charnov 1976b). Originally, the Marginal Value Theorem was framed to predict the optimal time to leave a patch as the rate of food intake decreased with respect to time spent foraging in the patch. However, the same idea can be used to predict the optimal density *R* to leave a patch when the costs of foraging begin to outweigh the rewards.

The fitness benefits of foraging in the current patch can be represented by the current harvest rate within the patch (rate-of-gain *H* in Equation 4), but at the same time, there are generally three fitness costs associated with foraging in a patch (Brown and Kotler 2004). The first is the metabolic loss rate of foraging, C, as the forager spends time foraging in the patch. The second cost recognizes that a forager may be killed or injured by its predators while foraging in the patch with some probability *P* over time. One of the critical insights from foraging theory is the idea of **tradeoffs**. That is, a forager may face a tradeoff between foraging benefits and the benefits of doing something else. In this case, *P* reflects the tradeoff between food and safety (Brown and Kotler 2004). Abramsky et al. (2002) demonstrated that the presence of predators decreased the amount of time gerbils spend foraging; however, the addition of extra food, even a small amount, offset the risk of predation and increased foraging time. Thus, the gerbils were willing to trade some amount of safety for an increase in the amount of food. This same idea of tradeoffs is incorporated into the last foraging cost, which is the cost of missing out on the fitness benefits of doing something else that enhances fitness (e.g., other types of foraging or reproduction). This is what economists call "opportunity cost," and ecologists have incorporated it into their foraging models as the **missed opportunity cost** (*MOC*).

So, a forager should leave a depletable food patch and go somewhere else or do something else when the harvest rate is no longer greater than the sum of loss rates. For dropping harvest rates or increasing loss rates, this point occurs when the reward rate equals the loss rate, or when:

$$H = C + P + MOC \qquad \text{Equation 6}$$

where we can think of $H = \dfrac{aRE}{1+ahR}$ from Equation 4, although here we use *E* for the reward before subtracting metabolic cost, instead of e for the net reward after the metabolic cost has been subtracted.

The point at which a forager stops foraging or quits harvesting a given food patch is called its **giving-up density** (GUD; Brown 1988). We can measure the GUD in the laboratory or field as the density of leftover food in a food patch. This generates a number of testable predictions. Because the density of resource *R* is intimately tied to the harvest rate, if we substitute Equation 4 and flip things around in Equation 6, we get:

$$R = \frac{C+P+MOC}{a[E-h(C+P+MOC)]} \qquad \text{Equation 7}$$

We find that, given *R* is positive, GUDs should decline when both the resource value *E* and the probability of success *a* are greater, making the denominator larger and thus *R* smaller. But a forager's GUD should increase with greater handling time *h,* making the denominator smaller and thus *R* larger. Greater metabolic loss rate *C*, higher predation rate *P*, or higher missed opportunity cost *MOC* (fitness costs) makes both the numerator larger and the denominator smaller and thus *R* larger. Therefore, as *P*, *C*, *MOC*, and *h* increase, we would expect to see more berries in a patch after the forager leaves, but as *E* and *a* increase, those berries will get scarcer and scarcer.

Evidence supports this (Bedoya-Perez et al. 2013). Giving-up densities in artificial resource patches have been measured for several mammals of conservation interest, including hyraxes, klipspringers, springbok, ibex, white-tailed deer, squirrels, and even red foxes (Brown and Kotler 2004). Hyraxes (*Procavia capensis*) in Augrabies Falls National Park, South Africa, are site-attached herbivores restricted to the vicinity of kopjes and rocky substrate. GUDs in experimental-resource patches reveal that hyraxes prefer early morning and late afternoon foraging and respond strongly to predation risk factors (higher *P*), exhibiting increasing GUDs with distance from cover and lower GUDs under protective cover than in the open (Druce et al. 2006). Tree squirrels will also have higher GUDs as food patches are placed farther from the safety of a tree (Morgan et al. 1997). Cottontail rabbits have lower GUDs on food pellets that have a higher nitrogen content (higher *E*), such as alfalfa, versus pellets containing a lower amount of nitrogen, such as timothy hay (Abu Baker et al. 2021). Crows have been shown to have higher GUDs on consecutive cold days (higher *C*) relative to warmer days (Kilpatrick 2003), and if there is more food available in the environment (higher *MOC*), starlings will use food patches less thoroughly (higher GUD; Olsson et al. 2002).

Habitat Selection

Habitat selection defines the preference of organisms for a certain territory or home range over others. Foraging may be a large part of this decision. Here is an evolutionary ecology "chicken or egg" question: Does diet preference dictate where an organism lives? Or does habitat dictate diet preference? Or perhaps the organism can move among habitats, choosing when to leave or give up on a spot and move to another, like a "patch use" decision. Such decisions may be seasonally transient, as when migratory birds such as storks spend summers in Poland, in northern Europe, and winters as far south as Botswana, in southern Africa. Likewise, wildebeest seasonally migrate north and south with the rains in the Masai Mara, part of the Serengeti ecosystem in Kenya and Tanzania. Where an animal takes up residence may include diverse considerations such as overall opportunities for food, safety, homes, and mates. Another consideration is the presence and density of other individuals. For instance, if a wooded area is particularly rich in acorns relative to another (perhaps one has mostly oak trees and the other maple trees), then squirrels may prefer inhabiting oak groves, thus increasing crowding and competition. At some point, a squirrel may favor a less crowded maple grove over a more crowded oak stand. This is the topic of density-dependent habitat selection.

The theory of density-dependent habitat selection (Fretwell 1969) teaches us that at low population size (*N*), all individuals should occupy the best habitat, but at higher *N*, some individuals should spill over from the good habitat into more marginal habitats. Individuals should distribute themselves so that those using different habitats have roughly equal fitness. This is called an **ideal free distribution** or IFD (Fretwell and Lucas 1970). According to IFD, individuals are free to move among habitats and should assort themselves such that the expected reward (food, safety, etc.) is constant across the habitat spectrum. The reward in a rich habitat is depressed as population density increases, to the point at which the reward in an adjoining poorer habitat looks more enticing (the grass is greener…). As a consequence, unoccupied habitats should be colonized until the reward in those habitats is equal to that of the currently occupied habitat. At an IFD, all occupied habitats offer the same fitness, and this equilibration of fitness is achieved by having different numbers of individuals occupying each habitat.

Paterson and Blouin-Demers (2017) tested the IFD hypothesis with *Urosaurus ornatus*, the ornate tree lizard. In the Chiricahua Mountains of southeastern Arizona, ornate tree lizards prefer the thermally warm and arthropod (food) rich wash habitat to the upland habitat. They found the density of lizards to be higher in wash sites than in upland sites. Lizards chose the upland habitat when the density of lizards in the wash habitat was high, and using monthly survival as a proxy for fitness, they found no fitness differences between the two habitats.

One of the assumptions of IFD is that individuals are *free* to move to the habitat of their choice. However, the ideal despotic (or dominance) distribution model (IDD) predicts that some individuals will benefit from a higher fitness because of resource monopolization (e.g., territoriality) while subordinate individuals are constrained to poorer habitat, not by choice, but by dominant individuals (Fretwell and Lucas 1970). In other words, there are competitive asymmetries between individuals, and, on average, individuals in the best habitats will have a higher reproductive success because they have excluded others. IDD predicts that the fitness differences between the two habitats should be marked. When Morris (1989) tested whether white-footed mice (*Peromyscus leucopus*) exhibited IFD or IDD among forest, forest edge, and fencerow habitats, he found, as expected for both models, a negative relationship between population density and mean litter size (fitness). However, the overall fitness in the forest habitat was greater than either the fencerow or edge habitats, indicating IDD. The natural history of white-footed mice supports the IDD model because females are known to be territorial during breeding season (Morris 1989).

Applications of Foraging Theory to Conservation

Diet choice, patch-use, and habitat selection can be used as behavioral indicators for population conservation (Whelan and Jedlicka 2007). For example, a healthy population may exhibit seasonal fluctuations in diet choice according to dry versus wet seasons or cold versus warm seasons. Cottontail rabbits and many other lagomorphs in temperate climates feed on preferred grasses and forbs during the summer but strip bark from shrubs during the winter when preferred foods are absent. A change in diet to bark feeding during the summer, for such animals, could indicate a food-stressed population (Morris et al. 2009). The point at which animals leave natural or experimental food patches can also provide valuable behavioral indicators. The increase in the GUDs of ibex in the presence of people quantified the negative impact of large numbers of tourists at desert oases in Israel (Tadesse and Kotler 2012). At the habitat selection level, foraging locations may reveal human hunting pressure and disturbance. Hunting is banned in both Sagmartha (Mount Everest) National Park and the Annapurna Conservation Area in Nepal, but enforcement has been more successful at Sagmartha National Park. Consequently, Himalayan tahr (*Hemitragus jemlahicus*) in Sagmartha forage intensively both inside and outside of agricultural fields, even with humans passing by just meters away (Ale 2007). In contrast, in the Annapurna Conservation Area, Himalayan tahr remain 100 or more meters from humans, preferring the high ground or a separating ravine. These fear responses (predation-risk) can be quantified by measuring vigilance and approach distance (Frid and Dill 2002; Ale and Brown 2009), which is used to indicate poaching and hunting pressure, and more generally, the risks of predation faced by prey (Morris et al. 2009).

Finally, the way individuals distribute themselves across habitats can also be an important indicator of preferred habitat, helping to prioritize conservation areas. For instance, at low abundance, an animal's first choice of habitat should indicate the highest quality habitat. This holds true whether animals disperse according to IFD or IDD. Density-dependent habitat selection can also be used to monitor changes in habitat over time. By comparing plots over time of the density in habitat A versus the density in habitat B, changes in relative habitat quality over time can be detected quantitatively.

Plant Foraging Theory

If the concepts of diet choice, patch use, and habitat selection seem biased toward mobile foragers, you would not be mistaken. Recall, however, that we began this chapter with the idea that plants are also foragers. Plant foragers have been likened to foraging bees (McNickle et al. 2009) and ants (Lopez et al. 1994). One group of plants has even been compared to sit-and-wait predators. Ellison (2020) points out that carnivorous (insect-trapping) plants attract, capture, kill, and digest prey with the capability to adjust trap features, switch between different types of traps, activate traps in response to external cues, and even choose whether to use traps. However, most ideas about plant foraging have focused on "root strategies."

We may naively think that roots are "rooted" and immobile in the ground, but plant root growth can be highly **plastic** and responsive to nutrient and other cues (Cahill and McNickle 2011). Root strategies include differences in branching, biomass, length, uptake capacity, and longevity (McNickle et al. 2009). While stationary, plants are also modular organisms, which allows them to forage by growing organs in multiple locations at once. In fact, nutrient-foraging canopy roots have been found in both temperate and tropical rainforest trees, which take advantage of soils that develop in their leaf canopies (Nadkarni 1981; Hertel et al. 2011). While it is generally true that animals and plants differ in how they forage and the currency in which they forage (animals move around to gain energy and nutrients, whereas plants grow body parts to gain nutrients and water to make energy-capturing organs), despite these differences, McNickle et al. (2009) argue that plants should still behave in ways that enhance fitness.[1] If plant behavior can be defined as "phenotypic plasticity… produced by stimuli for which alternative responses would produce differential fitness" (McNickle et al. 2009), then plants can indeed exhibit foraging behavior.

We can apply a cost-benefit approach to plant diets, similar to the ways we study mobile foragers. Costs, which can be calculated in units of carbon expended by a plant, include growth and maintenance respiration whereas benefits can be calculated in units of carbon gain with respect to acquisition limiting nutrients or water. Efficiencies are gained when benefits with respect to costs are maximized. For example, optimal root longevity is linked to maximized efficiency in resource acquisition, which in turn, can be linked to individual plant fitness (Eissenstat et al. 2000).

As for the ranking of food items, one might ask whether plants can make a diet choice when their "diet" is made up of **essential nutrients** (e.g., nitrogen cannot substitute for phosphorus and is therefore essential) rather than **substitutable nutrients** (e.g., 2 bananas = 1 chocolate bar). However, plants do face diet choices because essential nutrients exist in multiple chemical forms. For instance, nitrogen exists as nitrate, ammonium, and amino acids (McNickle et al. 2009). These different forms are absorbed or taken up via different pathways. Thus, plants should uptake only high-quality forms and include other forms in their "diet" only when high-quality forms are difficult to obtain.

Patch use theory can also be applied to plants because resources essential to plant growth are usually heterogeneously distributed in space and time, and plants often have the plasticity to respond to this (Bell and Lechowicz 1994). While some plant species have low "precision" with respect to nutrients in a patchy environment and place their roots at random, other plant species are "precise" in their foraging and grow a large proportion of their roots in nutrient-rich patches (Kembel and Cahill 2005). The clonal dune plant *Hydrocotyle bonariensis* is capable of veering away from "less favorable" grass patches and sending its ramets into "favorable" empty patches, where it is suspected they are responding to higher nutrient levels (Evans and Cain 1995). Consistent with the marginal value theorem of patch use, the presence of nutrient-rich patches can cause plant roots to reduce their rate of soil exploration such that they "leave" nutrient-rich patches later than they leave nutrient-poor patches (McNickle and Cahill 2009). There is also evidence that plants exhibit density-dependent responses and are capable of choosing habitat patches in an ideal free distribution (Gersani et al. 1998).

As first recognized by Charles Darwin, because plants are modular, they face both evolutionary and behavioral foraging choices (Kutschera and Niklas 2009). Evolutionarily, the heterogeneity of plant resources can act as a major selective force on plasticity in plant growth. Above ground, plants can change shoot strategies in response to light flux, quality, soil nutrients, water status, competition, and stress. Belowground, plants can change root branching and elongation and can form associations with mycorrhizal fungi (Hutchings and DeKroon 1994). For both mobile and immobile foragers, life-history and behavioral traits that increase resource acquisition increase fitness. This, in turn, affects the impacts on the dynamics, coexistence, and evolution of their respective populations. "We are what we eat" is thus a profound summation of the evolutionary ecology of foraging.

[1] Jack Schultz, an evolutionary ecologist from University of Houston, was once heard to say, "plant are really animals, they just behave very slowly" (C. Whelan, University of Illinois Chicago, personal communication).

References

Abramsky, Z., Rosenzweig, M. L. and Subach A. 2002. The costs of apprehensive foraging. Ecology 83: 1330–1340.
Abu Baker, M. A., Emerson, S. E., Gorman, J. and Brown, J. S. 2021. Dietary choices in a generalist herbivore, the eastern cottontail *Sylvilagus floridanus*, in urban landscapes. Behaviour 158: 377–395.
Ale, S. B. 2007. Ecology of the snow leopard and the Himalayan tahr in Sagarmatha (Mt. Everest) National Park, Nepal. PhD thesis. University of Illinois Chicago, IL, USA.
Ale, S. B. and Brown, J. S. 2009. Prey behavior leads to predator: a case study of the Himalayan tahr and the snow leopard in Sagarmatha (Mt. Everest) National Park, Nepal. Israel Journal of Ecology and Evolution 55: 315–327.
Anderson, R. C. and Mather, J. A. 2007. The packaging problem: bivalve prey selection and prey entry techniques of the octopus *Enteroctopus dofleini*. Journal of Comparative Psychology 121: 300–305.
Bedoya-Perez, M. A., Carthey, A. J. R., Mella, V. S. A. McArthur, C. and Banks, P. B. 2013. A practical guide to avoid giving up on giving-up densities. Behavioral Ecology and Sociobiology 67: 1541–1553.
Begon, M., Harper, J. L. and Townsend, C. R. 1996. Ecology, 3rd edition. Blackwell Science. Cambridge, MA.
Bell, G. and Lechowicz, M. J. 1994. Spatial heterogeneity at small scales and how plants respond to it. pp. 391–411. *In*: Caldwell, M.M. and R.W. Pearcy [eds.]. Exploitation of Environmental Heterogeneity by Plants. Academic Press, San Diego, California.
Boback, S. M., Cox, C. L., Ott, B. D., Carmody, R., Wrangham, R. W. and Secor, S. M. 2007. Cooking and grinding reduces the cost of meat digestion. Comparative Biochemistry and Physiology, Part A 148: 651–656.
Brown, J. S. 1988. Patch use as an indicator of habitat preference, predation risk, and competition. Behavioral Ecology and Sociobiology 22: 37–47.
Brown, J. S. and Kotler, B. P. 2004. Hazardous duty pay: studying the foraging cost of predation. Ecology Letters 7: 999–1014.
Cahill, J.F., Jr. and McNickle, G. G. 2011. The behavioral ecology of nutrient foraging by plants. Annual Reviews of Ecology, Evolution, and Systematics 42: 289–311.
Carmody, R. N. and Wrangham, R. W. 2009. The energetic significance of cooking. Journal of Human Evolution 57: 379–391.
Charnov, E. L. 1976a. Optimal foraging: Attack strategy of a mantid. The American Naturalist 110: 141–151.
Charnov, E. L. 1976b. Optimal foraging: the marginal value theorem. Theor. Popul. Biol. 9: 129–136.
Davies, N. B. 1977. Prey selection and social behavior in wagtails (*Aves monticillidae*). Journal of Animal Ecology 46: 37-57.
Druce, D. J., Brown, J. S., Castley, J. G., Kerley, G. I. H., Kotler, B. P., Slotow, R. et al. 2006. Scale-dependent foraging costs: habitat use by rock hyraxes (*Procavia capensis*) determined using giving-up densities. Oikos 115: 513–525.
Eissenstat, D. M., Wells, C. E., Yanai, R. D. and Whitbeck, J. L. 2000. Building roots in a changing environment: implications for root longevity. New Phytol. 147: 33–42.
Ellison, A. M. 2020. Foraging modes of carnivorous plants. Israel Journal of Ecology and Evolution 66: 101–112.
Evans, J. P. and Cain, M. L. 1995. A spatially explicit test of foraging behavior in a clonal plant. Ecology 76: 1147–1155.
Fretwell, S. 1969. Dominance behavior and winter habitat distribution in juncos (*Junco hyemalis*). Bird-banding 40: 1–83.
Fretwell, S. D. and Lucas, H. L. Jr. 1970. On territorial behavior and other factors influencing habitat distribution in birds. I. Theoretical development. Acta Biotheoretica 19: 16–36.
Frid, A. and Dill, l. 2002. Human-caused disturbance stimuli as a form of predation risk. Conservation Ecology 6: 11 http://www.consecol.org/vol6/iss1/art11.
Gersani, M., Abramsky, Z. and Falik, O. 1998. Density-dependent habitat selection in plants. Evolutionary Ecology 12: 223–234.
Hertel, D., Kōhler, L. and Rillig, M. C. 2011. Mycorrhizal, endophytic and ecomorphological status of tree roots in the canopy of a montane rain forest. Biotropica 43: 401–404.
Hilborn, A., Pettorelli, N., Orme, C. D. L., and Durant, S. M. 2012. Stalk and chase: how hunt stages affect hunting success in Serengeti cheetah. Animal Behaviour 84: 701–706.
Hutchings, M. J. and DeKroon, H. 1994. Foraging in plants: the role of morphological plasticity in resource acquisition. Advances in Ecological Research 25: 159–238.
Kembel, S. W. and Cahill, J. F. 2005. Plant phenotypic plasticity belowground: a phylogenetic perspective on root foraging trade-offs. The American Naturalist 166: 216–230.
Kilpatrick, A. M. 2003. The impact of thermoregulatory costs on foraging behaviour: a test with American crows (*Corvus brachyrhynchos*) and eastern grey squirrels (*Sciurus carolinensis*). Evolutionary Ecology Research 5: 781-786.

Krebs, J. R., Erichsen, J. T., Webber, M. I. and Charnov, E. L. 1977. Optimal prey selection in the great tit (*Parus major*). Animal Behaviour 25: 30–38.
Kutschera, U. and Niklas, K. J. 2009. Evolutionary plant physiology: Charles Darwin's forgotten synthesis. Naturwissenschaften 96: 1339–1354.
Lopez, F., Serrano, J. M. and Acosta, F. J. 1994. Parallels between the foraging strategies of ants and plants. Trends in Ecology & Evolution 9: 150–153.
MacArthur, R. H. and Pianka, E. R. 1966. On optimal use of a patchy environment. The American Naturalist 100: 603–609.
McAuliffe, J. 1984. Sahuaro-nurse tree associations in the Sonoran Desert: competitive effects of sahuaros. Oecologia 64: 319–321.
McNickle, G. G. and Cahill, J. F. Jr. 2009. Plant root growth and the marginal value theorem. PNAS 106: 4747–4751.
McNickle, G. G., St. Clair, C. C. and J. F., Jr. Cahill. 2009. Focusing the metaphor: plant root foraging behaviour. Trends in Ecology & Evolution 24: 419–426.
Morgan, R., Brown, J. S. and Thorson, J. M. 1997. The effect of spatial scale on the functional response of fox squirrels. Ecology 78: 1087–1097.
Morris, D. W., Kotler, B. P., Brown, J. S., Sundararaj, V. and Ale, S. B. 2009. Behavioral indicators for conserving mammal diversity. Ann. N. Y. Acad. Sci. 1162: 334-56. doi: 10.1111/j.1749-6632.2009.04494.x.
Morris, D. W. 1989. Density-dependent habitat selection: testing the theory with fitness data. Evolutionary Ecology 3: 80–94.
Nadkarni, N. M. 1981. Canopy roots: convergent evolution in rainforest nutrient cycles. Science 214: 1023–1024.
Olsson, O., Bruun, M. and Smith, H. G. 2002. Starling foraging success in relation to agricultural land-use. Ecography 25: 363–371.
Paterson, J. E. and Blouin-Demers, G. 2017. Density-dependent habitat selection predicts fitness and abundance in a small lizard. Oikos 127: 448–459.
Shaw, B. 1903. Act I. Man and Superman. https://www.panarchy.org/shaw/maxims.1903.html. Retrieved 3 January 2025.
Somers, N., D'Haese, B., Bossuyt, B., Lens, L. and Hoffmann, M. 2008. Food quality affects diet preference of rabbits: experimental evidence. Belgian Journal of Zoology 138: 170–176.
Stephens, D. W., Brown, J. S., and Ydenberg, R. C. 2007. Foraging: Behavior and Ecology. University of Chicago Press, Chicago, IL.
Tadesse, S. A. and Kotler, B. P. 2012. Impact of tourism on Nubian ibex (*Capra nubiana*) revealed through assessment of behavioral indicators. Behavioral Ecology 23: 1257–1262.
Whelan, C. J. and Jedlicka, D. M. 2007. Augmenting population monitoring programs with behavioral indicators during ecological restorations. Israel Journal of Ecology and Evolution 53: 279–295.
Witham, T. G. 1977. Coevolution of foraging in *Bombus* and nectar dispensing in *Chilopsis*: a last dreg theory. Science 197: 593–596.
Young, T. P., Stanton, M. L. and Chrisian, C. E. 2003. Effects of natural and simulated herbivory on spine lengths of *Acacia drepanolobium* in Kenya. Oikos 101: 171–179.

CHAPTER 6

Life Histories and Life History Tradeoffs

"Youth is a blunder, manhood a struggle, and old age a regret" (Benjamin Disraeli)

What is a **life history**? Perhaps for some people, our life stages may be typified by youthful blunder, middle-aged struggle, and the waning regret phases, as British statesman Benjamin Disraeli wittily put it. However, evolutionary ecologists define life history in terms of growth and reproductive strategies, as well as the traits that characterize an organism as it passes through the different stages of its life cycle—from fertilization to embryonic development to maturation and reproduction, to senescence and eventually death. While we are familiar with the human life cycle and the growth and reproductive traits characterizing the life history stages we experience from conception to death, it is important to recognize that life history traits and life cycles are not the same for all species. They are, in fact, wildly varied.

The American beech (*Fagus grandifolia*), a dominant component of deciduous forests in southeastern Canada and northeastern United States, is a large, abundant species in Warren Woods, a temperate deciduous forest in Michigan that has been studied by scientists since the early 1900s. It can grow up to 30 meters tall and survive up to 400 years. It bears seeds around age 40, producing thousands of seeds in a "mast" every two to eight years during its lifespan (Burns and Honkala 1990). However, the likelihood of any beech nut finding a space, germinating, and starting its climb to the canopy is very small. The forest world is so perilous and competitive that of the millions of seeds the tree produces, only a few survive to reproductive age. Many of its seeds and seedlings will be consumed as delicacies by slugs, birds, rabbits, deer, and squirrels. Those that survive the feeding frenzy must then fortuitously arrive in an open space with enough light to grow to maturity. As a mature beech ages, like many trees, its bark wrinkles, and its trunk becomes weak and dry, making it susceptible to damage or toppling in a storm. Its life ends, but the opening its death creates will quite likely provide space for one of its seedlings to grow to maturity.

In contrast, another Warren Woods denizen, a female mayfly (*Hexagenia limbate*), reproduces a short time after she transforms from the aquatic stage to flying adult. Mayflies live in a stream as nymphs for a year, feeding on various plant cells, diatoms, algae, and even small insects. Incredibly, nymphs emerge from the water as adults (sans functioning mouthparts!) and live just one day to mate, and if female, to deposit eggs in the stream before dying. Thus, a mayfly has just one chance to produce offspring. As mayflies fly around looking for mates, they will be eaten by birds, bats, frogs, and even spiders. Unlike many of her siblings, this particular mayfly, perched on a giant old maple tree, survived the egg stage and the nymph stage. During her one day of adulthood, she successfully mated and laid up to 8,000 eggs in the Galien River, whose calm waters pass through Warren Woods on their way to Lake Michigan.

The mayfly life cycle, from stream-deposited egg to aquatic nymph, through about 30 molts to terrestrial adulthood, and finally to the deposition of new eggs into the stream again, is complex

DOI: 10.1201/9781003008538-6

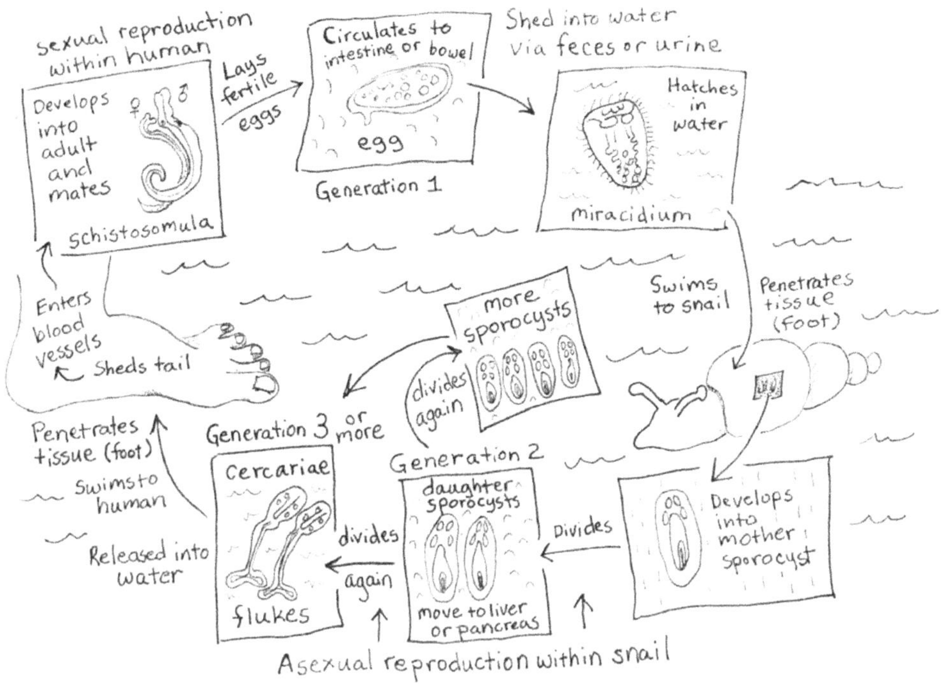

Figure 1. Conceptualized sketch of the life cycle of *Schistosoma*.

compared to the life cycle of a beech tree, but life cycles can get even more involved. In parts of Africa, South America, and Asia, parasitic worms (blood flukes) of the genus *Schistosoma* cause schistosomiasis, a waterborne disease. Over 200 million people are infected annually, challenging the global health community to develop methods to control the parasite. Schistosomiasis is difficult to control because the parasite *Schistosoma* has such a complicated life cycle. In fact, its life cycle—from egg to adult to egg again—spans multiple generations (Figure 1). *Schistosoma* eggs from infected individuals are released into water through feces or urine, where they hatch into free-swimming larvae called miracidia. The miracidia enter freshwater snails by piercing the snail's foot tissue, where they transform into sporocysts. These then divide into thousands of free-swimming cercariae larvae (the second generation). Cercariae emerge daily from snails and penetrate any human skin exposed to infested water. Once cercariae enter a human body, they travel through the bloodstream to the liver, where they begin feeding on red blood cells until they mature into adults. Reproduction occurs monogamously when the female is enveloped by the male. Their eggs then enter the intestines or bladder through blood vessels and ultimately exit the body through feces or urine, thus completing the *Schistosoma* life cycle.

Life cycles in eukaryotic organisms describe a species' development from zygote (fertilized egg) to maturity and ultimately to the production of new zygotes. Trees such as beech or maple, as well as birds and mammals, show relatively simple life cycles. However, complex life cycles enable species to exploit multiple habitats and niches. Like *Schistosoma,* most primitive plants have multigenerational life cycles. Insects, parasites, and amphibians also exhibit complex life cycles, which consist of two or more distinct morphological stages that inhabit different habitats. Each of these different generations or stages has different physiologies and shows different behaviors. But why is there this complexity? And why is there such a diversity of other life history traits, such as size at birth, age, and size at maturity, and number and size of offspring? All these life history traits, like many other physiological, morphological, and behavioral traits, have genetic bases and are therefore subject to natural selection (Stearns 1977). In the evolutionary game of life, "winning," however, comes down to one thing: maximizing the number of offspring that successfully reproduce.

So, if organisms are all trying to maximize the number of "successful" offspring, why is the solution to the basic problem of reproduction so varied and at times quite complex?

If we were to construct a hypothetical life history that yielded the greatest lifetime reproductive output, we would end up with an organism that begins reproducing as soon as it is born, produces offspring infinitely, and lives forever. This "Darwinian demon" would eventually take over the world (Law 1979). Fortunately, in the real world, no super-species exists that is optimized to do everything, has access to unlimited resources, and whose fitness is high in all possible circumstances. This is because, when it comes to life histories, there are always tradeoffs: a jack of all trades is a master of none.

Life History Tradeoffs

To grow, survive, and produce offspring, all organisms must obtain energy and nutrients from their environment. However, it is a fundamental constraint that no resource is available both instantaneously and in unlimited supply. As a result, there are evolutionary tradeoffs where allocation to one life history trait comes at a cost to another. One way to think about life history tradeoffs is to imagine a species' life history as a finite pie, with slices representing how the organism divides its resources among growth, survival, and reproduction. For example, devoting more of the pie to early reproduction comes at the cost of growth, which might reduce the potential for larger clutch sizes in the future. Time, energy, and nutrients spent on one goal cannot support another. The life histories that organisms possess represent evolutionary solutions to these life history tradeoffs within the environment they inhabit.

Three basic questions related to the allocation of resources in the life history pie are:

1. How many offspring should an organism produce during each reproductive episode?
2. How often should it breed?
3. When should it begin to reproduce?

We know the presence of predators, availability of nutrients, etc., have distinct effects on an organism's life history pie, so what is the best way to apportion the pie under any given ecological circumstance?

Offspring Numbers vs. Offspring Size (Reproductive Allocation)

How many offspring should an organism produce during each reproductive episode? Godfray et al. (1991) note that the evolution of clutch size can "claim to be one of the oldest topics in life history theory." The term clutch size refers to the number of eggs laid in a single brood (or nest) by a female bird (Lack 1947), but it can be more generally defined across taxa as the number of offspring produced by a female in a single reproductive bout. Clutch size in fishes ranges from one or two large live young produced by mako sharks to 600 million eggs per clutch laid by the ocean sunfish (Winemiller 1995). However, it is worth pointing out that no fish produces millions of large live young. This pattern extends across multiple taxa. In the 1960s, ornithologist David Lack (1967) hypothesized a fundamental trade-off between clutch size and egg size. He reasoned that producing offspring requires resources, meaning an organism can either produce many smaller offspring or few large offspring for a given amount of resources. Large offspring are considered to be of higher "quality" because they are often less likely to be consumed or are better provisioned at birth for faster growth or growth under extreme environmental or competitive conditions. In terms of our life-history pie, this is an allocative tradeoff within the reproduction slice.

Several studies support the presence of this tradeoff, although it can be difficult to detect with statistical rigor due to other life history and ecological correlates (for example, when the size of the mother correlates with the size of the clutch), which tend to mask the allocative tradeoff between egg or newborn size and clutch or litter size. Blackburn (1991) controlled for these confounding variables and found that, among 1,530 bird species, there was indeed a negative relationship (tradeoff) between egg size and clutch size. Additionally, he found that a small increase in egg size was "balanced" by a large decrease in clutch size. When Charnov and Ernest (2006) corrected for size-related correlations, they found that mammals also have an inverse proportional relationship between the number of offspring and offspring size.

Plants also vary widely in the number of seeds they produce and the size of those seeds. Consider a coconut palm tree that produces the world's largest seeds but only 30 to 40 coconuts per reproductive event, whereas orchids produce the world's smallest seeds but as many as four million of them per fruit (Arditti and Ghani 2000). Plant biologists long ago recognized a negative relationship (i.e., a tradeoff) between seed size and seed number in different species of plants (Stevens 1932). Šera and Šerý (2004), using a measure of reproductive capacity of the population based on the number of seeds, confirmed a tradeoff between seed number and seed weight from 500 wild herbaceous species from Central Europe (Figure 2).

Clutch size is related to evolutionary fitness, as noted by Lack (1947). In general, organisms are expected to evolve a clutch size that maximizes the number of young surviving to maturity, averaged over the parent's lifetime. A small clutch size might not be evolutionarily advantageous on its own, yet many organisms produce small clutches, broods, or litters. Horses generally produce only one foal, while pigs usually have a litter of four to six piglets. Why is this the case? Under what conditions is it better to have many small offspring rather than a few large ones, and vice versa?

In plants, Šera and Šerý (2004) found that species with an annual life cycle produce more seeds than perennials, and that plants with more upright than lateral spread tend to have larger seeds. Annual plants are specialists in disturbed and marginal habitats without a lot of other competing plant species, so among annuals, high numbers of small seeds improve a species' dispersal ability. However, this is at the expense of the ability of seedlings to establish successfully, which is an advantage of larger seeds that can be packed with extra resources. Thus, the tradeoff between seed size and seed number leads to differing advantages depending on the habitat.

In environments with a high risk of predation, having a larger clutch can be both beneficial and detrimental. Large broods of birds may make more noise and attract predators (Skutch 1949), but in invertebrates, large clutches are often advantageous (Godfray et al. 1991). For example, Ascalaphid larvae (*Acaloptynx furciger*) form a mass to rebuff an attacker (Henry 1972), and some moths lay eggs in a way that the outer eggs protect the interior eggs from parasitism (Dowden 1961). In

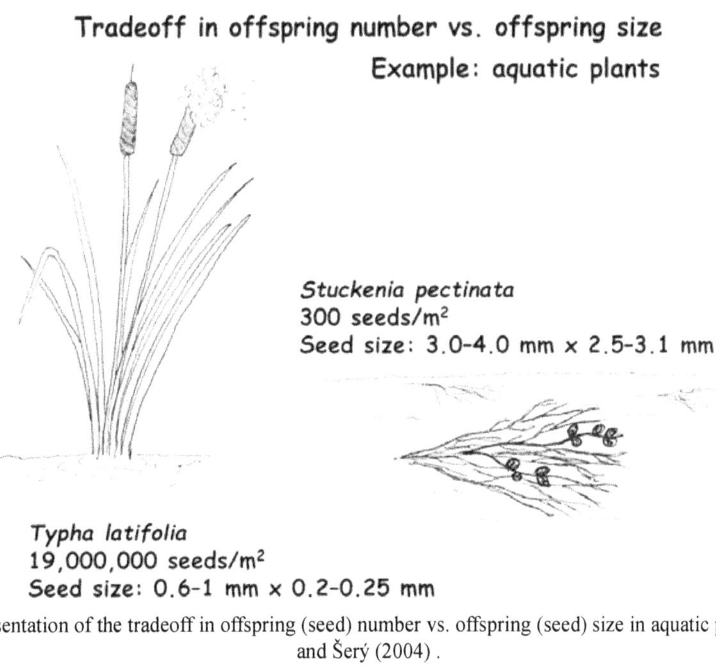

Figure 2. Representation of the tradeoff in offspring (seed) number vs. offspring (seed) size in aquatic plants. Source: Šera and Šerý (2004).

contrast, long-tailed skuas, a type of seabird, have a clutch size of no more than two because they incubate their eggs on their feet, which experiments reveal may be an evolutionary response to increased predation on eggs in nests (Andersson 1976).

Two hypotheses have been proposed for the evolution of clutch size in birds exhibiting parental care: food limitation (when the amount of resources in the environment limits the amount of food parents can gather; Lack 1948) and the frequency of nest predation (when predators use parental activity to find nests and therefore constrain parental food delivery rates [see the costs of foraging in Chapter 5 *Foraging ecology*]; Skutch 1949). Lack (1948) thought food limitation was important because he noticed that the clutch size of breeding birds varies with latitude: at higher latitudes, birds have larger clutches (generally twice the size). The reason for larger clutches at higher latitudes, according to Lack, is that higher latitudes provide longer periods of daylight during the breeding season, which may allow parents more time for foraging and feeding greater numbers of offspring than their counterparts in the tropics, supporting the food limitation hypothesis. However, since Lack first proposed this hypothesis, it has been tested a number of times by artificially enlarging brood size (ecologists added eggs to nests). An analysis of 42 independent brood-enlargement studies shows significantly more fledglings were produced in enlarged broods than in normal-sized broods, indicating parents were not food-limited in the amount of young they could raise (Vander Werf 1992).

Considering the alternative nest predation hypothesis, Martin et al. (2000) found a strong negative relationship between daily predation rate and nestling visitation and food delivery rates by parents in subtropical North and South American bird communities. This finding supports the idea that predation imposes a foraging cost—higher predator presence results in reduced food delivery. Interestingly, they also found that clutch size was positively related to food delivery rate but even more negatively related to daily predation rate, indicating predation rate had more of an effect on clutch size than food delivery rate.

When considering the tradeoff between the number and size of offspring, it may be tempting to link this energy investment by the parent to the distinction between precocial and altricial young. **Precocial** young are born or hatched well-furred or feathered, with eyes and ears open, and with a developed thermoregulatory and nervous system. This tends to be associated with longer gestation in mammals and longer incubation times in birds, resulting in a reduced need for parental care (Augustine et al. 2019). On the flip side, **altricial** young are born or hatched essentially naked, with eyes and ears closed, limited mobility, and poor thermoregulation. This developmental state typically requires less parental investment before hatching or birth, but more parental care afterward (Augustine et al. 2019). However, it turns out attempts to link the developmental state of the offspring upon birth or hatching to clutch or litter size have been fraught with exceptions. In mammals, the stage of neonatal development is largely a function of time, so longer gestations tend to have precocial neonates in smaller litters (Derrickson 1992). Yet, numerous exceptions prevent this from being a strict rule: capybaras, coypus, and hares all produce precocial offspring without significantly increasing gestation length or decreasing litter size, while primates—despite having the longest relative gestation periods—do not produce very precocial young at all (Derrickson 1992). This is also true for birds, where the difference in developmental state at hatching may be more closely related to whether birds lay eggs on the ground or in trees, rather than to clutch size (Starck and Ricklefs 1998). What this highlights is that ecological conditions after birth or hatching may be what drives the evolution of clutch size.

Lack's (1947) tradeoff between number and size in the production of offspring is assumed to be allocative because producing offspring requires resources, which are limited. However, Lack's hypothesis does not apply if there is a tradeoff between the number of offspring in a clutch and future reproductive success (Godfray et al. 1991). In some cases, it is better to wait to have offspring.

Offspring Now vs. Offspring Later (Reproduction vs. Parental Growth and Survival)

When should an organism start reproducing? Some organisms reproduce early in life, while others reproduce later. Starting to reproduce early may seem advantageous in terms of greater lifetime reproduction, but those who adopt a late-reproduction strategy can invest their energy into growing and developing before reproduction begins. If natural selection maximizes an individual's relative fitness, then the age and size at first reproduction are optimal when the benefits outweigh the costs. If delaying maturity results in a larger body size or greater access to resources at the age of first reproduction, and if larger body size or increased resources translate to a greater fecundity, then delaying reproduction can be advantageous.

Consider a hypothetical mammal that grows until sexual maturity and then begins to reproduce. Suppose a female of this mammal begins reproducing at age three, and she is capable of producing 15 offspring per year throughout her lifetime. But what if a different individual delayed reproduction until age four, when she was big enough to produce 20 offspring per year (a 33% greater annual fecundity)? By comparing the accumulated number of offspring (Table 1), we can see that until age five, the individual maturing early has a greater accumulated fecundity. At age six, both have the same accumulated fecundity, but from age seven onward, the individual maturing later has greater accumulated reproduction (Table 1).

Evidence for a tradeoff between reproduction and growth has been found in herbaceous plant species, conifers, and angiosperm tree species. The thin growth rings in trees in mast years indicate a drain on resources and suggest a tradeoff in the allocation of energy between growth and reproduction (Koenig and Knops 1998). **Mast-seeding** is a reproductive strategy that results in the synchronized production of large crops in some years and small crops in others. In Douglas fir trees, as the production of cones increases, the radial growth of the adult tree decreases (Eis et al. 1965). In the southern hemisphere, in New Zealand, the hard beech, *Nothofagus truncata*, also exhibits mast seeding, and Monks and Kelly (2006) showed that when these trees reproduced the most, they grew the least, and vice versa (Figure 3).

Evidence for the reproduction vs. growth tradeoff is also found in animals. In sardinella fish, as fecundity decreases, the growth rate has been shown to increase (Tsikliras et al. 2007). In an experiment with the freshwater clam *Anodonta piscinalis* (Jokela and Mutikainen 1995), individuals moved into high-density cages had the same high reproductive output as those at the control density, but shell growth was slower, and somatic (body) mass was lower. This suggests that when resources are limited, resources are allocated more to reproduction than to growth.

All else equal, selection should favor later maturity if it results in larger clutch sizes, as the Table 1 exercise reveals. However, we have not accounted for the probability of adult survival. In our example, natural selection should favor later maturity only if there is a high enough probability of adult survival to age 7; otherwise, it should favor earlier maturity. Thus, the answer to when an organism should begin to reproduce, or whether to devote resources to the reproduction slice or

Table 1. Cumulative lifetime reproduction of two female mammals differing by one year in age at first reproduction (maturity), where a delay allows 33% greater annual fecundity (increase from 15 offspring to 20 offspring per year).

	M_x = Birth Rate		Accumulated Offspring	
Age (year x)	**Mature at 3**	**Mature at 4**	**Mature at 3**	**Mature at 4**
1	0	0	0	0
2	0	0	0	0
3	15	0	15	0
4	15	20	30	20
5	15	20	45	40
6	15	20	60	60
7	15	20	75	80
8	15	20	90	100
9	15	20	105	120
10	15	20	120	140

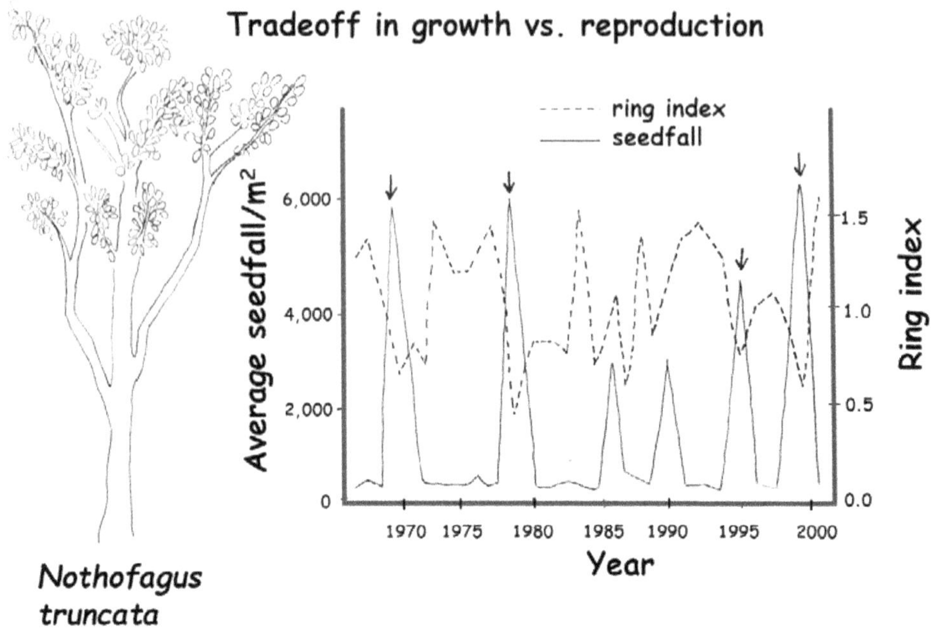

Figure 3. Illustrated example of the tradeoff in growth (tree ring width as index) versus number of seeds produced (seedfall/m^2). Arrows indicate high seedfall years in which the ring index was depressed relative to the previous year (interpreted from Figure 1 in Monks and Kelly 2006).

the growth slice of the life history pie, should also depend on the likelihood of survival to the next reproductive bout.

In a review of published data, Shine and Charnov (1992) found that as survival of adult lizards and snakes increases, their age at maturity also increases. The annual rate of adult survival among snakes ranged from 35% to 85% of the population, while the corresponding age at reproductive maturity ranged from two to seven years, respectively. Similarly, the annual rate of lizard survival ranged from 8% to 67%, and age at first reproduction ranged from eight months to six and a half years, respectively (Shine and Charnov 1992).

Guppies in Trinidad (Trinidad and Tobago) show variations in life history characteristics associated with the risk of demise due to the type of predator in the habitat. Where the chief predators are cichlids, which eat mature guppies, the guppies mature earlier (because the probability of adult survival is low). But where the chief predators are killifish, which eat small immature guppies, they mature later (because the probability of adult survival is high). In one experiment (Reznick 2016), guppies were transplanted from a site with cichlids to a site with killifish, but where no guppies had been before. After 11 years, the age of maturity had increased in the guppy population, female size at the age of maturity had also increased, and the guppy population had fewer but larger offspring.

The selection on life history traits due to predators in the aforementioned examples is similar to the unintended impact of human selection on harvested wild animals and plants. Combining data from 29 wild species (21 fish, 4 invertebrates, 2 ungulates, and 2 plants) harvested by humans, Darimont et al. (2009) found that in 282 of 297 case studies, body size declined, with an average decrease of 18%, and in 173 of 178 case studies, reproduction occurred at an earlier age or smaller size, with a decrease of 25% on average. They extended their analysis to show that these are relatively large magnitudes of change in short amounts of time.

In a study using quantitative genetic analysis and pedigree reconstruction, Pigeon et al. (2016) tested whether the significant decline in average horn size of male bighorn sheep in Alberta (Canada) during a strong trophy hunting period between 1973 and 1995 was due to the effects of human-caused selection. In wild sheep, horn size is an important determinant of reproductive success, but it can take 8–10 years to grow advantageously large horns. Because horn size is an inherited trait, rams with large horns tend to have offspring with large horns, while rams with small horns tend to have offspring with small horns. Thus, when rams with larger horns are removed, smaller-horned males, which larger-horned males would have outcompeted, can become reproductively successful and pass on the "smaller-horned" trait. Pigeon et al.'s (2016) analysis provides compelling evidence that the observed decline in male horn length was due to an evolutionary response to human-caused selection, rather than random genetic drift. Although their study controlled for age (using sheep just 2–4 years old) to examine the evolutionary effects on horn size, it follows that males in the population were also reproducing at a younger age, since it takes nearly a decade to grow large horns. This again highlights the tradeoff between age at reproduction and growth.

Commercial fishing, as a source of decreased probability of adult survival, would seem to impose substantial selection pressure on fished populations. In the 1930s, the northeast Atlantic cod (*Gadus morhua*) had a median age at maturity ranging between nine and 11 years. By the 1970s, age at maturity had declined to between seven and nine years (Law 2000). Overfishing did not just shrink cod numbers, but also their size (about 20%). As fishing intensified, the animals, which once matured at 80 centimeters in length, began to reproduce at younger ages, at about 65 centimeters. At first, biologists feared the exhausted populations had permanently evolved these new traits, which would complicate restoring the population and size-age composition. But recently, Pinsky et al. (2021) examined the entire cod genome and found no sign of sweeping changes, suggesting natural selection was not the cause of these phenotypic changes. One possibility, according to Pinsky and colleagues, is that phenotypic plasticity rather than genetic evolution allowed for the adaptive change in maturation timing, although they could not eliminate the possibility of low-intensity selection on hundreds or thousands of genes (termed polygenic selection) or the evolution of highly repetitive regions of the cod genome. This is an important consideration when studying the evolutionary ecology of life history traits because adaptive characteristics may not necessarily be "hardwired" by genetic selection but can be brought about by phenotypic or behavioral plasticity. But in the end, the decision to reproduce now versus growing to reproduce later is subject to the same advantages and constraints, whether the life-history trait is genetic, environmentally plastic, or behavioral.

All Eggs in One Basket vs. Bet Hedging (Reproduction and Offspring Survival vs. Parental Growth and Parental Survival)

How often should an organism breed? Should it breed once and die (**semelparity**), or should it breed multiple times (**iteroparity**), and for how long? Charnov and Schaffer (1973) proposed that relatively low adult survival and high juvenile survival are expected to result in the evolution of annual semelparous species, whereas the reverse should select for perennial iteroparous species. Many small, short-lived semelparous plants, such as ragweed (*Ambrosia artemisiifolia*), live in ephemeral habitats like deserts, where good conditions (rain) come in bursts, then disappear. Their best strategy is to expend all their energy in one bout of reproduction instead of storing it for future reproduction, as their chance of surviving to the next period of abundance is slim. However, offspring survivorship is high because they remain as seeds until the next period of good environmental conditions arrives. In contrast, in wet equatorial regions with little seasonal variation in temperature and rainfall, several species of fig (*Ficus*) bear fruit continuously, forming a reliable year-round food supply for birds and primates. Stable environments promote iteroparity, where adult survival is high, but the probability of any one offspring growing to maturity is low.

For semelparous species, which breed only once, physiological decline and death follow rapidly. In truly semelparous species, death follows a single bout of reproduction, whether after one year or many years of growth. Because they reproduce only once, individuals invest all available resources into maximizing reproduction in this single event, effectively "putting all their eggs in one basket" at the expense of longer lives. Examples of semelparous species include Pacific salmon, bamboo, annual plants, many insects, and octopuses.

For iteroparous species, which breed many times, the physiological decline is more gradual, but the tradeoff between fecundity and parental survival still exists. The European kestrel (*Falco tinnunculus*), which feeds on voles and shrews, provides an example. European kestrels lay between three and seven eggs per brood. In one experimental study (Korpimaki and Rita 1996), when researchers experimentally altered brood size by one chick, the number of fledged chicks was not affected, but the future survival of the female parent was reduced by brood enlargements and increased by brood reductions, indicating a tradeoff between the number of offspring and parent survival. Many organisms, including perennial plants and most vertebrates, including humans, are iteroparous, reproducing multiple times during their lifetime. Iteroparous species breed repeatedly, devoting some of their resources during each breeding episode to their survival for later breeding events. By breeding repeatedly, iteroparous species "hedge" against the possibility that any one reproductive bout might occur during a period of poor offspring survival.

The concept of *r*- and *K*-selected species is one of the earliest attempts to categorize the great variety of life-history diversity (MacArthur and Wilson 1967). In general, *r*-selected species, like annual plants or mice, are under selection for a high intrinsic rate of increase with early reproduction and small bodies, which typically favor semelparity. *K*-selected species, like oaks and elephants, are selected for traits that are adaptive at population carrying capacity, such as late reproduction and large bodies, which typically favor iteroparity. However, this pattern does not always fit. Sometimes larger, longer-lived organisms—such as salmon, agaves, and bamboo—are semelparous. Under what conditions is it advantageous to risk survival until reproduction and offspring survival during a single reproductive bout?

Although semelparous organisms, such as *Arabidopsis thaliana* (a flowering weed plant from the mustard family), are often short-lived with lifespans of a year or less (making them ideal for genetic studies), many semelparous organisms, like bamboos, live for several years before they reproduce and die. Once a bamboo plant establishes itself in a disturbed habitat, it spreads by vegetative growth, continually sending up new stalks until the habitat becomes densely packed with bamboo shoots. Some bamboo species grow for 12 to 60 years before synchronously flowering once and dying as an adaptive response to "swamping" the predators of their vulnerable seeds (Gadgil and Prasad 1984).

In a comparative phylogenetic analysis of salmonid fishes, Crespi and Teo (2002) found that having fewer reproductive episodes was linked to higher reproductive investment (egg weight) per reproductive episode. This higher reproductive investment is made possible by growing larger and waiting to reproduce. Pacific salmon (*Oncorhynchus* spp.) grow rapidly for several years after migrating to productive oceanic waters, culminating in a return to freshwater spawning grounds and a single enormous reproductive effort, in which a large portion of the body is converted to eggs, and after which they subsequently die. Other salmonids, such as rainbow trout (*Oncorhynchus mykiss*), can spawn multiple times in their lifetime. Still, their reproductive investment is much smaller than that of semelparous salmon, and their bodies are not consumed by their reproductive effort. The evolution of semelparity in salmonids, according to Crespi and Teo (2002), was due to the long-distance migration to productive ocean waters, where salmon can grow rapidly. This migration results in both an increased probability of adult mortality between breeding episodes and an increase in juvenile survivorship due to the higher egg weight in larger fish. Their result supports Charnov and Schaffer's (1973) idea that low adult survivorship, coupled with high offspring survivorship, drives the evolution of semelparity.

Agave plants thrive in arid climates with sparse and erratic rainfall. Most agaves reproduce only once; when the plant is ready, it sends up a huge flowering stalk producing a large number of seeds. When ready to reproduce, the parent plant cannot obtain the nutrients and water necessary for stalk growth from the environment and instead draws those resources from its leaves, which have been accumulating water and other resources for years. This results in the parent plant dying soon after the seeds are produced (Nobel 1977). Selection for tall, showy inflorescences by pollinators

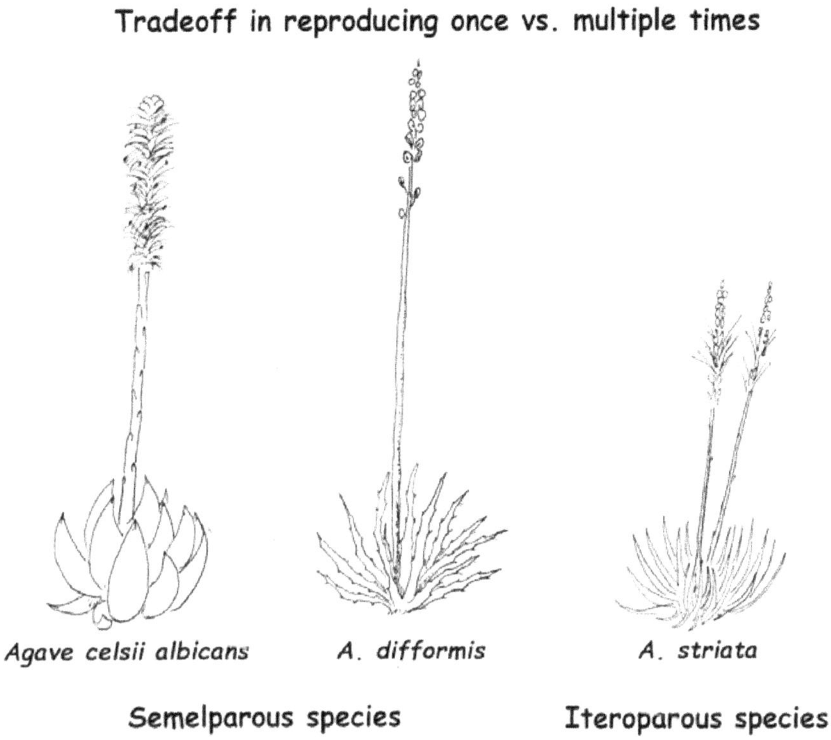

Figure 4. Rendering of three agave species. Most agaves, like *Agave celsii albicans* and *A. difformis*, are semelparous, producing an extremely tall inflorescence to attract bat pollinators, which are highly effective at spreading agave pollen, and then dying soon after. *A. striata*, however, is iteroparous, continuing to live after reproduction and using a greater percentage of diurnal pollinators (bees and bumblebees), which can pollinate without the need for taller, more costly floral stalks (Rocha et al. 2005).

drove the evolution of this trait in agaves, and, by necessity, a semelparous life history (Schaffer and Schaffer 1979). Although most agave species are semelparous, some reveal an iteroparous life history strategy (Figure 4). In a study of the reproductive ecology of five sympatric species in central Mexico (four semelparous and one iteroparous), Rocha et al. (2005) found that the four semelparous species compete for the same bat pollinators, and that fruit set (an indicator of successful pollinator attraction) was positively correlated with inflorescence height. This supports Schaffer and Schaffer's (1979) theory that pollinators select for taller inflorescence, thus driving the evolution of a semelparous life history. In contrast, the iteroparous species had limited overlap in shared pollinators, utilizing day-time pollinators instead, and showed no relationship between inflorescence height and fruit set.

Reconciling Life Histories With Population Dynamics

As we have seen, the bewildering variety of life histories can be explained by tradeoffs in just a few parameters, such as fecundity (number of offspring), survival, and age at reproductive maturity, coupled with the ecological conditions that make certain traits adaptive. Now, suppose we want to take what we have learned here and apply it to what we have learned from Chapter 2 *Population Dynamics*. To characterize the life history strategies of a population, these three parameters can be tabulated in a **life table**. A life table summarizes fecundity (m) and survival (l) in relation to age, size, or life stage (x) of the individuals in the population of interest. Although insurance companies originally developed life tables to provide a means of determining how long a person of a particular age could be expected to live, they have also proven to be invaluable to ecologists.

Life Table and Survivorship

To construct a life table, ecologists track the survival and reproduction of a cohort of individuals until no survivors remain. A **cohort** is a group of individuals born during the same period. For example, imagine a cohort of 1,000 gray squirrels all born in the spring of a single year. As time progresses, some individuals die, so there are fewer and fewer individuals of the original cohort present each following year. All sorts of biological phenomena vary with age. Reproduction, for instance, begins at puberty (in humans), but the reproductive rate is rarely constant and can differ between young and older adults. Likewise, the probability of living from one instant to the next is not only a function of an organism's age but also the conditions it encounters in its immediate environment.

Plotting a survivorship curve can help us visualize how survival or mortality rate changes over an organism's lifetime. It can be plotted as (l_x), which represents the probability of surviving to age x as a function of age x.

Three idealized types of survivorship curves are recognized: Type I, Type II, and Type III (Figure 5).

Type I Survivorship Curve: Species with Type I survivorship exhibit high survivorship during much of their youth and adult lives, but survivorship decreases with age (x). The curve may display a relatively shallow slope early in life (i.e., high survivorship during juvenile and reproductive years) but then steeply decline as mortality rates increase at ages when reproduction no longer occurs. In general, *K*-selected species—like most large mammals and large birds whose population sizes fluctuate at or near their carrying capacity—reveal a Type I curve. The death rate in such individuals increases with age (x).

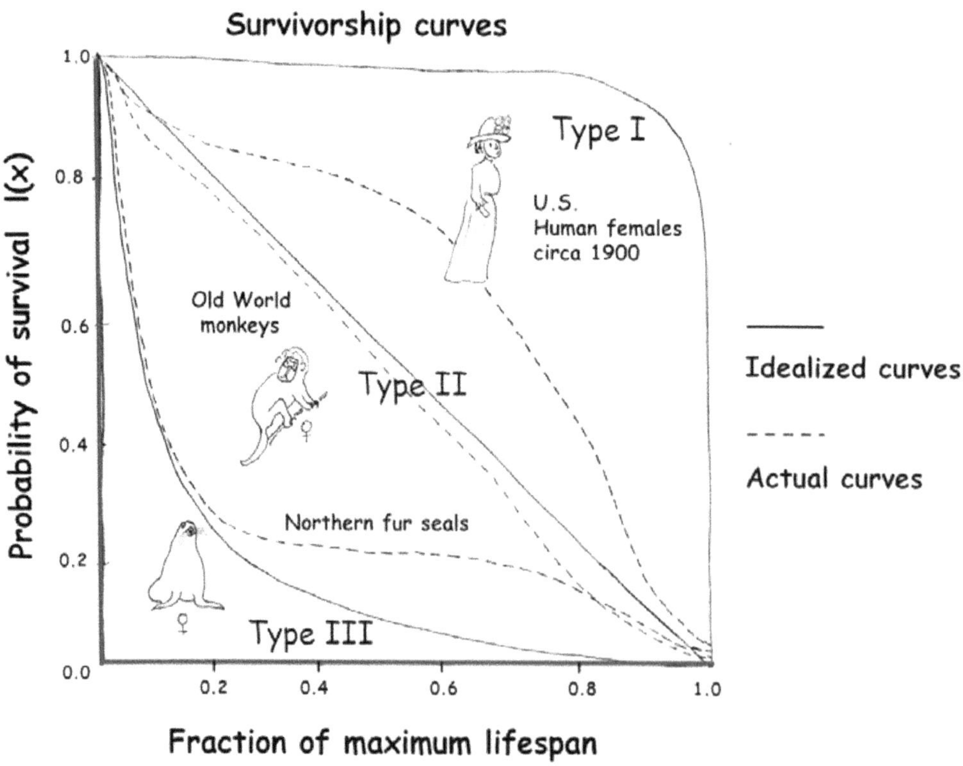

Figure 5. Idealized Type I, Type II, and Type III survivorship curves (solid lines) and actual survivorship curves (dashed lines; interpreted from Figure 1 in Barlow and Boveng 1991).

Type II: In species exhibiting Type II survivorship, individuals are lost mostly due to accidents and predation. The death rate is independent of age, meaning there is a constant probability of survival (l_x) throughout their life. Species of many small birds, small mammals, and insects fall into this category.

Type III: A highly concave curve, Type III, results when organisms have low survivorship (high mortality rate) during juvenile stages. Oysters, trees, sea turtles, and other large reptiles, and large fish produce a large number of progeny and face high mortality when young. The free-swimming larval stages of large fish or the acorn and seedling stages of oaks are particularly vulnerable to predators. But once they survive their youth, they can put significant energy into continued survival and reproduction. The longer they survive, the more progeny they will produce. The death rate in such species declines with age.

The shape of the survivorship curve is often related to the degree of parental care. Most birds and mammals that protect their young show a less concave survivorship curve than fish and most insects, which do not protect their young. In general, *K*-selected species (whose population dynamics remain near carrying capacity) show a Type I survivorship curve, while *r*-selected species (species whose populations are governed by their biotic potential, *r*) reveal Type II and Type III curves.

Like survival, fecundity (*m*) can also vary with age over an organism's lifetime. To understand how both survival and fecundity affect the population growth of our imaginary squirrel population, we can construct a life table (Table 2) to follow the survival and reproductive fates of a cohort of 1,000 individuals. In this table, *Nx* represents the number of individuals alive at age *x* from the original cohort of 1,000 individuals; *lx* is the fraction of the original cohort surviving to age *x*, which we will assume follows a Type III survivorship curve; and m_x is the age-specific fecundity, or the average number of offspring per individual at age *x* (note that because we are averaging over all individuals in the population, the number of babies produced per female would be about twice this if the sex ratio is 50:50).

Recall from Chapter 2 *Population Dynamics*, we assumed individuals in a population do not vary in their birth and death rates (i.e., *b* and *d* were constants). However, this is not what happens in real populations, and it certainly was not true for our hypothetical squirrel cohort in Table 2. So, how can we link this lifetime variability in fecundity and survival to population dynamics?

Table 2. A life table of hypothetical gray squirrel cohort of 1,000 individuals, where N_x is the number of individuals in a cohort alive at age *x*, l_x is the probability a newborn will survive to age *x*, m_x is the average number of offspring produced per *surviving* individual at age *x*, $l_x m_x$ is the average number of offspring produced per *original* individual at age *x*, and $x l_x m_x$ weights the average number of offspring produced by an individual by the age of that individual.

Age(*x*)	*Nx*	l_x	m_x	$l_x m_x$	$x l_x m_x$
0	1000	1	0	0	0
1	300	0.3	3	0.90	0.9
2	250	0.25	3	0.75	1.5
3	200	0.2	2	0.40	1.2
4	150	0.15	3	0.45	1.8
5	100	0.1	2	0.20	1.0
6	0	0	0	0	0
			$\sum m_x = 13$	$\sum l_x m_x = 2.7 = R_0$	$\sum x l_x m_x = 6.4$

Note: Most life tables consider just the females of the population so that average age-specific fecundity is not biased by sex ratio or the fact that fecundity is determined by the number of eggs or seeds a female produces, not the number of sperm or pollen a male produces. Note that if we had applied a "females only" life table to the population dynamics equations, we would be following just the number of females over time. You might debate whether this is a better way to model the dynamics of a population with two sexes.

Population dynamic equations are used to predict the change in the numbers of individuals over time. To do this, we need to determine the ***net reproductive rate*** (R_0) of the population, which represents the individual lifetime reproductive potential adjusted for survival. The net reproductive rate can be thought of in this way: if survivorship of the squirrel cohort was 100% to age 6, then the net reproductive rate (R_0) would be the average number of offspring produced by an individual at each age (m_x), summed over all ages (0 through 6). From Table 2, this sum of m_x is $\sum m_x = 13$. However, in real populations and our hypothetical case, some individuals die at every age. Therefore, the net reproductive rate (R_0) for a given cohort is obtained by multiplying the proportion of individuals surviving to each age (l_x) by the average number of offspring produced at each age (m_x), and then adding the products across all age groups: .

The **net reproductive rate** is formally defined as the average number of offspring produced by an individual in a population over its entire lifetime. If we assume the survival and fertility schedule remains constant over time, then when the net reproductive rate is 1.0, a given population is neither increasing nor decreasing, but rather replacing its numbers exactly; in other words, the population size remains constant. If the net reproductive rate is below 1.0, the population will decrease, and if it is above 1.0, the population will increase. The net reproductive rate thus represents the growth rate of a population *per generation*. In our hypothetical population of gray squirrels, the net reproductive rate (R_0) is 2.7, which means that the population can more than double its size each generation—an astonishingly large population growth rate for our hypothetical population.

The value of R_0 is a function of the age-specific patterns of birth and survivorship. It is a product of an organism's life history traits: the allocation of resources to reproduction, the timing of reproduction, the trade-offs between the size and number of offspring produced, and the extent of parental care. R_0 therefore provides a means of evaluating specific life history characteristics both for the individual (evolutionary fitness) and the population to which it belongs (dynamic consequences).

However, the net reproductive rate (R_0) should not be confused with the coefficient of population growth (*r*, or intrinsic growth rate). R_0 is related to generation time and, as such, is not suitable for comparing different populations unless their generation times are similar. Noting that the mean generation time is related to both R_0 and r, if we know the generation time T, we can relate R_0 to r.

Generation time is defined as the average time between the birth of an individual and the birth of its offspring. The average generation time, T, which is also the average age at which an individual reproduces, can be found as follows:

$$T = \frac{\sum x l_x m_x}{R_0} \qquad \text{Equation 1}$$

Here, the denominator represents the net per capita reproductive contribution of a single individual, and the numerator of this equation weights the average number of offspring produced by an individual at each age by the time since it was first born. In Table 2, we see that for organisms that live only one year, the numerator would equal the denominator ($xl_1m_1 = l_1m_1 = 0.9$) and thus the generation time T would equal one year. But, for long-lived species, T depends on the relative reproductive contribution of each age over a lifetime. So, if we sum up the reproductive contribution at age x, multiplied by the time it takes to make that contribution, and then divide this by the net per capita reproduction over an individual lifetime, we can obtain the average age at which an individual in this population reproduces an offspring or the average time between generations. In our squirrel example, $T = \frac{64}{27} = 2.4$.

We see now that R_0 is related to r by *time,* so it is possible to incorporate life history with simple population dynamics. We do this by assuming that the arbitrary time t in our population growth equation is the generation time T.

Recall the finite (discrete) population growth rate (λ) from Chapter 2 *Population Dynamics*, equals $\frac{N_t}{N_0}$, which can be approximated by e^{rt}: If we make $t = T$, then:

$$\frac{N_T}{N_0} \approx e^r T \qquad \text{Equation 2}$$

Now, assuming that the original population started with one individual ($N_0 = 1$), we know from Table 2 that the number of offspring produced by a single individual after a single generation (N_T) will be the same on average as R_0. So, substituting $N_0 = 1$ and $N_T = R_0$ into Equation 2, we get:

$$R_0 \approx e^r T \qquad \text{Equation 3}$$

Taking the natural log of both sides, we can approximate the population's per capita growth rate (r):

$$\ln (R_0) \approx rT \qquad \text{Equation 4}$$

And rearranging, we get:

$$r \approx \frac{\ln(R_0)}{T} \qquad \text{Equation 5}$$

In our hypothetical squirrel cohort, the net reproductive rate (R_0) is 2.7 and the generation time (T) is 2.4. Now we can calculate the per capita population growth rate (r):

$$r \approx \frac{\ln(2.7)}{2.4} \approx \frac{0.993}{2.4} \approx 0.41$$

The positive value of r in this case indicates that birth rates are higher than death rates, and the population is increasing. A value of r lower than zero would indicate a declining population, while a value of zero would indicate a constant population size. This hypothetical squirrel cohort, based on the lifetime survivorship and reproduction reported in Table 2, is doing quite well indeed.

Life tables allow us to understand why, in some situations, it is better to delay reproduction until an older age. But why do we age at all? It would seem adaptive to live and reproduce for as long as possible. An oft-cited evolutionary explanation for aging was put forth by Peter Medawar (1952), who suggested that the force of natural selection remains strong until the age of first reproduction and then declines with age. Medawar used this theory, called the "mutation accumulation theory of aging," to explain why deleterious mutations expressed later in life—such as those causing familial Alzheimer's disease—escape the purifying effects of natural selection. He proposed that such mutations accumulate in populations over time, leading to the evolution of senescence via genetic drift. Can you apply what you have learned about life history tradeoffs and this theory to explain why birds and bats are exceptionally long-lived (Munshi-South and Wilkinson 2010), or why human females live decades past reproductive age (Hawkes 2003)? The evolution of aging is a vast and fascinating field, perhaps driven in part by our human desire for immortality. It encompasses a body of research showing that longevity is regulated by the ability of somatic cells to repair damage, and is influenced by genes acting at both the population and individual levels to shape the aging phenotype (Kirkwood and Austad 2000).

References

Andersson, M. 1976. Clutch size in the long-tailed skua *Stercorarius longicaudus*: some field experiments. Ibis 118: 586–588.

Arditti, J. and Ghani, A. K. A. 2000. Tansley Review No. 110. Numerical and physical properties of orchid seeds and their biological implications. The New Phytologist 145: 367–421.

Augustine, S., Lika, K., and Kooijman, S. A. 2019. Altricial-precocial spectra in animal kingdom. Journal of Sea Research 143: 27–34.

Barlow, J. and Boveng, P. 1991. Modeling age-specific mortality for marine mammal populations. Marine Mammal Science 7: 50–65.

Blackburn, T. M. 1991. An interspecific relationship between egg size and clutch size in birds. Auk 108: 973–977.

Burns, R. M. and Honkala, B. H. 1990. Silvics of North America: 1. Conifers; 2. Hardwoods. Agriculture Handbook 654. U.S. Vol 2. Department of Agriculture, Forest Service, Washington, DC.

Charnov, E. L. and Ernest, S. M. 2006. The offspring-size/clutch-size trade-off in mammals. The American Naturalist 167: 578–582.

Charnov, E. L. and Schaffer, W. M. 1973. Life-history consequences of natural selection: Cole's result revisited. The American Naturalist 107: 791–793.

Crespi, B. J. and Teo, R. 2002. Comparative phylogenetic analysis of the evolution of semelparity and life history in salmonid fishes. Evolution 56: 1008–1020.

Darimont, C. T., Carlson, S. M., Kinnison, M. T., Paquet, P. C., Reimchen, T. E. and Wilmers, C. C. 2009. Human predators outpace other agents of trait change in the wild. Proceedings of the National Academy of Science 106: 952–954.

Derrickson, E. M. 1992. Comparative reproductive strategies of altricial and precocial eutherian mammals. Functional Ecology 6: 57–65.

Dowden, P. B. 1961. The gypsy moth egg parasite, *Ooencyrtus kuwanai*, in southern Connecticut in 1960. Journal of Economic Entomology 54: 876–878.

Eis, S., Garman, E. H. and Ebel, L. F. 1965. Relation between cone production and diameter increment of Douglas fir (*Pseudotsuga*), grand fir (*Aibes grandis*), and western white pine (*Pinus monticola*). Canadian Journal of Botany 43: 1553–1559.

Gadgil, M. and Prasad, S. N. 1984. Ecological determinants of life history evolution of two Indian bamboo species. Biotropica. 16(3): 161–172.

Godfray, H. C. J., Partridge, L. and Harvey, P. H. 1991. Clutch size. Annual Review of Ecology and Systematics 22: 409–429.

Hawkes, K. 2003. Grandmothers and the evolution of human longevity. American Journal of Human Biology 15: 380–400.

Henry, C. S. 1972. Eggs and rapagula of *Ululodes* and *Ascaloptynx* (Neuroptera: Ascalaphidae): a comparative study. Psyche 79: 1–22.

Jokela, J. and P. Mutikainen. 1995. Phenotypic plasticity and priority rules for energy allocation in a freshwater clam: a field experiment. Oecologia 104: 122–132.

Kirkwood, T. B. and Austad, S. N. 2000. Why do we age?. Nature 408: 233–238.

Koenig, W. D. and Knops, J. M. H. 1998. Scale of mast-seeding and tree-ring growth. Nature 396: 225–226.

Korpimaki, E and Rita, H. 1996. Effects of brood size manipulations on offspring and parental survival in the European kestrel under fluctuating food conditions. Ecoscience 3: 264–273.

Lack, D. 1947. The significance of clutch-size. I and II. Ibis 89: 302–352.

Lack, D. 1948. The significance of clutch size. III. Ibis 90: 25–45.

Lack, D. 1967. The significance of clutch-size in waterfowl. Wildfowl, 18(18): 125–128.

Law, R. 1979. Optimal life histories under age-specific predation. The American Naturalist 114: 399–417.

Law, R. 2000. Fishing, selection, and phenotypic evolution. CIES Journal of Marine Science 57: 659–668.

MacArthur, R. H. and Wilson, E. O. 1967. The Theory of Island Biogeography. Princeton University Press, Princeton, NJ.

Martin, T. E., Martin, P. R., Olson, C. R., Heidinger, B. J. and Fontaine, J. J. 2000. Parental care and clutch sizes in North and South American Birds. Science 287: 1482–1485.

Medawar, P. B. 1952. An Unsolved Problem in Biology: An Inaugural Lecture Delivered at University College, London, 6 December, 1951. H. K. Lewis and Company.

Monks, A. and Kelly, D. 2006. Testing the resource-matching hypothesis in the mast seeding tree *Nothofagus truncata* (Fagaceae). Austral Ecology 31: 366–375.

Munshi-South, J. and Wilkinson, G. S. 2010. Bats and birds: exceptional longevity despite high metabolic rates. Ageing research reviews 9: 12–19.

Nobel, P. 1977. Water relations of flowering of *Agave deserti*. Botanical Gazette 138: 1–6.

Pigeon, G., Festa-Bianchet, M., Coltman, D. W. and Pelletier, F. 2016. Intense selective hunting leads to artificial evolution in horn size. Evolutionary Applications 9: 521–530.

Pinsky, M. L., Eikeset, A. M., Helmerson, C., Bradbury, I. R., Bentzen, P., Morris, S. et al. 2021. Genomic stability through time despite decades of exploitation in cod on both sides of the Atlantic. PNAS 118: e2025453118, https://doi.org/10.1073/PNAS2025453118.

Reznick, D. 2016. Hard and soft selection revisited: How evolution by natural selection works in the real world. J. Hered. 107: 3–14.

Rocha, M., Valera, A. and Eguiarte, L. R. 2005. Reproductive ecology of five sympatric *Agave Littaea* (Agavaceae) species in central Mexico. American Journal of Botany 92: 1330–1341.

Schaffer, W. M. and Schaffer, M. V. 1979. The adaptive significance of variations in reproductive habit in the Agavaceae II: Pollinator foraging behavior and selection for increased reproductive expenditure. Ecology 60: 1051–1069.

Šera, B. and Šerý, M. 2004. Number and weight of seeds and reproductive strategies of herbaceous plants. Folia Geobotanica 39: 27–40.

Shine, R. and Charnov, E. L. 1992. Patterns of survival, growth, and maturation in snakes and lizards. The American Naturalist 139: 1257–69.

Skutch, A. F. 1949. Do tropical birds rear as many young as they can nourish? Ibis 91: 430–455.

Starck, J. M. and Ricklefs, R. E. 1998. Patterns of development: the altricial-precocial spectrum. pp. 1–28. *In*: Starck, J. M. and R. E. Ricklefs [eds.]. Avian growth and development: evolution within the altricial-precocial spectrum (No. 8). Oxford University Press, USA.

Stearns, S. C. 1977. The evolution of life-history traits: A critique of the theory and a review of the data. Ann. Rev. Ecol. Syst. 8: 145–171.

Stevens, O. A. 1932. The number and weight of seeds produced by weeds. American Journal of Botany 19: 784–794.

Tsikliras, A. C., Antonopoulou, E. and Stergiou, K. I. 2007. A phenotypic trade-off between previous growth and present fecundity in round sardinella *Sardinella aurita*. Population Ecology 49: 221–227.

Vander Werf, E. 1992. Lack's clutch size hypothesis: an examination of the evidence using meta-analysis. Ecology 73: 1699–1705.

Winemiller, K. O. 1995. Fish ecology. pp. 49–65. *In*: Encyclopedia of Environmental Biology. Vol 2. Academic Press, New York, NY.

CHAPTER 7

Light Competition in Forests and Evolutionary Game Theory

No organism is free from the evolutionary meddling of others (Vincent and Brown 2005)

Here we are in the woods. Maybe it is Henry Longfellow's primeval forest, or maybe it is the snowy woods of Robert Frost, or maybe it is Edward Thomas's forest of aspens, or perhaps it is the woods around Walden Pond where Henry Thoreau lived for two years. The captivating allure of the woods evokes the romanticism of poets and writers. But if we, too, found ourselves in such enchanting places on our evolutionary ecology journey, we would surely stop for a moment in wonder, gazing up at those far-off leaves perched atop a long, long pole, and ask, "Why is there wood?" Or, more specifically, "Why do trees have woody trunks?"

Is it for transporting water and soil nutrients to the leaves? Or for sending photosynthates (carbohydrates, proteins, and lipids) back to the roots? Is it to create habitat for squirrels and other wildlife? Is it to protect against pathogens? Or is it to provide humans with a wonderful building material? Nope. None of these is correct. Wood is an adaptation that allows plants to grow taller and raise some, if not all, of their leaves high above the ground. They have wood because it enables them to play an evolutionary game.

What? Wait. What is an evolutionary game?

Evolutionary Game Theory

Game theory is a branch of mathematics first formalized by John von Neumann and Oskar Morgenstern (1944) in their book *Theory of Games and Economic Behavior*. Games have players (agents), strategies (choices), payoffs (money, fame, utility), and rules (how the players' strategy choices translate into payoffs). In games, the success of a strategy depends on which other strategies are present. In other words, an individual's best choice depends on the choices of others.

Game theory originated in simple two-player matrix games that attempted to explore the consequences of conflict and cooperation in the context of strategies and payoffs. The game of chicken is a simple example. Urban legend has it that children played this game in the middle of the last century, and it goes like this: two boys riding bicycles speed toward each other, heading for a collision. If they both stay the course, both they and their bikes will suffer debilitating injury (at a costly $100 each). However, if one chickens out and swerves first, the loser of the game will forever be branded "chicken" and must pay $10 to the one who stayed the course longer. If both swerve, they remain unhurt ($0), but each bike home in disgrace for their cowardice.

This narrative can be formalized into a matrix with players and their strategies (on the exterior) and payoffs with respect to each strategy (in the interior; Figure 1).

DOI: 10.1201/9781003008538-7

		Player 2	
		Not Swerve	Swerve
Player 1	Not swerve	Player 1 -$100 Player 2 -$100	Player 1 +$10 Player 2 -$10
	Swerve	Player 1 -$10 Player 2 +$10	Player 1 $0 Player 2 $0

Figure 1. An example of a payoff matrix for the game of chicken. *(Note: The illustration was created by T.L.S. Vincent circa 1983, toured the world with T. L. Vincent at game theory conferences, and was first published in Vincent and Brown [2005]).*

John Nash, winner of the Nobel Prize in economics in 1994 and romanticized in the 2001 movie *A Beautiful Mind*, was the first to come up with solutions (eponymously called Nash equilibria) to these types of games. A **Nash equilibrium** is one in which no player can gain an advantage by changing their strategy. For the game of chicken, there are three Nash solutions. In two of the solutions (where player 1 always swerves and player 2 always does not *or* player 2 always swerves and player 1 always does not), players must know whether they are player 1 or player 2 (e.g., for the chickens depicted in Figure 1, the left chicken will do best always to swerve, and the right chicken will do best always to stay the course). Note that the payoffs are asymmetric, making the situation "unfair," as the "bully" always wins. There is a third Nash solution that does not require the players to know whether they are player 1 or 2. This solution yields equal payoffs to both players and involves playing a given strategy with some probability, as a mixed strategy. Riechert and Hammerstein (1983) derived the mixed solution to our particular payoff matrix. It turns out the Nash equilibrium is to swerve with a probability of 0.9 and not to swerve with a probability of 0.1. The equilibrium payoff is −$1 for each player. They parenthetically note, "truly rational beings would not play this Game of Chicken at all, since the equilibrium payoff is negative."

Imaginary games aside, what makes game theory interesting is that its formal structure allows one to predict optimal strategies in game-like situations: business, economics, finance, supply chains, marketing, sports, politics, traffic, trade, war, negotiations… and, as we see here, biology.

Evolutionary games are a spin-off of game theory. George Price and John Maynard Smith, in a classic paper in the journal *Nature* in 1973, were among the first to explore evolution as a mathematical game, and they were the first to conceive of the evolutionarily stable strategy (Maynard Smith and Price 1973; Maynard Smith 1982). It turns out that evolution by natural selection can be thought of as an evolutionary game: it has players (individual organisms), strategies (heritable traits), strategy sets (evolutionarily feasible strategies), and payoffs (fitness in terms of per capita growth rate; Vincent and Brown 2005). Evolutionary games provide a logical framework that can be used to understand the evolution of a trait or biological strategy when success depends on which other strategies are present in the "game." The biggest difference between classical game theory, which focuses on payoffs to players, and evolutionary game theory is that the latter focuses on strategies that will persist over time (Vincent and Brown 2005). Players in the evolutionary game cannot choose their strategies, and they come and go through births and deaths. However, evolutionary strategies (if successful) persist over time as they are passed from generation to generation. The solution to the evolutionary game is the **evolutionarily stable strategy** (ESS). An ESS is a strategy that can resist invasion by rare alternative strategies when it is common. Once fixed in a population, natural selection will maintain this strategy as an adaptation and prevent any alternative strategy from invading.

Maynard Smith and Price (1973; Maynard Smith 1974) first conceived of the ESS in terms of animal contests, such as the sparring "tournaments" of deer. Like the game of chicken, the biggest payoff goes to the individual who continues the contest longer. The ESS solution to the game of chicken is the same as the mixed Nash solution (Hummert et al. 2014) because if most of a population plays the mixed solution, no other strategy can invade, and no other strategy can perform better.

Maynard Smith and Price's ESS made a big splash because they used game theory to model evolution from an **individual** selection point of view—that individual strategies are subject to natural **selection**—rather than the **group selection** argument popularized by Lorenz (1963) and widely accepted at the time, which held that animals limit their aggression to avoid endangering the survival of the species (Leimar and McNamara 2023).

Evolutionary games sometimes reflect classical game theory. For example, in classical game theory, there is the **cooperative advertising game**, which explores when members of a supply chain might cooperate to pay for advertising (Jøgensen and Zaccour 2014). While not the same in detail, chorusing frogs also play a cooperative advertising game. Male frogs expend an enormous amount of energy calling for females during the breeding season. They typically join other males to call for females because, by calling together, they can be heard over greater distances. If you have ever sat by a pond at dusk and listened to a frog chorus, you know how loud it can get! Female arrival rates per caller are correlated with the number of calling males in the chorus, so it pays to advertise. On the other hand, loud noises attract predators, and time spent calling is not time spent foraging (see Chapter 5 *Foraging Dynamics*), so calling is risky and lowers energetic reserves.

Lucas and Howard (1995) consider frog choruses to be a game because the fitness payoff to any individual's behavior depends on the behaviors of other individuals in the population. In the cooperative advertising game, individuals are free to make decisions that are in their own best interest and need not be concerned about others' interests (Jøgensen and Zaccour 2014). In the classical advertising game, members of a supply chain (the players) decide when to pay for advertising and when to free-ride on the advertising of others in the supply chain. In the frog chorus game, the players have three strategies: call (which is energetically expensive), satellite (an inexpensive strategy in which satellites intercept females attracted by calling males), and leave (to forage or hide; Lucas and Howard 1995). The payoffs to the frog game are in terms of lifetime mating success, so the best strategy is one that maximizes the lifetime mating success of a single male. When this "best strategy" is used by most of the population and cannot be invaded by any other strategy, it is an ESS. Lucas and Howard (1995) modeled this frog advertising game with a density-dependent component (where the density of males determines female arrival rate) and a frequency-dependent component (where the female arrival rate decreases as the frequency of satellite strategies increases). They also considered factors such as physiological state, predation risk, and age-dependent mating success (older frogs are better at it). Their model predicted that males should call in stable environments, but in unstable environments, males should call only under favorable energetic conditions and forage when conditions are less favorable. They also predicted that first-year frogs should adopt satellite strategies, while older frogs should call.

But we were talking about wood. How is tree height in forests an evolutionary game?

Well, it turns out wood has its roots (pun intended) in another type of game, which in classical game theory is called the prisoner's dilemma. This is couched as a game of cooperation where one can either cooperate with the other player or defect and be selfish. The game is typically set up

		Prisoner 2 Cooperate	Prisoner 2 Defect
Prisoner 1	Cooperate	Prisoner 1's sentence 2 years Prisoner 2's sentence 2 years	Prisoner 1's sentence 10 years Prisoner 2's sentence 0 years
	Defect	Prisoner 1's sentence 0 years Prisoner 2's sentence 10 years	Prisoner 1's sentence 8 years Prisoner 2's sentence 8 years

Figure 2. An example of a payoff matrix for the prisoner's dilemma. *(Note: The illustration was created by T.L.S. Vincent and first published in a journal of high school math problems in Western Australia in 1983. It subsequently toured the world with T. L. Vincent at game theory conferences and was first published in Vincent and Brown [2005]).*

thus: two people are arrested for a minor crime with a two-year punishment, but are suspected of committing a major crime bearing a ten-year prison term. A confession is essential to the prosecutor's case, so the suspects are placed in separate cells where each is given a choice: confess to the crime and expose their partner (defect) or stay silent and protect their partner (cooperate). The prosecutor, to elicit the needed confession, makes the following offer. If they both confess (defect against their partner), each will be given a lighter sentence of eight years in prison. If only one of them confesses (defects on their partner) but the other stays silent (cooperates with their partner), the confessor will go free, and the other will get a ten-year prison sentence. However, if both stay silent (cooperate), there is only enough evidence to send them to prison for two years for the minor crime. The payoff matrix is in Figure 2, and the only Nash equilibrium and the ESS is to defect. However, you will notice that the cooperative solution is better for both players.

Expanded to multiple players, the prisoner's dilemma becomes the "tragedy of the commons" (Hummert et al. 2014). *The Tragedy of the Commons* first appeared in the journal *Science* in a 1968 essay by Garrett Hardin. In it, Hardin proposed that when resources are held in common, it becomes a rational decision to exploit the resource to the fullest. His example is a shared pasture (the commons) used by herders to graze their cattle. When a herder adds an animal to the herd, they eventually gain from the sale of the additional meat (a gain of +1 for the herder). The only loss to the herder from adding an animal is the additional contribution to overgrazing, but all herders share this loss. It amounts to only a fraction of −1. Thus, the rational decision is to keep adding cattle until the land is overgrazed and the cattle starve, even though a cooperative solution of limiting the number of cattle would give the highest combined payoff for all herders. This idea has been applied to evolutionary games and is used to describe evolution toward the inefficient use of a common resource (Hummert et al. 2014). It can be argued that global climate change is the greatest tragedy of the commons yet played by humans.

Back to trees in the forest. Competition for light should cause significant selection pressure on leaf height, creating an evolutionary game with outcomes depending on the strategies and densities of competitors (Givnish 1982). In competition for light, plants face a dilemma similar to that of the prisoners and the herdsmen. If they cooperated and shared the incident sunlight equally (similar to individually potted plants in a greenhouse, or fruit trees planted in an orchard), each individual would fare better than if it competed for the shared resource. However, a plant benefits from being taller and shading its neighbor, so, like the prisoners each cheating on their partner, each plant will try to "win" the game of light competition, attempting to grow taller than its neighbor. This generates a special form of the tragedy of the commons known as an "arms race" (a reference to the buildup of military arms between hostile countries; Falster and Westoby 2003). Trees, often tall ones, are the ESS outcome in this game. Such plants grow tall and large, have delayed maturation, and result in dense shading (Falster and Westoby 2003). This is why there is wood in the forest biome.

The Forest Biome

A **biome** describes communities of flora and fauna that share adaptations particularly suited to the climate (temperature and precipitation; see Chapter 12 *Climate change*) and circumstances (marine or terrestrial) they live in. The term was first proposed by Frederic Clements in 1916 (Carpenter 1939). While apt for delineating groups of organisms based on adaptations to the physical environment, its usefulness is limited by the fact that ecologists often differ on how coarsely or finely to draw these groups. However, lists of terrestrial biomes typically include tropical rain forest, tropical dry forest, tropical savanna, desert, temperate grassland, temperate savanna and shrubland, temperate forest, temperate coniferous rainforest, boreal coniferous forest, and tundra. In general, forest biomes, whether tropical or temperate, are found in areas with greater precipitation (see Figure 1 in Chapter 12 *Climate Change*).

By definition, forests contain woody plants that reach certain heights by virtue of stems and trunks composed of cellulose and lignin. Generally, to be considered a forest, there must be a nearly complete canopy of leaves. When viewed from a satellite or aerial photo, one sees only occasional gaps among a "sea" of green during periods when the trees are leafed. Gaps are created when trees fall over or snap off due to storms or diseases, and such gaps are temporary because seedlings and saplings then enter a race to fill the gap. The only exceptions to this rule happen when soil moisture or topography—such as bare rock, cliffs, shallow soils, or waterlogged soils—is unsuitable for trees. Hence, forests typically have what is referred to as a closed canopy. A forest grades into a wooded savanna when the trees do not form a complete canopy, but rather appear as solitary or several aggregated individuals within an overall sea of grass and non-woody plants. The conceptual dividing line in the continuum from forest to woody savanna is somewhat arbitrary, but in general, the greater the percent coverage of an area by a tree-supported canopy, the more likely it is considered a forest.

Only a subset of biomes is forests. The trees that make up a forest are a special kind of plant with a woody trunk. Most plants are **autotrophs** (they make their own sugars and carbohydrates) and require light, carbon dioxide, nutrients, and water as essential ingredients for sugar production (see Chapter 5 *Foraging Dynamics*). Plants convert this production into leaves (to collect more light and carbon dioxide), roots (to collect more water and nutrients), and reproductive parts (to produce offspring). About 500 million years ago, land plants evolved from their aquatic ancestors, solving the evolutionary problem of terrestrialization with the development of cell walls and roots (see Chapter 11 *Biodiversity*). Wood, or **xylem**, first appeared early in the evolution of land plants and functionally provided water transport, structure, and storage of water, minerals, and carbohydrates (Baas et al. 2004). The woody stems and trunks of trees and shrubs do not absorb water and nutrients (roots do). Likewise, most stems and trunks do not **photosynthesize** (produce sugars from water and carbon dioxide using sunlight energy), although there are exceptions. For instance, the palo verde tree (*Parkinsonia* spp.) grows in the deserts of North America. It is deciduous and grows leaves following periods of rain and loses them during intervening dry spells; however, its bark, by containing cells with chlorophyll and pores, can also photosynthesize and contributes significantly to reducing drought-related mortality (Ávila-Lovera et al. 2024). Although a successful strategy in dry habitats, these trees are not found in forest biomes (Berry et al. 2021), so wood and tree trunks are not primarily adaptations for photosynthesis. Wood does, however, play a role in the transport of water and nutrients from roots to leaves (as **xylem**), and its associated inner bark (as **phloem**) takes sugars from leaves to roots. But why transport anything in the first place? That is because wood, in its structural role, is an adaptation for making plants taller in response to competition for light with their neighbors.

In trees, photosynthesis (fueled by sunlight) occurs in the crown. However, the tree's crown also shades everything below it, leading to competition for light. Wood enables height in shrubs and trees, and in some cases, allows for the lateral growth of branches into areas of sunlight unobstructed by other trees. Height in trees and shrubs helps avoid shading from the leaves of other plants while casting shade on those same plants. Shrubs and trees are woody plants that engage in competition for light.

Light Competition as an Evolutionary Arms Race

So, how tall should a tree be? The obvious answer is just a bit taller than its neighbors. But what if the neighboring trees respond by growing a bit taller as well? Should the original tree keep one-upping its neighbors? Can this "arms race" of tree height continue indefinitely? Before we consider light competition as an evolutionary game, it is important to understand the tradeoffs and constraints that plants face when building a wooden pole to outgrow their neighbors.

When comparing canopy heights across forest biomes, we might observe that some forests, like the redwoods of California, USA, or the Kauri of New Zealand, have extremely tall canopies held aloft by gigantic woody trunks, while others, like the black pine forests of Alaska and Canada, are much shorter. Why might this be so? The oft-heard aphorism "there is no free lunch" applies here. It costs a plant to build a tall pole, and these costs are not spread evenly across the planet.

The Cost of Wood

First, there is the material and allocative cost of wood. Wood is made up largely of cellulose, hemicellulose, and lignin, but its structure and components can differ strikingly between species of trees (Baas et al. 2004). The greatest difference is between conifers ("softwoods") and angiosperms ("hardwoods"). All wood is made up of xylem and phloem, but conifers have simpler wood than angiosperms; their xylem is made up of long cells (**tracheids**) with tiny holes (**pits**). In addition to tracheids, angiosperms also have wide, tube-like **vessels** made up of cells called **vessel elements**. The cells that make up both tracheids and vessels are connected axially, and they are dead when they function as conduits for the transport of water and mineral salts from roots to the canopy. Xylem makes up the inner portion of a woody trunk, and it is separated from the outer covering of bark by a layer of living cells called the **vascular cambium** (Figure 3). The inner portion of bark is made up of phloem, consisting of living cells called **sieve-tube members** and associated **companion cells**, which work together to transport sugars produced by leaves to other parts of the plant. The outer portion of bark is made up of another kind of dead-when-functional cell called **cork**, created from a layer of living cells called **cork cambium** lying outside of the phloem (Figure 3). Connecting the xylem to the phloem is another kind of cell called **xylem parenchyma,** which is also a living cell that primarily functions as storage tissue (Słupianek et al. 2021).

The cambium continuously produces both xylem and phloem, consequently thickening the tree trunk and allowing trees to obtain extraordinary heights in the competition for light. Because wood is largely made up of dead cells (from 60–90%; Słupianek et al. 2021), it would seem that construction costs are up front and maintenance costs are minimal when compared to leaves; however, xylem becomes old and compressed with time, creating non-conducting **heartwood**, so trees must continue to expand their girth with functional **sapwood** created from the vascular cambium layer. The production and maintenance costs of woody tissue are also an allocative cost because photosynthetic products used to construct a woody trunk do not contribute to greater production of leaves and roots, which would increase growth, or to greater production of reproductive tissue, which would increase the number, quality, or frequency of offspring (Becker et al. 2000).

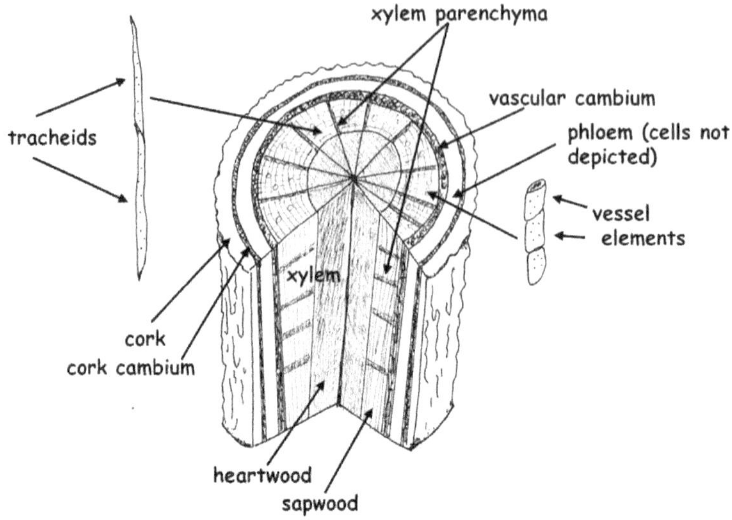

Figure 3. Idealized sections showing the important elements of wood.

The Cost of Height

In addition to the material and allocative costs of wood, there are also costs associated with height. Huge trees may be more likely to be blown down during extreme wind events (Peterson 2007), and taller trees may be more vulnerable to drought (Araújo et al. 2024). This latter problem is related to how trees move water against gravity from their roots to their canopy leaves.

The movement of water from soil to roots to leaf canopy is explained by the **cohesion-tension theory** of water transport, first proposed by Boehm (1893) and Dixon and Joly (1894). This controversial theory, which is notoriously difficult to evaluate (e.g., Zimmermann et al. 1994; Melcher et al. 1998), suggests that **transpiration** is the main force that pulls water from roots to leaves. Transpiration refers to the loss of water vapor from plant surfaces, mostly through plant-controlled gateways called **stomata** in leaves. According to the cohesion-tension theory, the transport of water against gravity is entirely driven by tension (negative pressure) gradients created by transpiration: the water potential created at the surface of leaves by evaporation is more negative than the potential in stems and roots, which is more negative than soil water potential (water moves from high to low potential). As radiation (light) and temperature increase, so too does the negative potential and "pull" on leaf water. The reason water moves upward with this tension is because of cohesion, which is the tendency of water molecules to stick together, and because of the adherence of water molecules to the hydrophilic walls of xylem cells. For this to work, xylem cells must resist **cavitation** (the breaking of the water column) by the formation of air bubbles (**embolisms**), which become more and more likely with increasing transpiration (Tyree and Zimmermann 2002), increasing friction due to path length (Koch et al. 2004), and with increasing conduit width of xylem cells (Tyree and Sperry 1989).

There is no controversy that the cohesive strength of water is adequate to lift water to the tallest trees (Loomis et al. 1960), and there is no doubt that water moves passively in plants, that xylem water is under tension, and that evaporation can act as a pump to move water (Tyree and Zimmermann 2002). However, much debate exists on how trees manage to avoid cavitation under the large negative pressures needed to move water against the assumed increasing frictional resistance of lengthening xylem or under arid conditions (Schenk et al. 2017). Hydraulic safety (the avoidance of cavitation) is adaptive, and variation in xylem architecture, which can be categorized by plant habitat, may attest to adaptive modifications (Baas et al. 2004). Furthermore, plants appear to be able to repair embolisms by the expulsion of gas via root pressure (works for small plants) or the dissolution of bubbles via diffusion. However, there is a tradeoff because making xylem safer from cavitation reduces the efficiency of water flow (Tyree and Sperry 1989; Tyree and Zimmermann 2002). There is even evidence that plants use surfactants, likely originating from living xylem parenchyma, to coat the conduits to help prevent embolisms from forming (Schenk et al. 2017; Słupianek et al. 2021). The simplest way plants can control embolism-causing negative pressures is by shutting down the "pull" of transpiration through stomatal closure. However, this means that carbon dioxide can no longer enter leaves for photosynthesis. While this may protect a tree from cavitation, it has negative impacts on plant growth for obvious reasons (Tyree and Zimmermann 2002).

This debate aside, it is clear that taller trees can suffer from hydraulic issues. Too much tension, and the water column is compromised, leading to hydraulic failure. Araújo et al. (2024) found that, within species of tropical trees, taller trees have lower resistance to embolism than shorter conspecifics, which explains patterns of taller tree mortality during drought. Thus, as trees grow taller, their costs are expected to increase. Furthermore, we expect that trees can respond adaptively to these costs, allowing tree height to vary accordingly.

Breaking Even

If costs increase with height, then at some point there will be a "break-even" height, beyond which it is no longer economical for the tree to produce more wood to gain additional height (Ryan and Yoder 1997). Differences in hydraulic resistance and evaporative pulling force may explain variations in maximal tree heights across habitats in relation to precipitation and temperature (Ryan and Yoder 1997). For example, black oak forests occur in drier, colder areas of the eastern and central United States and obtain canopy heights of 12–15 m, whereas cedar-hemlock forests occur in wetter, cooler areas of the Pacific Northwest and can attain heights of 45–50 m. Other costs of height can vary heterogeneously at relatively small scales. Higher winds, steeper slopes, and shallower soils all increase the costs of being tall. Soil composition is also important, with sand providing little anchorage against tipping, whereas aggregate soils with a mix of particle sizes, including pebbles and small stones, better anchor trees. Thus, we expect break-even heights to vary with habitat conditions, even within species. Additionally, evolutionary adaptations to particular habitats will likely play a role in creating different break-even points among species.

Light Competition and the Best Response Curve

Imagine a tree without a trunk. Without the costs of wood and height, this plant can devote all of its photosynthate to leaves, roots, and reproduction. However, what happens if it has a taller neighbor shading its energy and manufacturing sectors (leaves)? Given the reduced growth and reproduction, it might be beneficial to invest a little of that photosynthate into lifting its leaves higher.

Competitive interactions among individual trees to reach the canopy—that single layer of sun-exposed leaves—are a fact of life in forests. Competition for light is an asymmetric game in which taller individuals "win" a disproportionate share of the incoming radiation, thereby increasing their fitness relative to others when they intercept more light, both through higher photosynthetic production and by shading out their neighbors (Falster and Westoby 2003). Because this is an evolutionary game, we can attempt to find the height (or heights) that are an evolutionarily stable strategy (ESS).

One way to find the ESS is to use the concept of "best response." Imagine the height that maximizes the fitness of a focal tree (y-axis of Figure 4) in response to the height of its neighboring trees (x-axis of Figure 4). The line where $y = x$ (45° line) defines the focal tree's height with respect to the height of the neighboring trees when both are equal. When competing for light, the optimal height for our focal tree is just a little taller. Thus, the **best response curve** begins as a line slightly above the 45° line (Figure 4). As the height of the neighboring trees increases (along the x-axis), the benefit of being taller remains the same (being taller than one's neighbor always gives the same maximum amount of available light). However, the cost of height increases, as we saw in the previous section. So, there comes a point along the best response curve where the focal tree cannot

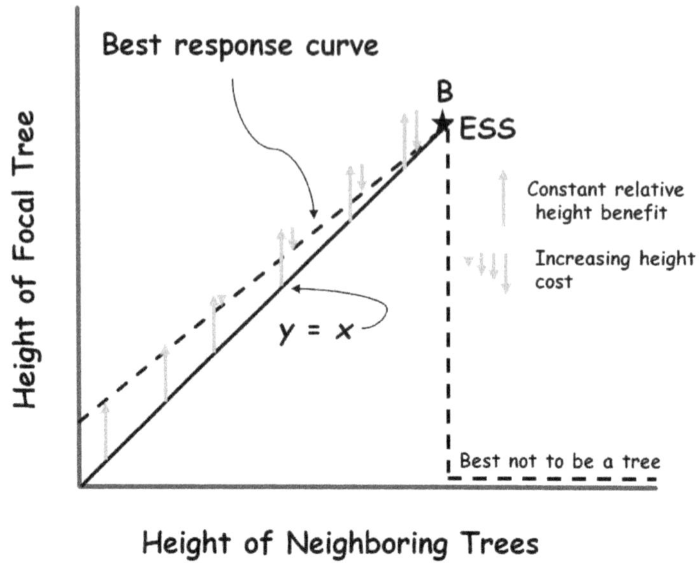

Figure 4. Best response curve for light competition. The 45° line shows all cases where the height of the focal tree equals the height of neighboring trees. The best response curve indicates the height that maximizes fitness. Point B is the break-even point, where the costs of height (which increase with height) equal the benefits of height (which are constant). The ESS corresponds to the height of the tallest neighboring tree at the break-even point. If the neighboring trees are taller, it is best to opt out of the game.

benefit by being taller than its neighbors (point B–the break-even point). If the neighboring trees are taller (beyond point B on the *x*-axis), it no longer benefits the focal tree to be slightly shorter; in fact, the best response drops to zero at point B. Beyond this point, the greatest fitness for the focal tree is to "opt out" of the height game and become a plant that is not a tree.

The ESS (and Nash solution) for the height game can be found at the intersection of the best response curve with the $y = x$ line (or 45° line). At this point, the best response of the focal individual is to do what everyone else is doing, because no individual can benefit from unilaterally changing their strategy (Nash), and if B is the break-even limit, no other strategy can invade (ESS).

The best response curve predicts an "arms race" for light that results in a single ESS height for any particular forest. All trees in that forest must either conform to this height, drive the height even higher (excluding other potential canopy tree species that have lower break-even heights), or opt out of the height game. Because the best response curve predicts a single height strategy, it does not explain why some forests have multiple heights (see *Détente* section below). However, assuming light competition is the prevailing limit to growth, and the break-even height (where cost equals benefit) is environmentally determined, it does explain why some canopies (i.e., the height of the tallest leaves) tend to have uniform height in particular habitats.

Various environmental factors are expected to shift the break-even tree height between forests. These include the amount of incident radiation (sunlight), temperature, precipitation, and rates of soil nutrient fluxes. Thus, we expect to see differences in canopy heights among forests (Figure 5).

Considering the sensitivity of tall trees to drought (see *The Cost of Height* section above), we might expect some of the tallest canopies in the world to occur in tropical and temperate rainforests where there is abundant rainfall. Indeed, rainfall is the main predictor of height in plants (Moles et al. 2009). But surprisingly, although rainforests cover large portions of the world, tree species exceeding 70 m in height are rare globally, and they do not necessarily grow in rainforests (Tng et al. 2012). Some of the tallest forests occur in what might be described as Mediterranean ecosystems, which are typically characterized by moderate winter rainfall, hot and dry summers, and often hilly and even mountainous terrain. These conditions hardly seem propitious for tall forests, and indeed, most of the vegetation of Mediterranean ecosystems is composed of woodland scrub,

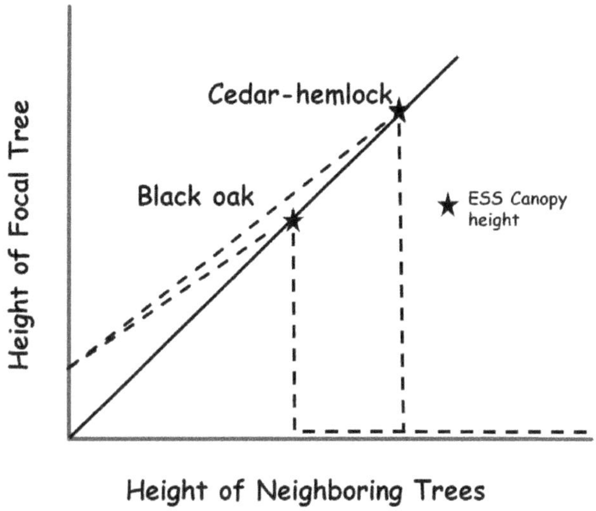

Figure 5. Imagined the best response curves and ESS canopy heights for two forests due to differing break-even points.

thorn scrub, savanna, grassland, or desert. Yet in some cases, we find the very tallest forests, with trees exceeding 80 m in height, in this climate. These forests include the redwoods of California (*Sequoia sempervirens*, *Pseudotsuga menziesii*, *Sequoiadendron giganteum*), USA, and the karri of Western Australia (*Eucalyptus diversicolor*). The conditions that enable the growth of these exceedingly tall forests do not appear to be linked to the Mediterranean climate in general, but rather to specific thermal and moisture conditions. In an analysis of average annual temperature with respect to seasonal temperature variation, Larjavaara (2014) found the world's largest trees occur under relatively rare thermal conditions, where the annual average temperature is between 7–10°C and the seasonal variation is less than 20°C. When these conditions were mapped globally, they corresponded to regions with the tallest forests on the North American continent (redwoods and Sitka spruce [*Picea sitchensis*] forests from California to Washington, reaching up to 100 m), the South American continent (Patagonia cypress forests in Chile, reaching up to 70 m), and the Australian continent (karri forests in southwestern Australia, eucalyptus forests of Tasmania and Victoria in eastern Australia, also growing to 100 m). However, these rare conditions also occurred in areas where the tallest forests do not exist, possibly due to factors such as human history (e.g., Western Europe and Great Britain), arid conditions (e.g., South Africa), or biogeographic history (e.g., New Zealand, where the tallest trees reaching 70–80 m are introduced species).

These special temperature conditions occur where the climate is mild and strongly maritime (Larjavaara 2014), but a lack of wind may also be important. Within these select climatic areas, extraordinarily tall forests grow on the level bottoms of ravines and escarpments affording deep, aggregate soils built up from erosion of hillsides and slopes; they occur in virtually windless spots, where topography and circumstances typically suppress gusts or high winds; and they grow relatively close to cold oceans offering drifting fog that provides drizzling rain, even in the summer, well above the average regional rainfall. However, some of the world's largest trees also occur in tropical Borneo, and these may be the exception that proves the rule. Many tropical areas between 7–20° latitude experience regular cyclones and hurricanes that can be particularly damaging to trees (Laurance and Curran 2008), but analysis of spatial patterns of tree heights suggests that the tallest trees in Borneo (> 70 m) only reach these heights by being sheltered from prevailing winds (Shenkin et al. 2019).

Even under the best conditions on Earth, trees are expected to encounter physiological constraints that limit how tall they can grow. For example, Jensen and Zwieniecki (2013) found that in tropical SE Asia, the leaf size of angiosperm trees becomes increasingly constrained with height, resulting from physical constraints imposed when pumping sugars from leaves to the rest of the tree. Given these constraints, they predict the maximum height that angiosperm trees can attain is 1046 m. The world's tallest angiosperm is a yellow meranti (*Shorea faguetiana*), which was measured at 100.8 m using direct tape (Shenkin et al. 2019).

For conifers, there also appear to be constraints related to the vascular system. Coastal redwoods (*Sequoia sempervirens*) are long-lived evergreen conifers that grow in the wet temperate forests of northern California, and among them is one of the tallest know trees on earth (112.7 m; Koch et al. 2004). Coastal redwoods can exceed 2,000 years in age and commonly exceed heights of 90 m (Wilson et al. 2016). Koch et al. (2004) studied sixteen trees between 85 m and 113 m tall and found that as the trees become taller, they suffer greater water stress, and the photosynthetic capacity per unit leaf mass decreases. Based on their physiological and structural measurements, Koch et al. (2004) calculated that the maximum height limit for coastal redwoods in their current environment would be 122–130 m or 7–13% taller than the tallest living tree.

Why are the tallest trees not at their eco-physiological limit? Note that reaching that limit would be disadvantageous; it is like living paycheck to paycheck. You lack the resources for emergencies, cannot expand your abode, and have enough to raise kids; you only have enough to break even with your bills. In tree terms, trees cannot invest in reproduction or vegetative growth—they exist, photosynthesizing as much as they respire. Therefore, the fitness-maximizing (ESS) break-even point cannot be at the limits set by physiologically driven habitat constraints, because trees need enough photosynthate to grow and reproduce. So, what is the ESS break-even height?

David King (1990) built a game-theoretical model of tree growth that answers this question. His model, which considered trees of even-aged stands and linearly rising costs with height, predicts that the ESS for trees of even-aged stands should be 87% of their theoretical height limit (or 13% smaller), which is similar to what we calculated for redwoods.[1] In testing this model with data from forestry yield tables, King (1990) found that average-sized trees in even-aged stands follow the predicted ESS, reaching at least 70–90% of their maximum theoretical height. The resulting evolutionarily stable strategy maximizes an individual's competitive ability for light. However, the model also shows that production is maximized at a shorter height—basically at the point between where the canopy closes and the ESS. In other words, the redwood's strategy is an ESS, but it is not the optimal strategy. This has been observed in other systems as well. The green revolution of the 1960s saw up to 10-fold increases in rice and grain yields due to new strains. A key feature of these high-yielding crops was their stature. Before this, wheat, barley, and rice stalks were much taller, likely due to competition for light. Cultivars from the 1960s were artificially bred for shorter crop strains. By breeding out the arms race, material "wasted" in the stems could be diverted to grain yield.

In our best response curve light competition game, we predicted a single ESS outcome, that is, a single optimal height. If we assume strategies correspond to species, then the best response curve predicts a single winner. Redwoods grow in warm (10–16°C annual mean), moist (640–3100 mm annual precipitation), temperate climates. They avoid wind, drought, and cold, and thrive in soils that are deep, moist, well-drained, and of poor to moderate fertility (Wilson et al. 2016)—seemingly optimal conditions for tree growth! However, despite these optimal conditions for any tree, coastal redwood forests growing on alluvial flats (broad, flat areas of soils built-up by flowing water) can dominate the forest, accounting for more than 90% of the stand basal area (Lorimer et al. 2009). In these places, they have "won" the arms race against competing species. However, because the light competition game is about strategies rather than species per se, our best response curve also predicts that any species that can attain the ESS height could potentially occupy forest canopy (Figure 6). In contrast to California's temperate redwood forest, the canopy layer on Barro Colorado Island in Panama, is made up of a high diversity of species (about 135 species per 1,000 trees [Bohlman 2015]). But can we really say that the trees in both California and Panama are playing the same light competition game? Why do the outcomes differ so markedly?

1 We note that King (1990) uses the term "break-even height" for the height at which no production occurs due to maintenance costs; we call this the height limit, whereas what we call the "break-even height," he calls the "critical height," which is the ESS in both cases.

When the outcome of our evolutionary game model matches predictions, we may blithely allow that our assumptions are correct; in this case, that an increasing cost to height and a closed-canopy competitive game strictly among adults leads to a single ESS height. However, when our model predictions diverge from observations, this is more useful because it allows us to explore ignored complexity that may be important in real systems, thereby advancing our understanding of the evolution and ecology of trees and forests.

Détente–Cooling the Arms Race

So far, we have only considered adult trees. We have completely ignored how recruitment affects the game of light competition. Though forests may appear ancient and untouched, they are ecologically dynamic, both in space and time. Trees do not simply appear tall *de novo* and remain there for eternity. All canopy trees must start life as seedlings that germinate from seeds. Yet the deep, dark, quiet majesty of a closed-canopy forest is not conducive to seedling growth. Fortunately for the seedling, death in the form of windfall, disease, and decay occasionally befalls the ancient trees, creating small to sometimes large gaps of light. If a seed or seedling is lucky and "wins the lottery," it happens to be in the right place at the right time when this occurs.

The Lottery Game—Coexistence Due to Chance

Lotteries can be characterized as low-probability, high-payoff games where randomness plays an important role in competition for space. Once a gap opens, the chance of any given seedling reaching the canopy is virtually zero, but the chance of some seedling growing to replace a canopy gap is nearly 100%! The **unified neutral theory of biodiversity** (Hubbell 2001) models ecological communities as a lottery game where pure chance, rather than species-specific traits, determines who fills a vacated gap. This theory suggests that the number of coexisting species (biodiversity) is controlled mostly by the neutral drift of species abundances, which fluctuate based on the probability that an individual of a particular species fills a randomly occurring gap at a particular time. While empirical tests have often found that the neutral theory fails to fully explain natural communities (e.g., Yu et al. 1998; McGill 2003), its proponents claim that it remains a valuable tool. They argue its failure is informative, revealing the potential roles of dispersal limitation, speciation, and ecological drift. As such, it serves as a useful null model against which adaptation and natural selection may be tested (Rosindell et al. 2011).

While the neutral theory dismisses differences between species as unimportant (a dubious assumption at best–see Ale et al. 2019), evolutionary games must account for them in full. This is because, in an evolutionary game, an individual's fitness depends on its own strategy as well as other strategies present in the community. Even prior to Hubbell's neutral theory, Chesson and Warner (1981) were among the first to consider a "lottery" for space as a means toward coexistence among competing species.[2] Although not an evolutionary game per se, their model takes into account both the stochastic nature of a lottery game as well as the critical differences in species' strategies. Their "lottery competitive system" demonstrated that two species could coexist provided the following conditions were met:

1. Adults remain in the population after reproducing (which buffers population growth when environmental conditions are poor for recruitment).
2. Each species differs in their response to environmental conditions such that one has greater recruitment (via germination and seedling competition) in some years and the other in different years.
3. The environmental variability is large enough.
4. Space (i.e., the proportion of adults dying) becomes available randomly.

[2] Chesson and Warner (1981) cite Sale's (1977, 1978) idea of a coral reef fish lottery as inspiration.

When all of these conditions are fulfilled, Chesson and Warner demonstrated that stochastic environmental variability allows competing species to coexist stably.

The lottery competitive system results in something called the **storage effect**. This occurs when a species' average population growth rate is more strongly affected by the benefits of a good year than by the costs of a bad one. Essentially, strong recruitment events are "stored" in the adult population (Warner and Chesson 1985). The key to the storage effect is a strong relationship between environmental conditions and a species' competitive ability (Johnson and Hastings 2022). The mechanism works as follows: when a dominant species is common and another is rare, random adult mortality is more likely to remove individuals from the dominant species. This creates gaps for recruitment. If environmental conditions favor the recruitment of the dominant species, it will outcompete the rare species to fill these gaps, and community's composition will remain unchanged. However, if environmental conditions allow the rare species to outcompete the dominant species, then all the gaps will be filled by rare species' recruits, thereby increasing its relative abundance in the community. It is important to note that Stump and Vasseur (2023) point out the storage effect only works under strict assumptions. These include temporal partitioning of density-independent factors that vary in time, buffered population growth, and a covariance between the environment and density-dependent competition. These assumptions are often violated in real-world communities.

Gap dynamics—where only a few trees die, leaving a gap in the forest that may be filled by other trees—have been a focus for ecologists debating how much ecological communities consist of coexisting competing species or chance collections of species behaving individualistically (Brokaw and Busing 2000). We noted Hubble's neutral theory treated species as groups without distinguishing features—pure chance determines who "wins" a gap. Chesson and Warner's lottery competitive system considered differences between species, but only in terms of their varying recruitment success under randomly fluctuating environmental conditions. Note that the differing strategies are not specified or evaluated relative to one another—only with respect to the environment. The death of an adult is not treated as an opportunity for individuals of either species to compete, but rather as a high-reward recruitment event for a rare species that happens to do better, for example, in a rainy year. In a lottery game, chance plays a big role. But what if, instead, we thought of a gap not as a chance event, but as a light resource—one that differs in abundance and frequency from the surrounding forest? If gaps occurred frequently enough, wouldn't trees evolve strategies to exploit this resource? What if we viewed gaps through the lens of an evolutionary game framework instead of a lottery game framework? Evolutionary games involve strategies whose fitness depends on the other strategies present in the community. Differing strategies generally evolve when there are major life history tradeoffs; for example, height vs. growth rate (Ryan and Yoder 1997), seed mass vs. seed output (Westoby et al. 2002), high survival vs. rapid growth (Gilbert et al. 2006), shade tolerant vs. shade intolerant (Grubb 2016), and disturbance-dependent vs disturbance-averse (Pavlovic 1994). We shall keep these in mind as we explore possible evolutionary games that lead to multiple ESS strategies in forest biomes.

The Gap-Capture Game–Coexistence Due to Environmental Heterogeneity in Space and Time

Most sessile organisms (those that do not move, like trees) depend on gaps for regeneration (Sousa 1984). Given enough time, a gap is a 100% certainty, but the frequency and size of gap formation can vary. For instance, Sousa (1984) describes extreme examples of frequency and area differences from the forests of southern Appalachia, where scattered deaths of single trees occur frequently, primarily due to windthrow, lightning, or glaze storms, and the average gap is only 31 square meters. In contrast, in the forests of northern Minnesota, where infrequent fires are the primary disturbance, gaps can range from 400 to 4,000 hectares in area. Gaps generally exhibit pronounced differences in light and moisture levels compared to the surrounding vegetation, and these differences, along with the size and frequency of gaps, strongly impact the types of species that can take advantage of gaps for regeneration. Furthermore, gaps are not necessarily created at random. For example, spruce bark

Figure 6. Games trees might play.

beetles preferentially attack their host tree, Norway spruce (*Picea abies*), resulting in 3 to 6 times more gaps than would be expected by chance, while not attacking co-occurring broad-leafed trees (Orman and Dobrowolska 2017). Depending on conditions, older (Runkle 2013), diseased (Lorimer et al. 2009), and taller (Stovall et al. 2019) trees may also be more likely to die and create gaps. Additionally, gaps can occur with predictable frequency or under certain conditions. For example, there are greater mortality rates in canopy trees during strong El Niño–Southern Oscillation events (see Chapter 12 *Climate Change*; Johnson et al. 2017), and fire frequency is predictably high for yellow pine forests compared to spruce-hemlock forests (Van de Water et al. 2011). Because gaps are often not random in frequency, size, or environmental conditions, strategies can evolve to increase the probability of "winning" the gap—more like a game of blackjack, or even poker, than pulling a lever on a slot machine.

***Live Fast–Die Hard Strategies*:** One way to take advantage of frequently occurring small to medium-sized gaps is to play the game of life within the gap before the light-winning ESS strategy takes over. To succeed, the "live fast–die hard" strategy must complete its life cycle before the canopy closes (Figure 6). Trees that use this strategy are often called **pioneer trees**. Pioneer trees are characterized by prolific seed production, high seed dispersibility (by animals or wind), and disturbance-cued (high light, low moisture) germination from soil seed banks, and they are often short-lived (Swaine and Whitmore 1988; Dalling and Brown 2009). Their fast growth can be attributed to the rapid deployment of large and relatively cheap leaves (Popma et al. 1992).

In the few decades it takes for successional processes to close the gap, pioneer species can produce thousands of seeds annually. For example, the **neotropical** (new-world tropics) pioneer genus *Cecropia*, which is restricted to germinating in gaps due to its light requirements, can produce a seed density of 66,000 seeds per square meter annually during its 20 to 40-year lifespan (Dalling et al. 1998; Zalamea et al. 2012). For *Cecropia*, this strategy produces abundant propagules, but these do not last long on the surface (due to ants) nor in the seed bank (due to fungal pathogens), nor do its seeds appear to be spread very widely (Dalling et al. 1998). Therefore, it is a strategy that appears most successful under conditions of frequent tree-fall gaps in nearby vicinities. Although *Cecropia* seeds do not appear to be spread widely, high seed dispersibility (though not necessarily seed longevity) is a characteristic of tropical tree pioneers (Swaine and Whitmore 1988) and would be expected to take advantage of frequent but widely dispersed gaps.

Note that live fast–die hard strategies exist on the seed mass vs. seed output tradeoff as well as reflecting fast growth vs. survival tradeoffs.

***Sit-And-Wait Strategies*:** It is possible that the frequency and size of gaps may play a role in influencing whether a live fast–die hard strategy or a sit-and-wait strategy is more successful. Plants can "sit and wait" for good growing conditions either as a seed or as a juvenile (seedling or non-reproducing sapling) understory plant.

For plants, **seed banks** (the accumulation of nongerminating seeds in the soil matrix) and between-year **seed dormancy** provide a "bet-hedging" strategy that ensures some seeds avoid germinating in "bad" years. This strategy is adaptive (increases a plant's fitness) in environments where conditions for germination fluctuate strongly from year to year, or as a way to avoid competition during times of high seed production by waiting until seed density is lower (Venable and Brown 1988). Dalling and Brown (2009) used carbon dating to measure the age of seeds from three neotropical pioneer species. They found germinable seeds between 18 and 38 years old, suggesting that these seeds have adaptive longevity allowing them to sit and wait to take advantage of gaps that occur in their location within that time period. The "youngest" seeds (i.e., 18-year-olds) came from the fastest-growing species in the forest, while the "oldest" seeds (i.e., 38-year-olds) came from a rare species restricted to huge treefall gaps that occur infrequently. Because the three pioneer species varied in their growth strategies, Dalling and Brown (2009) inferred that seed banks may allow pioneer trees to coexist by utilizing different gap conditions as they appear over time.

Strong competition can occur in gaps, as evidenced by the increased probability of mortality in seedlings when close to conspecific neighbors (Johnson et al. 2017). However, Johnson et al. (2017) also found that there was less mortality response to "other" (heterospecific) species' tree and seedling densities. Recall from Chapter 3, *Species Interaction: Competition*, that coexistence is promoted by greater competitive effects on conspecifics than on competitor species, perhaps implying mechanisms for coexistence among pioneer species even within the same gap.

Unlike the seed sit-and-wait strategy, which allows plants to wait until optimal conditions for growth arise, the sapling sit-and-wait strategy faces a notorious shade-tolerance tradeoff. That is, trees that do well under low light levels do poorly under high light levels, and vice versa (Grubb 2016). Plants that do well under high light levels are characterized as pioneer species, which are obligate gap specialists, completing their entire life cycle relatively quickly after a gap appears (live fast–die hard). In contrast, plants that do well under low light levels are characterized as shade-tolerant species, which are understory specialists (see the light-harvesting game below; Popma et al. (1992)). However, in an analysis of trees in a lowland tropical forest in Mexico, Popma et al. (1992) found a third strategy: some species can survive prolonged periods as juveniles in the shade, waiting for a small gap to appear, which they can then exploit to complete their life cycle.

The key leaf traits in the sapling sit-and-wait strategy appear to be thicker leaves with higher leaf mass per unit area, containing higher nutrient concentration than the pioneer or understory strategies. These trees also seem to generate highly significant differences between their sun and shade leaves, more so than either gap specialists or understory trees, indicating greater phenotypical plasticity than the other two strategies (Popma et al. 1992). In other words, trees that play the sapling sit-and-wait game are not bound by the shade-tolerance tradeoff. In fact, as indicated by the higher leaf nutrient concentration, it appears that the sit-and-wait strategy succeeds by accumulating resources within their leaves, allowing them to achieve a higher maximum rate of photosynthesis than other species once a light gap appears (Popma et al. 1992).

Gravel et al. (2010) studied the growth of understory saplings in maple–beech forests "released" (from shade) by the appearance of gaps. The maple *Acer saccharum* and the beech *Fagus grandifolia* differ in their response to light availability. The maple performs better than the beech under high-light conditions and frequent gaps because of its greater fecundity and ability to persist at a minimal seedling height (Poulson and Platt 1996; Marks and Gardescu 1998). However, beech seedlings can persist longer in the understory (more than 100 years, compared to 20 for the maple; Poulson and Platt 1996), which may explain why Gravel et al. (2010) found that beech outperformed maple using the sit-and-wait strategy after extended periods of low-light conditions. They found that despite high densities of maple seedlings, after a decade of low-light, slow-growth conditions, maples suffered greater recruitment failure, allowing the beech saplings to capitalize once the canopy opened.

***Vegetative Regeneration Strategies*:** Sprouting in downed trees from stems, base of the trunk, or tipped-up roots (Figure 7) is a form of post-disturbance reproduction from trees damaged by the

Figure 7. Conceptual rendering of vegetative regeneration via sprouting.

creation of a gap. Dietze and Clark (2008) studied whether vegetative reproduction played a role in gap dynamics in 18 replicated experimental gaps in forests of the southern Appalachians and Carolina Piedmont, USA. After monitoring damaged adult survival, sprout initiation, and growth and mortality of gap-colonizing sprouts and preexisting understory saplings, they found that recruitment from sprouts constituted 26–87% of early gap regeneration, largely because sprouts from recently damaged trees grew significantly faster than competing saplings. They concluded from this study that consistent differences among species in sprouting can affect competitive outcome among species in gaps (thus, vegetative reproduction may not necessarily be "neutral" with respect to maintaining diversity). They also found that sprouting occurs early in gap formation and tapers off as the gap ages. Importantly, they discovered that species differences in sprouting do not conform to the usual plant tradeoffs, such as those related to growth rates, shade tolerance, or seed abundance and survival. Thus, sprouting may represent a different tradeoff axis from the live fast–die hard or sit-and-wait strategies, adding complexity to the gap capture game.

The Light Harvesting Game—Coexistence at Different Light Levels

Plants have two different strategies to deal with shading by their neighbors: avoid it or tolerate it (Gommers et al. 2013). So far, most of the strategies we have considered are avoidance strategies: either grow as tall as possible to out-shade one's neighbors (best response curve ESS) or utilize light gaps in space and time until the canopy inevitably closes. However, waiting for a gap to appear and taking advantage of it is not the only strategy available to forest trees. Sometimes it is possible to make a living existing in the understory using **shade-tolerant strategies**. Shade-tolerant strategies allow multiple species to play the light capture game, resulting in greater forest biodiversity.

Taking advantage of a thin canopy: One way in which light competition shifts from an arms race with a single strategy ESS (i.e., "tallest tree wins") to the coexistence of multiple species (a multi-strategy ESS) is when conditions reduce the density of leaves in the canopy. While still lowering light levels, thin canopies can allow at least some light to penetrate. Iwasa et al. (1984) modeled light competition using assumptions similar to those in King's (1990) model and the concepts illustrated by the best response curve—specifically, that the benefit-to-cost ratio of increasing height declines in adult plants. However, Iwasa et al. (1984) relaxed the assumption that a focal individual could be completely shaded by neighboring trees. They found that when the canopy is sufficiently open to allow some incident light to reach lower layers, multiple height strategies can coexist as ESSs. While shorter strategies result in lower photosynthetic gains, they also incur lower costs related to height, so the fitness of both tall and short trees can be equal. In such cases, taller strategies will still dominate in density, while the abundance of shorter strategies declines with decreasing height.

***Cheating in the Height Game*:** Iwasa et al.'s (1984) model depended on the presence of a thin canopy, but it did not consider shade-tolerant strategies per se. Trees can essentially cheat at the height game if they have shade-tolerant adaptations. Popma et al. (1992) found that gap-independent plants (those that spend their entire life history in the understory) have elongated, thin leaves with fewer but larger stomata, suggesting gap-independent species maximize their carbon gain by producing "cheap" leaves. Pioneer species are also characterized as having cheap leaves (Popma et al. 1992), but for them, there is enough light to gain plenty of photosynthate before the leaf is damaged or desiccated and must be replaced. Westoby et al. (2002) argue that when a plant grows a leaf, it must last long enough to pay back initial investment costs. However, under low light conditions, photosynthetic rates can be very low, and therefore leaves must last a very long time to acquire a net gain of photosynthate. However, Westoby et al. (2002) point out that species characteristic of shaded understories usually have leaves with long lifespans, and the reason they do, despite flimsy, cheap leaves, is because understories are characterized by low wind, high humidity, and reduced risk of wilting under dense shade.

Grubb (2016) suggests that there may be a tradeoff between exploiting growth under high light conditions (like pioneer species) and surviving in deep shade (like shade-tolerant species), because rapid growth incurs high respiration costs, which could lead to starvation in plants adapted to deep shade. Thus, many shade-tolerant plants cannot play the gap game and should therefore have a phenotype adapted to permanently shaded environments.

Based on molecular evidence, Gommers et al. (2013) hypothesize that shade-tolerant plants use molecular antagonists to keep light-seeking traits, such as rapidly elongating stems and petioles, from being turned on under low light conditions so they can persist as shorter plants in the understory. That a molecular switch (or switches) is involved in the production of a shade-avoiding or shade-tolerant phenotype would go far in explaining the shade-avoiding vs. shade-tolerant tradeoff in both seedlings and saplings of different species.

Other light-harvesting strategies under low-light conditions may include differences in growth form, such as those that can take advantage of light occurring in sun flecks scattered in different places at different points in time throughout the sun's passage above the canopy. Poulson and Platt (1996) found evidence that strong **apical dominance** (where more growth happens vertically than horizontally) in maples leads to fast upward growth in vertical light gradients (which occur in gaps), whereas the long lateral branches of beech trees are advantageous in capturing scattered light flecks. In harvesting light flecks, plants must be able to tolerate sudden changes in photon flux density (the amount of photons per unit area per time) and capitalize on the sudden wealth without harming their photosystems (Ruban 2015). Efficient photosynthesis requires different chloroplast qualities and repair mechanisms under high light and low light conditions, so some plants have evolved sophisticated and varying strategies to cope with rapid variations in light fluctuation, though there is much to be learned about them (Mathur et al. 2018).

Suppressing Overstory Regeneration: In a twist, smaller trees can sometimes outcompete canopy trees if they can prevent canopy trees from regenerating. New Zealand lies in the middle of the favorable, mild, and strongly maritime climatic conditions that favor giant trees (Larjavaara 2014); however, New Zealand forests are predominantly composed of two heights, emergent but sparse conifer species and a dense lower canopy of angiosperms. McGlone et al. (2010) argue that these diverse angiosperm trees—which are small in size (likely an adaptation to small island conditions of Miocene-era archipelago) and have tiny leaves (most likely an adaptation against moa herbivory [Ciarle et al. 2025])—allow enough light to reach the understory for their own regeneration. However, the light levels are not enough for large conifers to grow from seed without a canopy gap. This dynamic keeps the conifers sparse enough to prevent them from excluding the angiosperms via light competition.

Opting Out of the Arms Race: Of course, one way to win the arms race is to opt out of it altogether by having no wood at all! Instead of investing photosynthate into woody plant parts, bryophytes such as mosses, liverworts, and hornworts take advantage of the moist, benign understory found in temperate rainforests to carpet the forest floor. Without the need to invest in woody structures, these plants can complete their life cycles under low light conditions, as long as the habitat remains suitably moist.

Assuredly, another game is being played among these nonvascular plants because, as long as there is more than one type of player, there is always an evolutionary game. By framing light competition as an evolutionary game, we can understand the evolutionary constraints faced by certain strategies and explore why particular strategies might be adaptive in a forest biome. Above all, we were able to see how the presence of other individuals with different strategies can impact a focal individual's fitness. Light competition is not the only evolutionary game that plants engage in. For example, there are root-shoot allocation games, root competition games, herbivore defense games, and even mutualist games (McNickle and Dybzinski 2013). As we hinted in the introduction, there are also games for other types of organisms: habitat selection games, foraging games, conflict games, sexual selection games, etc. Once we embark on this path, we may come to realize that the key to understanding not only the evolution of various strategies but also the diversity of coexisting strategies lies in thinking that "all of life is a game" (Vincent and Brown 2005).

References

Ale, S. B., Halloway, A., Mitchell, W. A. and Whelan, C. J. 2019. Does God roll dice? Neutrality and determinism in evolutionary ecology. Biology & Philosophy 34: 1–16.

Araújo, I., Marimon, B. S., Junior, B. H. M., Oliveira, C. H., Silva, J. W., Beú, R. G. et al. 2024. Taller trees exhibit greater hydraulic vulnerability in southern Amazonian forests. Environmental and Experimental Botany 226: 105905.

Ávila-Lovera, E., Haro, R., Choudhary, M., Acosta-Rangel, A., Pratt, R. B. and Santiago, L. S. 2024. The benefits of woody plant stem photosynthesis extend to hydraulic function and drought survival in *Parkinsonia florida*. Tree Physiology 44: p.tpae013.

Baas, P., Ewers, F. W., Davis, S. D. and Wheeler, E. A. 2004. Evolution of xylem physiology. pp. 273–295. *In*: Hemsley, A. R. and Poole, I. [eds.]. The Evolution of Plant Physiology. Academic Press.

Becker, P., Meinzer, F. C. and Wullschleger, S. D. 2000. Hydraulic limitation of tree height: a critique. Functional Ecology 14: 4–11.

Berry, Z. C., Ávila-Lovera, E., De Guzman, M. E., O'Keefe, K. and Emery, N. C. 2021. Beneath the bark: assessing woody stem water and carbon fluxes and its prevalence across climates and the woody plant phylogeny. Frontiers in Forests and Global Change 4: 675299.

Boehm J. 1893. Capillarität und saftsteigen. Ber. Deut. Bot. Ges. 11: 14–23.

Bohlman, S. A. 2015. Species diversity of canopy versus understory trees in a neotropical forest: Implications for forest structure, function and monitoring. Ecosystems 18: 658–670.

Brokaw, N. and Busing, R. T. 2000. Niche versus chance and tree diversity in forest gaps. Trends in Ecology & Evolution 15: 183–188.

Carpenter, J. R. 1939. The biome. The American Midland Naturalist 21: 75–91.

Chesson, P. L. and Warner, R. R. 1981. Environmental variability promotes coexistence in lottery competitive systems. The American Naturalist 117: 923–943.

Ciarle, R., de Lange, P. and Burns, K. C. 2025. Ancestral state reconstruction sheds new light on the loss of divarication hypothesis on New Zealand's outlying islands. Journal of Ecology 113: 1293–1303.

Dalling, J. W., Swaine, M. D. and Garwood, N. C. 1998. Dispersal patterns and seed bank dynamics of pioneer trees in moist tropical forest. Ecology 79: 564–578.

Dalling, J.W. and Brown, T. A. 2009. Long-term persistence of pioneer species in tropical rain forest soil seed banks. The American Naturalist 173: 531–535.

Dietze, M. C. and Clark, J. S. 2008. Changing the gap dynamics paradigm: vegetative regeneration control on forest response to disturbance. Ecological Monographs 78: 331–347.

Dixon H. H. and Joly, J. 1894. On the ascent of sap. Proceedings of the Royal Society B 57: 3–5.

Falster, D. S. and Westoby, M. 2003. Plant height and evolutionary games. Trends in Ecology & Evolution 18: 337–343.

Gilbert, B., Wright, S. J., Muller-Landau, H. C., Kitajima, K. and Hernandéz, A. 2006. Life history trade-offs in tropical trees and lianas. Ecology 87: 1281–1288.

Givnish, T. J. 1982. On the adaptive significance of leaf height in forest herbs. The American Naturalist 120: 353–381.

Gommers, C. M., Visser, E. J., St Onge, K. R., Voesenek, L. A. and Pierik, R. 2013. Shade tolerance: when growing tall is not an option. Trends in Plant Science 18: 65–71.

Gravel, D., Beaudet, M. and Messier, C. 2010. Large-scale synchrony of gap dynamics and the distribution of understory tree species in maple-beech forests. Oecologia 162: 153–161.

Grubb, P. J. 2016. Trade-offs in interspecific comparisons in plant ecology and how plants overcome proposed constraints. Plant Ecology & Diversity 9: 3–33.

Hardin, G. 1968. The tragedy of the commons: the population problem has no technical solution; it requires a fundamental extension in morality. Science 162: 1243–1248.

Hubbell, S. P. 2001. The Unified Neutral Theory of Biodiversity and Biogeography. Princeton University Press, NJ.

Hummert, S., Bohl, K., Basanta, D., Deutsch, A., Werner, S., Theißen, G. et al. 2014. Evolutionary game theory: cells as players. Molecular BioSystems 10: 3044–3065.

Iwasa, Y., Cohen, D. and Leon, J. A. 1984. Tree height and crown shape, as results of competitive games. Journal of Theoretical Biology 112: 279–297.

Jensen, K. H. and Zwieniecki, M. A. 2013. Physical limits to leaf size in tall trees. Physical Review Letters 110: 018104.

Jøgensen, S. and Zaccour, G. 2014. A survey of game-theoretic models of cooperative advertising. European Journal of Operational Research 237: 1–14.

Johnson, D. J., Condit, R., Hubbell, S. P. and Comita, L. S. 2017. Abiotic niche partitioning and negative density dependence drive tree seedling survival in a tropical forest. Proceedings of the Royal Society B 284: 20172210.

Johnson, E. C. and Hastings, A. 2022. Towards a heuristic understanding of the storage effect. Ecology Letters 25: 2347–2358.

King, D. A. 1990. The adaptive significance of tree height. The American Naturalist 135: 809–828.

Koch, G. W., Sillett, S. C., Jennings, G. M., and Davis, S. D. 2004. The limits to tree height. Nature 428: 851–854.

Larjavaara, M. 2014. The world's tallest trees grow in thermally similar climates. New Phytologist 202: 344–349.

Laurance, W. F. and Curran, T. J. 2008. Impacts of wind disturbance on fragmented tropical forests: A review and synthesis. Austral Ecology 33: 399–408.

Leimar, O. and McNamara, J. M. 2023. Game theory in biology: 50 years and onwards. Philosophical Transactions of the Royal Society B 378: 20210509.

Loomis, W. E., Santamaria-P, R. and Gage, R. S. 1960. Cohesion of water in plants. Plant Physiology 35: 300.

Lorenz, K. 1963. Das Sogenannte Böse. Wien, Austria: Dr. G. Borotha-Schoeler Verlag. *In*: German; transl. M. K. Wilson. 1966. On Aggression. London, UK: Routledge.

Lorimer, C. G., Porter, D. J., Madej, M. A., Stuart, J. D., Veirs Jr., S. D., Norman, S. P. et al. 2009. Presettlement and modern disturbance regimes in coast redwood forests: implications for the conservation of old-growth stands. Forest Ecology and Management 258: 1038–1054.

Lucas, J. R. and Howard, R. D. 1995. On alternative reproductive tactics in anurans: dynamic games with density and frequency dependence. The American Naturalist 146: 365–397.

Marks, P. L. and Gardescu, S. 1998. A case study of sugar maple (*Acer saccharum*) as a forest seedling bank species. Journal of the Torrey Botanical Society 125: 287–296.

Mathur, S., Jain, L., and Jajoo, A. 2018. Photosynthetic efficiency in sun and shade plants. Photosynthetica 56: 354–365.

Maynard Smith, J. 1974. The theory of games and the evolution of animal conflicts. Journal of Theoretical Biology 47: 209–221.

Maynard Smith, J. 1982. Evolution and the Theory of Games. Cambridge University Press.

Maynard Smith, J. and Price, G. R. 1973. The logic of animal conflicts. Nature 246: 15–18.

McGill, B. J. 2003. A test of the unified neutral theory of biodiversity. Nature 422: 881–885.

McGlone, P. S., Richardson, S. J. and Jordan, G. J. 2010. Comparative biogeography of New Zealand trees: species richness, height, leaf traits and range sizes. New Zealand Journal of Ecology 34: 137–151

McNickle, G. G. and Dybzinski, R. 2013. Game theory and plant ecology. Ecology letters 16: 545–555.

Melcher, P. J., Meinzer, F. C., Yount, D. E., Goldstein, G. and Zimmermann, U. 1998. Comparative measurements of xylem pressure in transpiring and non-transpiring leaves by means of the pressure chamber and the xylem pressure probe. Journal of Experimental Botany 49: 1757–1760.

Moles, A. T., Warton, D. I., Warman, L., Swenson, N. G., Laffan, S. W., Zanne, A. E. et al. 2009. Global patterns in plant height. The Journal of Ecology 97: 923–932.

Orman, O. and Dobrowolska, D. 2017. Gap dynamics in the Western Carpathian mixed beech old-growth forests affected by spruce bark beetle outbreak. European Journal of Forest Research 136: 571–581.

Pavlovic, N. B. 1994. Disturbance-dependent persistence of rare plants: anthropogenic impacts and restoration implications. pp. 159–193. *In*: Bowles, M. L. and C. J. Whelan [eds.]. Restoration of Endangered Species: Conceptual Issues, Planning and Implementation. Cambridge University Press, New York.

Peterson, C. J. 2007. Consistent influence of tree diameter and species on damage in nine eastern North America tornado blowdowns. Forest Ecology and Management 250: 96–108.
Polis, G. A., Anderson, W. B. and Holt, R. D. 1997. Toward an integration of landscape and food web ecology: the dynamics of spatially subsidized food webs. Annual review of ecology and systematics, 28: 289–316.
Popma, J., Bongers, F., and Werger, M. J. A. 1992. Gap-dependence and leaf characteristics of trees in a tropical lowland rain forest in Mexico. Oikos 63: 207–214.
Poulson, T. L. and Platt, W. J. 1996. Replacement patterns of beech and sugar maple in Warren Woods, Michigan. Ecology 77: 1234–1253.
Riechert, S. E. and Hammerstein, P. 1983. Game theory in the ecological context. Ann. Rev. Ecol. Syst. 14: 377–409.
Rosindell, J., Hubbell, S. P. and Etienne, R. S. 2011. The unified neutral theory of biodiversity and biogeography at age ten. Trends in Ecology & Evolution 26: 340–348.
Ruban, A. V. 2015. Evolution under the sun: optimizing light harvesting in photosynthesis. Journal of Experimental Botany 66: 7–23.
Runkle, J. R. 2013. Thirty-two years of change in an old-growth Ohio beech-maple forest. Ecology 94: 1165–1175.
Ryan, M. G. and Yoder, B. J. 1997. Hydraulic limits to tree height and tree growth. BioScience 47: 235–242.
Sale, P. F. 1977. Maintenance of high diversity in coral reef fish communities. The American Naturalist 111: 337–359.
Sale, P. F. 1978. Coexistence of coral reef fishes—a lottery for living space. Environmental Biology of Fishes 3: 85–102.
Schenk, H. J., Espino, S., Romo, D. M., Nima, N., Do, A. Y., Michaud, J. M. et al. 2017. Xylem surfactants introduce a new element to the cohesion-tension theory. Plant Physiology 173: 1177–1196.
Shenkin, A., Chandler, C. J., Boyd, D. S., Jackson, T., Disney, M., Majalap, N. et al. 2019. The world's tallest tropical tree in three dimensions. Frontiers in Forests and Global Change 2: 32.
Słupianek, A., Dolzblasz, A. and Sokołowska, K. 2021. Xylem parenchyma—role and relevance in wood functioning in trees. Plants 10: 1247. doi: 10.3390/plants10061247.
Sousa, W. P. 1984. The role of disturbance in natural communities. Annual review of ecology and systematics 15: 353–391.
Stovall, A. E., Shugart, H. and Yang, X. 2019. Tree height explains mortality risk during an intense drought. Nature Communications 10: 4385.
Stump, S. M. and Vasseur, D. A. 2023. Reexamining the storage effect: Why temporal variation in abiotic factors seems unlikely to cause coexistence. Ecological Monographs 93: e1585.
Swaine, M. D. and Whitmore, T. C. 1988. On the definition of ecological species groups in tropical rain forests. Vegetatio 75: 81–86.
Tng, D. Y. P., Williamson, G. J., Jordan, G. J. and Bowman, D. M. J. S. 2012. Giant eucalypts-globally unique fire-adapted rain-forest trees? New Phytologist 196: 1001–1014.
Tyree, M. T. and Sperry, J. S. 1989. Vulnerability of xylem to cavitation and embolism. Annual Review of Plant Physiology and Plant Molecular Biology 40: 19–36.
Tyree, M. T. and Zimmermann, M. H. 2002. Xylem Structure and the Ascent of Sap, 2nd edition. Springer, New York.
Van de Water, K. M. and Safford, H. D. 2011. A summary of fire frequency estimates for California vegetation before Euro-American settlement. Fire Ecology 7: 26–58.
Venable, D. S. and J. S. Brown. 1988. The selective interactions of dispersal, dormancy and seed size as adaptations for reducing risk in variable environments. The American Naturalist 131: 360–384.
Vincent, T. L. and Brown, J. S. 2005. Evolutionary Game Theory, Natural Selection, and Darwinian Dynamics. Cambridge University Press, Cambridge.
von Neumann, J. and Morgenstern, O. 1944. Theory of games and economic behavior. Princeton University Press, Princeton, NJ.
Warner, R. R. and Chesson, P. L. 1985. Coexistence mediated by recruitment fluctuations: a field guide to the storage effect. The American Naturalist 125: 769–787.
Westoby, M., Falster, D. S., Moles, A. T., Vesk, P. A. and Wright, I. J. 2002. Plant ecological strategies: some leading dimensions of variation between species. Annual Review of Ecology and Systematics, 33: 125–159.
Wilson, S. M., Mason, B., Jinks, R., Gil-Moreno, D. and Savill, P. 2016. The redwoods and red cedar. Quarterly Journal of Forestry: 110: 244–256.
Yu, D. W., Terborgh, J. W., and Potts, M. D. 1998. Can high tree species richness be explained by Hubbell's null model? Ecology Letters 1: 193–199.
Zalamea, P. C., Heuret, P., Sarmiento, C., Rodríguez, M., Berthouly, A., Guitet, S. et al. 2012. The genus Cecropia: a biological clock to estimate the age of recently disturbed areas in the Neotropics. Plos One 7: e4263.
Zimmermann, U., Meinzer, F. C., Benkert, R., Zhu, J. J., Schneider, H., Goldstein, G. et al. 1994. Xylem water transport: is the available evidence consistent with the cohesion theory? Plant, Cell & Environment 17: 1169–1181.

CHAPTER 8

Food Webs

"Food webs are the road maps through Darwin's famous 'entangled bank'..." (Pimm et al. 1991)

A biological community is an association of interacting species in a defined area where boundaries and habitats are broadly set. While it is often convenient for ecologists to circumscribe a community for ecological study, these defined communities are almost always linked to other communities (Polis et al. 1997). For example, if we were interested in the inhabitants of a particular stretch of stream, the stream—though defined—is linked to the surrounding forest community by decomposing leaves falling into the water, by animal populations such as salamanders and mayflies that utilize both the water and the forest, and by a host of other animals (plants, fungi, and bacteria) that cross the boundaries we have set. Migratory birds may link temperate and tropical forests, while oceans and freshwater systems may be connected through the movement of anadromous fish and nutrient cycling. Likewise, within a community, no pair of predator and prey, parasite and host, or grazer and plant lives in isolation. Each is embedded in a complex web of interactions with other predators, parasites, prey, and competitors. Even a simple ecosystem, such as a desert, may harbor a diverse web of relationships. Polis (1991) describes a sand community in California's Coachella Valley desert interconnected by about 175 plants, over 100 vertebrates, and thousands of insects, plus arachnids, nematodes, and microorganisms.

A complex web of interactions is a "system." To understand it fully, ecologists need to have some knowledge of the parts and how they affect the "whole." How can we make sense of it all?

Although "follow the energy and matter" is not quite as catchy as "follow the money," it works equally well as a method for untangling complex relationships and laying bare underlying community structure. The flow of energy and matter through an ecosystem involves feeding relationships, which in turn affect the abundance of each species involved in the relationships. During photosynthesis, energy and matter are transferred from the abiotic environment to plants. When animals eat plants, energy and matter are passed on to the animals. Similarly, when one animal eats another, the energy contained in the first animal is also passed on. This flow of matter and energy can be conveniently depicted as the links between organisms in a **food chain**. Although matter and energy flow in parallel processes, they differ in that energy must obey the laws of thermodynamics and can flow only unidirectionally along food chains, with losses to heat along the way. In contrast, matter, such as chemicals like carbon and nitrogen, can cycle from the bottom to the top level and back down again. Food chains depict a single, linear path representing the consumption of one species by another and the transfer of energy and matter (chemicals or nutrients) between them. A food chain is just one pathway (typically consisting of only 3 or 4 links long; Pimm and Kitching 1987). However, if an organism is eaten by multiple types of consumers, energy and matter can follow multiple paths. Most feeding relationships in a community are not simple food chains, but rather multiple interconnected food chains that form a food web.

A **food web** is a diagram of interlocking food chains depicting interactions among prey, predators, and competitors, and these depictions often become quite complex. It is well understood

DOI: 10.1201/9781003008538-8

that food webs describe communities from the standpoint of predator–prey (or consumer–resource) interactions. However, they also encompass competitive interactions among consumers that share the same food sources, and they can reveal indirect interactions that may lead to surprising competitive or even mutualistic relationships.

Direct interactions influence the presence and abundance of organisms within communities. They modify each other's fundamental niches, relative abundance, and distributions along gradients, thereby influencing species composition and community structure. However, when species are interconnected by a web of trophic (food-related) relationships, the direct effect of one species on another often sets off a chain of events that affects additional species, setting the stage for indirect interactions to play out in the community. **Indirect interactions** occur when one species does not interact directly with a second species but instead influences the second species via interaction with a third species. In other words, two species interact in a way that involves one or more intermediate species. In complex networks of interacting species, the number of indirect interactions—requiring paths with more than one segment—grows exponentially with species diversity. This poses challenges in recognizing how the abundance of one species is affected by others through feeding relationships.

One approach to simplifying the complexity of interactions is to place species into broad categories called **trophic levels**. This is achieved by grouping organisms that obtain their energy in similar ways into a single trophic level and looking at the relationships between these feeding levels.

The first trophic level is made up of autotrophs or **producers**, which include photosynthesizing and chemosynthesizing organisms. Photosynthesizing organisms such as phytoplankton, macroalgae, and plants make their own "food" by converting the energy from light and using the matter from carbon dioxide to form organic molecules. Chemosynthesizing bacteria convert the energy from the oxidation of iron or sulfur and use the matter from carbon dioxide or methane to form organic molecules. In subsequent trophic levels, organisms called "consumers" do not make their food but derive energy and matter from consuming others. **Primary consumers** are species that feed on producers, while **secondary consumers** are those that feed on other consumers. **Decomposers** represent a terminal level essential to the "food cycle" because they break down matter from other organisms and return it to the first trophic level as chemicals that producers can use. Any self-sustaining ecosystem must contain producers and decomposers, but consumers are not essential to complete a "food cycle."

Sometimes, grouping organisms by trophic level is too simple for the questions we want to explore, and we may instead want to examine groups of organisms within a given trophic level based on similarities in food type or feeding strategy. Richard Root (1967) was the first to call this grouping a guild. Root defined a **guild** as "a group of species that exploit the same class of environmental resources in a similar way," and he was particularly interested in how several species of birds all utilize the same food resource in the same habitat. In this case, he studied the "foliage-gleaning guild" in an oak woodland in California (USA), which included members such as vireos, warblers, and a gnatcatcher (Root 1967). As the concept of guilds gained traction, it became apparent that members of a guild need not belong to the same taxonomic group. For example, ants and desert rodents that feed on the same seed species can be considered members of the same guild (Brown and Davidson 1977). Alternatively, though less commonly, members of different guilds may be closely related. For example, Höfer and Brescovit (2001), studying Brazilian spiders, found that the spider Family Ctenidae has members in both the nocturnal aerial ambushers guild and the ground ambushers guild. However, they also found that, across most other spider Families, most members were restricted to a single guild. Ecological researchers often determine guilds by assigning diet categories and ranking their importance for each species. For example, Kissling et al. (2012) divided all terrestrial bird species into nine broad guilds: herbivores, granivores, frugivores, nectivores, aerial insectivores, terrestrial insectivores, carnivores, scavengers, and omnivores. Another approach is to use a multivariate statistical cluster analysis based on food types in the diet. In this way, Lešo and

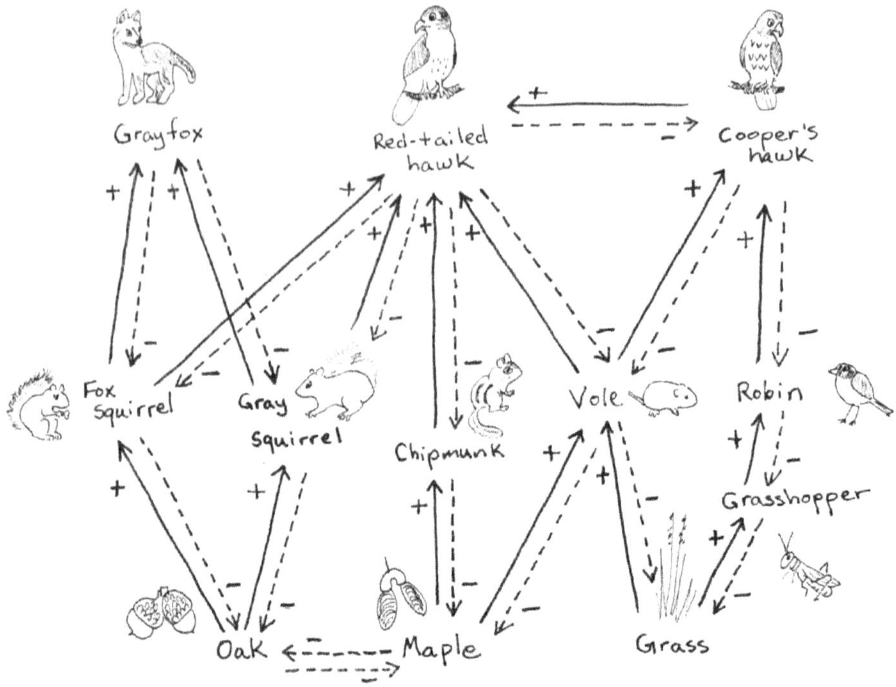

Figure 1. A subset of a food web in a typical oak–maple forest community in Illinois (Midwestern United States). Arrows indicate the effect of one species' density on the growth rate of another (note: magnitude of effect is not shown here). Paired positive and negative arrows indicate the effect of prey density on the growth rate of its predator and vice versa. Paired negative arrows indicate the mutual negative effect on the growth rate of competing species that directly interfere with one another.

Kropil (2007) divided the insectivorous forest birds of the West Carpathians into foliage gleaners, bark foragers, aerial insectivores, and ground insectivores.

Guilds are used to group consumers by diet to study competition, abundance, and diversity, while food webs, which consider all members of all trophic levels, allow ecologists to tease apart complex relationships between species. Consider a typical oak–maple forest community in Midwestern North America (Figure 1). Plants, including oaks, maples, and grasses, serve as the producers. Squirrels, voles, and grasshoppers feed on maple seeds, acorns, and grass, making them primary consumers. These primary consumers are then eaten by secondary consumers, such as the Cooper's Hawk. Red-tailed Hawk preying on Cooper's Hawk (secondary consumers) typify a tertiary consumer. In this forest ecosystem, we consider only 12 species to illustrate a food web. Of course, we could go on adding many more species—such as white-footed mice and ants—but doing so would make the food web too complex for our illustrative purpose here. Note that, in addition to producers and consumers, food webs also include consumers of dead organic matter, such as scavengers, detritivores, and decomposers (not shown in Figure 1). In our example of a food web, only the Red-tailed Hawk feeds at two trophic levels. In this case, the carnivorous Red-tailed Hawk consumes not only most of the primary consumers, but another carnivore (Cooper's Hawkas well. However, feeding at two trophic levels is not unusual; foxes and coyotes, for example, are opportunistic feeders, consuming mice, marmots, and other small animals, in addition to fruits and seeds. When organisms eat both producers and consumers, they are called **omnivores**. Omnivores do not fit conveniently into simple trophic-level feeding categories, and therefore, they tend to make food webs more complex with additional connections.

Direct and Indirect Interactions and Their Effects

Organisms within a food web affect each other's per capita growth rate. The simplest way that one species can affect the abundance of another is through a direct effect, which occurs when two species interact without involving other species. However, this is not always the case in nature. If we consider communities with multiple species, we can trace not only direct interactions but also indirect interactions between species at the same or other trophic levels.

How many pairwise interactions (direct effects) can be recognized in the diagram (Figure 1)? The answer is 17 (you may wish to count the pairs of arrows). Direct effects occur when individuals from different species influence each other's population growth rates, and these effects have both signs and magnitudes. For example, oak and maple directly compete for soil nutrients and light, negatively influencing seed production, which in turn decreases the population growth rate of each. Voles directly affect grass through herbivory (primary consumption), enhancing the population growth of the former while decreasing the population growth of the latter. Gray foxes consuming gray squirrels represent a direct positive effect on the fox's per capita growth rate and a negative effect on the squirrel's per capita growth rate.

In our example food web (Figure 1), gray squirrels and fox squirrels do not compete directly. There is no positive or negative sign between them. However, they compete indirectly via an intermediate species. Because they both consume acorns (oak seeds), they are competing for the same resource. Indirect effects, which are ubiquitous in nature, are simply a product of two or more linked direct effects.

Five important indirect interactions are recognized: Resource competition, apparent competition, keystone species, trophic cascade, and indirect mutualism.

Resource Competition

Resource competition is a common form of indirect interaction. In resource competition, also called exploitative competition, two consumers negatively affect each other's growth rate through a shared prey species. The negative effect of fox squirrels on the growth rate of gray squirrels is readily observed (Figure 2). When fox squirrels consume acorns, they decrease the acorn density. Because oak seeds are a shared resource, the reduction in acorn density lowers the growth rate of gray squirrels. The reverse is also true, so they interact indirectly through resource competition. Likewise, the Cooper's Hawk and Red-tailed Hawk compete because both consume voles (Figure 1). Such resource competition between two species is mediated by a third species.

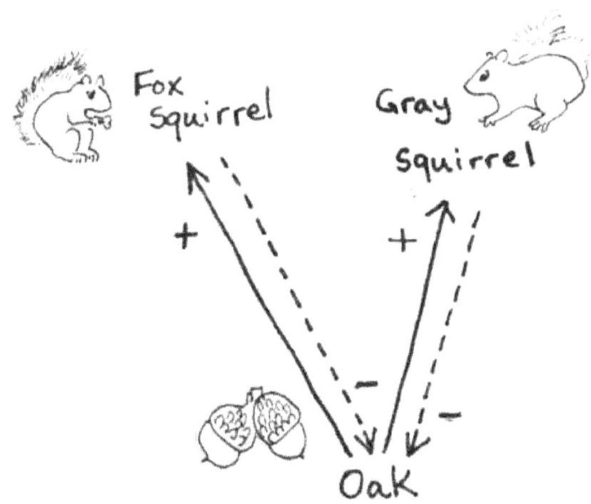

Figure 2. Resource (exploitative) competition: Two consumers compete for the same prey. A solid line represents a positive effect on growth rate, and a dashed line indicates a negative effect on growth rate; hence, the effect of fox quirrel density on the per capita growth of the gray squirrel population is negative, and vice versa. (Follow the direction of arrows from one squirrel to another and use the logic of simple algebra: negative times positive is negative).

A classic study of resource competition in the Sonoran Desert near Tucson, Arizona (USA), showed that seed-eating desert rodents and harvester ants, despite differences in their sizes and evolutionary lineages, compete indirectly for the same seeds (Brown and Davidson 1977). The negative impact of one competitor on another is mediated by the abundance of a third species (seeds). In experimental enclosures where rodents were excluded, the number of ant colonies increased by 71%, and when ants were excluded, rodent biomass increased by 24%—indicating their population sizes had each been decreased by the presence of the other (the definition of competition). Both rodents and ants increased and maintained higher densities on plots from which the other taxon had been excluded compared to unmanipulated control plots where the two taxa coexisted. In the case of the ants, this increase was particularly dramatic. Rodents and ants compete primarily by exploiting common food resources, thereby reducing their availability. However, they also compete directly by actively interfering with each other. Seed-eating rodents have been observed eating ants and stealing seeds stored in ant granaries. Ants also interfere with rodents' foraging at seed clumps by biting them.

Sometimes the effects of indirect competition can come from multiple sources. In the Trans-Himalayan zone in Nepal, people pasture domestic goats and sheep in the nutritionally rich alpine meadow habitat in spring and summer and move them to a lower elevation for winter pasture (Shrestha and Wegge 2008). Similar to free-range cattle, there are also free-range yaks that utilize the alpine meadows in spring and move to higher elevation areas for the rest of the year. Wild blue sheep (*Pseudois nayaur*) also inhabit this alpine habitat along with these domesticated ungulates. Shrestha and Wegge (2008) explain the low birth rate of blue sheep in this system as a result of resource competition. This begins in the spring when the blue sheep overlap in diet (forbs) and elevation with domestic sheep and goats, then continues in the summer when blue sheep move higher and overlap in diet (grasses) with yaks. Finally, in winter, after the sheep and goats are moved to lower levels, blue sheep again overlap in diet (shrubs) and elevation with yaks.

Apparent Competition

Interspecific resource competition occurs when the individuals of two species have a negative impact on each other's fitness. In some cases, this negative impact is the result of "apparent competition" (Holt 1977), in which two species share a common predator. For instance, fox squirrels and gray squirrels share a common predator, the gray fox (Figure 3). The same predator also hunts Cooper's Hawk and voles, the Red-tailed Hawk (Figure 1). Apparent competition happens between the fox and grey squirrels because an increase in the number of fox squirrels has a positive effect on the number of grey foxes, which then has an increased negative effect on the number of grey squirrels. The reverse is true for the fox squirrel, creating apparent competition between the two species.

Apparent competition, also called competition for "enemy free space" (Jeffries and Lawton 1984), is distinguished from resource competition because it does not involve negative interactions between species competing for a shared limited resource. Instead, safety from predators becomes the limiting factor. However, the consequence of apparent competition—a decrease in the abundance of one or both prey species—resembles the population effects that would result from classical resource competition. Apparent competition can occur when prey species contribute to an increased predator growth rate, attract an aggregation of predators (increasing predator immigration), or enhance the predator's capture and consumption rate (increasing predator efficiency).

Arctic hares *(Lepus arcticus)* once roamed the boreal forests of Newfoundland, Canada. However, after the introduction of the snowshoe hare (*Lepus americanus*) in the latter half of the 19th century (Strong and Leroux 2014), Arctic hares decreased to such an extent that they are now confined to mountain and coastal barrens (Small and Keith 1992). Lynx, the natural enemy of the introduced snowshoe hare, was initially suspected to be the cause of the Arctic hare's decline. However, subsequent detailed studies found that it was not the lynx in this "apparent competition

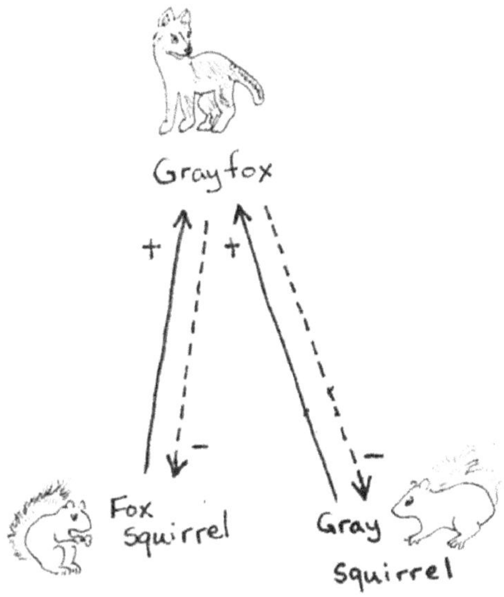

Figure 3. Apparent competition: Two prey experience negative effects on each other's fitness via a common predator. A solid line represents a positive effect, while a dashed line indicates a negative effect. The effect of the fox squirrel population on the per capita growth of the gray squirrel population is negative, and vice versa. (Remember, as in Figure 2, follow the direction of arrows from one squirrel to the other and use the logic of simple algebra: Positive times negative is negative).

tale," but the red fox, which turned out to be the common predator of both species of hare (Holt and Bonsall 2017). The red fox population rose abundantly by feeding on the introduced snowshoe hare population. In sufficient numbers, they not only kill snowshoe hares but also Arctic hares. The Arctic hare, which is much more susceptible to red fox predation than the snowshoe hare, is now restricted to rocky refugia in the tundra, where they can escape fox predation (Small and Keith 1992). The introduced snowshoe hare proved to be a superior competitor to the Arctic hare in boreal forests due to both its consumption of common food resources and its positive effect on fox population numbers, causing the decline of Arctic hares.

The indirect interactions of apparent competition can have practical implications. For example, biological control may be made more effective by creating situations where predators introduced to control a pest maintain higher densities by feeding on multiple prey species. This contrasts with the conventional strategy of seeking a very specific predator for a particular prey. Field experiments have shown that under some circumstances, predators are more effective at controlling a certain target prey when they are released together with an alternative prey. In California, the introduction of the predatory mite along with an alternative prey species, the Willamette mite (*Eotetranychus willamettei*), resulted in a much greater reduction in the abundance of grapevine-infesting Pacific mites (*Tetranychus pacificus*) than when the predator was introduced without its alternative prey (Karban et al. 1994).

For the Arctic hare and Pacific mite, competing prey species increased the risk of predation, but sometimes predation risk for one species can be reduced (a positive effect!) if the predator prefers the competing species. In western India, chital deer (*Axis axis*) utilize similar habitat in both Gir Sanctuary (where they coexist with domestic cattle and buffalo) and Gir National Park (where domestic livestock are excluded). Both the deer and the livestock share a common predator, the Asiatic lion (*Panthera leo*). Vijayan et al. (2012) found that although the Sanctuary is a quantitatively inferior habitat for deer due to the high livestock grazing intensity, the chital deer preferred using the Sanctuary because lions preferentially sought the easier-to-catch livestock, thereby reducing the predation risk for the chital deer. It was only when the food resources of the Sanctuary were too limited to support a larger deer population that chital deer moved to the riskier habitat of the National Park (Vijayan et al. 2012).

Keystone Species

The chital deer–livestock example shows that instead of causing mutual negative effects, the presence of predators can sometimes be beneficial to a less preferred species when predator effort is directed toward a more preferred species (apparent commensalism). But what if predation is directed toward a competitively superior species to the advantage of a competitively inferior species? This describes the special case of "keystone predation."

Paine (1966) described a food web from the intertidal zone at Mukkaw Bay, Washington (USA), a typical rocky shore community along the west coast of North America. This food web consisted of 15 species, including chitons, limpets, mussels, acorn barnacles, gooseneck barnacles, and two predators: the sea star *Pisaster* and a whelk (*Thais*, a type of sea snail). *Pisaster* (the top predator) obtained 90% of its energy from a mixture of chitons, mussels, and barnacles. Paine removed the sea star from certain locations and monitored the response of the community. In two years, the diversity of intertidal invertebrates in the control plot remained at 15 species, whereas the diversity within the experimental removal plot declined from 15 to eight species. The decline in community diversity occurred because the removal of the sea star from experimental plots allowed the mussels and gooseneck barnacles (strong competitors) to increase substantially, thereby outcompeting the inferior species such as limpets, chitons, and acorn barnacles.

Paine (1969) termed the sea star a **keystone species** because it exerted substantial control over community composition and stability. A keystone species is defined as one whose impact on species composition and community processes is disproportionately large relative to its abundance or biomass (Power et al. 1996). Put another way, its removal would lead to a decrease in at least half the species in a community (Mills et al. 1993). Although the term was originally applied only to predators, it is now widely accepted that keystone species can occur at any trophic level. Most food web models assume the "importance" of each species is either equal or follows a normal distribution; however, the keystone species concept introduces the idea of "interaction strength," which is assumed to be stronger for keystone species relative to other species (Mills et al. 1993).

The loss of a keystone species leads to a reduction in biodiversity. Eurasian beavers (*Castor fiber*) are considered keystone species because their dam-building activities support a higher diversity of invertebrates, increase fish density and biomass, provide breeding habitat for amphibians and reptiles, and provide forage and shelter for mammals (Janiszewski et al. 2014). Other examples of keystone species include sea urchins in kelp communities, pocket gophers in mountain meadows, termites in Brazil, and palm nuts in the tropics (Mills et al. 1993).

Trophic Cascade

A "trophic cascade" refers to the effect of a top predator on a lower trophic level via an intermediate species. A predator (carnivore) reduces the abundance of its prey (herbivore), which cascades down to the trophic level below the prey, increasing the abundance of the prey's resources (such as plants in terrestrial ecosystems or phytoplankton in aquatic habitats).

Consider populations of gray foxes, fox squirrels, and oaks (Figure 4). The effect of the gray fox population on the per capita growth of oaks is positive because it decrease the density of fox squirrels, which feed on oak seeds.

A long-term study of sea otters (*Enhydra lutris*) in the North Pacific (Estes and Duggins 1995) showed that their position as a top predator helped maintain a rich and stable kelp (large brown algae) forest community. Sea otters feed on sea urchins, which graze on the kelp that harbors many species. As in the gray fox example (Figure 4), the interaction between sea otters and sea urchins cascaded down along the food chain, with a positive impact on the kelp forest. However, since the 1990s, increased predation of otters by killer whales has resulted in a precipitous decline in sea otter abundance at several islands in the Aleutian archipelago. By 1997, sea urchin biomass and its grazing intensity on kelp forests increased eightfold, and kelp density decreased by a factor of twelve (Estes et al. 1998).

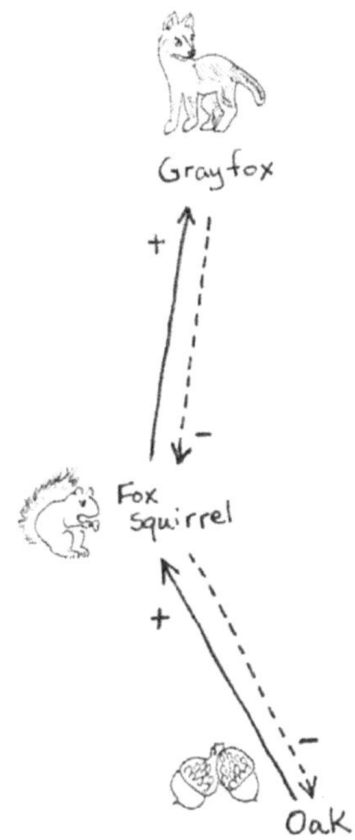

Figure 4. A trophic cascade occurs when a top predator indirectly influences lower trophic levels via intermediate species. A solid line represents a positive effect, and a dashed line indicates a negative effect. The effect of the gray fox population on the per capita growth of oaks is positive. (Follow the direction of arrows from fox to oak and apply simple algebra: negative times negative equals positive). The positive impact of the gray fox on the oak population is a top-down effect. The impact of the oak population on the gray fox is also positive. (Positive times positive equals positive.) This is a bottom-up effect.

In the three-trophic-level coastal community, the positive impact of sea otters on the growth of kelps was evident (Estes and Duggins 1995). However, when the coastal community changed to a four-trophic-level system because of increased predation by the topmost carnivore (killer whales), the abundance of carnivores in the next lower trophic level (sea otters) decreased. The cascading effects continued down to the herbivores in the lower level (urchins), which increased, and to the producers in the lowest level (kelp), which decreased. This cascading effect is not unique and has been seen in a variety of other communities.

In an experimental study in the tropical lowland forests of Costa Rica, specialist *Tarsobaenus* beetles were added experimentally to a three-trophic-level ant-herbivore-plant system (Dyer and Letourneau 1999). The average abundance of the herbivore-eating *Pheidole* ant was reduced fivefold, the average herbivory to *Piper* tree leaves almost tripled, and the tree leaf area was reduced by nearly half. Thus, the indirect impact of beetles on the bottom trophic level was negative $[(-) \times (-) \times (-) = (-)]$. From this and other examples, we can posit that plant biomass will be greater in odd-numbered trophic-level systems (even-numbered links) and less in even-numbered trophic-level systems (odd-numbered links).

As you might expect, humans, as "predators," can also cause trophic cascades. In the Canadian provinces of Newfoundland and Nova Scotia, annual Atlantic cod catches between 1850 and 1960 increased from 100,000 to 300,000 tons. Starting in the 1970s, larger fishing trawlers and other new technologies led to a rapid increase in the number of cod caught, with a peak catch of 800,000 tons. However, the cod population soon crashed, leading to the closure of the fishery in 1992 and the loss of a livelihood for over 30,000 Canadian fishermen. The ramifications of cod overfishing cascaded down through the ecosystem as well (Frank et al. 2005). The cod's prey (small fish and bottom-dwelling invertebrates) increased, zooplankton prey decreased, and phytoplankton producers increased. The rise in phytoplankton eventually lowered the ocean nitrate concentration, an essential nutrient. This chain event is a clear example of how human actions can induce a trophic cascade.

When predator-mediated effects on food webs were initially explored in the 1960s, indirect effects were thought to propagate only when changes in the population size of one species altered the density of another species through consumption (direct mortality). This, in turn, altered the densities of species in other trophic levels. This kind of effect is termed **density-mediated indirect effects**. For example, decreasing the density of sea stars in an intertidal community caused an increase in mussels, which excluded other species, such as snails, from occupying the limited open space on the rocks. In the unmanipulated community, the positive indirect effect of sea stars on snails occurs because the sea stars reduce the density of mussels (Paine 1966). However, more recent research has shown that trophic cascades also occur when one species causes changes in the *behavior* of another. This is called **trait-mediated indirect effects**. When a predator causes its prey to change its feeding behavior, it in turn alters how much food the prey consumes.

The reintroduction of wolves into Yellowstone National Park in the winter of 1995–1996 dramatically changed the behavior of elk, causing them to feed in safer areas and avoid habitats with a high risk of predation. This change in foraging behavior decreased plant consumption, although the density of elk did not change. Female elk and bison showed significantly higher vigilance in areas with wolves than in areas without wolves. The result was that elk behavior in the presence of wolves may have more far-reaching consequences for elk and bison ecology than the actual killing of individuals by wolves (Laundre et al. 2001, 2010). Trait-mediated indirect effects are aptly understood in the limelight of the "ecology of fear" (Brown et al. 1999), where elk are so scared by the presence of wolves that they change their foraging patterns and habitat choice.

The reintroduction of wolves into Yellowstone National Park has led to significant changes in the park's vegetation. Ripple and Beschta (2012) showed that the reintroduction of wolves restored a trophic cascade that existed before the extermination of wolves 70 years earlier, with woody browse species growing taller and canopy cover increasing in some areas. After wolf reintroduction, elk populations decreased, while both beaver and bison numbers increased, possibly due to the increased availability of woody plants and herbaceous forage resulting from less competition with elk (Ripple and Beschta 2012).

Trait-mediated indirect effects are not limited to mammals. Schmitz et al. (1997) conducted a unique experiment in a community of spiders (predator), grasshoppers (herbivore), and grasses (producer) to understand whether the fear instilled by the predator in their prey could produce trophic-level effects similar to those generated by direct predation. They glued the lethal spiders' mouthparts, rendering them ineffective, and observed the degree of herbivory in cages with spiders (without manipulation) and in cages where spiders had their mouthparts glued. In the presence of spiders, regardless of whether the spiders had their mouths glued shut, grasshoppers changed their diet to avoid areas that spiders liked to lurk and spent less time eating. The result was lower grasshopper densities, irrespective of whether the spiders could kill grasshoppers. Even with their mouths glued shut, the mere presence of spiders changed the behavior of the grasshoppers, which led to less foraging and a decrease in their populations. These effects trickled down to the plant community, increasing plant biomass in comparison to cages where only grasshoppers were present. Fear matters.

Two general patterns emerge from the principle of trophic cascades.

First: The top-down (predator) effect is usually negative on the lower trophic level because, by consuming (or scaring) prey, a predator reduces the number or biomass of prey. The bottom-up (prey) effect is always positive on the higher trophic level because prey supply nutrition and energy to predators, thereby helping to increase predator numbers or biomass.

Second: An odd trophic number (1, 3, etc.) tends to create a "green" world, whereas an even number of trophic levels (2, 4, etc.) tends to create a "brown" world. The odd trophic-level world is green because the trophic cascade produces a positive effect on producers. The even trophic-level world is brown because the trophic cascade produces a negative effect on producers.

When we revisit our Midwest oak-maple community (Figure 1), we observe that an increase in the abundance of robins causes a decrease in the population of grasshoppers, increasing the abundance of grasses. This is known as "the world is green" hypothesis, where the community is structured by top-down control. In 1960, Hairston, Smith, and Slobodkin surmised that, because most communities contain an abundance of vegetation, trophic groups must be controlled from the top of the food web; predators must keep the number of herbivores in check (Hairston et al. 1960). A "firestorm" of controversy followed their publication. Murdoch (1966) pointed out that not all "green" is edible, and Ehrlich and Birch (1967) argued that there are no universals in biological regulation. However, Slobodkin et al. (1967) rebutted that certain biological methods of regulation are characteristic of specific trophic levels (not populations). Their tentative answer to the question "Why is the world green?" was that the herbivore trophic level does not turn the world brown largely because there is a predator trophic level. Cases where human actions or natural events remove predators demonstrate they can control rodent plagues or insect outbreaks that destroy vegetation. One of the first studies of unambiguous top-down effects came from a bird-insect-plant association. Marquis and Whelan (1994) excluded birds from some white oak trees by netting (cage treatment), while letting other trees remain available to the birds (control treatment). Birds significantly reduced the abundance of herbivorous insects on the white oaks to such an extent that oaks had less leaf damage from insects and subsequently attained a higher biomass.

However, another valid way of looking at our food web in Figure 1 is that an increase in grasses causes an increase in the grasshopper population, which in turn causes an increase in the number of robins. This is the viewpoint of bottom-up control over the abundance of trophic groups. As Murdoch (1966) pointed out, the alternative view of bottom-up control asserts that not all plant material is edible: The nutritional quality or anti-herbivore structural and chemical defenses of plants suggest a bottom-up form of regulation that limits herbivore resources. It would be naïve to view plants as passive, waiting for animals to devour them. We must remember that evolution plays a role here. Any plant that is slightly less desirable will have an advantage. Plants have evolved physical, chemical, morphological, and behavioral defenses against being attacked, including leaves modified into spines (e.g., holly plants, cotoneaster bushes), leaf surfaces clothed in epidermal hairs (e.g., African violet), leaves steeped in toxins (e.g., tobacco plant), or parts with little nutritional value (e.g., woody stems). Thus, even without the help of predators to keep herbivore numbers low, the world would still be filled with populations of plant species that can persist despite being consumed by voracious herbivores. As Pimm et al. (1991) aptly wrote: "The world is prickly and tastes bad."

Clearly, the dichotomy between top–down and bottom–up control of abundance is a simplification of the complexities of the real world. In a recent revisit of "the world is green hypothesis", Wilkinson and Sherratt (2016) explain that this "marvel of nature" (where plants are not consumed to the point where they can't support herbivores; Crawford 1989) is possible because many plant communities are simultaneously controlled both from the top down by predators (herbivores) and from the bottom up by resources. They conclude that bottom-up processes are important for explaining the presence of vegetation and that top-down forces explain the type of vegetation (Wilkinson and Sherratt 2016).

In eastern Venezuela, Terborgh et al. (2006) investigated six islands created as a result of a large hydroelectric project to provide electricity for millions. Not long ago, most of this area was an unbroken expanse of green dominated by top predators such as jaguars and harpy eagles. But after the project, the newly created islands were too small to support viable populations of medium-sized mammals, such as armadillos and various primate species, that consume invertebrate herbivores such as leaf-cutter ants. The six small islands supported much lower densities of tree saplings than larger islands because there were no predators to reduce the defoliating ant populations. However, the predator-free islands with higher leaf-cutter ant densities, according to Terborgh and the team, would eventually become dominated by ant-resistant lianas. Their results are consistent with the view that top-down forces (the absence of major predators) do not lead to a plant-free system (a brown world), but one that is still green, albeit dominated by very different types of plants.

Indirect Mutualism

An indirect mutualism is the positive indirect effect that two predators have on each other via distinct prey species that happen to compete for a common resource (Figure 5).

Indirect mutualism occurs, for example, when gray squirrels and chipmunks feed on different species—oak acorns and maple seedlings, respectively—but the maple trees and oak trees compete for space, soil nutrients, water, and access to sunlight through the forest canopy. By eating oak acorns, gray squirrels negatively affect the oak population, which in turn has a positive effect on the maple population (fewer oaks mean more resources for maples). An increase in the population size of maples benefits chipmunks, because more food is available to them. The reverse is also true; the chipmunk benefits from the gray squirrel eating the oak acorns because it allows the competing maple population to increase.

Overall, in nature, direct interactions are readily understood, while indirect interactions can seem surprising or even counterintuitive.

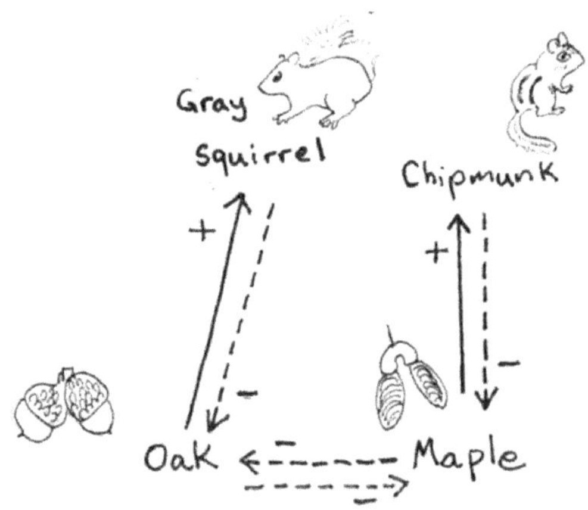

Figure 5. Indirect mutualism is manifested when there are two predators and two competing prey species. A solid line represents a positive effect, and a dashed line indicates a negative effect. The effect of the gray squirrel population on the per capita growth of the chipmunks is positive (negative times negative times positive equals positive) and vice versa.

Indirect Effects and Evolution

Largely because it is simpler and less costly in time and resources, most studies exploring the role of species interactions on community structure focus on interactions between two or a few species. Because only a few species are involved, these types of studies underestimate the importance of multiple interactions responsible for shaping overall community structure and dynamics. Food webs can contain strong direct interactions, such as those between plants and their pollinators, but many direct interactions can be diffuse, where one species (like the snowshoe hare) is influenced by interactions with many different species, such as the lynx, coyote, and horned owl (e.g., Tyson et al. 2010). Food webs also create indirect interactions, and as we've seen, changes in the abundance of one species can lead to changes in the abundance of others outside the direct link, as well as shifts in habitat use and behavior. The effects on one population's density extend in all directions through its direct interactions with predators, competitors, and prey. However, a population may also be influenced by its predator's predator, competitor's competitor, prey's prey, and so on. If population size is affected through indirect interactions, do these interactions also influence the evolution of that population?

Miller and Travis (1996) argue that evidence to evaluate indirect evolutionary effects would ideally come from evolutionary changes in a focal species when subjected to different combinations of direct and indirect effects. Walsh (2013) presents a collection of this evidence, including genetic responses to indirect effects found in experiments involving peas, aphids, and ladybird beetles

(Astles et al. 2005). Selective responses to indirect effects are also found in wild cotton, with selection for extrafloral nectaries that attract ants but exclude herbivores (Rudgers and Strauss 2004). This even includes evolutionary responses to indirect effects of predators and competitors observed in a laboratory setting with protozoa (terHorst 2010). The evidence thus indicates that indirect food web interactions can also influence evolution.

Food Webs and Community Structure

Communities, unlike individuals, are not directly under the influence of natural selection. They do not adapt like individual organisms and pass on selected traits. However, we can view them as entities with distinct structural characteristics, including species richness, productivity, food web complexity, and food chain length. These in turn might be expected to influence the stability, resiliency, and functional efficiency of the community as a whole.

There are different ways to consider and measure the structural complexity of food webs. One way is by the mean food chain length. The length of any one food chain within a food web is the number of links between a top predator and the base of the food web (basal species). If, for instance, there is a plant, an herbivore, and a carnivore, there are two links between three species, and the chain length is two. The mean chain length is the average of the lengths of all the (unlooped) chains in a food web (Williams and Martinez 2000). A second measure of structural complexity of food webs is by something called "connectance" (Williams and Martinez 2000). Intuitively, the maximum number of links that can occur in a food web will be a direct function of species richness (S). A theoretical maximum number of links for a food web that has S species, assuming that each species may link to every other species (including itself), is S^2. Connectance (C) is the ratio of the actual number of links (L) to the theoretical maximum number (S^2) or $C = L/S^2$. For example, if a community has $S = 7$ species, the maximum number of links is $S^2 = 49$, but if it has $L = 8$ links, then its connectance $C = L/S^2 = 8/49 = 0.16$. This can be compared to a community with 4 species ($S^2 = 16$) and 5 links, which has a connectance $C = 5/16 = 0.31$. How does connectance matter in food webs? Does a greater connectance lead to more stability?

Stability is the ability of a system to return to an equilibrium point. Think of stability as a ball in a bowl—move the ball up the side and it will roll back to the bottom, which is the stable point. **Resiliency** is the speed with which the system returns to its initial state (think of a ball on a platter having low resilience and a ball on a mixing bowl having high resilience). Are some food web structures more stable or resilient than others? This question is not only of theoretical interest (e.g., are there limits to food chain length or web complexity?), but it also has conservation relevance because some types of communities may be more susceptible to human and natural disturbance or perturbations than others, and therefore may require more conservation efforts.

Classical ecological wisdom held that increased complexity within a community leads to increased stability (MacArthur 1955, Elton 1958). In more complex communities, with more species and more interactions, there are more possible pathways by which energy can pass through the community. So, if, for instance, a disturbance changed the density of one species, this would have relatively little effect on the densities of other species in the community because energy can still pass using other pathways. A complex community would therefore be resistant to change. However, this classical view was challenged by later attempts to mathematically explore the question of community complexity versus community stability (May 1972; Pimm 1982). In models where predator density influences prey density and vice versa (i.e., Lotka–Volterra models), disturbed food webs with higher connectance generally had lower stability. However, there are a couple of interesting exceptions to this when the model had bottom-up control, such that the food supply was not affected by the consumers (nectar for hummingbirds is a good example), and increased connectance enhanced stability (DeAngelis 1975; Pimm 1982). Another exception occurred in stable modeled communities, where increased connectance improved *resilience* (Pimm 1982). Furthermore, in a study of marine food webs off the coast of Africa, in the Caribbean, and the coastal shelf in the northeastern United States, Dunne et al. (2004) found that communities with increased connectance exhibited greater robustness to species loss. **Robustness** refers to the resistance of a system to collapse when species are removed.

Food web complexity is not the only feature of a community that may impact its stability and functionality. Over one hundred years ago, Elton (1927) introduced the term **food chain** in his book *Animal Ecology* and emphasized the use of food webs and food chains as important summaries of community patterns, posing questions about the limits of food-chain length that still intrigue ecologists. Elton also observed that, in general, typical predators are larger than their prey. Predators, he reckoned, need large habitats, which could ultimately impose limits on the length of food chains. Elton noted that small organisms at the base of the food chain are more numerous than their larger predators, and these predators are more numerous than their predators further up the chain, and so on. This is known as the Eltonian pyramid. While this generalization does not apply to systems where large primary producers, such as trees, are foraged by much smaller and more numerous herbivores, including aphids or other insects, similar pyramids can be produced when considering biomass or productivity.

In highly productive systems where carbon accumulation per area per unit time is very high, inverted pyramids of numbers or biomass can occur. In these cases, the abundance or biomass of a lower trophic level is *less* than that of an adjacent higher trophic level. For example, in relatively clear oligotrophic lakes, herbivorous zooplankton reduce phytoplankton to very low levels of abundance or biomass (Del Giorgio et al. 1999). These inverted pyramids occur due to efficient energy transfer, where high turnover rates of phytoplankton support a large standing biomass of consumers. Inverted pyramids are also reported in pristine coral reef ecosystems, where apex predators (sharks) outnumber their fish prey (Friedlander and DeMartini 2002). Predator-prey models of these communities reveal the importance of fear responses as a way for prey to cope with high biomass of big, fierce predators. Sharks play an important role in coral reef community structure not only because they eat prey, but also because they scare their prey (Malone et al. 2020).

A part of Elton's (1927) ecological legacy was establishing the notion of the food web and its role in community structure, based primarily on the direct predator-prey interactions between species. However, Lindeman (1942; see also Cook 1977), in an important contribution to the understanding of food webs in relation to community structure and functioning, viewed the interacting species embedded in a food web from the perspective of energy and nutrient flow, thereby creating a common currency for studying and comparing disparate ecosystems. In Cedar Creek Bog, a senescent lake north of Minneapolis, Minnesota (USA), Lindeman quantified energy flow through the producers, primary consumers, and secondary consumers in the system. Lindeman introduced the idea of ecological efficiency, a measure of the fraction of energy entering one trophic level that is passed on to the next higher trophic level. Energy transfer between trophic levels, according to Lindeman, is inefficient, typically ranging from 5–15%. This inefficiency of energy transfer between trophic levels provides one possible explanation for the limited length of food chains, as relatively little energy remains after passing through four or five trophic levels (Slobodkin 1960). We, therefore, might expect that if energy limits food chain length, increasing productivity should lead to an increase in the number of trophic levels.

In experimental manipulations of productivity, Pimm and Kitching (1987) tested the effects of variation in productivity on the relatively simple food webs that develop in water-filled tree holes in tropical Australia: Detritus (leaf litter)—larval mosquitoes—larvae of predatory midge—predator tadpoles. They concluded that increasing productivity decreases food-chain length. In the early 1970s, Rosenzweig (1971), analyzing several predator-prey models, suggested that population dynamics become less stable as systems become more productive and prey populations increase. His models revealed the "paradox of enrichment," explaining why productive systems might have shorter rather than longer food chains (see Chapter 4 *Species Interaction: Predation*).

An alternative to the productivity–chain length hypothesis argues that food chains should be shorter in unpredictable environments. Pimm and Lawton (1977) base this argument on the return times (resiliency) following disturbance of modeled food chains. The longer the food chain, the longer the return time, implying an increased chance of extinction in an unpredictable environment that may not allow time for recovery. This might explain why, in a survey of natural lakes, Post et al. (2000) found no relationship between food-chain length and productivity, but that larger lakes tended to support longer food chains independent of any differences in productivity. Is it possible that larger lakes have more predictable environments than smaller ones, which can be subject to extreme heat, cold, and evaporation?

If it feels like there is no simple rule of thumb to understand the structure of food webs, it is because there is not any. Today's ecologists realize that a complex and contingent framework of interacting constraints governs food chain length and complexity. This framework includes resource availability (energy-flow), the history of community organization, the type of predator–prey interactions, disturbance, and ecosystem size. Instead of searching for singular explanations, the trend now is to identify which constraints are in operation within particular food webs (Post 2002). For example, Ward and McCann (2017), in considering aquatic food chain length, show that productivity and ecosystem size drive food chain length, but their effects are context dependent in lake and marine ecosystems. They demonstrate that ecosystem size governs food chain length in low-productivity ecosystems, while productivity is the driver in large and highly productive ecosystems.

There also appears to be no simple rule, like the number of connections, explaining stability in complex food webs. Drossel et al. (2001) posit that food web "evolution" is the missing explanation. They consider the long-term development of food web structure over time (food web evolution) to be a function of "true" biological evolution occurring at the level of individuals. In real food webs, chain link strengths and linkage patterns can change through adaptive evolutionary processes, which can happen quite quickly (in the span of 100 years; Thompson 1998). Drossel et al.'s (2001) population dynamic models with an evolutionary component show that this process results in complex food webs with a large proportion of weak links, which help stabilize the population dynamics.

Other food web complexities not shown in Figure 1 are parasites and pathogens. How do parasites fit into food webs? Can you guess how parasites might change relationships within a food web? Lafferty et al. (2008) call our attention to evidence that parasites can alter food-web characteristics such as chain length, connectance, robustness, interaction strength, and energy flow. Food-web structure, in turn, is expected to affect infectious disease dynamics. While adding parasites and pathogens to a food web might seem straightforward, Lafferty et al. (2008) suggest that parasites may affect hosts differently than predators affect prey, and that parasites, especially with their complex life cycles, may act differently than pathogens in food webs. Regardless of how they are incorporated, the authors suggest that considering parasites and pathogens can improve our understanding of both food webs and infectious disease dynamics.

Since the time of Carl Linnaeus, who, in his essay *The Economy of Nature*, briefly itemized the stages of two food chains (one terrestrial and another aquatic), unraveling the patterns in food webs has been an unrelenting quest to understand their nature and origin (Egerton 2007). Despite the failure to find an all-encompassing explanation, the search for simple patterns in seemingly complex food webs has yielded a wealth of practical knowledge, including trophic control of pests, indirect effects of increased harvest, nutrient cycling, and the bioaccumulation of environmental contaminants. Moreover, it has inspired the union of disparate disciplines such as population dynamics, community ecology, energy flow, foraging ecology, and eco-evolutionary dynamics, to name a few, in the contemplation of an entangled bank—an idea Darwin (1859) illustrates so aptly: "clothed with many plants of many kinds, with birds singing on the bushes, with various insects flitting about, and with worms crawling through the damp earth, and to reflect that these elaborately constructed forms, so different from each other, and dependent on each other in so complex a manner, have all been produced by laws acting around us."

References

Astles, P. A., Moore, A. J. and Preziosi, R. F. 2005. Genetic variation in response to an indirect ecological effect. Proceedings of the Royal Society B 272: 2577–2581.
Brown, J. H. and Davidson, D. W. 1977. Competition between seed-eating rodents and ants in desert ecosystems. Science 196: 880–882.
Brown, J. S., Laundré, J. W. and Gurung, M. 1999. The ecology of fear: optimal foraging, game theory, and trophic interactions. Journal of Mammalogy 80: 385–399.
Cook, R. E. 1977. Raymond Lindeman and his trophic-dynamic concept in ecology. Science 198: 22–26.
Crawford, R. M. M. 1989. Studies in Plant Survival. Blackwell Scientific Publications, Oxford.
Darwin, C. 1859. On The Origin of Species by Means of Natural Selection. John Murray, London, UK.
DeAngelis, D. L. 1975. Stability and connectance in food web models. Ecology 56: 238–243.
Del Giorgio, P. A., Cole, J. J., Caraco, N. F. and Peters, R. H. 1999. Linking planktonic biomass and metabolism to net gas fluxes in northern temperate lakes. Ecology 80: 1422–1431.
Drossel, B., Higgs, P.G. and McKane, A. J. 2001. The influence of predator-prey population dynamics on the long-term evolution of food web structure. Journal of Theoretical Biology 208: 91–107.
Dunne, J. A., Williams, R. J. and Martinez, N. D. 2004. Network structure and robustness of marine food webs. Marine Ecology Progress Series 273: 291–302.
Dyer, L. A. and Letourneau, D. K. 1999. Trophic cascades in a complex terrestrial community. PNAS 96: 5072–5076.
Egerton, F. N. 2007. Carl Linnaeus and the economy of nature. ESA Bulletin 88: 72–88.
Ehrlich, P. R. and Birch, L. C. 1967. The "balance of nature" and "population control." The American Naturalist 101: 97–107.
Elton, C. S. 1927. Animal Ecology. Sidgwick and Jackson, London, UK.
Elton, C. S. 1958. The Ecology of Invasions by Animals and Plants. University of Chicago, Chicago, IL.
Estes, J. A. and Duggins, D. O. 1995. Sea otters and kelp forests in Alaska: generality and variation in a community ecological paradigm. Ecological Monographs 65: 75–100.
Estes, J. A., Tinker, M. T., Williams, T. M. and Doak, D. F. 1998. Killer whale predation on sea otters linking oceanic and nearshore ecosystems. Science 282: 473–476.
Frank, K. T., Petrie, B., Choi, J. S. and Leggett, W. C. 2005. Trophic cascades in a formerly cod dominated ecosystem. Science 308: 1621–1623.
Friedlander, A. M. and DeMartini, E. E. 2002. Contrasts in density, size, and biomass of reef fishes between northwestern and the main Hawaiian islands: the effects of fishing down apex predators. Marine Ecology Progress Series 230: 253–264.
Hairston, N. G., Smith, F. E. and Slobodkin, L. B. 1960. Community structure, population control, and competition. The American naturalist 94: 421–425.
Höfer, H. and Brescovit, A. D. 2001. Species and guild structure of a Neotropical spider assemblage (Araneae) from Reserva Ducke, Amazonas, Brazil. Andras 15: 99–119.
Holt, R. D. 1977. Predation, apparent competition, and the structure of prey communities. Theor. Popul. Biol. 12: 197-229.
Holt, R. D. and Bonsall, M. B. 2017. Apparent competition. Annu. Rev. Ecol. Evol. Syst. 48: 447-71.
Janiszewski, P., Hanzal, V., and Misiukiewicz, W. 2014. The Eurasian beaver (*Castor fiber*) as a keystone species-a literature review. Baltic Forestry 20: 277–286.
Jeffries, M. J. and Lawton, J. H. 1984. Enemy free space and the structure of ecological communities. Biological Journal of the Linnean Society 23: 269–286.
Karban, R., Hougen-Eitzmann, D. and English-Loeb, G. 1994. Predator-mediated apparent competition between two herbivores that feed on grapevines. Oecologia 97: 508–511.
Kissling, W. D., Sekercioglu, C. H. and Jetz, W. 2012. Bird dietary guild richness across latitudes, environments and biogeographic regions. Global Ecology and Biogeography 21: 328–340.
Lafferty, K. D., Allesina, S., Arim, M., Briggs, C. J., De Leo, G., Dobson, A. P. et al. 2008. Parasites in food webs: the ultimate missing links. Ecology Letters 11: 533–546.
Laundre, J. W., Hernandez, L. and Altendorf, K. B. 2001. Wolves, elk, and bison: reestablishing the "landscape of fear" in Yellowstone National Park, U.S.A. Canadian journal of Zoology 79: 1401–1409.
Laundre, J. W., Hernández, L. and Ripple, W. J. 2010. The landscape of fear: ecological implications of being afraid. The Open Ecology Journal 3: 1–7.
Lešo, P. and Kropil, R. 2007. A comparison of three different approaches for the classification of bird foraging guilds: an effect of leaf phenophase. Folia Zool. 56: 51–70.
Lindeman, R. L. 1942. The trophic-dynamic aspect of ecology. Ecology 23: 399–417.
MacArthur, R. 1955. Fluctuations of animal populations and a measure of community stability. Ecology 36: 533–536.
Malone, M. A., Halloway, A. H. and Brown, J. S. 2020. The ecology of fear and inverted biomass pyramids. Oikos 129: 787–798.

Marquis, R. J. and Whelan, C. J. 1994. Insectivorous birds increase growth of white oak through consumption of leaf-chewing insects. Ecology 75: 2007–2014.
May, R. M. 1972. Will a large complex system be stable? Nature 238: 413–414.
Miller, T. E. and Travis, J. 1996. The evolutionary role of indirect effects in communities. Ecology 77: 1329–1335.
Mills, L. S., Soulé, M. E. and Doak, D. F. 1993. The keystone-species concept in ecology and conservation. BioScience 43: 219–224.
Murdoch, W. W. 1966. Community structure, population control and competition—a critique. The American Naturalist 100: 219–226.
Paine, R. T. 1966. Food web complexity and species diversity. The American Naturalist 100: 65–75.
Paine, R. T. 1969. A note on trophic complexity and community stability. The American Naturalist 103: 91–93.
Pimm, S. L. 1982. Food webs. Population and community biology series. Chapman and Hall, New York.
Pimm, S. L. 1991. The Balance of Nature: Ecological Issues in the Conservation of Species and Communities. University of Chicago Press, Chicago.
Pimm, S. L. and Kitching, R. L. 1987. The determinants of food chain lengths. Oikos 50: 302–307.
Pimm, S. L. and Lawton, J. H. 1977. Number of trophic levels in ecological communities. Nature 268: 329–331.
Pimm, S. L., Lawton, J. H. and Cohen, J. E. 1991. Food web patterns and their consequences. Nature 350: 669–674.
Polis, G. A. 1991. Complex trophic interactions in deserts: an empirical critique of food-web theory. The American Naturalist 138: 123–155.
Polis, G. A., Anderson, W. B. and Holt, R. D. 1997. Toward an integration of landscape and food web ecology: the dynamics of spatially subsidized food webs. Annual Review of Ecology and Systematics 28: 289–316.
Post, D. M. 2002. The long and short of food-chain length. Trends in Ecology & Evolution 17: 269–277.
Post, D. M., Pace, M. L. and Hairston, N. G. Jr. 2000. Ecosystem size determines food-chain length in lakes. Nature 405: 1047–1049.
Power, M. E., Tilman, D., Estes, J. A. Menge, B, A., Bond, W. J., Mills, L. C. et al. 1996. Challenges in the quest for keystones. BioScience 46: 609–620.
Ripple, W. J. and Beschta, R. L. 2012. Trophic cascades in Yellowstone: The first 15 years after wolf reintroduction. Biological Conservation 145: 205–213.
Root, R. B. 1967. The niche exploitation pattern of the blue-gray gnatcatcher. Ecological Monographs 37: 317–350.
Rosenzweig, M. L. 1971. Paradox of enrichment: destabilization of exploitation ecosystems ecological time. Science 171: 385–387.
Rudgers, J. A. and Strauss, S. Y. 2004. A selection mosaic in the facultative mutualism between ants and wild cotton. Proceedings of the Royal Society B 271: 2481–2488.
Schmitz, O. J., Beckerman, A. P. and O'Brien, K. M. 1997. Behaviorally mediated trophic cascades: effects of predation risk on food web interactions. Ecology 78: 1388–1399.
Shrestha, R. and Wegge, P. 2008. Wild sheep and livestock in Nepal Trans-Himalaya: coexistence or competition? Environmental Conservation 35: 125–136.
Slobodkin, L. B. 1960. Ecological energy relationships at the population level. The American Naturalist 94: 213–236.
Slobodkin, L. B., Smith, F. E. and Hairston, N. G. 1967. Regulation in terrestrial ecosystems, and the implied balance of nature. The American Naturalist 101: 109–124.
Small, R. J. and Keith, L. B. 1992. An experimental study of red fox predation on arctic and snowshoe hares. Canadian Journal of Zoology 70: 1614–1621.
Strong, J. S. and Leroux, S. J. 2014. Impact of non-native terrestrial mammals on the structure of the terrestrial mammal food web of Newfoundland, Canada. PLoS ONE 9: e106264.
Terborgh, J., Feeley, K., Silman, M., Nuñez, P. and Balukjian, B. 2006. Vegetation dynamics of predator-free land-bridge islands. The Journal of Ecology 94: 253–263.
terHorst, C. P. 2010. Evolution in response to direct and indirect ecological effects in pitcher plant inquiline communities. The American Naturalist 176: 675–685.
Thompson, J. N. 1998. Rapid evolution as an ecological process. Trends in Ecology & Evolution 13: 329–332.
Tyson, R., Haines, S., and Hodges, K. E. 2010. Modelling the Canada lynx and snowshoe hare population cycle: the role of specialist predators. Theoretical Ecology 3: 97–111.
Vijayan, S., Morris, D. W. and McLaren, B. E. 2012. Prey habitat selection under shared predation: tradeoffs between risk and competition? Oikos 121: 783–789.
Walsh, M. R. 2013. The evolutionary consequences of indirect effects. Trends in Ecology & Evolution 28: 23–29.
Ward, C. L. and McCann, K. S. 2017. A mechanistic theory for aquatic food chain length. Nature Communications 8: 2028. doi: 10.1038/s41467-017-02157-0.
Wilkinson, D. M. and Sherratt, T. N. 2016. Why is the world green? The interactions of top-down and bottom-up processes in terrestrial vegetation ecology. Plant Ecology & Diversity 9: 127–140.
Williams R. J. and Martinez, N. D. 2000. Simple rules yield complex food web. Nature 404: 180–183.

Chapter 9

Ecological Succession

"...I have great faith in a seed" (Thoreau 1860)

"One thing leads to another" (The Fixx 1983)

Ecological succession is chiefly about vegetational communities, about relative abundance, and most essentially about "change," though not random change. It explains patterns of change that seem tantalizingly predictable. If a fire burns in a lodgepole pine (*Pinus contorta*) forest in the Rocky Mountains (USA), the bare ground is first covered by fire grass (*Agrostis* sp.), then dominated by fireweed (*Epilobium angustifolium*), followed by aspen (*Populus* sp.), and finally returns to a lodgepole pine forest (Clements 1916).

Observations of the natural replacement of one plant community by another are not new. Daubenmire (1968) credits the first written observations of this phenomenon (vegetation sequences in fields and flood plains) to Plato's student Theophrastus in the third century B.C., but it was not until a couple thousand years later that Clements (1916; quoted in Egerton 2015) credited Jean André Deluc (1727–1817) with first coining the term **succession**. The naturalist philosopher Henry David Thoreau (1817–1862) was clearly aware of this idea and used the term succession to explain the transition to forests from abandoned farms and pastures in New England (USA). Thoreau also observed that when pines were logged, oaks replaced them, and vice versa. Pines could not replace cut pines, Thoreau explained, because oak seedlings were already growing there. Furthermore, since pine seeds blew annually into oak forests, they were ready to germinate when oaks were cut. In an address to members of the Middlesex Agricultural Society, Massachusetts, Thoreau stated: "Though I do not believe that a plant will spring up where no seed has been, I have great faith in a seed" (Thoreau 1860). However, it was not until the late 19th and early 20th centuries that the nascent ecological sciences benefited from the rapidly accumulated observations, interpretations, and theories of succession by the foundational powerhouses H. C. Cowles (1899) and F. E. Clements (1916). Cowles formalized the study of vegetational succession using Lake Michigan's sand dunes, while Clements introduced the concept of succession as an orderly process, moving from bare earth through a sequence of "seres" (plant communities as units of succession) to a "climax"—the stable vegetational sere at the "end" of succession.

In describing succession as a vegetational sequence beginning with bare earth, Clements (1916) observed that the initial state of succession can arise in habitats that are either "originally bare" or "bare by denudation." We now describe these as **primary** and **secondary succession**. Both cases rely on some form of disturbance, which can be brought about by abiotic factors (e.g., retreating glaciers, drought, flood, fire, volcanism, etc.) or biotic factors (e.g., digging gophers, grazing elephants, human agriculture, etc.). Disturbances can also be described in terms of amplitude (severity) and frequency. In general, severe disturbances are rarer and often obliterate the existing organic matter, while modest disturbances are more frequent and often leave organic matter intact despite removing the existing vegetation. However, this is not a general rule, as severe disturbances such as hurricanes or floods often enhance the organic content of the "bare earth" they leave behind, and the simple

DOI: 10.1201/9781003008538-9

creation of a child's sandlot has no organic content at all. Thus, the type of disturbance does not matter as much as the type of "bare earth," which sets the initial conditions that ultimately lead to either primary or secondary types of succession. Daubenmire (1968), typical of 20th-century textbooks, defines "primary bare areas" as unweathered parent material, the surface of which has never supported plants, while "secondary bare areas" retain the soil ecosystem (humus) and, in some cases, the seeds and roots of the previously existing vegetation. In both cases, "succession" refers to the colonization of this bare earth and its subsequent change in biotic communities over time.

Primary Succession

Primary succession is distinct from secondary succession in its progression from a nonliving substrate to living soil with the accumulation of organisms and organic matter. For primary succession, we define the initial phase of "bare earth" as a mineral substrate devoid of organic matter. Bare mineral soil may seem like a rather extreme condition on Earth, with its teeming biosphere, but there are at least four main types of disturbances can completely obliterate the physical environment (organic soil) and all life on it, creating conditions for primary succession to occur.

Extraterrestrial Impact Craters

When sizable asteroids, comets, or meteorites hit Earth's surface, they kill all life in the vicinity of the impacted area, and the impact crater becomes sterile due to temperatures reaching up to 15,000 K (Cockell and Lee 2002). Listing well over 50 impact craters in a suite of biomes, Cockell and Lee (2002) proposed that such impacts result in a generalized sequence of primary succession, beginning with the sterilizing impact event, followed by a first phase of hydrothermal microbial activity. During this phase, water (preferentially trapped by crater walls) often forms lakes. This is a transitory phase, as the crater cools, forming the second phase, which may be complex and involve succession to a climax lake community containing life forms such as diatoms, snails, and fish. This is followed by a subsequent succession and climax community of vegetation after the lake waters recede. In general, these climax vegetation communities will form an intra-crater "oasis" because of the moist hydrologic depression created by the crater (Cockell and Lee 2002). In the final phase, however, the crater erodes and fills, losing the hydrological depression, resulting in an ecosystem that becomes indistinguishable from the surrounding region. This reversion occurs over a long period and may be seen as part of Earth's geological dynamics.

Volcanoes

Newly erupted lava—creating both new land and ash that buries and smothers existing vegetation—consists of sterile volcanic rock that may contain nutrient-rich metals and minerals but is otherwise devoid of organic matter. Primary succession has been documented around the world where active volcanism has occurred. One of the longest-studied tropical successions is that of the Krakatau Islands in Indonesia (Whittaker et al. 1989). These islands were first "sterilized" in 1883 by a series of eruptions, and scientists of the period documented the new soils and followed the colonization of barren ground on three of the islands. About 100 years later, each of the islands had experienced similar vegetational successions, starting with an initial colonization by wind-dispersed photosynthetic bacteria and ferns on a surface of little-altered ash. This was followed by "pioneer" grasses taking advantage of changing soils and microclimates, the development of a savanna grassland, and then the accumulation of bird or bat-dispersed trees, forming a forest community of more shade-tolerant and water-loving species (Whittaker et al. 1989). Whittaker et al. (1989) found that where there were differences in succession, these could be accounted for by the presence of nearby sources of propagules and animal dispersal agents.

The deadliest and economically most destructive volcanism in the continental United States during recent times was the eruption of Mount St. Helens on 18 May 1980. The initial eruption produced both air-fall and pyroclastic flows of hot volcanic gases, ash, and pumice, which killed 57 people, destroyed an estimated billion dollars in infrastructure, and obliterated 600 square kilometers of forests (Peterson 1990). In areas of pyroclastic flow and avalanche debris, almost no organisms survived (Dale et al. 2005a).

A major factor impeding early primary succession on the sterile ash-fall on Mount St. Helens was the distance from potential colonists. Once propagules arrived, colonists were able to establish themselves only in favorable microsites where more nutrients and moisture collected (del Moral and Wood 1993). Lupines (*Lupinus*), legumes with roots harboring a special association with nitrogen-fixing bacteria, were critical to the enrichment of the nutrient-poor pyroclastic ash deposits, allowing other species to colonize (Halvorson et al. 2005). Comparisons of soil carbon and nitrogen, twenty years after the eruption, showed significantly increasing differences in nutrient concentrations and microbial biomass between lupine-colonized and uncolonized soils (Halvorson et al. 2005). Because lupines can survive low nutrient conditions by fixing nitrogen, they subsequently acted as "resource islands," which accumulated and built soil, while also providing ameliorating local conditions, trapping seeds, and providing habitat for other organisms (Halvorson et al. 2005).

The year 2015 marked the 35th anniversary of the Mount St. Helens eruption, and reflecting on this, Dale and Crisafulli (2018) summarized what had returned and what had been learned about primary succession up to that point. First, the explosive event was complex, and the mechanisms of its disturbance varied in type, intensity, and severity, resulting in a mosaic of initial conditions. These included survivors such as *Chamerion* fireweed (from seed), *Ambystoma* salamanders and *Anaxyrus* toads (in burrows or frozen ponds), *Thomomys* pocket gophers (in burrows), and *Camponotus* ants (in logs or underground). Second, the ecological response varied spatially and temporally across landscape features, with early plant colonists including nitrogen-fixing lupine (*Lupinus lepidus*), penstemon (*Penstemon cardwellii*), paintbrush (*Castilleja miniata*), *Racomitrium* moss, willow (*Salix sitchensis*), alder (*Alder viridis*), and Douglas fir (*Pseudotsuga menziesii*). Similarly, early animal colonists included elk (*Cervus elaphus*), deer mice (*Peromyscus maniculatus*), white-crowned sparrows (*Zonotrichia leucophrys*), *Melanoplus* grasshoppers, and carabid beetles (*Calostoma tepidum*). Finally, some areas of vegetation have returned to forest, vertebrate species have increased in diversity and population sizes, birds have recolonized, and nearly all species of small mammals have returned (Dale et al. 2005a). The Mount St. Helens disaster taught ecologists that succession is complex, proceeding at varying paces along diverse paths, checkered with periodic disturbances and random events. All of these factors impact species accrual and community development occurring throughout succession (Dale and Crisafulli 2018).

Sand Dunes

One of the first studies of primary succession (Cowles 1899) was conducted on the dunes of Lake Michigan (USA) in what is now Indiana Dunes National Park. Although the accumulation and drifting of sand is not as dramatic a disturbance as a meteorite or volcano, it provides an example of primary succession because piles of windblown sand can smother preexisting habitat. A sand dune is itself a barren medium that requires organisms to colonize it from elsewhere. In 1899, Henry Cowles recorded differences in vegetation on dunes at different stages and inferred successional patterns from them. Plant succession on dunes, according to Cowles, could be characterized as stages of development, passing from (1) bare sand colonized by annual and biennial succulents, grasses, and forbs, to (2) stationary beach dune, which forms as perennial grasses and certain shrubs or cottonwoods colonize the area and can tolerate the dry, exposed habitat, thereby securing the sand from movement, to (3) wandering dunes, which occur when the dune-forming plants, although initially stabilizing the sand, collect more sand and build the dunes so high that they can no longer hold the sand in place. The process finally evolves to (4) the final stage, the established dune, which occurs when vegetation starts taking hold in lower places protected from wind and where there is higher moisture, allowing it to grow up the slope and "capture" the dune. Once a dune has reached the "established dune" stage, vegetative communities then pass through successive associations, culminating in the "normal climax" forest of oak and pine.

Glaciers

Glaciers are the result of a net accumulation of snow during cooler global temperatures, generally known as "Ice Ages." Given enough time, the snow gradually compacts and forms colossal rivers of ice, the weight of which scours anything underneath down to bare rock. When warmer temperatures prevail, as for example, during the "interglacial" periods of the Ice Age, there is a net loss of this frozen water due to this warming, and melting glaciers "retreat," exposing the barren substrate or scoured rock where no life has existed for thousands of years. When a glacier retreats, the colonization of life ensues. As ecologists studying Mount St. Helens attest (Dale et al. 2005a), primary succession after a disturbance may take more than just a decade or two to play out, influencing ecological communities for centuries to come and easily exceeding the life of an ecologist. Fortunately, ecologists can gain information over longer time scales if the different stages of the succession can be observed in terms of community gradients in space. This is called a "space-for-time" substitution or **chronosequence**. Glaciers, due to their linear direction of retreat, often appear to display a chronosequence of successional stages, from newly bare earth at the terminus to areas now kilometers away, first exposed hundreds of years earlier and now covered by vegetation.

William Cooper, one of Henry Cowles' students, was the first to study succession at Glacier Bay National Park, Alaska, where bare areas were rapidly becoming exposed in the years following its maximum in 1750 during the Little Ice Age (Connor et al. 2009). In 1916, Cooper established a series of geographically clustered quadrats on similar gently sloping glacial tills and revisited them every five to ten years until the 1930s (Cooper 1923, 1939). During each visit, Cooper estimated the percentage of area covered by each species and mapped the location of individual plants within each quadrat. Other studies followed, but instead of creating a long-term study of colonization at sites of initially bare rock, these took advantage of the glacial chronosequence to examine soil development (e.g., Crocker and Major 1955), patterns of vegetation succession (e.g., Reiners et al. 1971), and mechanisms of change (e.g., Chapin, III et al. 1994). In general, the soil environment changed greatly over 200 years of succession, increasing in soil organic matter, moisture, and total nitrogen while decreasing in phosphorus (Crocker and Major 1955; Chapin III, et al. 1994). The youngest sites ($\leq$ 10 years old) supported a pioneer community of early dominants that fix nitrogen (cyanobacteria, lichen, the mat-forming shrub *Dryas*) as well as scattered fireweed and young shrubby willows, cottonwoods, and spruce. The sites, approximately 30 years and older, supported a mat of *Dryas* with persistent individuals of willow, cottonwood, alder, and spruce. After 50 years, alder formed dense thickets, and *Dryas* disappeared. The sites over 100 years old contained alder and willow with encroaching spruce overtopping. A site over 200 years old supported a spruce-dominated forest.

Early colonists in primary succession are often nitrogen-fixing "soil builders" that may provide ameliorating microsites for later colonists, thus leaving a persistent notion that primary succession is dependent on facilitation by early species for the colonization of the last species. However, like Whittaker et al. (1989) at Krakatau, Chapin, III et al. (1994) found that dispersal ability was critical to early-successional dynamics. Similar to Dale and Crisafulli (2018) at Mount St. Helens, they found that succession is complex, with both facilitative and inhibitory effects playing a role. The idea of facilitation and "predictability" in primary succession was further challenged by a recent revisiting of Cooper's classic study sites by Buma et al. (2019). Based on 100 years of data, Buma and the team observed that clear differences in the most recent vegetation composition resulted more from early establishment patterns (i.e., who arrived first) rather than from a specific sequence of species change. Amazingly, they found the effect of early nitrogen fixers was insignificant on subsequent trajectories (although nitrogen did accumulate in the soil over time; Buma et al. 2017), and light emerged as the single limiting resource, with intense competition among plants and near-complete monopolization by dominant species (Buma et al. 2019; Figure 1).

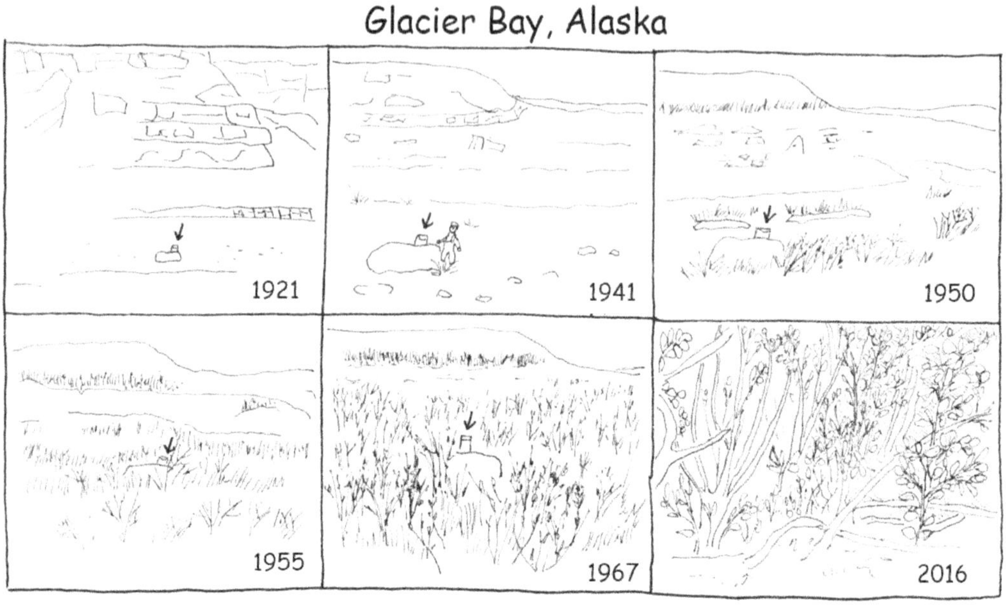

Figure 1. A foundation of ecology rediscovered: 100 years of succession on the William S. Cooper plots in Glacier Bay, Alaska (pen and ink interpretation of photos in Buma et al. 2017). Depiction of "cross rock" glacial erratic (arrow) easily visible until it was obscured by vegetation in 2016.

If primary succession is so complex, with a mishmash or mixture of possible trajectories, what is it about primary succession that makes it distinct from other kinds of vegetative change?

Common to all our examples of primary succession is the initial "sterile" conditions, yielding to colonization by organisms, followed by a change in the vegetation community over time. Rejmanek et al. (1982) compared the colonization curves for seven successional plant communities, and what made primary succession distinct from secondary succession, according to them, was two things: first, primary successions start with zero species (instead of typically 30–40 in secondary successions), and second, they tend to accumulate species more slowly than secondary successions. Thus, as Dale et al. (2005b) say, "primary succession is…an endpoint along a continuum of abundance of residual organisms and biological structures left by a disturbance." In other words, in the continuum from zero to "a lot" of organisms and organic matter left after a disturbance, primary succession represents the singular endpoint (zero), while secondary succession encompasses everything else.

Secondary Succession

Secondary succession occurs in habitats where soil and sometimes organisms are left intact following a disturbance. Disturbances creating such conditions include low-impact fires, floods, hurricanes, storms, agricultural clearings, clearcuts, and mining, among others. After a hurricane, flood, or fire hits an area, seeds, spores, or residual individuals remain. The amount of these biological legacies varies depending on the disturbance. An abandoned pasture starts with an intact community of plants and soil. Clearcutting a forest leaves only the soil and understory plants intact. Fire may remove the entire above-ground plant community, leaving only fire-tolerant seeds and fire-resistant roots. The initial conditions for secondary succession are therefore a continuum of "bare earth" and residual organisms, distinct only from primary succession in that the environment is not sterile. The pace of secondary succession is thus expected to be faster than primary succession because secondary succession is more about reestablishment and "recovery" of the biotic community rather than colonization of an abiotic environment.

The notion of "recovery" or the idea that "natural" areas have an underlying "climax" plant community following a disturbance, has been used to reconstruct the transitional stages of secondary succession. For example, Bard (1952) considered the 250-year-old forest called Mettler's Woods in Somerset County, New Jersey (USA) to be a vegetational climax. Using this "climax" community as the endpoint and a chronosequence of 27 previously cultivated fields abandoned between one and 60 years prior to his study, Bard reconstructed the sequence of secondary succession. He proposed nine stages of transition, starting from abandoned fields dominated by annuals such as ragweed (*Ambrosia*) and biennials such as evening primrose (*Oenothera*), progressing to perennial herbs like goldenrod (*Solidago*), to short-lived conifers such as juniper (*Juniperus*), and culminating in a final stage where the oak-hickory hardwood forest eventually replaced the conifers.

At Cedar Creek Natural History Area in Minnesota (USA), Richard Inouye and his team sampled a chronosequence of 22 old fields of known ages, ranging from one to 56 years after abandonment. Their study (Inouye et al. 1987) found a general trend of annual and nonnative species decreasing over time, while native prairie and woody species increased. They also observed that resource concentrations changed over time, with light penetration to the soil surface decreasing and soil nitrogen concentration increasing as the old fields aged. As the fields became older, they became more heterogeneous. By using this heterogeneity and looking at the effects of light and soil on species cover separately from field age, Inouye et al. (1987) concluded that nitrogen availability appeared important in influencing local species composition and abundance. However, they also found that time, acting through differences in dispersal and colonization rates, was crucial in determining which species were present in particular fields. In other words, on a very broad scale, the old fields were gradually reverting to the surrounding vegetation, but on a fine scale, the replacement of species over time appeared to be influenced by dispersal and colonization.

While it is clear community structure changes depending on time since disturbance, what about the size of the disturbance? Bob Holt and colleagues designed an experiment that manipulated the size of isolated disturbance patches (small, medium, and large) by mowing and disking in a Kansas old field (Holt et al. 1995). They found after six years that while the largest patches had greater species richness, this richness consisted of unique but rare species. Overall, however, early successional processes led to similar outcomes across all patches—specifically, the dominance of Canada goldenrod (*Solidago canadensis*), which they attributed to a common initial seed bank among the experimental patches.

Examples of old field secondary succession make "succession" a tantalizingly predictable sequence of vegetative change: it has a beginning, a middle, and often, what appears to be an end. However, a survey of the literature finds researchers repeatedly concluding that the process is disappointingly complex. This raises the following question: Are there predictable mechanisms driving successional change, or is this process driven more by random events and happenstance?

Models of Succession

Danish botanist Johannes Warming (1895), who studied interactions between plants and their environment in Denmark, was the first to introduce the concept of plant "communities" and their change over time. Although influenced by Warming, Cowles (1899) formalized the study of succession at the geologically dynamic Indiana Dunes. While Cowles believed that succession tended toward a stable equilibrium, he did not believe that the equilibrium state was ever reached. He suggested that the successional stages leading up to it were never quite a straight line (Kingsland 1995).

Following on the heels of Cowles, the ideas of botanist Frederic Clements emerged, with his studies of the more stable grasslands and conifer forests of the western prairie. These plant communities influenced Clements (1916, 1936) to see stability. He postulated that complex communities represent vegetation types that reach a **climax state** or "equilibrium," governed by local climate, topography, and interacting biotic and abiotic factors. Climax communities could include tall-grass species in a prairie or mature trees in a forest. After a disturbance, according to Clements, any given ecosystem would eventually return to its characteristic assemblage of species.

For instance, an oak-hickory forest after a devastating fire would return to the climax community defined by the dominant oak and hickory species, rather than one dominated by spruce, pine, or birch. Just as we expect a chick to develop into a chicken—not a dog or mushroom—Clements proposed that a community functions like an "organism," with stages of growth (which he called **seres**) leading to an inevitable outcome. The annual plant sere leads to a perennial forb sere, followed by a shrub sere, and then finally the climax oak-hickory sere. One thing leads to another until a climax is reached.

Theories of succession did not end with Clements (1916). Sometimes, great debates raged. Henry A. Gleason's *Individualistic Concept of Plant Association* argued that plant communities are not cohesive seres that inevitably transition to a stable climax community. Instead, he believed plant species behaved *individually,* and if species happened to form associations, it was merely a "coincidence" (Gleason 1926). Gleason did not outright deny associations between plants: "Plant associations exist; we can walk over them, we can measure their extent, we can describe their structure in terms of their component species, we can correlate them with their environment, we can frequently discover their history and make inferences about their future" (Gleason 1926). However, the concept of a climax community was a northern latitude bias, Gleason further noted, due to its wide diversity in environmental conditions with relatively few species. In the tropics, where the opposite occurs, he argued, it may be "impracticable or even impossible" to define a climax community. Gleason's individualistic "non-equilibrium" view of plant associations maintained that communities were not nearly as predictable as Clements proposed.

Other arguments followed. British ecologist Tansley (1935) disagreed with the organismic metaphor of a community as well, though he defended Clements' idea of the plant formation as a natural unit (the sere) and argued that ecosystems develop toward a state of equilibrium. Elton (1930), another influential British ecologist, did not believe in equilibrium or any "balance of nature." In the aftermath of the prolonged debate between Clements-type ideas and Gleason-type ideas, Frank Egler (1954) suggested a breakthrough. Contrasting the more Clementsian succession of "relay floristics" (Figure 2a), he introduced the "initial floristic composition" hypothesis (Figure 2b). Its basic tenet was that succession at a site is largely determined by the species composition of plant propagules already present at the site after a disturbance. Because some species

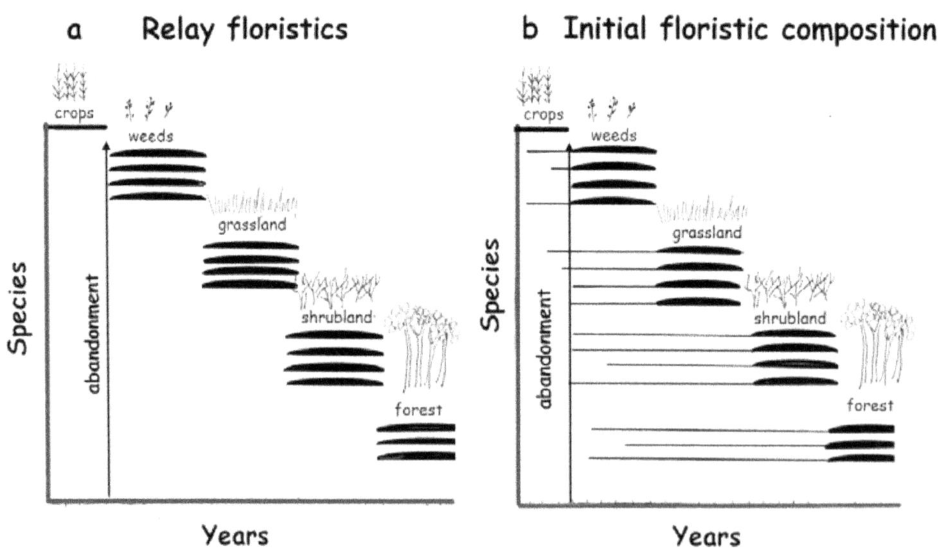

Figure 2. Two successional hypotheses (interpretation of Figures 1 and 2 in Egler 1954). (a) Egler (1954) calls the hypothesis of successional seres "relay floristics." Groups of species appear, making conditions unsuited for themselves but suitable for the next invading group until a stable endpoint is obtained. (b) Egler (1954) proposed a new hypothesis of succession called "initial floristic composition," where the propagules of most species are already present, but the dominance of species changes over time due to differential growth rates.

live longer than others but grow slowly, they take time to become dominant and gradually replace short-lived species. So, one may see a gradual transition in the identity of the dominant species in a community.

The "relay floristics" hypothesis (Figure 1a) was specifically applied to secondary succession in abandoned old fields (e.g., Clements 1916) and does not address the initial conditions of primary succession. However, Egler's (1954) "initial floristic composition" hypothesis (Figure 2b) was a step on the road toward an evolving general consensus of Gleason's idea that plants act on their own, rather than as a unit, during the process of plant succession.

In an attempt to organize the complexity of successional ideas into a coherent framework that considers the range of individualistic species responses throughout the course of succession, Connell and Slatyer (1977) offered three models of succession. These models were based on whether the first species to colonize or emerge from bare ground after a disturbance facilitates, tolerates, or inhibits the subsequent colonization or growth of other species.

Facilitation

This model proposes that while several species may attempt to colonize the disturbed area, only some with unique characteristics can occupy the "bare earth" and establish populations. These early occupants, through litterfall and the accumulation of carbon and nitrogen resources, stabilize soil, sand, or protect from harsh conditions for later plants. These colonizers thus modify the environment such that it becomes more suitable for later-successional species than for themselves. The facilitation model seems to explain many examples of primary succession. Cowles (1899) credits the colonization by perennial grasses and certain shrubs, which can tolerate the dry, exposed habitat, with stabilizing the dunes so that later-successional species can colonize. At Glacier Bay, Chapin III et al. (1994) found that earlier "pioneer" stages, such as *Dryas* and alder, can ameliorate the harsh conditions left by retreating glaciers by reducing pH, increasing nitrogen, and adding a layer of organic soil, allowing the seedlings of later successional spruce to germinate and grow. Once a spruce forest is established, however, the later environmental conditions (e.g., low light, less phosphorus) make it less suitable for the growth of earlier colonizing species.

Tolerance

The tolerance model proposes that late-successional plants are largely unaffected by plants at earlier stages of succession. Egler's (1954) initial floristic composition idea is an example of this model. This model posits that the initial stages of colonization are not just limited to a few pioneer species but also include species that are dominant in later-successional stages. These later successional species can survive and grow despite the presence of other species that established more quickly; they were too slow-growing to be dominant first. The tolerance model predicts a succession of species with increased ability to tolerate environmental factors that change as the vegetative community develops, such as lowering light levels, decreasing soil moisture and nutrients, and increasing allelochemicals and grazing. For example, in forest successions, species would be successively replaced by those that are increasingly more shade-tolerant. The end point of a shade-tolerance succession occurs when the most shade-tolerant species (e.g., beech in a beech forest) dominates the site and casts shade that even its offspring cannot survive, until a disturbance creates an opening. This points to the tolerance model's idea that the climax species essentially outcompete other species for limited resources, which can also be a feature of the inhibition model (described below). Connell and Slatyer (1977) admit that the tolerance model is an intermediate case between the facilitation and inhibition models, representing a single point (neutrality) on a continuum from initial colonists helping (facilitation) to initial colonists hindering (inhibition) subsequent species in the succession. Perhaps because of this, while the tolerance model is theoretically plausible, Connell and Slatyer (1977) found no convincing examples of it.

Inhibition

Many species arriving early after a disturbance, according to Connell and Slatyer (1977), may hinder rather than facilitate or contribute neutrally to the establishment of others. The possession of space is also important. The inhibition model proposes that early colonists dominate and persist vegetatively until a disturbance or mortality removes them, releasing space and resources for other occupants. As a result, one may never see a traditional successional sequence. For example, Facelli and Facelli (1993) removed the litter of *Setaria faberii* (giant foxtail, an Asian grass, often an early-successional species in old fields of New Jersey) and noted that the removal increased the biomass of a typically later-successional species, *Erigeron annuus* (daisy fleabane, a native North American herbaceous plant). Phytotoxic compounds from *Setaria* litter or physical obstruction by the litter may have hindered the growth of *Erigeron* and maintained *Setaria*, the early-successional species, as the dominant species.

However, if there is a life-history tradeoff that affects colonization rate versus lifespan, then the inhibition model can produce a traditional successional sequence where the first colonizers are short-lived annuals and biennials. Because they are short-lived, their deaths provide more chances to be replaced by longer-lived species than vice versa (Connell and Slatyer 1977).

Although Connell and Slatyer's three alternative models focused research on plant succession, not all successions fit neatly into these models. For example, Chapin III et al. (1994) found both facilitative effects and inhibitory effects from the pioneer species on the subsequent colonization and growth of the climax spruce species. Furthermore, facilitative or inhibitory effects may be important at many stages of succession, not just the initial phase. For example, alders are a mid-successional species that may facilitate spruce growth by adding soil nitrogen, but they also grow thick canopies, which inhibit the germination and establishment of spruce seedlings (Chapin III et al. 1994).

While facilitation, inhibition, and tolerance may be useful for understanding succession to a certain degree, without specifying the mechanisms that cause them, these models are too general to be useful in creating predictive models of succession. Is a mechanistic model of succession possible?

The Resource Ratio Hypothesis of Succession

David Tilman (1985) created a mechanistic model of plant succession by combining the uniqueness of species' traits with the uniqueness of habitat conditions and stitching them together with a graphical theory of plant competition to create the "resource-ratio" hypothesis of plant succession. To grasp this concept, we must first understand his theory of plant competition.

Consider just a single limiting resource, such as nitrogen, and a population of, say, little bluestem grass (*Schizachyrium scoparium*). As the little bluestem population grows, it depletes the availability of nitrogen in the soil. This depletion continues until the per capita growth rate of the population, dN/dtN (which is a function of resource R), balances its loss rate m (due to senescence, herbivory, etc.).

Because plants consume resources, the availability of resources in the habitat changes depending on both the supply rate of the resource and the consumption rate by the plant population. Therefore, the rate of change in resource availability (dR/dt) equals the supply rate (S) minus the consumption rate. In this model, S represents a habitat-specific "carrying capacity" of the resource in the absence of consumption.

If we let our plant population and resource availability reach equilibrium, the growth rate of the plant population will balance the mortality rate, and the supply rate of the resource will equal the consumption rate by the plant population. From this balance, both the equilibrium biomass (N^*) and the equilibrium resource availability (R^*) can be determined (Figure 3).

Plant foraging differs from animal foraging in that most plant resources are "essential." In other words, a plant can't substitute nitrogen for light or phosphorus for water. In contrast, animal resources are often "substitutable." For example, when a rabbit population declines, a hawk can switch to hunting mice. Because plant resources are essential, once resource availability is reduced

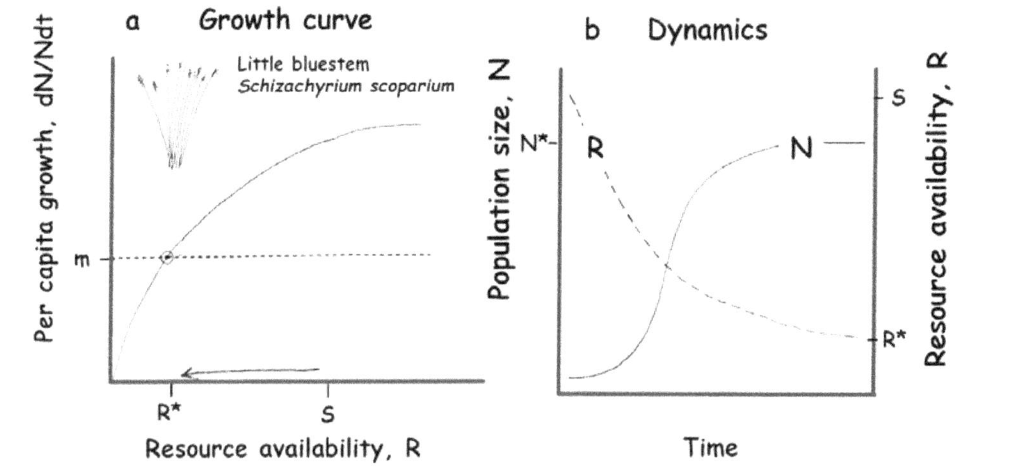

Figure 3. Resource consumption and growth dynamics for one species (interpreted from Tilman's [1982] Figure 10). (a) Hypothetical growth curve of little bluestem where the solid line is the per capita rate of growth with respect to a limiting resource and the dashed line is its mortality rate (m, which is unrelated to resource availability or growth rate). "*S*" indicates the habitat supply rate of the resource (e.g., nitrogen) without consumers, and R^* is the equilibrium value of resource availability if the habitat had a monoculture of little bluestem. (b) If the functions for both population growth and resource supply are specified, then solving the equations over time gives the dynamics. Little bluestem eventually reaches equilibrium biomass (N^*), and the resource availability is drawn down to R^*.

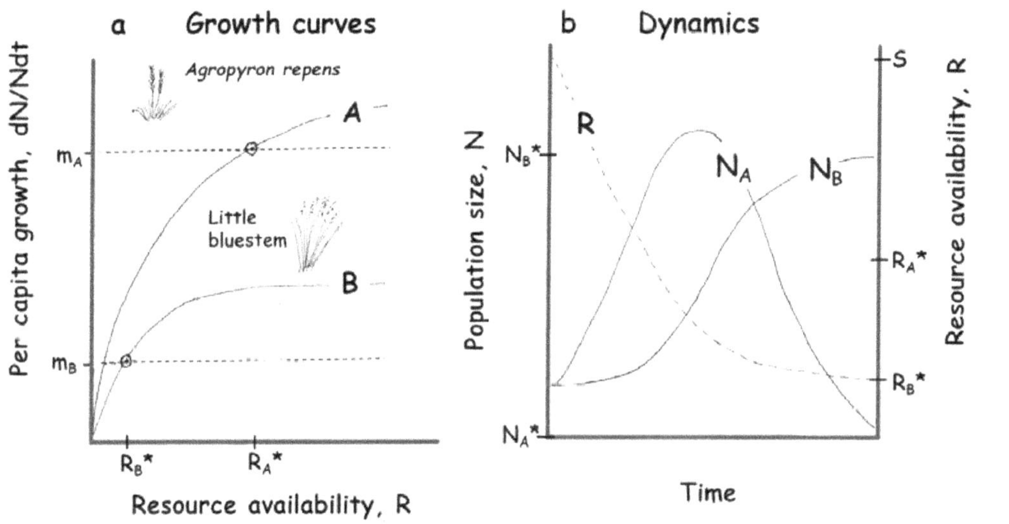

Figure 4. Resource consumption and growth dynamics for each of two species growing by itself (interpreted from Tilman's [1982] Figure 11). (a) Hypothetical growth curves of *Agropyron* (species A) and little bluestem (species B) on a single limiting resource (e.g., nitrogen) where the R_B^* for little bluestem is lower than the R_A^* for *Agropyron*. (b) Little bluestem (species B) outcompetes *Agropyron* (species A) in this scenario; over time, the population of little bluestem arrives at N_B^*, and the population of *Agropyron* goes to zero.

to the R^* level, the plant population can no longer increase, because plants generally cannot switch to another resource to compensate. The R^* for a particular resource varies by species. Tilman and Wedin (1991) found that little bluestem has a lower R^* for nitrogen than the early-successional grass *Agropyron repens*. According to Tilman's (1982) theory, little bluestem should outcompete *Agropyron* for nitrogen, which it did in experimental trials (Tilman and Wedin 1991). By reducing nitrogen levels to a point where *Agropyron* could no longer grow and compensate for its biomass losses, little bluestem caused *Agropyron* to die off, allowing little bluestem to "win" (e.g., Figure 4).

If species do not have tradeoffs, then the species that is best at drawing down the most limiting resource to its lowest level will "win," and the habitat will become a monoculture. Plant competition, however, is rarely that simple. Plants *do* have tradeoffs. As we saw in Chapter 6, *Life Histories and Life History Tradeoffs*, plants are modular organisms with life histories and morphological strategies that vary in their allocation of leaves, stems, roots, or seeds, as well as in their growth and senescence rates. This means plant species that put most of their production into roots may be better at nitrogen uptake but cannot devote as much production to stems and leaves, which are important for competing for light. These morphological and life-history tradeoffs are critical when plants compete for resources because they can strongly influence which plant strategy will "win" or whether they can coexist in particular types of habitats.

In Chapter 3, *Species Interaction*: *Competition*, we created zero-growth isoclines in a state space that depicted the combinations of species 1 and species 2 densities where the first (or second) species neither increased nor decreased (zero growth). Here, we will also depict zero-growth isoclines, but in this case, instead of a state space with species densities, we will use a state space of the two most limiting resources. This is because we want to understand how a particular plant grows (based on its traits) in response to changes in habitat conditions (based on limiting resources).

Using the single-resource competition theory presented above, Tilman (1982, 1988) created zero-net-growth isoclines for a hypothetical species consuming two limiting resources (Figure 5). The availabilities of the two limiting resources (e.g., light and nitrogen) are scaled on the R_1 and R_2 axes and are depicted as points in the state space (several numbered points are shown in Figure 5; these could represent either different habitats or different resource levels in the same habitat at different points in time).

If the resources R_1 and R_2 are **perfectly essential** (i.e., to grow, a plant can't substitute one resource for the other), then the dashed "A" lines in Figure 5 would represent the zero-growth isocline. Population growth relative to the A isocline can be understood in the following way. If the habitat contains more light than R_1* and more nitrogen than R_2* (e.g., habitat 6), the plant population will increase. In habitat 2, the plant population cannot grow because it is limited by

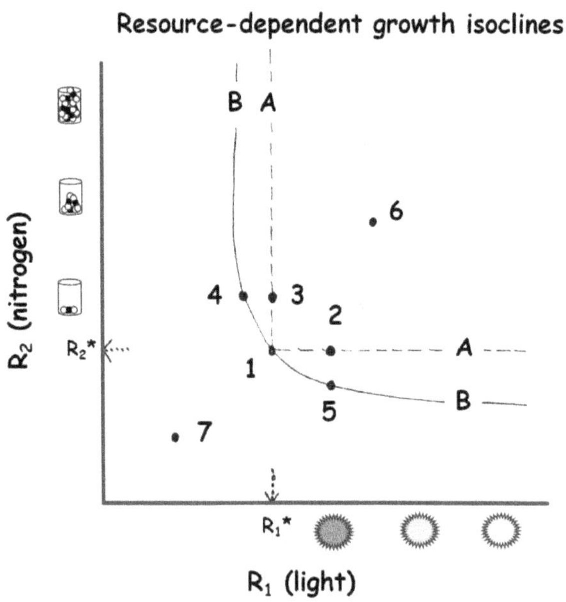

Figure 5. Resource-dependent growth isoclines for a single species in a habitat with two limiting resources (e.g., light and nitrogen) where a plant species has no foraging plasticity with respect to the resources (dashed line A), or a plant species has some foraging plasticity with respect to the resources (solid line B; interpreted from Tilman's [1988] Figure 6). Numbers depict different habitats with respect to nitrogen and light. The R*s depicted here are the lowest levels to which the species can drive each resource. Note that the dotted arrows simply point to the value of R* for each resource and are not part of the isocline.

nitrogen; in habitat 3, the plant population cannot grow because it is limited by light; in habitat 1, the plant population is equally limited by light and nitrogen. In habitats 1, 2, and 3, the population neither grows nor declines because these points lie on the zero-growth isocline. The plant population decreases in habitats 4, 5, and 7 because there is insufficient light or nitrogen to maintain it.

Plants face major foraging tradeoffs. For example, plants that are good light competitors because they have tall stems are generally not good nitrogen competitors, as this is because they allocate more growth to aboveground foraging structures rather than belowground foraging structures. The A isocline in Figure 5 assumes that plants are rigid in the way they allocate growth to their foraging structures. For plant A, light and nitrogen are perfectly essential resources; the most limiting resource will always hinder population growth. That is the case for habitats 2 and 3 in Figure 5. But what if plants have some degree of foraging plasticity? That is, what if they can adjust their consumption of resources by allocating more growth to stems or leaves in low light and allocating more growth to roots in poor soils? An optimally foraging plant, if capable of doing so, should adjust its morphology or physiology to be equally limited by all resources. For example, for plant A in habitat 2, nitrogen limits growth and light does not, but if this plant were capable of giving up some allocation to stem growth in favor of root growth, then a population of this plant could also thrive in habitat 5, which has much less nitrogen. Thus, if plants have foraging plasticity, light and nutrients are no longer perfectly essential, but are instead considered **interactively essential**. This results in the B isocline in Figure 5 (Tilman 1988).

The reason the B isocline has a rounded corner is that, near the bend, adding either resource can lead to an increase in population density. For example, in habitat 5, the increased light above R_1* can compensate for the decreased nitrogen below R_2* because light is not currently limiting growth. As a result, the plant can allocate less to growing leaves and more to growing roots to increase its ability to acquire nitrogen, which is the limiting resource for growth. The two essential resources thus become functionally substitutable over a small range of resources encompassing the plant's foraging flexibility.

Note that the *R**s for both resources remain the same for isocline A (perfectly essential resources) or isocline B (interactively essential resources). However, for isocline B, the plant population can now grow in habitats 2 and 3 (where isocline A could not), remains constant for habitats 4 and 5 (where isocline A would decline), but still cannot grow at resource levels below the isocline (e.g., habitat 7).

Now, we need to understand what happens to the availability of resources in the environment when plants consume them. Each habitat has an inherent rate of nutrient supply, which could be the rate at which nitrogen is fixed from the atmosphere, the rate at which phosphorus erodes from rocks, or the rate of disturbance that creates light gaps. We call this inherent supply rate "S." In Figure 3b, we saw how the growth of a single species reduced the availability of a single resource from its supply point down to *R**. The same principle applies to two limiting resources. There will be an environmental supply point (S_1, S_2), and for a particular species of plant, it will consume both resources until the population can no longer grow (where growth rate balances loss rate), and the consumption rate of both resources balances the supply rates. This point (R_1*, R_2*) represents the equilibrium value of resources on the species' isocline (the plant population will grow to N*, but this will not be depicted in the resource state space).

Figure 6a depicts a state space with three supply points $(S_1, S_2)^X$, $(S_1, S_2)^Y$, and $(S_1, S_2)^Z$ for resources 1 and 2 in habitats X, Y, and Z, respectively, and the zero-growth isocline of a species that is better at foraging for resource 2 than for resource 1. Plants are expected to optimize consumption such that both resources equally limit them. As Tilman (1988) says, "Excess consumption of a non-limiting resource is of no selective advantage…" Thus, a plant that is better at foraging for nitrogen will be less limited by nitrogen (*y*-axis, resource 1) than light (*x*- axis, resource 2) and will therefore consume light at a greater rate than nitrogen (the plant population needs more light than nitrogen to grow; $c_{R1} > c_{R1}$). In Figure 6a, this results in a consumption vector ($\vec{C}$) canted north of 45° (the line of equal consumption rates).

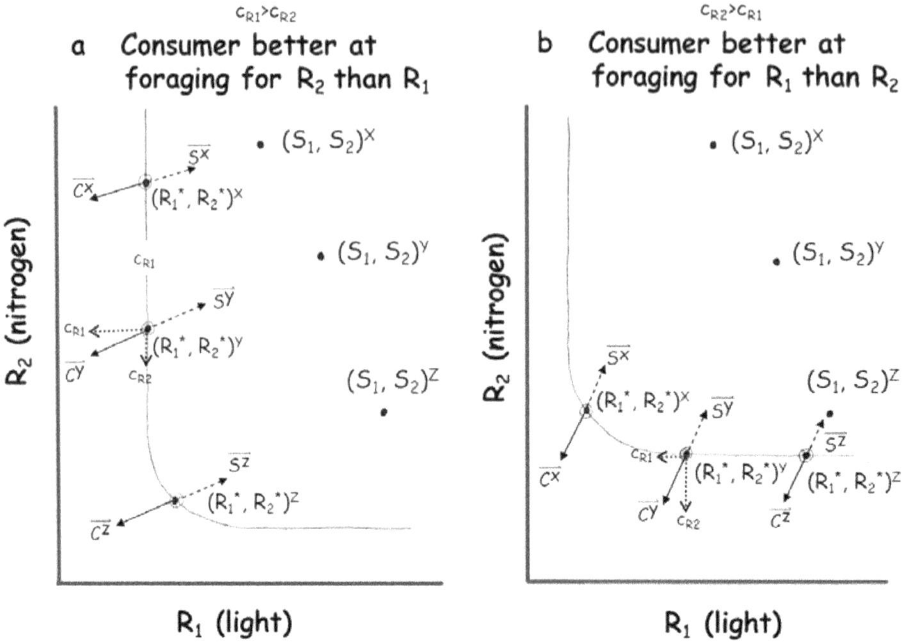

Figure 6. Foraging dynamics of two competitors (interpreted from Tilman's [1988] Figure 2.8 and narrative in Tilman [1988]). For a single plant consumer with two limiting resources (R_1 and R_2), its optimal foraging consumption vector ($\vec{C}$) is generated from the ratio of the resource consumption rates (c_{R1} and c_{R2}). At equilibrium, the supply vector ($\vec{S}$) is equal and opposite to the consumption vector and points back to the supply point. (a) The consumption vector for a consumer that is better at foraging for R_2 than for R_1 is shown at the equilibrium points (R_1*, R_2*) for three habitats (X, Y, and Z) with differing supply points. (b) The consumption vector for a consumer that is better at foraging for R_1 than R_2 is shown at the equilibrium points (R_1*, R_2*) for three habitats (X, Y, and Z) with differing supply points.

Figure 6b depicts an isocline of a plant that is better at foraging for light than nitrogen (i.e., it needs more nitrogen than light to grow), and its consumption vector ($\vec{C}$) is angled south of 45°. Although the state space is static, Tilman (1988) gives us some dynamical rules of thumb. The plant population will grow until growth equals loss (N*), and the resources will decline from the supply point to an equilibrium point on the isocline (R_1*, R_2*) where supply rate balances consumption rate. Graphically, the consumption vector will be equal and opposite to the supply vector, which will point back to the supply point (Figure 6a and Figure 6b). Thus, for a given consumption vector (foraging trait of the plant), and a given resource supply point (e.g., habitat X, Y, or Z), we can predict where the equilibrium point will lie on the isocline (see Figure 6a and 6b). Note that the equilibrium biomass (*N**) is not shown in this state space, but it can vary depending on the supply point, just as the equilibrium availability of resources (R* values) varies depending on the supply point.

Now that we understand plant growth and consumption in a state space with two resources, we can use Tilman's (1988) graphical rules of thumb for resource and population dynamics to predict the outcome of competition between plants based on the habitat resource availability. In Figure 7, there are four competitive scenarios. In scenario ***a***, species A "wins" in all habitats where the supply points are above its *R** for both resources because species A can still grow at resource levels where species B cannot. In scenario **b**, the opposite is true. Species B "wins" because its *R** for both resources is always lower than for A, regardless of the supply point.

In scenario **c**, there is an equilibrium point at which both species can coexist. But whether the dynamics lead to this equilibrium point depends on the location of the resource supply point. Remember, the system is dynamic, so as both species grow, the availability of resources will change from the supply point (maximal resource availability in the environment) to an equilibrium point on

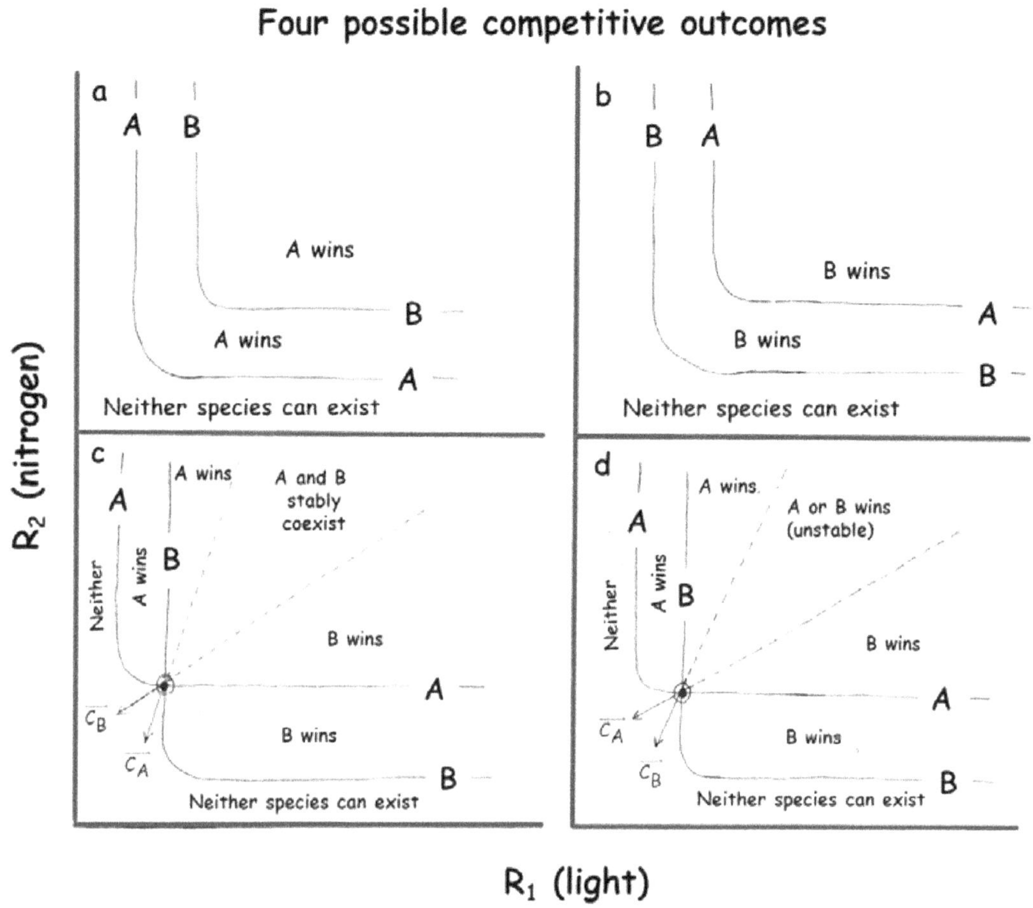

Figure 7. For two plant species competing for two limiting essential resources, competitive outcomes are given for the four possible kinds of isocline scenarios where supply points are located in the regions of the state space as indicated in relation to the dashed lines and isoclines (interpreted from Tilman's [1988] Figure 2.9). For example, supply points located in the space where "A wins" will result in equilibrium resource availability on the A isocline, which excludes the growth of B. (Note: Anyone interested in the math behind these graphical rules of thumb should consult Tilman [1982, 1988]).

one or both plant isoclines. In Figure 7c, a supply point in the regions where "A wins" will result in resource availability on A's isocline being below B's isocline for resource 1. A supply point in regions where "B wins" will result in resource availability on B's isocline below A's isocline for resource 2. However, coexistence is possible because the isoclines of A and B cross (creating an equilibrium point), and as long as the resource supply point lies between the dashed lines defined by the slope of each species' consumption vector (their respective resource ratios), they can coexist.

There is a caveat to this coexistence rule: each species must be an optimal forager, consuming proportionally more of the resource that most limits its growth. In Figure 7c, species A is more limited by nitrogen than B (isocline lies higher on the y axis), so its consumption vector has a steeper slope, indicating it consumes more nitrogen than light to grow. Species B is more limited by light than A (the isocline lies more to the right on the x-axis), so its consumption vector is shallower, indicating it consumes more light than nitrogen to grow.

However, if A and B were non-optimal foragers and consumed more of the resource that limited them *less*, this would result in scenario **d**, where the slopes of the consumption vectors are reversed, and coexistence is possible only if the supply point is exactly at the equilibrium point (Figure 7d). Otherwise, if the supply point is within the zone defined by the slopes of the consumption vectors, the dynamics of nonoptimal consumption would drive habitat resource levels into the zones where

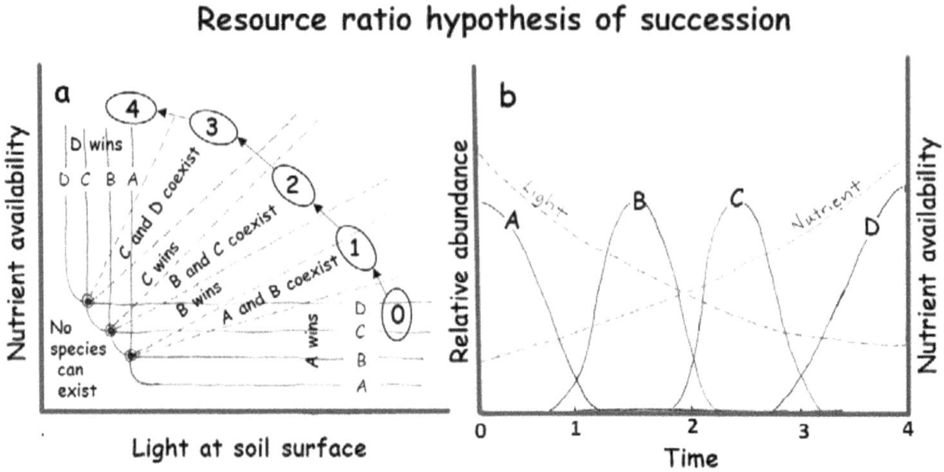

Figure 8. The resource ratio hypothesis of succession (interpreted from Tilman's [1988] Figure 7.1). Habitats do not usually have a single point of resource supply because there is always some heterogeneity; this is depicted by resource supply "ovals" zero through four. (a) During primary succession, the light-to-soil-nutrient supply ratio decreases, moving the habitat from zero to four. This results in a "succession" of species from A to D. (b) The population dynamics of species A through D and the dynamics of nutrient and light availability are depicted through time.

either A or B wins. Assuming nonoptimal foraging does not normally occur, we will dismiss scenario **d** as unimportant for our discussion here.

So, how does all this relate to succession? Tilman (1988) proposed the resource ratio hypothesis of succession (hereafter RRS) based on these mechanistic ideas of plant competition (Figure 8).

Suppose species A through D are optimally foraging plants that have differing resource-dependent growth isoclines, reflecting their requirements for a limiting soil resource and light. Since primary succession begins in a nutrient-poor but high-light environment (a set of supply points located in the oval "0"), as seen in Figure 8a, one would expect to see changing resource supply over time as nutrients and biomass accumulate while light levels decline. This scenario would result in the successional replacement of species due to the competitive tradeoffs discussed earlier. The successional dynamics with respect to species abundance and resource availability are shown in Figure 8b. This is the "resource ratio hypothesis" of succession. Changes in species dominance happen because of the changing ratio of limiting resources.

There are several implications of the RRS hypothesis. First, the RRS hypothesis does not assume succession must be directional or have a fixed endpoint. An essential feature of the RRS hypothesis is that changes in the relative supply of limiting resources should lead to changes in the composition of a plant community. Tilman (1988) gives an example. In the classical Glacier Bay succession, the habitat transitions from nitrogen-limited to light-limited, and species shift from nitrogen fixers to spruce trees. However, if the forest is invaded by *Sphagnum* moss, it causes the soils to acidify, lowering nitrogen mineralization rates, reducing plant biomass, and increasing light penetration. This results in muskeg-like vegetation dominated by plants that are better competitors under high-light, low-nitrogen mineralization soil conditions. The RRS hypothesis does not assume directionality or repeatability, nor does it state that succession occurs as light becomes a more-limiting resource than nitrogen. Instead, it proposes that if there is a temporal gradient in resources, it will sequentially favor species that are superior competitors at specific resource ratios.

The second implication of the RRS hypothesis is that the temporal gradient need not be one of productivity. In secondary succession, the "starting point" is usually considered to have high nutrients, high light—like an old field, which, when left fallow, will eventually be invaded by shrubs and trees, as Thoreau observed in New England. Tilman (1988) explains this type of succession in terms of a biomass "loss rate" gradient. In Figure 9a, we can see the tradeoff in species growth rates versus habitat loss rates. Species A has the slowest maximal growth rate, but under a low biomass

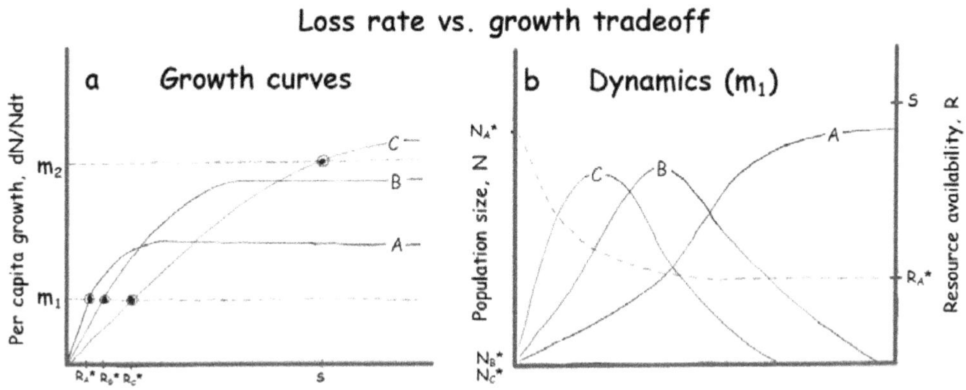

Figure 9. Successional dynamics for a growth vs. loss tradeoff (interpreted from Tilman's [1988] Figure 2.1). (a) Hypothetical growth rate versus habitat loss rate tradeoff where m_1 is a low rate of biomass loss, and m_2 is a high rate of biomass loss, and species A, B, and C have increasingly greater maximal growth rates. (b) Transient dynamics (succession) in a habitat with low loss rate m_1 where the limiting resource is drawn down to R^*_A.

loss rate (m_1), it will eventually outcompete B and C for the limiting resource (Figure 9b). But if the habitat has a high loss rate (m_2), species C will outcompete species B and A because they cannot grow fast enough to compensate for such a high loss rate. For example, pastures have high loss rates due to herbivory by grazers and periodic mowing. As a result, the species that thrive in such environments are short-lived grasses and forbs with high growth rates. However, when grazing and mowing are removed, the habitat transitions to one with a lower loss rate, and species A will eventually outcompete B and C for the limiting resource. The transient dynamics of this process are "succession."

In their study of old field ecosystems at the Cedar Creek Ecosystem Science Reserve in Minnesota (USA), Clark et al. (2018) found that replicated simulations of five major drivers of successional outcome—soil fertility, fire, climate, competition, and demography—followed similar successional trajectories due to "loss rate" compensatory tradeoffs. Species that are abundant late in succession have low colonization rates and high mortality rates early in succession, but become better competitors later in succession. Although trajectories differed among the drivers, stochastic processes contributed only short-term noise, while known life-history characteristics affected trajectory outcomes in a repeatable manner.

The third implication of the RRS hypothesis is that any kind of tradeoff gradient—not just temporal ones—can influence the type of vegetation that occurs. For example, plant morphologies and their abilities to garner resources under specific environmental conditions change along a tropical elevational gradient, from paramo species to rainforest species, or along a temperate North American moisture gradient, from desert species to mesophytic forest, or along a latitudinal temperature gradient from tundra to tropical forest. From this perspective, succession is essentially just a transition in vegetation along an environmental gradient in time, rather than a transition in vegetation along an environmental gradient in space.

In the examples above, we focused on terrestrial vegetation, but succession also happens in the marine environment. For example, in denuded patches near Caribbean coral reefs, succession in marine epibenthic communities (organisms that live on the surface of the bottom of the ocean) starts with an accumulation of biofilm made up of bacteria and microalgae adhering to particulate matter on surfaces. This biofilm provides a base for the settlement of opportunistic filamentous algae and cyanobacteria. Other species of algae appear later, gradually changing the community composition until competitive fleshy algal species dominate (Fricke et al. 2011). In fact, another type of succession does not involve plants at all. The death of an organism often sets off what is known as **heterotrophic succession**. The death of a whale that sinks to the sea floor (whale fall) is an excellent example, with communities changing over time from mobile scavengers to enrichment opportunists, and finally to a sulphur-loving assemblage that includes chemoautotrophs—this final stage can last for decades (Smith and Baco 2003).

Tilman (1988) argues that a mechanistic theory of resource supply rates and morphological tradeoffs can explain the transient dynamics of many types of successions and their multiple possible outcomes. In fact, this theory need not be restricted to plant succession. Tilman (1982) shows how the dynamics of competitors with substitutable resources or switching resources can also be modeled in resource state space. By understanding succession in a mechanistic way, that is, by the advantages of particular foraging traits in particular environments, we can see how evolution plays a central role—dictating the optimal foraging constraints and life history tradeoffs organisms must navigate as they grow in a dynamic world.

References

Bard, G. E. 1952. Secondary succession on the piedmont of New Jersey. Ecological Monographs 22: 195–215.

Buma, B., Bisbing, S., Krapek, J. and Wright, G. 2017. A foundation of ecology rediscovered: 100 years of succession on the William S. Cooper plots in Glacier Bay, Alaska. Ecology 98: 1513–1523.

Buma, B., Bisbing, S., Wiles, G. and Bidlack, A.L. 2019. 100 yr primary succession highlights stochasticity and competition driving community establishment and stability. Ecology 100: e02885.

Chapin III, F. S., Walker, L. R., Fastie, C. L. and Sharman, L. C. 1994. Mechanisms of primary succession following deglaciation at Glacier Bay, Alaska. Ecological Monographs 64: 49–175.

Clark, A. T., Knops, J. M. H. and Tilman, D. 2018. Contingent factors explain average divergence in functional composition over 88 years of old field succession. The Journal of Ecology 107: 545–558.

Clements, F. E. 1916. Plant succession: analysis of the development of vegetation. Carnegie Institution of Washington Publication Sciences 242: 1–512.

Clements, F. E. 1936. Nature and structure of the climax. The Journal of Ecology 24: 252–84.

Cockell, C. S. and Lee, P. 2002. The biology of impact craters-a review. Biological Review 77: 279–310.

Connell, J. H. and Slatyer, R. O. 1977. Mechanisms of succession in natural communities and their role in community stability and organization. The American Naturalist 111: 1119–44.

Connor, C., Streveler, G., Post, A., Monteith, D. and Howell, W. 2009. The Neoglacial landscape and human history of Glacier Bay, Glacier Bay National Park and Preserve, southeast Alaska, USA. The Holocene 19: 381–393.

Cooper, W. S. 1923. The recent ecological history of Glacier Bay, Alaska. Ecology 4: 93–128.

Cooper, W. S. 1939. A fourth expedition to Glacier Bay, Alaska. Ecology 20: 130–55.

Cowles, H. C. 1899. The ecological relations of the vegetation on the sand dunes of Lake Michigan. Part I.- Geographical relations of the dune floras. Botanical Gazette 27: 95–117, 167–202, 281–308, 361–391.

Crocker, R. L. and Major, J. 1955. Soil development in relation to vegetation and surface age at Glacier Bay, Alaska. The Journal of Ecology 43: 427–448.

Dale, V. H. and Crisafulli, C. M. 2018. Ecological responses to the 1980 eruption of Mount St. Helens: key lessons and remaining questions. pp. 1–18. *In*: Crisafulli, C. M. and Dale, V. H. [eds.]. Ecological responses at Mount St. Helens: revisited 35 years after the 1980 eruption. Springer New York, NY.

Dale, V. H., Crisafulli, C. M., and Swanson, F. J. 2005a. 25 Years of Ecological Change at Mount St. Helens. Science 308: 961–962.

Dale, V. H., Swanson, F. J. and Crisafulli, C. M. 2005b. Disturbance, survival, and succession: understanding ecological responses to the 1980 eruption of Mount St. Helens. pp. 3–11. *In*: Dale, V. H., Swanson, F. J. and Crisafulli, C. M. [eds.]. Ecological responses to the 1980 eruption of Mount St. Helens. Springer, New York, NY.

Daubenmire, R. 1968. Plant communities: a textbook of plant synecology. Harper and Row, New York, NY.

del Moral, R. and Wood, D. M. 1993. Early primary succession on the volcano Mount St. Helens. Journal of Vegetation Science 4: 223–234.

Egerton, F. N. 2015. History of ecological sciences, part 54: succession, community, and continuum. Bulletin Ecological Society of America 96: 426–474.

Egler, F. E. 1954. Vegetation science concepts I. Initial floristic composition, a factor in old-field vegetation development with 2 figs. Vegetatio Acta Geobot 4: 412–417.

Elton, C. 1930. Animal Ecology and Evolution, Methuen, New York.

Facelli, J. M. and E. Facelli. 1993. Interactions after death: plant litter controls priority effects in a successional plant community. Oecologia 95: 277–282.

Fricke, A., Teichberg, M., Beilfuss, S. and Bischof, K. 2011. Succession patterns in algal turf vegetation on a Caribbean coral reef. Botanica Marina 54: 111–126.

Gleason, H. A. 1926. The individualistic concept of the plant association. Bulletin of the Torrey Botanical Club 53: 7–26.
Halvorson, J. J., Smith, J. L. and Kennedy, A. C. 2005. Lupine effects on soil development and function during early primary succession at Mount St. Helens. pp. 243–254. *In*: Dale, V. H., Swanson, F. J. and Crisafulli, C. M. [eds.]. Ecological responses to the 1980 eruption of Mount St. Helens. Springer, New York, NY.
Holt, R. D., Robinson, G. R. and Gaines, M. S. 1995. Vegetation dynamics in an experimentally fragmented landscape. Ecology 76: 1610–1624.
Inouye, R. S., Huntly, N. J., Tilman, D., Tester, J. R., Stillwell, M. and Zinnel, K. C. 1987. Old-field succession on a Minnesota sand plain. Ecology 68: 12–26.
Kingsland, S. E. 1995. Modeling Nature: Episodes in the History of Population Ecology. University of Chicago Press, Chicago.
Peterson, D. W. 1990. Overview of the effects of influence of the activity of Mount St. Helens in the 1980s. Geoscience Canada 17: 163–166.
Reiners, W. A., Worley, I. A. and Lawrence, D. B. 1971. Plant diversity in a chronosequence at Glacier Bay, Alaska. Ecology 52: 55–69.
Rejmanek, M., Haagerova, R. and Haager, J. 1982. Progress of plant succession on the Paricutin Volcano: 25 years after activity ceased. American Midland Naturalist 108: 194–198.
Smith, C. R. and Baco, A. R. 2003. Ecology of whale falls at the deep-sea floor. pp. 319–333. *In*: Gibson, R. N. and R. J. A. Atkinson [eds.] Oceanography and marine biology. CRC Press.
Tansley, A. G. 1935. The use and abuse of vegetational terms. Ecology 14: 284–307.
The Fixx. 1983. Reach the Beach. MCA Records.
Thoreau, H. D. 1860. The Succession of Forest Trees; and Wild Apples. An address read to the Middlesex Agricultural Society, Concord (Massachusetts). Also published in a New York newspaper (Thoreau 1880).
Tilman, D. 1982. Resource competition and community structure. Monographs in Population Biology 17. Princeton University Press, Princeton, NJ.
Tilman, D. 1985. The resource-ratio hypothesis of plant succession. The American Naturalist 125: 827–852.
Tilman, D. 1988. Plant strategies and the dynamics and structure of plant communities. Monographs in Population Biology 26. Princeton University Press, Princeton, NJ.
Tilman, D. and Wedin, D. 1991. Dynamics of nitrogen competition between successional grasses. Ecology 72: 1038–1049.
Warming, E. 1895. Plantesamfund - Grundtræk af den økologiske Plantegeografi. P.G. Philipsens Forlag, Kjøbenhavn.
Whittaker, R.T., Bush, M. B. and K. Richards. 1989. Plant recolonization and vegetation succession on the Krakatau Islands, Indonesia. Ecological Monographs 59: 59–123.

CHAPTER 10

Nutrient Cycling

"...for dust thou art, and unto dust shalt thou return" – Genesis King James Bible

On Christmas Eve, 1968, one of the most famous and impactful photographs, both of its time and now, was taken by Apollo 8 astronauts as they endeavored to be the first humans to reach the Moon—it was a snapshot of Earth observed from space.

In this iconic photograph, dubbed "Earthrise," our planet looks strikingly blue, the color from oceans that dominate over 70% of its surface. This abundance of water, filling oceans, rivers, and lakes, constitutes the **hydrosphere** of our planet. The next outstanding feature is the scattered clouds, indicating Earth is surrounded by an **atmosphere**, including the blanket of gases composed primarily of nitrogen (N_2), oxygen (O_2), carbon dioxide (CO_2), and water vapor. Below the clouds, it is possible to discern continents, mountains, valleys, and other terrestrial features, constituting the **lithosphere**. Ninety-four percent of the lithosphere is composed of oxygen, iron (Fe), silicon (Si), and magnesium (Mg). Another feature strikingly visible to any astronaut arriving aboard the International Space Station is the Great Barrier Reef off the Australian continent. The distinctive feature of this colossal coral reef is that living organisms built it. The planet's 10 million different kinds of living things together constitute the **biosphere**. All four systems are interconnected and interact with one another.

Nearly 79 years earlier, on November 25, 1889, the passengers of the *Golcanda*, a British ship sailing in the Red Sea, saw a huge swarm of migrating locusts. The sight was so magnificent that a naturalist aboard aptly reported the event in the journal *Nature*, estimating that over 24 thousand trillion locusts, weighing about 43 billion tons, were migrating that day at a high altitude on a northwestern course. In hindsight, this was an exaggerated overestimate of locusts, but the point remains. When this report came to the attention of Russian geochemist Vladimir Vernadsky, he marveled at the grandeur of the observation, expressing it in the language of chemistry in his book *La Geochimie:* "...the swarm of locusts, expressed in terms of chemical elements and metric tons, may be seen as **analogous** to a moving rock formation" (Smil 1997). For Vernadsky, the elements embodied in the swarming locusts came from plants that originally assimilated those elements from the air (atmosphere), water (hydrosphere), and the very rocks of the Earth's crust (lithosphere).

Compared to spectacular volcanic eruptions, floods, or hurricanes that can move tons of materials from one place to another in catastrophic events, the transport of materials by organisms can seem subtle. However, the sheer ubiquity of living transport makes it vast and relentless. Many organisms do not stick to one place, and their movements transport chemical elements ingested at one location and deposited in another. Examples include herds of wildebeests making their annual migration from Tanzania's Serengeti to Kenya's Masai Mara, following the rains, and consuming and depositing organic material as they travel. The European stork travels from Africa to Europe every spring and every autumn returns to Africa, transporting materials in their feces, feathers, and, eventually, carcasses. Arctic terns make an amazing flight of over 18,000 kilometers from the Arctic Circle to the western coasts of Europe and Africa, then across the Antarctic Ocean to

DOI: 10.1201/9781003008538-10

Antarctica. A chick born and raised in the Arctic may die and decompose in the Antarctic, shifting small amounts of nutrients from one pole to the other. Pacific salmon are renowned for transporting ocean productivity, sometimes up to thousands of miles inland[1] via freshwater tributaries.

The details of migratory animals aside, Vernadsky's synthesis—that migratory locust swarms transport embodied elements much faster and farther than the weathering of a rock of equal mass—was visionary. His observation that atoms continuously migrate from inert matter (non-living environment) to living matter and back again was ahead of his time.

Today, the discipline of ecosystem ecology deals with the storage (**reservoirs**) and transfer (**flux**) of matter and energy through living systems. Tansley (1935) was the first to define an **ecosystem** as "including not only the organism-complex but also the whole complex of physical factors forming what we call the environment of the biome." He goes on to explain, "There is constant interchange of the most various kinds within each system, not only between the organisms, but between the organic and inorganic."

Energy vs. Matter

Ecosystems are made up of various compartments or reservoirs of energy and matter: primary producers, herbivorous and carnivorous consumers, decomposers and detritivores, a stockpile of dead organic matter, plus the physical and chemical environment that provides sources and sinks for energy and matter. All living components of the ecosystem are made up of elemental matter that was originally assimilated by primary producers from the physical and chemical environment. The energy used by plants, algae, and cyanobacteria to bind this matter together to form biological macromolecules (carbohydrates, proteins, lipids, DNA, etc.) primarily comes from the sun. This fixation of energy into biomass by primary producers is called **primary productivity**, defined as the amount of biomass produced per unit area per unit time. Gross primary productivity is the total fixation of energy by photosynthesis. However, most measures of productivity account for the portion of this energy that is respired away and lost as heat during biomass production; this is called **net primary productivity** or **NPP**.

Solar radiation is essentially a plant resource that varies in availability with latitude, providing 0.20 cal/cm^2/min at 60 degrees latitude and about 0.45 cal/cm^2/min in the tropics (Raschke et al. 1973). A forest-to-forest comparison of mean NPP of a tropical seasonal forest (1,600 g/m^2/year) to a boreal forest (800 g/m^2/year; Begon et al. 1996 after Whittaker 1975) shows how differences in solar radiation are reflected in primary production. However, this is only a crude relationship, because temperature and water are equally powerful drivers of global NPP. In fact, water is the most limiting factor for terrestrial NPP on 40% of Earth's land surface, temperature on 33% of the land surface (including boreal forests), and solar radiation in 27% of global vegetated areas—ironically, in moist tropical areas where temperature and water are adequate, but cloud cover dramatically reduces solar radiation (Running et al. 2004). Because photosynthesis is how primary producers capture the energy of the sun, this NPP trifecta—solar radiation, water, and temperature—is unsurprising. Photosynthesis requires both light and water in addition to carbon dioxide, and the capture of carbon dioxide by plants is affected by temperature. Specifically, lower temperatures decrease the kinetic energy of the critical photosynthetic enzyme Rubisco (Yamori et al. 2006).

The energy plants capture from the sun is used to produce plant biomass, which is then passed on to herbivorous consumers, who, in turn, transfer that energy to their consumers. Because of thermodynamic laws, each time this energy is used or transferred, a portion of it is lost as heat, increasing the entropy of the universe, albeit in a tiny fraction. However, much of the primary production on Earth is not limited by the input of energy. As long as the sun shines, Earth has a

---

[1] Some Chinook salmon (*Oncorhynchus tschawytscha*) travel from the Bering Sea, across the entire width of Alaska, into British Columbia, Canada.

continual source of energy, despite its thermal losses. It is an **open system** for energy because it both receives and loses energy to the surrounding universe.

On the other hand, Earth is a **closed system** for matter, and primary production is often more limited by the input of matter than by energy. NPP can vary on both broad and local scales due to limits in essential matter. Ocean NPP is nutrient-limited rather than energy-limited and can respond rapidly to influxes of nutrients. This is demonstrated, for example, by oceanic plankton blooms during upwelling events and near-coastal runoff (Berger et al. 1989), as well as during occasional atmospheric deposition of iron, phosphorus, and nitrogen by dust storms (Jickells and Moore 2015). Clearly, it is not just energy that is important to the production of life in the biosphere; certain types of matter are also important. But what are they, and why are they so limiting?

The "Stuff" of Life

Although energy is the driving force in living systems, it is only one requirement for living organisms to grow and multiply. Organisms are, after all, made up of matter. Even plants cannot use solar energy directly, as it must be converted into chemical energy that can be used by cells. Therefore, a supply of matter (materials) must be available not only to capture the sun's energy but also to convert that energy into useful biomass (everything needed to make an organism). These materials come from the physical world: the atmosphere, hydrosphere, and lithosphere.

The biosphere comprises organisms that are mostly made up of carbon-based macromolecules, proteins, carbohydrates, nucleic acids, and fats. Proteins, serving a diversity of structural and mechanistic cell functions, are built from chains of amino acids composed mainly of four basic elements: hydrogen (H), oxygen (O), carbon (C), and nitrogen (N). Perhaps it is no coincidence that these four elements are among the top five most abundant elements in our own Milky Way Galaxy, prompting Carl Sagan's 1970s aphorism that "we are made of star stuff." Though this idea may seem cliché now, recent space mission samples from comets Wild 2 and 67P have proven that these essential elements of life can and have been forged into simple amino acids in the stars (Altwegg et al. 2016). We may never know for sure whether the amino acids that formed the basis for life on Earth came from space or were synthesized under Earth's primitive conditions (either possibility is plausible; Naraoka et al. 2023), but it certainly opens the possibility that life on this planet is not unique in the universe.

The most abundant elements in our galaxy—hydrogen, oxygen, and carbon—also make up the most abundant biological macromolecules or **biomolecules** in Earth's biosphere: carbohydrates. Simple carbohydrates (sugars), ubiquitous to all cellular life, are the chemical "batteries" that cells use to store energy. These sugars are formed when molecules of carbon, entering the biosphere as CO_2, are bound to oxygen and hydrogen via photosynthesis. Plants also create complex carbohydrates in the form of chainlike molecules of cellulose and cross-linked molecules called lignin, which together make up most of the above-ground living and dead woody biomass in trees and shrubs. These biomolecules serve a vital function in plants' competition for light and structural elements that are resistant to herbivory or pathogens. Plants create not only simple and complex carbohydrates out of carbon, hydrogen, and oxygen, but in combination with other elements, they also produce fats and proteins, all of which become part of the net primary productivity available for consumption by the rest of the food chain.

Carbohydrates aside, the biosphere also utilizes other elements that happen to make up large percentages of the Earth's mass, including calcium (Ca; 1.5%), sulfur (S; 3%), magnesium (Mg; 14%), and iron (Fe; 32%) (Morgan and Anders 1980). Calcium is essential for the structure of bones, teeth, eggshells, and shells of mollusks and other shell-critters. All proteins require sulfur to function properly. Without magnesium, chlorophyll cannot capture the solar energy needed for photosynthesis, and iron (Fe) is a key constituent of hemoglobin, which transports oxygen in the blood of vertebrates.

Although phosphorus (P) is less abundant on Earth compared to these other elements, it is still about as abundant as the sodium found in table salt (Morgan and Anders 1980). Phosphorus is essential to living systems because it is required for the phospholipid bilayer that constitutes the cell membranes surrounding all cells. Within cells, phosphorus is essential for the synthesis of DNA, RNA, and the energy molecule ATP.

Nitrogen, although rarer than phosphorus, is still the most abundant element in Earth's atmosphere. Nitrogen is essential to all amino acids, the building blocks of proteins found in all organisms on Earth.

Not all abundant elements on Earth, however, are used by organisms. For example, silicon (Si) is the third-most abundant element on the planet, following iron and oxygen (Morgan and Anders 1980). However, the only organisms that use silicon extensively are plants such as horsetails and grasses, which deposit silica both intra- and extracellularly, effectively discouraging herbivory (Sapei et al. 2007; Hall et al. 2020), and diatoms, a ubiquitous photosynthesizing phytoplankton, which uses it to construct their silicified cell walls, or "frustules."

Approximately 30 elements are now known to be essential to life. We call these and their associated chemical compounds **nutrients**. They provide energy, give the body structure, and help regulate chemical processes. Nutrients are classified as either macronutrients or micronutrients depending on how much is needed for an organism to function. Macronutrients absorbed or consumed in large quantities include compounds such as carbohydrates, lipids, and proteins, as well as elements including calcium, sodium, potassium, and magnesium. In contrast, vitamins and dietary minerals such as copper and iron are micronutrients, required for life processes, but needed in only trace amounts. However, while all nutrients, both macro- and micro-, are essential to life on Earth, not all these nutrients are easily available to all organisms under all conditions.

Nutrient availability varies among environments, with some nutrients less available than others. For plants, copper deficiency is one of the most common micronutrient issues observed in Florida. Copper leaches easily from the root zone in Florida's acidic, sandy soils, which can be problematic for citrus fruits and vegetables grown on sawgrass peat in the Everglades (Schubert 1982). Phosphorus is highly limited in tropical forests and savannas with heavily weathered ancient soils, as in southwestern Australia and parts of South Africa (Laliberté et al. 2015). In contrast, nitrogen is more limited in deserts, tundra, and boreal forests, which are characterized by relatively young soils (Wang et al. 2010).

Nutrient limitation can be an important selective force shaping adaptations. For instance, most plants are limited to obtaining nitrogen from the soil, primarily in the form of ammonium or nitrate. However, certain bacteria can "fix" N_2 in the atmosphere into a form that plants can directly use. Alders (*Alnus* spp.) grow abundantly in nitrogen-poor boreal forests by forming root nodules that host a symbiotic association with a nitrogen-fixing actinobacterium in the soil. This symbiosis can supply almost all of the nitrogen needs of the host plant (Myrold and Huss-Danell 2002). Water can also be considered an essential nutrient. Adaptations to water limitation in deserts are myriad and include the capacity for water storage (e.g., camels), concentrated urine (e.g., kangaroo rats), and the ability to absorb CO_2 at night via open stomata (portals for gas exchange), then closing them during the day to prevent high evapotranspiration (loss of water) while completing the process of photosynthesis (e.g., cacti).

As we saw in Chapter 5, *Foraging Dynamics*, there is a whole science called **foraging theory** that seeks to quantify and explain the constraints and pressures faced by organisms foraging for limited energy and matter in the environment, and the adaptive strategies that evolved to overcome those limitations. Thus, understanding how global processes affect the availability of nutrients to organisms and how organisms affect these global processes is a necessary foundation to our understanding of biological consumers and their resources.

Biogeochemical Cycles

All scuba divers know that when their oxygen tank gets low, it is time to come to the surface. This is because we cannot store or reuse oxygen in our bodies, so another dive cannot be taken until the tank is refilled with oxygen. The same is true for all elements essential to living systems. Nutrients are consumed, and they must be made available again if life is to continue.

The recycling of nutrients is crucial to the biosphere because—except for small inputs from cosmic dust or an occasional asteroid—the planet is essentially a closed system with respect to matter. Elements required to build living structures and maintain internal homeostasis and support growth must come from Earth's crust, air, and water, from where they will eventually return. Because plants and animals only have access to nutrients that occur at or near the surface of the Earth (making them a finite resource), the continuity of life requires that the materials assimilated by organisms are eventually released, so they are available for other organisms to grow. If this did not occur, supplies of essential nutrients would have been depleted long ago as they became bound in the dead bodies of organisms that lived millions of years ago. Instead, the dead bodies of organisms and wastes from the living are broken down by weathering, detritivores, and decomposers, and their component parts are recycled. The dead leaves in autumn, rustling underfoot when we walk through a temperate deciduous forest, cover the decomposing remains of other plant matter and the dead tissues of organisms that once lived in the soil beneath. Soil organisms of different kinds transform the forest litter into a fine powder of decaying tissues, ultimately reducing them to smaller and smaller molecules, from which they were once synthesized, thereby returning elements to the Earth once again.

The movements of elements essential to life within and between their various biotic (living), and abiotic (non-living) reservoirs are aptly called **biogeochemical cycles**, as these include a variety of biological, geological, and chemical processes.

Three Common Characteristics of Any Biogeochemical Cycle

Reservoirs: All nutrient cycles have reservoirs. A reservoir or pool is a place where a biologically meaningful element resides, provided there is no energy source to move it to another reservoir. An element can reside in multiple locations distributed among Earth's four major storage reservoirs: the atmosphere (air), hydrosphere (water), lithosphere (rocks and soils), and organisms (biosphere). For instance, a large pool of nitrogen makes up 78% of the atmosphere, but nitrogen-fixing bacteria in the lithosphere can capture and convert atmospheric nitrogen to a form that is available to plants and animals in the biosphere.

States and Forms of Nutrients: Nutrients can occur in gaseous form (e.g., N_2, CO_2), mineral form (e.g., apatite, the main phosphorus-containing mineral), inorganic ionic form (NH_4^+, NO_3^-, SO_4^{--}, $H_2PO_4^-$), and organic form (various carbon-based compounds in living or dead organisms or their products). Plants take up nutrients mostly in simple ionic forms (chemical forms that are soluble in water), animals take in large and more complex organic forms by consuming other organisms or detritus, and microorganisms ingest nutrients in any mineral or organic form.

The chemical form of a nutrient often varies among different reservoirs. Carbon, for instance, occurs in the lithosphere as carbonates, in the atmosphere as carbon dioxide and carbon monoxide, and the hydrosphere mostly as carbonates and bicarbonates. In water, carbon occurs as dissolved carbon dioxide below pH 5, as bicarbonates between pH 5 and 7, and as carbonates above pH 10 (McNaughton and Wolf 1979).

The interconversion among forms is mediated by chemical, physical, and biological actions in the ecosystem. For instance, water vapor in clouds reacts with carbon dioxide in the atmosphere to form carbonic acid, which then falls to the ground and dissolves rocks, releasing calcium, magnesium, potassium, and other bound elements as soluble ions. These ions can then be taken up by plants and used to form complex macromolecules, such as cell walls (with calcium; Hirschi 2004) and chlorophyll (magnesium), as well as being essential for cellular transport, enzyme function (magnesium; Shaul 2002), and maintaining turgor pressure (potassium; Kumar et al. 2006).

Nutrient Cycles May Have a Fast Phase and a Slow Phase: The rates at which essential nutrients cycle in and out of the biosphere depend on their chemical reactivity and their physical state. Gas molecules, such as CO_2, N_2, and O_2, are transported quickly, whereas phosphorus, with no gaseous phase, is relatively inactive and moves slowly through its cycle, spending most of its time in the lithosphere. Most nutrients have a fast phase, which involves smaller and more active pools in the atmosphere, hydrosphere, or soil, whereas nutrients with a slow phase have larger, inactive geological pools (rocks and sediments). Once nutrients are locked in sediments, it can take thousands or even millions of years for chemical and physical weathering to release them into biologically usable forms.

The Phosphorus Cycle: Slow Cycling Limits NPP

Phosphorus, a key constituent of cell membranes, DNA, and the energy molecule ATP, is critical for all life. Although required in smaller quantities than oxygen, carbon, and hydrogen, phosphorus is an indispensable nutrient that can determine the survival and growth of organisms in any environment. Phosphorus is the 11th-most abundant element on Earth, but it is not readily available in a biologically usable form. Unlike carbon and nitrogen, phosphorus does not exist in a gaseous state, so the primary source of phosphorus for land plants is the lithosphere (rocks).

The global distribution of phosphate resources is uneven, with some regions possessing a larger share than others. For example, in 2010, only five countries—Morocco, China, South Africa, Jordan, and the United States—accounted for over 90% of the world's total phosphate reserves, with Morocco alone holding significantly more than all the other countries combined (Van Kauwenbergh 2010).

However, other places, such as the Amazon basin, have little to no phosphate-bearing rocks to supply soils with this essential nutrient. Because different types and ages of rocks contain different amounts of phosphorus, the availability of phosphorus to plants depends on the parent material in the soil where the plants are located. Compared to the pace of weathering, the uptake of phosphorus into the biosphere is rapid. Consequently, the weathering of continental rocks ultimately controls the pace of the global phosphorus cycle. Since phosphorus never enters gaseous form, it is deposited from the biosphere back into the lithosphere only by decomposers or when it is leached from land to sea, where it eventually becomes part of the ocean sediments. Once in the lithosphere, only tectonic uplift, volcanic activity, or other geological processes can expose it to surface weathering again. This explains why phosphorus has one of the slowest cycles of all the inorganic macronutrients (Figure 1) and why it often limits net primary production.

Phosphorus becomes available to the biosphere through weathering of a single rock mineral (apatite) or seabird guano, or the decomposition of dead tissues (Figure 1). Once phosphorus is released from these sources, it is typically in the form of phosphate (PO_4^-), which plants can absorb. Vitousek et al. (2010) suggest that one of the major means by which phosphorus limits primary productivity is through soil depletion. Phosphate, being in an ionic form (and thus transportable by water), is susceptible to leaching from terrestrial ecosystems to lakes and oceans. This leads to a gradual loss of phosphorus, which can occur over millions of years (Vitousek et al. 2010). Soil phosphate depletion of this form is evident from the Hawaiian archipelago chronosequence (with islands getting progressively older from southeast to northwest): on sites with the same type of soil parent material and similar forest vegetation, the oldest site (4.1 million years) lost 90% of the phosphorus from its parent material compared to the youngest site (300 years; Vitousek and Farringon 1997). Other sources of phosphorus limitation include soil barriers (e.g., permafrost, iron pans, and clay horizons), which form over hundreds to thousands of years and prevent plant roots from accessing deeper, nondepleted portions of the soil profile. Another major source of limitation is low phosphorus input from poor phosphorus-containing parent materials (Vitousek et al. 2010).

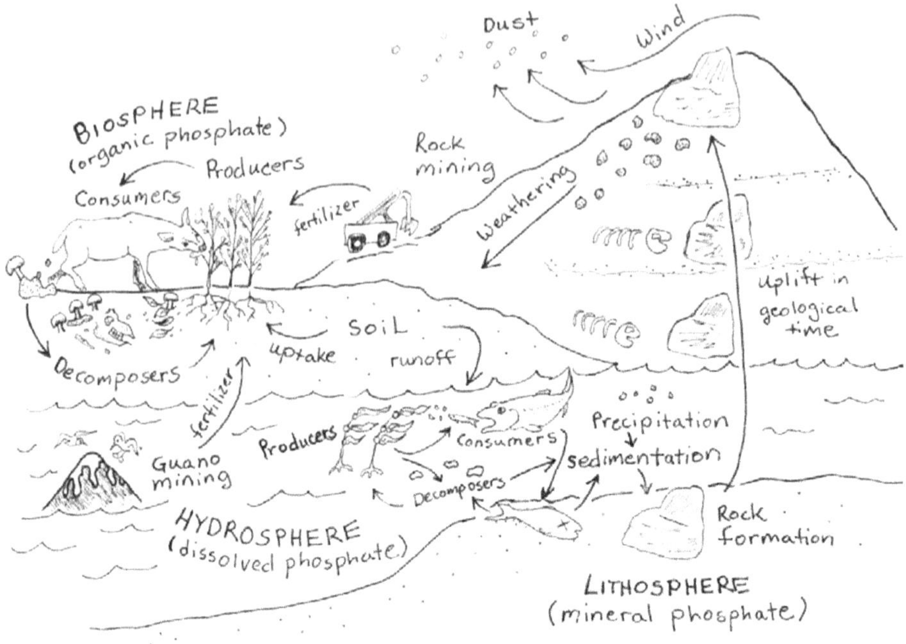

Figure 1. Phosphorus reservoirs, states, and fluxes.

In areas with phosphorus limitation, atmospheric transport of phosphorus, lifted by air via dust particulates, volcanic ejecta, or burned biomass, is critical to net primary productivity. In the incredibly phosphorus-limited Amazon basin, Barkley et al. (2019) estimate that about 10,000 to 30,000 metric tons of phosphorus emitted from biomass burning in Africa are deposited into the Amazon basin via prevailing winds, a distance of more than 5,000 km. However, compared to burning of fuels or volcanoes, the largest source of atmospheric phosphorus is via dust particles (Mahowald et al. 2008). Global dust emissions come largely from the deserts of North Africa, the Arabian Peninsula, and Central and South Asia, with smaller contributions from the deserts in North and South America and Australia (Wang et al. 2017). Each year, an estimated one to three billion metric tons of dust (containing nutrients and sometimes contaminants) are lofted from Earth's surface by wind and carried around the world, much of which ends up in the ocean (Wang et al. 2017). Oceans receive an average of over 550,000 metric tons of phosphorus annually, according to Mahowald et al. (2008). Because oceans are typically more phosphorus-limited than nitrogen-limited, this can have a big impact on their net primary productivity.

Phosphorus is also limiting in agricultural systems. This is because it is removed from soil through the harvest of biomass and by leaching into freshwater and marine ecosystems. Some of that leached phosphorus, which contributes to ocean primary productivity, supports secondary productivity in the form of marine fish and other marine organisms, which are then consumed by seabirds. By mining guano from seabird colonies, some of this marine phosphorus can be returned to agricultural soils. Until the 19th century, this commodity was so valuable that Chile, Bolivia, Peru, Spain, and even the United States fought wars over control of guano-rich islands (Cushman 2013).

We also mine phosphorus-bearing rocks for agricultural fertilizer. According to one estimate (Cordell et al. 2009), approximately 17.5 million tons of phosphorus were mined in 2005, with 14 million tons used in fertilizer. Much of the remaining phosphorus was used in cattle-feed supplements, food preservatives, and industrial cleaning agents. However, around 8 million tons of this phosphorus were lost directly from farms through soil leaching and erosion (Cordell et al. 2009).

Although we have outlined several examples of how phosphorus limits primary productivity and have emphasized its slow cycle and depletion from soils, surprisingly, nitrogen (N) is the more limiting nutrient in most productive terrestrial ecosystems. This is because dead biomass accumulates as a sink for nutrients, and in such cases, the biochemical regeneration of phosphorus occurs more rapidly than that of nitrogen, driving the system toward nitrogen limitation (Vitousek et al. 2010). This explains why Vitousek and Farringon (1997) found that while phosphorus was more limiting on 4.1-million-year-old Hawaiian soils, nitrogen was the more limiting nutrient in younger Hawaiian soils (300 years old). It is, therefore, important to understand N:P ratios in organisms and ecosystems, as well as the biogeochemical cycling of nitrogen.

The Nitrogen Cycle: "Fast" Yet Limiting to NPP

Nitrogen is an example of a nutrient that occurs abundantly in the gas phase, constituting 78.1% of our atmosphere. It is critically essential to life in the biosphere for, among other things, the synthesis of proteins and nucleic acids—the two most important macromolecules of life, without which cells could neither function nor reproduce. Despite its overwhelming abundance in the atmosphere, this nitrogen is not readily available to most life in the biosphere. It is inert due to the three covalent bonds that bind two nitrogen atoms together as N_2. It requires a significant amount of energy to dissociate atmospheric N_2 and convert it, or fix it, to a different chemical state, namely nitrate (NO_3^-) or ammonium (NH_4^+). Only in these forms can plants uptake and use nitrogen, allowing it to enter the biosphere. So, how does nitrogen move from the atmosphere to the biosphere in a form usable by plants?

Nitrogen fixation occurs in two processes: abiotic (physical) and biotic (biological). Most abiotic nitrogen fixation occurs when lightning breaks the bonds of atmospheric N_2. The visible flash during lightning accounts for 2% of its energy, while the rest of the lightning's energy heats the surrounding atmosphere enough to split N_2, forming nitrogen oxides and dioxides that dissolve in rain and enter the soil as NO_3^-. This form of nitrogen is available to plants through their roots, transported by the water that plants use to grow (Figure 2). Combustion from wildfires, fossil fuels burning, or volcanic eruptions also produces nitrates, which become suspended in the air and fall to the ground with precipitation. However, the total natural abiotic inputs of fixed nitrogen are low, amounting to about 5% of what is typically fixed by biological organisms (Galloway and Cowling 2002). This small amount of abiotic nitrogen input contrasts with the large abiotic inputs of phosphorus and most other essential nutrients, which are supplied by weathering of soil and rock (Vitousek et al. 2002).

Living organisms rather than geological processes largely mediate the nitrogen cycle. The only organisms capable of converting the vast supply of atmospheric N_2 to a biologically usable form are bacteria. In oceans, biologically useful nitrogen is fixed by free-living cyanobacteria, cyanobacteria in symbiotic relationships with unicellular alga, and many types of bacteria (Zehr and Capone 2020). Most terrestrial nitrogen fixation is performed by soil bacteria of two kinds: those living freely in the soil and those living in a symbiotic relationship in the root nodules of certain plants, such as clover, alfalfa, peas, soybeans, and alder. Symbiotic nitrogen fixation has evolved in only 10 of the 380 flowering plant families, with legumes (Fabaceae) containing by far the most genera (640; Soltis et al. 1995). Symbiotic nitrogen fixation is estimated to produce as much as 10 kg N/hectare per annum, compared to about 0.4 kg N/hectare per annum fixed by lightning strikes, a 25-fold increase (Smil 1997). Nitrogen-fixing symbioses also occur between *Azolla* ferns and cyanobacteria (Newton and Herman 1979), lichens and nitrogen-fixing symbionts (Millbank and Kershaw 1969), and between termites and nitrogen-fixing bacteria in their guts (Breznak et al. 1973).

The line between nitrogen-fixing symbionts and nitrogen-fixing free-living soil bacteria is a fuzzy one, as relationships between these microorganisms and plants come in a diversity of forms, but asymbiotic (free-living) nitrogen fixers are nearly ubiquitous in terrestrial ecosystems (Reed et al. 2011). The rate of nitrogen fixation by free-living nitrogen-fixers has been estimated to be

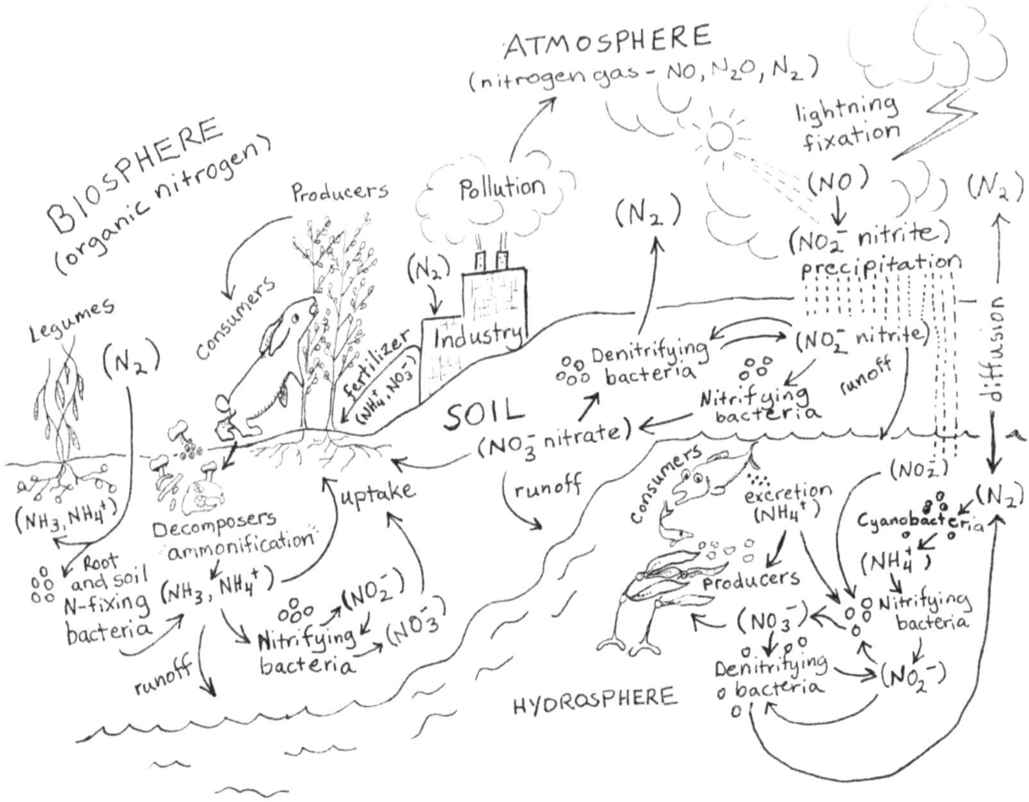

Figure 2. Nitrogen reservoirs, states, and fluxes.

greater than 150 kg N/hectare per year, which is comparable to the 100–500 kg N/hectare per year of fertilizer added to modern agricultural systems (Reed et al. 2011).

Within any particular ecosystem, fixed soil nitrogen can be assimilated into plants and passed along the food chain to consumers, residing in the biosphere for some time. Upon the death and decay of these organisms, nitrogen is returned to the soil as ammonia (NH_3). Most ammonia is converted to ammonium ions (NH_4^+) on the surface of soil particles. This process is called ammonification, and many kinds of decay-consuming microorganisms participate in it. Ammonia can be converted into nitrite (NO_2) by a variety of nitrifying bacteria, and then nitrobacteria oxidize these nitrites to nitrates (NO_3^-). Both ammonium and nitrate are readily taken up by plants again, completing what should be a relatively fast cycle. However, due to the nature of nitrogen's chemical bonding in organisms, organic nitrogen can often remain bound for long periods of time, even after most other biological material has decayed (Vitousek et al. 2002). This is because nitrogen is covalently bonded to carbon, which forms a stronger bond than both the ester bonds (phosphorus and sulfur) and the ionic bond (iron, magnesium, calcium, etc.) that other elements form with carbon. Additionally, at any given time, most of the decaying matter in an ecosystem is plant-derived. However, despite the uptake of nitrogen by plants, their carbon-to-nitrogen ratio is large, meaning they are composed mostly of carbon (e.g., cellulose and lignin), compared to animals, which are mostly made up of nitrogen-containing proteins. Thus, there is not a lot of nitrogen to be gained from plant decomposition. Because nitrogen is limiting to decomposing microbes, they tend to retain any nitrogen they obtain, further delaying the recycling of nitrogen into biologically available forms (Vitousek et al. 2002).

Another group of microorganisms, denitrifying bacteria, can remove biologically available nitrogen by reducing soil nitrates back to molecular nitrogen (N_2) under anoxic conditions such as those found in wet soils, bogs, marshes, and swamps. Once converted to a gas, the nitrogen escapes into the atmosphere. In addition, during both bacterial nitrification and denitrification, intermediary gaseous compounds such as N_2O, NO, and NO_2 are formed, which can also escape to the atmosphere. N_2O (nitrous oxide) is a greenhouse gas—the third most plentiful after carbon dioxide and methane—and it also destroys ozone when it reaches the stratosphere. Establishing rates of N_2O flux from soil to the atmosphere on an ecosystem scale can be challenging because ecosystem-scale controls are interactive and vary by system (forest, grassland, etc.). However, it is known that rainfall, temperature, and fertilizer application can all contribute to increased N_2O flux (Groffman et al. 2000).

In addition to nitrogen flux to the atmosphere, nitrogen can be easily removed from soils in a local area through soil leaching and biomass harvest (e.g., crops, timber). Because soil nitrogen occurs in a soluble ionic form, it is easily washed out of soils by rain and ultimately transported via freshwater to coastal areas and into the ocean. In intensive agricultural areas, crop harvests rapidly deplete nitrogen from the soil. Because of these losses, soil nitrogen must be replenished if plants are to continue growing there.

In preindustrial times, the annual flux of nitrogen from the atmosphere to land and aquatic ecosystems was roughly balanced by a reverse denitrification flux. Humans relied on lightning strikes, nitrogen-fixing bacteria, natural fertilizers (manure), and legumes with nitrogen-fixing symbionts to bring nitrogen to soils. In intensive agricultural areas, crop rotation, in which a leguminous crop is rotated with a cereal crop, was and still is a common practice among farmers worldwide to maintain soil fertility. Upon death, legume roots left in the soil decay, returning organic nitrogen compounds to the soil for uptake by the next generation of plants. It is an age-old practice—there is evidence that Egyptians of the Nile Delta, around 2300–2100 BCE, who practiced intercropping of cereal grains with legumes (such as clover, *Trifolum* sp.), understood that intercropping with legumes was more beneficial than cultivating a monoculture (Malleson 2016).

As agricultural production intensified in the 19th century, the nitrogen- and phosphorus-rich guano produced by millions of seabirds, such as nesting cormorants, became highly prized. According to recent estimates, the worldwide total nitrogen and phosphorus excreted in seabird breeding colonies were 591 and 99 gigatons per year, respectively (Otero et al. 2018). Because about 12% of that nitrogen and 22% of that phosphorus is readily dissolvable, seabird colonies are nutrient "hot spots," providing rich runoff to land and sea plants living down-current. This is especially true in the waters around Antarctica and its nearby islands, where seabirds are typically larger and have longer breeding seasons than seabirds elsewhere (Perkins 2018).

Because nitrogen is relatively limited in natural ecosystems, human alteration of the nitrogen cycle has large impacts on natural productivity and species composition (Vitousek et al. 2002). In the early 20th century, a German physical chemist, Fritz Haber, received the Nobel Prize in Chemistry for developing a high-pressure method of combining N_2 with hydrogen to produce ammonia. This "synthetic" nitrogen fertilizer revolutionized agriculture, but its abundance also created problems. Between 80 and 110 million tons of nitrogen fertilizers are applied to agricultural areas annually across the globe. Because of the mobility of nitrogen ions, 30% of these fertilizers eventually wind up in freshwater bodies, coastal areas, and oceans, causing **eutrophication**. This condition leads to an excessive abundance of photosynthetic algae, resulting in major aquatic species loss (Smil 1997).

Although nitrogen is a highly mobile element and can be easily leached from soils, nitrogen losses can vary significantly. The plants in contact with the soil may play a role in this variability. Lovett et al. (2002) found that nitrate export from forested watersheds in the Catskill Mountains of upstate New York is strongly influenced by the carbon-to-nitrogen (C:N) ratio of the soils. This, in turn, is influenced by the plant species present. They explain that when the C:N ratio is high, strong demand by soil microbes for that limited nitrogen causes less nitrogen leaching. They also found a strong relationship between higher C:N ratios and greater amounts of red oak (*Quercus rubra*), and lower C:N ratios and greater amounts of sugar maple (*Acer saccharum*). This suggests that a forest dominated by sugar maples will have more nitrate leached into the surrounding watershed than a forest dominated by red oak. They argue that because trees vary widely in their C:N litter chemistry,

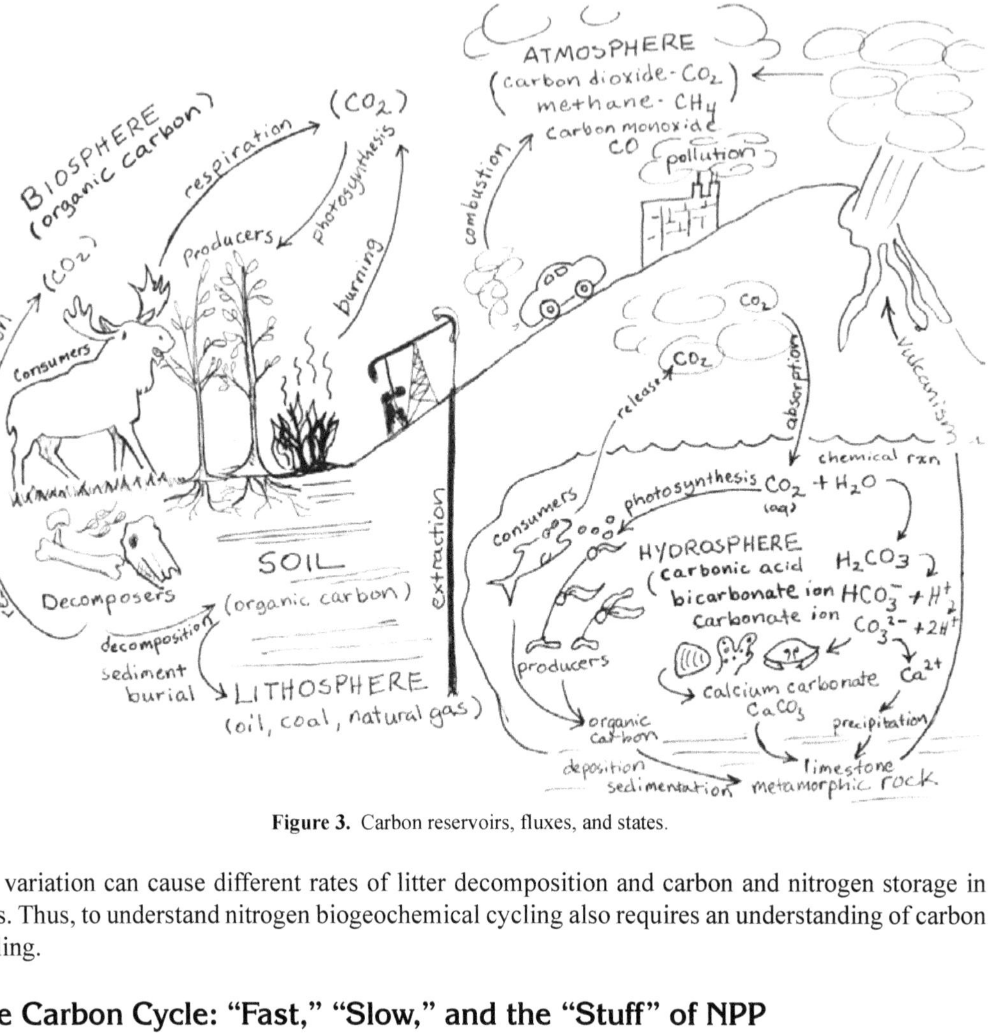

Figure 3. Carbon reservoirs, fluxes, and states.

this variation can cause different rates of litter decomposition and carbon and nitrogen storage in soils. Thus, to understand nitrogen biogeochemical cycling also requires an understanding of carbon cycling.

The Carbon Cycle: "Fast," "Slow," and the "Stuff" of NPP

Every single carbon atom that makes up an individual organism came from the environment, either from other organisms or fixed out of the atmosphere via photosynthesis.[2] Living and dead terrestrial organisms are a reservoir for about 2,000 gigatons (Gt) of carbon (Falkowski et al. 2000). Three other major reservoirs are the lithosphere (>75,000,000 Gt), the hydrosphere (38,400 Gt), and the atmosphere (720 Gt; Falkowski et al. 2000). Although the atmosphere contains only this seemingly small amount, this reservoir is still immeasurably important. Nearly all the flux of carbon into living systems comes from the atmosphere. Simply put, there would be no life as we know it without this source of carbon.

The biogeochemistry of carbon includes a **fast phase** and a **slow phase**. In the short term, elemental carbon in the form of CO_2 is removed from the atmosphere and enters living systems through photosynthesis. Producers such as plants, algae, and cyanobacteria remove carbon dioxide from the atmosphere or water during photosynthesis and use it to build organic carbon molecules. When consumers eat these products, the carbon is transferred to them. Both producers and consumers break down carbon compounds during cellular respiration, releasing both stored

[2] Note: This sweeping statement is overwhelmingly true, though not absolutely true, because chemosynthesis in the depths of the oceans fixes carbon from carbon dioxide or methane obtained from the hydrosphere.

energy and carbon dioxide that returns to the atmosphere. When plants grow, they tie up carbon in their tissues. When fire burns them, the carbon is released back into the environment as carbon compounds or carbon dioxide. When plants and animals die, they are decomposed, or broken down, by a host of decomposers—bacteria, protists, fungi, and soil detritivores such as earthworms and nematodes—releasing carbon into the soil and atmosphere and thereby completing the carbon cycle. This gaseous cycle is the fast phase of carbon cycling (Figure 3).

The carbon in living organisms is only a tiny fraction of the carbon that exists on Earth, and while some carbon molecules cycle rapidly between organisms and the atmosphere, a larger flux of carbon molecules exists between the atmosphere and the hydrosphere. These components of the carbon cycle exchange about 90 Gt of carbon per year in each direction via diffusion—without the involvement of any life in the process (Falkowski et al. 2000). This exchange is fast, attaining equilibrium rapidly. Nonetheless, because the ocean can hold about 50 times more dissolved inorganic carbon than the atmosphere, the ocean is a far greater reservoir.

In contrast to the fast phase of carbon cycling, the slow phase involves carbon spending most of its time locked in the lithosphere in sedimentary rocks such as calcite and limestone (> 60,000,000 Gt), which are largely composed of calcium carbonate ($CaCO_3$), and in kerogens (15,000,000 Gt), which are the transformed remains of ancient planktonic biomass in shales (sedimentary rocks). How did all that carbon get there?

When CO_2 moves from the atmosphere to the ocean, dissolved carbon is available to marine plants, algae, and bacteria for photosynthesis or is converted to carbonate and bicarbonate ions. These ions are used by certain marine organisms, such as corals and foraminifera plankton, to construct skeletons and shells. Carbonate ions can also combine with calcium in water to form calcium carbonate ($CaCO_3$). Calcium carbonate has a low solubility in water, so it precipitates out of the water and becomes part of the sediments at the ocean's bottom, along with the calcium carbonate skeletons from tiny marine organisms that died and sank to the ocean floor. Over millions of years, the calcium carbonate sediments that accumulate at the bottom of the ocean can develop into massive sources of carbon in the form of rocks known as dolomite and limestone. After being deposited on the ocean floor, it takes millions of years for that carbon to return to the biosphere via geotectonic processes. Plate tectonics, the movements of surface plates on Earth's crust, recycle carbon from the lithosphere. When a plate is thrust downward, the rocks and sediments are heated and transformed, releasing carbon dioxide into the atmosphere during volcanic eruptions. When a plate is thrust upward to Earth's surface, uplifted rocks and sediments are eroded by weathering, making carbon available in soils and freshwater. This is the slow, geological, or sedimentary phase of carbon cycling.

On land, carbon can be buried as organic matter before it fully decomposes. Over millions of years, some of this organic matter transforms into fossil fuels such as petroleum, natural gas, and coal, collectively forming the second-largest carbon pool or reservoir. Petroleum—used in the production of heating oil, diesel, and gasoline—forms when ancient organic material is subjected to heat and pressure over time. This process first degrades the organic material into kerogen, a waxy organic material. With more heat and pressure, kerogen converts into liquid petroleum, which seeps through rocks in the crust. Natural gas, often associated with petroleum, is mostly methane (CH_4), but may contain trace amounts of other gases, such as ethane (C_2H_6). Natural gas migrates through rocks in the crust and is trapped by geologic structures. Coal forms from decaying plant material that is buried, compressed, and lithified over time.

Fossil fuels are an important part of the modern carbon cycle. They represent carbon that was removed from the carbon cycle millions of years ago and stored or sequestered for long periods. When fossil fuels are burned as an energy source, the sequestered carbon is released into the atmosphere as carbon dioxide to become part of the fast cycle. While the slow cycle operates in rocks and sediments, the fast cycle operates in the biosphere. Carbon completes its fast cycle within years, moving carbon from the atmosphere to the biosphere and then back to the atmosphere. In the slow cycle, carbon can take millions of years to complete, moving carbon through the Earth's crust between rocks, soil, ocean, and atmosphere.

As carbon flows among these various reservoirs (atmosphere, hydrosphere, biosphere, and lithosphere), any change in the cycle that shifts carbon out of one reservoir adds more carbon to the others. For instance, the combustion of fossil fuels sends carbon dioxide from the lithosphere into the atmosphere, where it can also enter the biosphere and hydrosphere. Because CO_2 is a greenhouse gas and carbon enters the ocean as carbonic acid, this dramatic change in the carbon cycle has led to global warming and changes in surface ocean chemistry (IPCC 2021). This human-caused shift in biogeochemical cycling prompted ecologist Eugene Stoermer to coin the term **Anthropocene** to describe the geological epoch in which humans began significantly affecting biogeochemical cycling.

Putting the "Bio" in Geochemical Cycles

Nutrient cycling—the flux between reservoirs—can occur with very little involvement in the biosphere (e.g., phosphorus), or up to a great deal of involvement (e.g., nitrogen and carbon). And over the Earth's existence, biogeochemical cycles have affected not only the evolution of life, but they have become more and more dominated by the "bio" part of the cycle.

About 4.5 billion years ago, proto-Earth was struck by another planet, which led to the formation of the modern Moon and Earth. The collision caused the spin and tilt affecting diel and seasonal changes, and created heat that powers plate tectonics and volcanism, which now supplies essential elements to the biosphere (Nisbet and Sleep 2001). Teasing out the composition of Earth's environment as it transitioned from this Hadean eon impact to the Archaean eon when primitive life emerged has been challenging. Evidence indicates there was liquid water, but there is also evidence that the Sun was 20–25% dimmer. This presents a paradox: a dimmer Sun would suggest frozen water. Adding to the intrigue, evidence suggests that impacts from extraterrestrial objects created a chemically reducing environment, with a methane-rich atmosphere and vaporized water, leading to periods of high temperatures due to greenhouse effects. This itself is complicated by further evidence that these periods were transient (Zahnle et al. 2020). Nevertheless, regardless of the exact conditions, primitive life likely emerged as early as 4.1 billion years ago (Bell et al. 2015).

Evidence from resurrected proteins indicates that primitive life originated as thermophilic bacteria, surviving in hot ocean environments ("hot house"), possibly created by frequent extraterrestrial impacts. By 2.7 billion years ago, microbial mats and stromatolite-forming cyanobacteria were life *du jour*, indicating life had evolved photosynthesis and adapted to a cooler ("greenhouse") environment (Gaucher et al. 2008). Prior to photosynthesis, the earliest microbial ecosystems were probably made up of unicellular organisms that used **redox potential** as an energy source for biomass production, with methane as a metabolic waste. Productivity would have been low and energy-limited rather than nutrient-limited. These organisms may have influenced carbon cycling by releasing methane into the atmosphere, creating a low carbon monoxide-to-methane ratio, which would not have been possible under lifeless conditions (Sauterey et al. 2020).

The evolution of photosynthesis in oxygenic phototrophs, such as cyanobacteria, had a fundamental impact on Earth's history—paving the way for atmospheric oxygen and the "explosion" of animal life beginning in the Cambrian period (Sánchez-Baracaldo and Cardona 2020). The first rise in oxygen in the atmosphere, known as the Great Oxidation Event, occurred between 2.32 and 2.4 billion years ago at the beginning of the Proterozoic eon (Sánchez-Baracaldo and Cardona 2020). Oxygenic phototrophs were the first organisms to convert carbon dioxide and water into food (carbohydrates) using solar energy. The byproduct of photosynthesis was oxygen, which was added to the atmosphere for the first time. Much of this oxygen did not remain long in the early atmosphere, as it quickly combined with metals to form oxides, such as iron oxide. Later, a second major oxygenation event occurred between 800 and 600 million years ago, significantly increasing atmospheric oxygen (the Neoproterozoic Oxidation Event). This event coincided with the evolution of planktonic photosynthetic organisms, just prior to the Paleozoic era, when the Cambrian period began. Over geological time, oxygen accumulated in the atmosphere because carbon fixation exceeded the respiration of organic matter, which drew down carbon in the atmosphere and buried it in marine sediments (Sánchez-Baracaldo and Cardona 2020). Carbon found a new reservoir–the living world.

Terrestrial plants, which evolved in the Paleozoic from the Ordovician to the Carboniferous period (485–298.9 million years ago), are linked to a stepwise increase in Earth's oxygen levels, a drawdown of atmospheric CO_2, and a change from "greenhouse" to cooler conditions. The Carboniferous is known for its coal-bearing rocks, which resulted from the sequestering of atmospheric carbon into the newly innovated lignified, or "woody," plant biomass (more resistant to decay than non-lignified plant parts) and the swampy conditions under which this biomass was buried. The evolution of land plants—from rootless plants to lignified plants, to vascular plants with shallow roots, to tall perennial trees with deep roots, and finally seed plants—corresponds to changes in Earth's climate and atmosphere. Plant evolution reduced albedo, accelerated the water cycle, enhanced chemical weathering and clay formation in soils, decreased CO_2 and increased O_2 in the atmosphere, and promoted fire regimes, which contributed to the creation of an ozone layer (Dahl and Arens 2020).

Just prior to the evolution of land plants, high atmospheric CO_2 concentrations created a greenhouse effect that countered a dimmer sun (4% dimmer than now). This concentration was a stable balance between volcanism and carbonate sinks due to silicate weathering, a process whereby CO_2 reacts with water in the atmosphere to form carbonic acid (H_2CO_3), the primary chemical agent responsible for the weathering of calcium silicate rocks. During this period, carbonic acid reached the ground in rainfall and eroded calcium silicate rocks, thereby releasing calcium ions (Ca^{++}) and bicarbonate ions (HCO_3^-). Streams and rivers transported these ions to the oceans, where they reacted to produce sediment-forming calcium carbonates, CO_2, and water. The emergence of land plants contributed to greater silicate weathering in soils by acidifying them with organic matter and root exudates, increasing water flow and oxygen content in soils (Dahl and Arens 2020). The removal of CO_2 since the emergence of land plants has reduced its concentration in the atmosphere to the present-day level of 0.04%, cooled the planet, and produced a giant reservoir of carbon (100 million gigatons of carbonate sediments).

Shell-critters, such as crustaceans (krill, shrimps, and crabs) and mollusks (clams and oysters), as well as reef-building corals, evolved to use this abundant aquatic calcium carbonate to make shells and reef structures. These marine creatures, when deceased, harden and turn into limestone (a sedimentary rock), making the carbon of living systems geologically important when their deposited calcium-remains accumulate in such colossal amounts that they become geological features, such as the Great Barrier Reef off Australia, the coral islands of the Pacific and southern Florida, and the White Cliffs of Dover in England. The limestone can also become marble under increased temperature and pressure, providing the ancient Greeks with the material to build the Parthenon. In a modern spin of the cycle, the Parthenon is now subject to human pollution-caused chemical weathering, by which it will eventually be weathered back into the ocean.

In addition to changes in carbon cycling, nitrogen cycling was also transformed by the evolution of life. The nitrogen-fixing bacteria may have evolved 3.2 billion years ago, but this fixation (which removes N_2 from the atmosphere) would have been balanced by anaerobic prokaryotes, which used fixed nitrate to convert CO_2 into organic carbon, a process that returns fixed nitrogen to the atmosphere as N_2 (Lammer et al. 2018). As oxygen in the atmosphere increased upon the evolution of photosynthetic organisms during the Great Oxidation Event, oxygenated forms of nitrogen became easier to create, while also allowing for a greater flux of nitrogen back into the atmosphere as N_2. This set the stage for the expansion of nitrogen-fixing bacteria and the evolution of other microorganisms with the ability to oxidize ammonium to nitrite (NO_2^-) and nitrate (NO_3), a stepwise process of nitrification that allows this mobile form of nitrogen to be assimilated by plants (Lammer et al. 2018). As bacterial diversity increased, the atmosphere became dominated by nitrogen and oxygen, reaching present-day levels of 78% and 21%, respectively. The subsequent formation of ozone (O_3) shielded Earth from ultraviolet radiation, creating more benign conditions for the evolution of NPP-consuming land animals, including, much later, the evolution of another organism that has dramatically impacted biogeochemical cycles–*Homo sapiens* or humans.

References

Altwegg, K., Balsiger, H., Akiva Bar-Nun, A., Berthelier, J. J., Bieler, A., Bochsler, P. et al. 2016. Prebiotic chemicals—amino acid and phosphorus—in the coma of comet 67P/Churyumov-Gerasimenko. Science Advances 2: 5, e1600285. doi: 10.1126/sciadv.1600285.

Barkley, A. E., Prospero, J. M., Mahowald, N., Hamilton, D. S., Popendorf, K. J., Oehlert, A. M. et al. 2019. African biomass burning is a substantial source of phosphorus deposition to the Amazon, Tropical Atlantic Ocean, and Southern Ocean. PNAS 116: 16216–16221.

Begon, M., Harper, J. L. and Townsend, C. R. 1996. Ecology. 3rd edition. Blackwell Science, Cambridge, MA.

Bell, E. A. Boehnke, P. and Harrison, T. M., and Maob, W. L. 2015. Potentially biogenic carbon preserved in a 4.1 billion-year-old zircon. PNAS 112: 14518–14521.

Berger, W. H., V. S. Smetacek, and G. Wefer. 1989. Ocean productivity and paleoproductivity—An overview. pp. 1–34. *In*: Berger, W. H., Smetacek, V. S. and Wefer, G. [eds.]. Productivity of the Ocean: Present and Past. John Wiley, Hoboken, N. J.

Breznak, J. A., Brill, W. J., Mertins, J. W. and Coppel, H. C. 1973. Nitrogen fixation in termites. Nature 244: 577–579.

Cordell, D., Drangert, J. O. and White, S. 2009. The story of phosphorus: Global food security and food for thought. Global Environ. Change 19: 292–305.

Cushman, G. T. 2013. Guano and the Opening of the Pacific World: A Global Ecological History. Cambridge University Press, Cambridge.

Dahl, T. W. and Arens, S. K. M. 2020. The impacts of land plant evolution on Earth's climate and oxygenation state - an interdisciplinary review. Chemical Geology 547: 119665 doi.org/10.1016/j.chemgeo.2020.119665.

Falkowski, P., Scholes, R. J., Boyle, E., Canadell, J., Canfield, D., Elser, J. et al. 2000. The global carbon cycle: a test of our knowledge of earth as a system. Science 290: 291–6. doi: 10.1126/science.290.5490.291.

Galloway, J. N. and Cowling, E. B. 2002 Reactive nitrogen and the world: 200 years of change. Ambio 31: 64–71.

Gaucher, E. A., Govindarajan, S. and Ganesh, O. K. 2008. Paleotemperature trend for Precambrian life inferred from resurrected proteins. Nature 451: 704–707. https://doi.org/10.1038/nature06510.

Groffman, P. M., Brumme, R., Butterbach-Bahl, K., Dobbie, K. E., Mosier, A. R., Ojima, D. et al. 2000. Evaluating annual nitrous oxide fluxes at the ecosystem scale. Global Biogeochemical Cycles 14: 1061–1070.

Hall, C. R, Dagg, V., Waterman, J. M. and Johnson, S. N. 2020. Silicon alters leaf surface morphology and suppresses insect herbivory in a model grass species. Plants 9: 643. doi: 10.3390/plants9050643.

Hirschi, K. D. 2004. The calcium conundrum. Both versatile nutrient and specific signal. Plant Physiology 136: 2438–2442.

IPCC. 2021. Summary for Policymakers. Climate Change 2021: The Physical Science Basis. Contribution of Working Group I to the Sixth Assessment Report of the Intergovernmental Panel on Climate Change, Cambridge University Press, Cambridge. doi:10.1017/9781009157896.00.

Jickells, T. and Moore, C. M. 2015. The importance of atmospheric deposition for ocean productivity. Annual Review of Ecology Evolution and Systematics 46: 481–501. doi:10.1146/annurev-ecolsys-112414-054118.

Kumar, A. R., Kumar, N. and Kavino, M. 2006. Role of potassium in fruit crops - a review. Agric. Rev. 27: 284–291.

Laliberté, E., Lambers, H., Burgess, T. I. and Wright, S. J. 2015. Phosphorus limitation, soil-borne pathogens and the coexistence of plant species in hyperdiverse forests and shrublands. New Phytologist 206: 507–21.

Lammer, H., Zerkle, A. L., Gebauer, S., Tosi, N., Noack, L., Scherf, M. et al. 2018. Origin and evolution of the atmospheres of early Venus, Earth and Mars. Astron Astrophys Rev 26: 2. https://doi.org/10.1007/s00159-018-0108-y.

Lovett, G. M., Weathers, K. C. and Arthur, M. A. 2002. Control of nitrogen loss from forested watersheds by soil carbon:nitrogen ratio and tree species composition. Ecosystems 5: 712–718.

Mahowald, N., Jickells, T. D., Baker, A. R., Artaxo, P., Benitez-Nelson, C. R., Bergametti, G. et al. 2008. Global distribution of atmospheric phosphorus sources, concentrations and deposition rates, and anthropogenic impacts. Global Biogeochemical Cycles 22: GB4026.

Malleson, C. 2016. Informal intercropping of legumes with cereals? A re-assessment of clover abundance in ancient Egyptian cereal processing by-product assemblages: archeobotanical investigations at Khentkawes town, Giza (2300-2100 BC). Veget. Hist. Archeobot. 25: 431–442.

McNaughton, S. J. and Wolf, L. L. 1979. General Ecology. Syracuse University. New York.
Millbank, J. W. and Kershaw, K. A. 1969. Nitrogen metabolism in lichens. The New Phytologist. 68: 721–729.
Morgan, J. W. and Anders, E. 1980. Chemical composition of Earth, Venus, and Mercury. PNAS 77: 6973–6977.
Myrold, D. D. and Huss-Danell, K. 2002. Alder and lubine enhance nitrogen cycling in a degraded forest soil in Northern Sweden. Plant and Soil 254: 47–56.
Naraoka, H., Takano, Y., Dworkin, J. P., Oba, Y., Hamase, K., Furusho, A. et al. 2023. Soluble organic molecules in samples of the carbonaceous asteroid (162173) Ryugu. Science 379: p.eabn9033.
Newton, J. W. and Herman, A. I. 1979. Isolation of cyanobacteria from the aquatic fern, *Azolla*. Arch. Microbiol. 120: 161–165.
Nisbet, E. G. and Sleep, N. H. 2001. The habitat and nature of early life. Nature 409: 1083–1091
Otero, X.L., De La Peña-Lastra, S., Pérez-Alberti, A., Ferreira, T. O., and Huerta-Diaz, M A. 2018. Seabird colonies as important global drivers in the nitrogen and phosphorus cycles. Nature Communications 9: 246. https://doi.org/10.1038/s41467-017-02446-8.
Perkins, S. 2018. Bird poop brings 3.8 million metric tons of nitrogen out of the sea each year. Science. 23 Jan 2016. https://www.science.org/content/article/bird-poop-brings-38-million-metric-tons-nitrogen-out-sea-each-year.
Raschke, E., Vonder Harr, T. H., Bandeen, W. R. and Pasternak, M. 1973. The annual radiation balance of the earth-atmosphere system during 1969-70 from Nimbus 3 measurements. Journal of the Atmospheric Sciences 30: 341–364.
Reed, S. C., Cleveland, C. C. and Townsend, A. R. 2011. Functional ecology of free-living nitrogen fixation: a contemporary perspective. Annual Review of Ecology, Evolution, and Systematics 42: 489–512.
Running, S. W., Nemani, R. R., Heinsch, F. A., Zhao, M., Reeves, M. and Hashimoto, H. 2004. A continuous satellite-derived measure of global terrestrial primary production. BioScience 54: 547–560.
Sánchez-Baracaldo, P. and Cardona, T. 2020. On the origin of oxygenic photosynthesis and Cyanobacteria. New Phytol. 225: 1440–1446. doi: 10.1111/nph.16249). Epub 2019 Nov 6.
Sapei, L., Gierlinger, N., Hartmann, J., Noske, R., Strauch, P. and Paris, O. 2007. Structural and analytical studies of silica accumulations in *Equisetum hyemale*. Anal Bioanal Chem 389: 1249–1257.
Sauterey, B., Charnay, B., Affholder, A., Mazevet, S. and Ferriere, R. 2020. Co-evolution of primitive methane-cycling ecosystems and early Earth's atmosphere and climate. Nature Communications 11: 2705. https://doi.org/10.1038/s41467-020-16374-7.
Schubert, T. S. 1982. Copper deficiency of plants. Plant Pathology Circular Contribution No. 528. Bureau of Plant Pathology. Dept. Agric. & Consumer Serv. Division of Plant Industry, FL.
Shaul, O. 2002. Magnesium transport and function in plants: the tip of the iceberg. BioMetals 15: 309–323.
Smil, V. 1997. Cycles of life. Civilization and the Biosphere. Scientific American Library. New York
Soltis, D. E., Soltis, P. S., Morgan, D. R., Swensen, S. M., Mullin, B. C., Dowd, J. M. et al. 1995. Chloroplast gene sequence data suggest a single origin of the predisposition for symbiotic nitrogen fixation in angiosperm. PNAS 92: 2647–2651.
Tansley, A. G. 1935. The use and abuse of vegetational concepts and terms. Ecology 16: 284–307.
Van Kauwenbergh, S. J. 2010. World phosphate rock reserves and resources: IFDC Muscle Shoals. Paper presented at Fertilizer Outlook and Technology Conference, November 16–18, 2010.
Vitousek, P. M. and Farringon, H. 1997. Nutrient limitation and soil development: experimental test of biogeochemical theory. Biogeochemistry 37: 63–75.
Vitousek, P. M., Hättenschwiler, S., Olander, L. and Allison, S. 2002. Nitrogen and Nature. Ambio 31: 97–101.
Vitousek, P. M., Porder, S., Houlton, B. Z. and Chadwick, O. A. 2010. Terrestrial phosphorus limitation: mechanisms, implications, and nitrogen-phosphorus interactions. Ecological Applications 20: 5–15.
Wang, F., Zhao, X., Gerlein-Safdi, C., Mu, Y., Wang, D. and Lu, Q. 2017. Global sources, emissions, transport and deposition of dust and sand and their effects on the climate and environment: a review. Front. Environ. Sci. Eng. 11: 13. DOI 10.10007/s11783-017-0904-z.
Wang, Y. P., Law, R. M. and Pak, B. 2010. A global model of carbon, nitrogen and phosphorus cycles for the terrestrial biosphere, Biogeosciences 7: 2261–2282. https://doi.org/10.5194/bg-7-2261-2010.
Whittaker, R. H. 1975. Communities and Ecosystems. 2nd edition. Macmillan, London, UK.
Yamori, W., Suzuki, K., Noguchi, K., Nakai, M. and Terashima, I. 2006. Effects of Rubisco kinetics and Rubisco actation state on the temperature dependence of the photosynthetic rate in spinach leaves from contrasting growth temperatures. Plant, Cell, and Environment 29: 1659–1670.
Zahnle, K. J., Lupu, R., Catling, D. C. and Wogan, N. 2020. Creation and evolution of impact-generated reduced atmospheres of early Earth. The Planetary Science Journal 1:11. https://doi.org/10.3847/PSJ/ab7e2c.
Zehr, J. P. and Capone, D.G. 2020. Changing perspectives in marine nitrogen fixation. Science 368: eaay9514.

CHAPTER 11

Biodiversity

"The biosphere is diverse. Really diverse. You just won't believe how numerously, prodigiously, mind-bogglingly diverse it is..." (parody of Douglas Adams' Hitchhiker's Guide to the Galaxy)

Biodiversity first appeared in print in its titular role in a compendium edited by the eminent biologist Edward O. Wilson. He introduces *Biodiversity* (1988) with a dramatic opening: "The diversity of life forms, so numerous that we have yet to identify most of them, is the greatest wonder of this planet." He goes on to assert that biodiversity "must be treated more seriously as a global resource to be indexed, used, and above all, preserved." Along with other contributing authors, Wilson makes the case for why we should promote biodiversity as a valuable commodity and brings our attention to the benefits of biodiversity and the costs of losing it.

What is Biodiversity?

Suppose we followed Wilson's cue and put a value on biodiversity. In any analysis of costs and benefits, we must first take inventory of what we have ("index" it, as Wilson suggests). However, to do that, we need to understand what "it" is. The First International Convention on Biodiversity (CBD 1992) defined "it" as:

> "the variability among living organisms from all sources. Such sources include terrestrial, marine, and other aquatic ecosystems, plus the ecological complexes of which they are a part—including diversity within species, between species, and of ecosystems."

But if biodiversity is "all sources of variability" in living organisms and ecosystems, we are still left in a bit of a pickle when trying to answer, "How much of it is there?" "Where can we find more of it?" "Is there more of it here than there?" or "How bad is it if we lose it?" This is because, to assess, understand, and preserve "biological variability," we need a measurable unit. Unfortunately, we have not yet come up with a single measurement that encompasses all of biological variability. The best "unit" of variability often depends on the type of question being asked and by whom, which can lead to heated disagreements about whether something belongs in the unit. Despite this, the most common way to measure biodiversity is by using units called "species" (Barraclough 2019).

Taxonomic Species

Taxonomy, from which the term "species" is derived, is the science of naming and classifying organisms. However, the naming and classifying of organisms is not unique to science. It is difficult to find a human society that does not identify, name, and classify living organisms. Berlin (1992) explains this in terms of "human beings' similar perceptual and largely unconscious appreciation of the natural affinities among groupings of plants and animals in their environment." The fundamental unit in folk taxonomies, reflected in distinctive names, often corresponds to a scientific genus (Rea 1997). Once an organism is named, it can, of course, be distinguished from other organisms and

DOI: 10.1201/9781003008538-11

therefore be "counted" as a unique component of the ecosystem and part of its biological variability. For example, the "Pima" (O'odham) people in southern Arizona (USA) recognize about half as many plant species in their local flora as are recognized by Western science (Rea 1997). This illustrates the innate difficulty in measuring biodiversity—you will get a different count depending on how you define the fundamental unit.

If ecologists consider the fundamental unit of biodiversity to be the "species," what exactly is a species? So far, we have established that a species may be something with a name that is generally agreed upon; this is called a **taxonomic species**. But on what criteria do we name a species, and how do we all agree on it? The former comes from the discipline of taxonomy, and the latter comes from established naming rules that all scientists follow (e.g., the International Code of Zoological Nomenclature, the International Code of Nomenclature for Algae, Fungi, and Plants, the International Code of Nomenclature of Bacteria, etc.). Both taxonomy and these naming codes can ultimately be traced to the Swedish botanist Carl von Linné, who is considered the father of modern taxonomy and known by his latinized name, Carolus Linnaeus.

Linnaeus' life spanned most of the 1700s, a period of intense European exploration into what were considered novel and exotic lands. Linnaeus, who measured his prestige by the number of correspondents he kept (Jönsson 2011), had a large international network of about 600 people (similar to modern "Facebook friends"), including his students, whom he sent on expeditions all over the world. Many of these correspondents sent Linnaeus organisms collected from these far-flung corners of the globe. As a result, during his long career, Linnaeus' collection amassed over 14,000 plant specimens, 4,000 of which are type specimens for their species; 3,200 insect specimens representing 109 genera; 168 fish specimens representing 97 genera; and 1,500 shells representing 34 genera.

Linnaeus's collection was not the only collection at the time. Because of all this exploration, one of the greatest challenges for Linnaeus and other collectors was naming and classifying the growing number of organisms being collected from around the world (Jönsson 2011). By the end of the 1700s, multiple systems of biological classification had arisen, but Linnaeus' system rose to the top because it was simple to use and had international appeal (Jönsson 2011). Linnaeus originated the *binomial* system of naming, giving every organism a unique species name while grouping similar species into a genus based on key morphological traits. He also standardized his terminology in Latin, an unbiased language (because it was no country's native tongue) yet universal (at least to Christendom).

Following Linnaeus's ideas, there are general rules that taxonomists use when deciding whether to name something a "species." First, there must be a **type specimen** that demonstrates the key (and unique) features used to identify the species. Second, a description of this species' unique features is required; this description and its associated data must be published in a manner that is internationally accessible and in line with the International Code of Nomenclature appropriate for the taxon. For example, *Geospiza fortis* is a species of ground finch from the Galapagos archipelago. The type specimen for *Geospiza fortis* was collected by Charles Darwin in 1835 and described by John Gould. The description was published in 1837 in *The Zoology of the Voyage of H. M. S. Beagle, Part III Birds*, along with the description of seven other species that Gould grouped into the genus *Geospiza.*

Despite the naming of species following these agreed-upon rules, taxonomic species do not always last. Gould identified eight species in Darwin's finch collection. Today, only three of these eight species are recognized, while three others have been named and added to the genus by different researchers.[1] Clearly, John Gould's "unique" morphological differences in Darwin's specimens were

[1] *Geospiza fulginosa, G. fortis, G. magnirostris* (still named as such); *G. dubia, G. dentirostris* (both considered G. fortis); *G. strenua* (now considered *G. magnirostris); G. nebulosa* (now considered *G. difficilis*); and *G. parvula* (now called *Camarhynchus parvulus*).

not considered unique by later taxonomists. This presents a fundamental dilemma when defining a species by unique morphology: we impose human categories on a nature that is inherently variable. Open any drawer of taxonomic specimens, and you will find natural variation. Often, this variation makes it difficult to distinguish one species from another. For example, the drab blackish-brown *Geospiza* finches found on the Galapagos Islands are distinguished largely by differences in the depths of their beaks (Grant and Grant 2008). However, if you were to sort them based on this trait, you might lump *Geospiza difficilis* from Genovesa Island with *G. fulginosa* from Pinta Island, because their beak depths are identical (e.g., Schluter and Grant 1984: 184). Because of the often continuous nature of nature's variability, it is possible—and even desirable—to define a species by something other than its morphological uniqueness.

Biological Species

Reproductive isolation was introduced as a species concept by Ernst Mayr (1904–2005) in the 1940s (e.g., Mayr 1948). It was, in a way, his answer to the difficulty of distinguishing between dead *Geospiza* specimens. Let us instead examine Galapagos finches living on different islands. We can clearly see different species, reproductively isolated by their distinct mating calls (Grant and Grant 2008). Mayr (2000) defines a **biological species** "as groups of interbreeding natural populations that are reproductively isolated from other such groups." Mayr was an important figure in *the Modern Synthesis* (Huxley 1942), in which Charles Darwin's theory of evolution and the rediscovery of Gregor Mendel's theory of heredity were integrated into a unified framework. By using the biological species definition, Mayr sought to emphasize genetic relationships and descent through common ancestry. He also aimed to challenge traditional taxonomy. Instead of grouping similar individuals into a "species" and similar species into a "genus," he wanted populations themselves to determine species status (Mayr 2000). In other words, he wanted biology, rather than the human eye, to define what a species is.

Mayr's biological species concept was embraced because it aimed to remove human bias from the naming of species, but it was also widely criticized for its difficulty in application. For instance, it cannot be applied to asexual organisms, which form clones rather than populations. How does one apply the biological species concept to extinct organisms? It also faces challenges in determining what constitutes "reproductive isolation." The salamanders of California are a one-dimensional chain of reproductively linked populations (Jackman and Wake 1994), but is there enough genetic distance between each end of the chain to classify them as one species or two? And how much hybridization is allowed, for example, in European oaks (Gomory et al. 2001), before two populations are no longer considered separate "biological species"?

Grant and Grant (2008) astutely point out that, for Galapagos finches, "speciation is a process and not an event," meaning that reproductive isolation is not a "one-and-done" phenomenon. More modern views of what constitutes a species take this into account, as advances in genetics have revealed another layer of biodiversity at the genetic level.

Genetic Variation

Population genetics is the study of genetic variation within populations of organisms. Genetic variation, or within-species biodiversity, arises from allelic differences among individuals across their chromosomes and the genes they carry. **Genes** code for traits, such as blue eyes, while **alleles** are alternative versions of a particular gene, such as those for brown or green eyes. Populations can exhibit high or low genetic variation. High genetic variation occurs where there are frequent differences in alleles between individuals, while low genetic variation occurs when individuals rarely differ in their alleles. The relative frequencies of specific alleles in a population (e.g., 50% brown, 25% blue, and 25% green) will remain constant from one generation to the next, assuming generations are non-overlapping, mating is random, the population is vast, mutation is negligible, and there is no natural selection. This is called the **Hardy-Weinberg equilibrium**. But this rarely occurs in nature.

Species are often composed of small, localized populations (sub-populations, races, or demes), which tend to consist of individuals with alleles unique to that population and adapted to corresponding local environments (known as ecotypes). These genetic differences may also result from **population bottlenecks** and **founder effects** (see Chapter 13 *Evolutionary Ecology* for a detailed discussion). Such populations often differ from each other in genetic composition. For example, Cook Inlet, Alaska (USA), is home to 55 distinct genetic lineages of Chinook salmon (*Oncorhynchus tshawytscha*; Barclay and Habicht 2015). These distinct genetic lineages, or stocks, originate from different river drainages that all flow into the Cook Inlet. Some of the larger river drainages, such as the Kenai River, even contain multiple lineages (the Kenai River has thirteen). These genetic lineages reflect differences in the tributaries where they spawn, as well as the seasonal timing when they migrate to their spawning grounds.

Genetic variation can be compared between lineages or between populations within a species using a measure called the **fixation index.** Population geneticists use the term F_{ST} to represent this index, which is expressed as a proportion. When F_{ST} is close to zero, the compared lineages or populations are considered genetically similar, implying complete genetic mixing (**panmixis**). As F_{ST} gets larger, it signifies greater genetic differentiation between lineages or populations. When F_{ST} is close to one, there is almost complete divergence, and the lineages or populations are considered **genetically isolated** from each other.

It turns out that measuring the relative genetic isolation (F_{ST}) among populations within a species is very useful for managing an exploited resource like Cook Inlet Chinook salmon. Chinook salmon, like other salmon, are unique in that they are **semelparous** (see Chapter 6 *Life Histories and Life History Tradeoffs*), returning to their natal stream—where they hatched as fry three to seven years ago—to spawn for the first and last time in their lives. Most streams host populations with a high degree of reproductive isolation, which allows fish returning to spawn in different tributaries and at different times to be distinguished into genetic clusters (called stocks)[2] using F_{ST} (Barclay and Habicht 2015). This, in turn, enabled the Alaska Department of Fish and Game to collect genetic samples from the commercial harvest of Cook Inlet Chinook salmon to determine which stocks were being captured at various locations and times in Cook Inlet (Eskelin and Barclay 2018). This information allowed managers to set fishing limits appropriately, distinguishing between stocks suffering population declines and those that were not. The same concept of using F_{ST} to separate populations into genetic clusters can also be applied to separate populations into species and to construct **phylogenetic trees** (similar to family trees) that represent various populations or species (Mallet 1995).

Phylogenetic Species

Phylogenetics is the scientific study of evolutionary lineages and is used to distinguish and organize populations of organisms into lineages using genetics and statistics. Phylogenetic species are identified as genotypic clusters (Mallet 1995) that are linked into lineages by genetic similarities (de Queiroz 1988; Zachos 2016). The **phylogenetic** or **evolutionary species** concept reflects ancestry and descent with modification. It recognizes that reproductive isolation is a process that occurs over time (e.g., Figure 1). Phylogenetics also helps us understand how the two levels of biodiversity—within-species genetic variation and between-species differences (e.g., in morphology, behavior)—are approximately paralleled by microevolution and macroevolution (Wilson 1999). **Microevolution** refers to the small, gradual, adaptive changes occurring in isolated populations that can be attributed

[2] Note: Chinooks salmon stocks are not considered species but can be considered populations with some amount of genetic isolation from one another. A small number of salmon do not return to their natal streams to spawn but instead "stray," moving genes between populations.

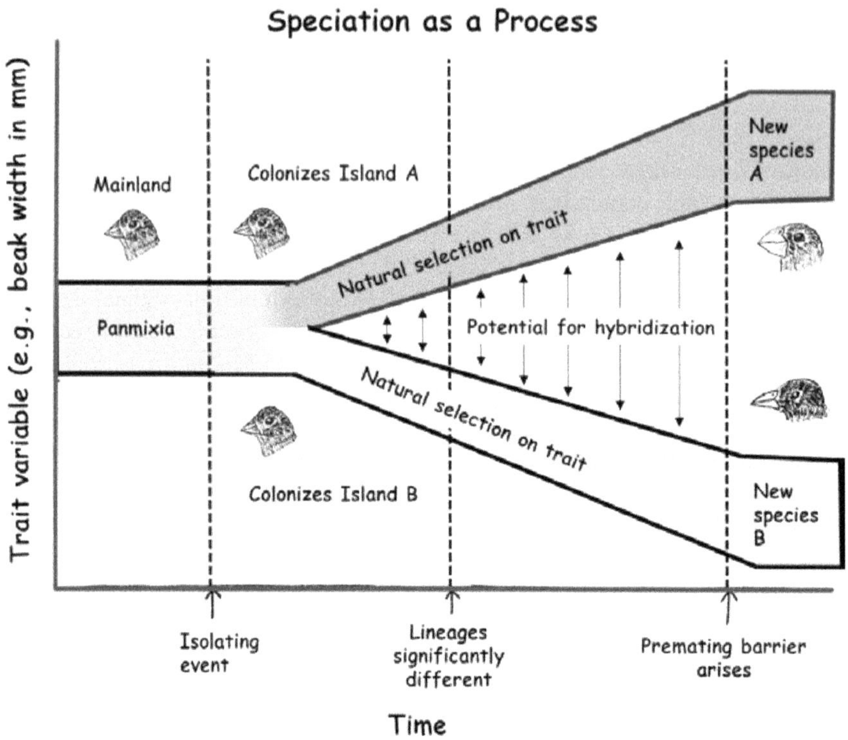

Figure 1. Phylogenetic speciation as a process from panmictic ancestors to reproductively isolated descendants (based on Grant and Grant 2008: pg. 109).

to changes in genes and chromosomes. Over time, these changes may eventually result in speciation caused by the emergence of reproductive barriers. **Macroevolution** refers to more complex and profound changes or, more specifically, to the processes that can result in novel species derived from ancestral ones (see Chapter 13 *Evolutionary Ecology* for further discussion).

Interestingly, this modern view of phylogenetic species brings us back to Linnaeus' binomial system of naming, which groups species into genera. Although Linnaeus grouped species based on morphological similarities, we now consider genera to be groups of species with a common ancestor. Linnaeus's (1758) *Systema Naturæ*,[3] in which he created nested groupings of species into genera, genera into orders, orders into classes, and classes into kingdoms, was prescient as a way to organize biodiversity in the age of evolutionary biology. Darwin's (1859) theory of evolution proposed that various life forms evolved from shared common ancestors that could be traced back to a presumed single organism. Darwin's macroevolutionary model, visualized as a branching tree, contrasted with the then prevalent "chain of being" idea of a linear succession of organisms from simple to complex. However, Darwin's idea, following Linnaeus by a century, falls neatly into the lap of Linnaeus' *Systema Naturae* because taxonomic groupings often reflect the morphological modifications that occur via evolutionary descent. Haeckel (in his 1866 *Generelle Morphologie der Organismen*) was among the earliest to combine the two ideas—evolution by descent and nested groupings—into what is known as the "Tree of Life" model of biodiversity. Haeckel's model depicts organisms in the Plant, Protist, and Animal kingdoms arising from a single common ancestor in the Monera (bacteria) kingdom (Figure 2a; Kutschera 2016).

[3] *Systema naturææ per regna tria naturæ, secundum classes, ordines, genera, species, cum characteribus, differentiis, synonymis, locis.*

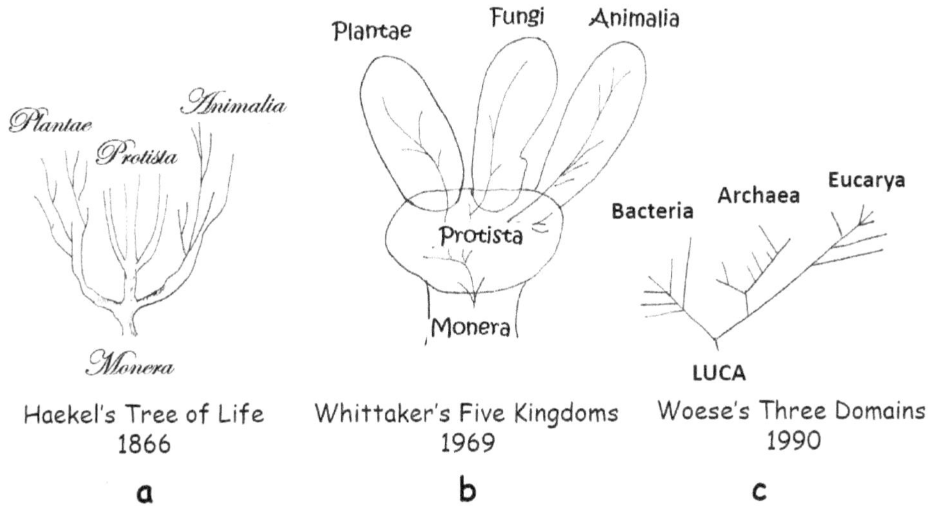

Figure 2. Three models of Earth's biodiversity are depicted as descending from a common ancestor. a) Haekel (1866), who was the first to represent biodiversity in the context of evolution, coined the term "phylogeny." b) Whittaker (1969) separated Earth's biodiversity into 5 kingdoms, depicting the multicellular kingdoms arising from the unicellular kingdoms. c) Woese et al. (1990) separate biodiversity into 3 domains by similarity in ribosomes, depicting them as arising from the last universal common ancestor (LUCA).

Whittaker (1969) improved Haeckel's tree of life by dividing it into five kingdoms: Animals, Plants, and Fungi, all branching from the Protists (simple eukaryotes with nucleated cells), which arose from the Monera (prokaryotes, one-celled organisms without nuclei; Figure 2b). Modifications to this basic tree of life have been made as technology has improved our ability to read the genetic code, and phylogenetics has grown as a science accordingly. Today, biologists recognize another configuration of the tree of life based on genetic distinctions in small subunit ribosomal RNAs, first proposed by Carl Woese et al. (1990): Eukarya, Archaea, and Bacteria—groups called "domains," the highest-level taxon (Figure 2c). The RNA of ribosomes—protein factories of cells common to all life—have small subunit structural features that are shared within each domain but distinct between domains, creating a natural distinction between these three groups. This RNA-sequenced-based phylogenetic tree was a breakthrough that could be used to relate all organisms and reconstruct the history of life (Pace 1997).

Origins of Species

Biodiversity is the product of past evolution, just as future biodiversity will be a product of contemporary evolution. Speciation, at least in eukaryotes, is a process of lineage splitting that begins with genetic variation and ends when enough differences accumulate over time, preventing interbreeding and leading to the formation of biological species (e.g., Figure 1). Grant and Grant (2008) make a strong case that speciation in Galapagos finches occurred because populations on different islands experienced adaptive changes in beak size due to selection via food resources, which resulted in changes to song quality, creating pre-mating barriers to interbreeding. The key components of speciation are thus population variation (via adaptation, etc.) and barriers to mating. As Earth's history attests, given these conditions, the accumulation of species (biodiversity) is simply a matter of time.

There appears to be geological evidence of life as far back as 4.1 billion years ago (Bell et al. 2015), but finding the last universal common ancestor of extant life (LUCA) that links the Eukarya, Archaea, and Bacteria is a tricky business. Sometimes, structures that resemble fossil microbes can arguably be created by geological processes (Donoghue 2020). The oldest widely accepted evidence comes from 3.43–3.35 billion-year-old microbial fossils found in the Strelley Pool Formation in

Western Australia. However, molecular clock methodology (calibrating molecular changes in the genome with time) has suggested that LUCA may have existed as soon as Earth became habitable around 4.52–4.47 billion years ago (Donoghue 2020). The transformation from organic soup to microbes marked the **first great evolutionary step** of four in the history of life (Wilson 1999; Figure 3).

In the beginning, "sex" (meiosis) had not yet evolved, so the means of generating genetic variation were limited. However, prokaryotes—the earliest microbes—were able to exchange genetic material through a process known as horizontal gene transfer (Gogarten and Townsend 2005). Horizontal gene transfer resulted in increased variation within cell lines, providing fodder for evolution. These single-celled organisms ebbed and flowed in response to climate, nutrients, sunlight, and other factors. At some point during the first billion years of life, cyanobacteria (also known as blue-green algae), using energy from light, began absorbing water molecules, assimilating the hydrogen, and releasing oxygen (the elixir of most modern life forms) as waste, thereby "inventing" photosynthesis. As cyanobacteria proliferated, oxygen began to accumulate in the atmosphere. About 3.4 billion years ago, in shallow seas, the cyanobacteria became slightly sticky, trapping micro-particles of dust and sand, which were bound together to form solid structures—the stromatolites (Collins and Jahnert 2014). This marked the world's first collective venture, as cyanobacteria formed colonies (billions per square meter of rock) resembling stacks of mattresses for reasons still debated. Today, we can observe modern versions of living stromatolites in extremely salty lagoons, bays, or shallow seas at Shark Bay (Australia), the Bahamas, and Baja California (Mexico), or in lakes in the East African Rift system and interior western Argentina (Landing and Johnson 2024).

As time passed, groups of primitive organisms emerged, constructing the world's first ecosystem. By 2.2 billion years ago, these ancient microbes had pumped so much oxygen into the atmosphere that its concentration rose dramatically to between 1% and 10% of present atmospheric levels during the Great Oxygenation Event (Mills et al. 2022). Following this oxygenation event, and continuing until the second great oxygenation event between 500 and 800 million years ago (the Neoproterozoic Oxygenation Event, which brought the atmosphere to its current composition of about 20% oxygen), a new development occurred—the eukaryotic cell. This marks what Wilson (1999) calls the **second great evolutionary step** in the diversification of life (Figure 3).

We can imagine that the genesis of eukaryotic life—cells containing mitochondria, from which protists, animals, plants, and fungi are formed—occurred when two adventurous prokaryotic cells collided and became one. This fanciful hyperbole may not be too far off the mark. Lynn Margulis (1970) was the first to propose that eukaryotic life arose from a symbiotic relationship between two types of formerly free-living bacterial cells. Recent phylogenetic analyses suggest that after the Bacteria and Archaea lineages split, the ancestral mitochondrion from the Bacteria lineage and the ancestral eukaryotic cell from the Archaea lineage became mutually dependent on one another and evolved into the last eukaryotic common ancestor (LECA: Mills et al. 2022). Descendants of the ancestral mitochondrion now generate energy for all eukaryotic cells using genes inherited from their former bacterial lifestyle, while descendants of the archaea cell grew larger to contain a DNA-filled nucleus and other small bodies collectively called organelles, including the mitochondria. That is how an entirely new type of cell—the eukaryotic cell that forms multicellular organisms like us—arose from prokaryotic cells. A similar process, according to Margulis (1970), also gave rise to the first "photosynthetic" eukaryote, which originated from a heterotrophic eukaryote that captured a coccoid cyanobacterium about 1.6 billion years ago, forming an endosymbiotic relationship. Over time, descendants of these symbionts evolved into photosynthetic eukaryotic cells with a new organelle called a plastid, eventually giving rise to algae and plants (Yoon et al. 2004). It seems likely that the energy advantage provided to eukaryotic cells by their mitochondria and plastids set the stage for the evolution of multicellular life.

The Neoproterozoic Oxygen Event, occurring 500–800 million years ago and bringing atmospheric oxygen to current levels, ushered in an era dominated by increasingly more complex

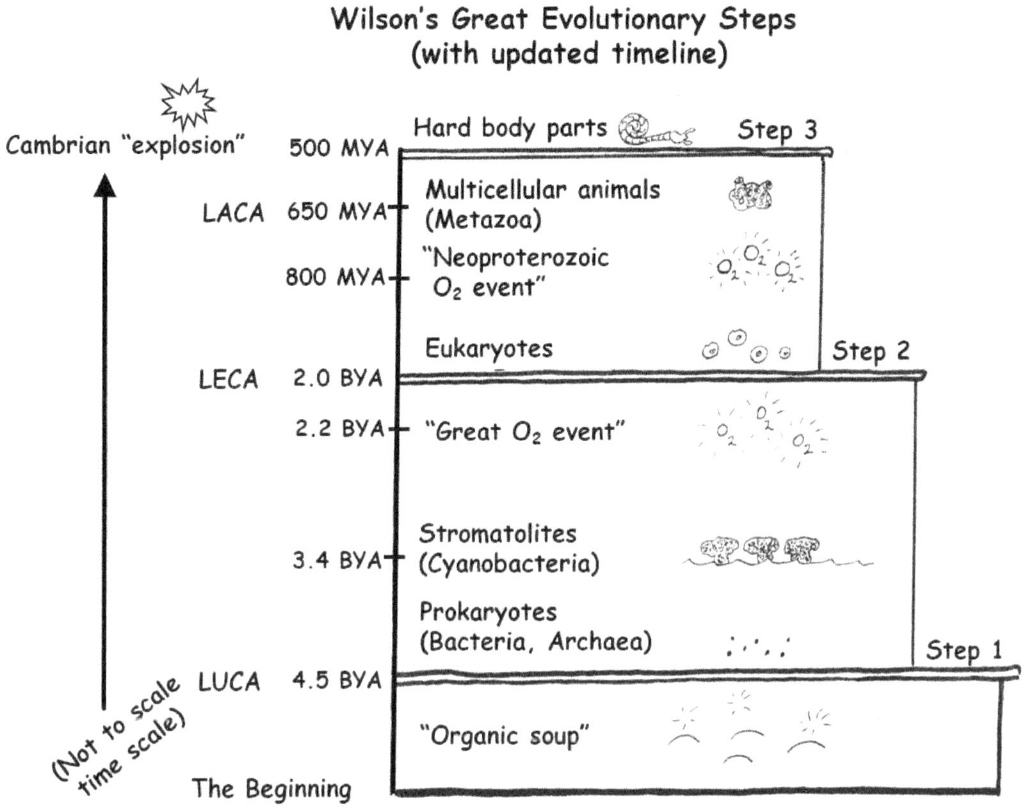

Figure 3. Wilson's (1999) great evolutionary steps are depicted along with the appearances of the last universal common ancestor (LUCA; Donoghue 2020), the last eukaryotic common ancestor (LECA; Mills et al. 2022), and the last animal common ancestor (LACA; Ros-Rocher et al. 2020).

organisms that depended on oxygen to survive (Edwards et al. 2017). The first multicellular animal, known as the last animal common ancestor (LACA), probably appeared between 800 and 650 million years ago at the start of the oxygenation event. The Cambrian "explosion" (named for the fast and extensive diversification of multicellular animals, or **metazoa**) occurred toward the end of the event, between 540 and 500 million years ago (Ros-Rocher et al. 2020). This explosion marked the **third great evolutionary step** identified by Wilson (1999), when animals evolved hard body parts (Figure 3). These hard-bodied animals (as well as those with soft bodies; Cole et al. 2020) proliferated in diverse and complex forms.

The Cambrian explosion is documented as one of the most important adaptive radiations in the history of life. But no one knows for sure why it occurred. One hypothesis proposed for the evolution and diversification of hard body parts is that predators evolved enhanced skills around this time, making animals with hard exoskeletons advantaged over their soft-bodied predecessors. The innovation of a hard skeleton ensured a relentless evolutionary game: the rule was either make your own tough mineralized shield or perish, because predators would target your easy-to-consume soft body instead (see Chapter 4 *Species Interaction: Predation*).

The Cambrian explosion took place in shallow marine waters, while simultaneously, green photosynthetic algae—the closest relatives of land plants—began to leave their watery homes. They evolved new strategies (genes related to multicellularity, such as cell walls, root development, and "terrestrialization," such as embryo development and cell signaling; Bowles et al. 2020) to gain a firm footing on swampy land, greening the continents for the first time. These newcomers included the bryophytes (hornworts, liverworts, and mosses) and the tracheophytes (vascular plants; Puttick et al. 2018). It took another 100 million years for terrestrial animal life to appear. Of the over 30 marine animal phyla, only the vertebrates, arthropods, mollusks, and annelids have significant numbers of terrestrial representatives. Interestingly, no phylum originated on land (Selden 2016). Nonetheless, these few terrestrial animal phyla radiated into the diversity of invertebrate and vertebrate animals we see on land today. Although the colonization and establishment of terrestrial plants paved the way for a diversity of terrestrial animals, the radiation of fungi onto land probably coincided with plants. Based on molecular evidence, Heckman et al. (2001) speculate that the first terrestrial eukaryotes were lichen-like algal-fungal symbionts, which may have been present on land before the Cambrian explosion. One of the key morphological transitions in fungi associated with the colonization of land is the shift to hyphal growth (e.g., the white fuzz we see on rotten food) and the loss of flagella (organs used for swimming, also seen in microscopic animals; Naranjo-Ortiz et al. 2019).

Today, the world supports an astonishing variety of species occupying an immense number of different niches. **Adaptive radiation** is the evolution of species with common ancestry into different niches in a relatively short evolutionary time. Adaptive radiations at the species level are dramatically displayed in distant, isolated archipelagoes, including the finches of the Galapagos Islands (e.g., Grant and Grant 2008), honeycreepers (e.g., Lerner et al. 2011), and silverswords (e.g., Baldwin et al. 1991) of the Hawaiian Islands in the Pacific Ocean, and the stick insects (e.g., Bradler et al. 2015) of the Mascarene Islands in the Indian Ocean. Anolis lizards on the less isolated islands of the Caribbean (and the adjacent Central American mainland) also exhibit adaptive radiation at the species level (e.g., Pinto et al. 2008). In freshwater environments, lakes act like islands, and adaptive radiations of amphipod species (e.g., Naumenko et al. 2017) in Lake Baikal in southern Siberia (Russia) and cichlid species in the Great Lakes of East Africa (e.g., Seehausen 2006) are remarkable examples.

Mainland relatives of the last common ancestor in these radiations—whether birds, insects, fish, or plants—tend to face pressure from competition, predation, and disease in their mainland habitats. It is thought that release from these pressures is a major driver of these radiations when ancestral immigrants first arrive and take full advantage of abundant, unutilized resources, resulting in a radiation into unoccupied niches (e.g., Seehausen 2006). However, there are indications that a propensity to speciate—such as the formation of mating barriers or the possession of a key innovation that others lack—may also play a role in determining which types of immigrating organisms ultimately form populations that adaptively radiate in these island opportunities (Grant and Grant 2008; Pinto et al. 2008).

Older adaptive radiations can also be seen at taxonomic levels above the species level. For instance, the enormous diversity of angiosperms (over 250,000 species) is largely due to the evolution of the flower—an innovation that opened up new means to exploit animals for dispersal. The diversification of mammals after the demise of dinosaurs is another example of adaptive radiation at a higher taxonomic level. Dinosaurs themselves can also be considered one of the most spectacular and successful examples of adaptive radiation at a higher taxonomic level. While evolutionary innovations (e.g., chlorophyll, hard body parts, lignin, and flowers) can generate adaptive radiations, ecological voids created by the formation of islands and mass extinctions do so as well.

Extinctions of Species

Although Earth's history shows how easily species originate and accumulate, it also reveals that species go extinct. Extinction is a natural part of life and evolution. No one knows how many species of organisms have existed since life began—perhaps several billion, maybe vastly more—but we know with certainty that almost all species that have ever lived (some estimates suggest at least 99.9%) have gone extinct (Newman 1997; Barnosky et al. 2011). Extinctions are natural and have occurred long before humans walked on Earth. Species go extinct, while new ones evolve.

This give-and-take of extinction and evolution is referred to as **background extinction**, and it is estimated that the rate of background extinctions is around 0.1 extinctions per million species per year (De Vos et al. 2015). *C'est la vie!*

The "grand procession of life" was obstructed, experts widely agree, by at least five mass extinctions. These extinctions, which resulted in a loss of over 75% of estimated species, are recorded in the fossil record at the ends of the Ordovician, Devonian, Permian, Triassic, and Cretaceous periods (Barnosky et al. 2011; Figure 4). All of these major extinctions occurred after the "third great evolutionary step" of hard body parts (this must be kept in mind, as fossilization occurs best with hard body parts). The best-known major extinction is the Cretaceous event (66 million years ago [Ma]) when the dinosaurs, which dominated Earth for over 150 million years, went extinct in a global catastrophe caused by an extraterrestrial impact (Kring 2007) and massive volcanism (Ma et al. 2022). These events paved the way for mammalian dominance and, eventually, the evolution of *Homo sapiens*.

The end-Triassic event (201 Ma), just before the dinosaurs' rise to prominence, was a drawn-out process of extinctions in diverse marine groups, including bivalves, brachiopods (which resemble bivalves but are not), ostracods (tiny crustaceans), calcareous algae, and the last conodonts (eel-like vertebrates), as well as several kinds of tetrapods (land animals whose descendants include the amphibians, reptiles, dinosaurs, and mammals). These extinctions appear to have been driven by a

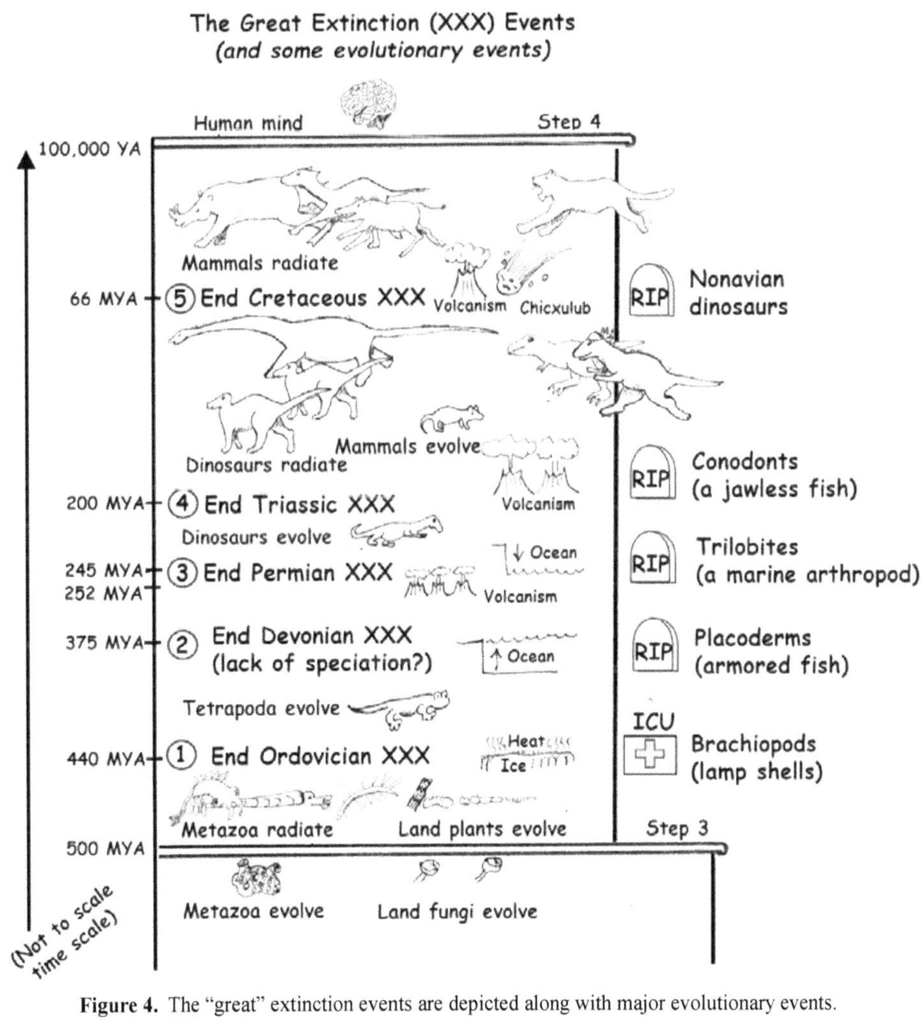

Figure 4. The "great" extinction events are depicted along with major evolutionary events.

"double-punch" of two large-scale volcanic eruptions occurring within a short geologic time span (Wignall and Atkinson 2020).

The Ordovician event (440 Ma), the first of the five mass extinctions, was probably caused by a geologically sudden glaciation event that changed the global environment. This was followed by an equally sudden warming event that impacted the surviving species, which had been selectively adapted to glacial conditions. These two extreme global events are estimated to have wiped out about 85 percent of the marine species (Sheehan 2001), which made up most of Earth's biodiversity at the time.

The biggest mass extinction, however, was the end-Permian (also called Permo-Triassic) event (245 Ma). This event, occurring just before the first dinosaurs evolved (230 Ma; Marsicano et al. 2016), obliterated an estimated 96% of marine species, representing approximately 54% of marine families, including trilobites (a whole class), which disappeared forever (Erwin 1990). The event is associated with an estimated sea level drop of 280 meters, reducing continental shelf habitat by 40%. This dramatic event may have been related to the formation of the supercontinent Pangea, causing environmental extremes such as global cooling and glaciation (Erwin 1990). The devastation of this period was further worsened by the massive volcanism of the Siberian Traps Large Igneous Province, which is implicated in the extinction of 70% of the terrestrial vertebrate families just prior (252 Ma) to the marine collapse (Del Corso et al. 2022).

A look at species turnover in the fossil record reveals that species also go extinct between these major events, but at a much lower rate and generally without wiping out an entire taxon (e.g., all trilobites at the end of the Permian). New species, however, are always arising in response to the ecological opportunities that arise after these losses. The estimated background rate of extinction (0.023 to 0.135 extinctions per million species per year) is lower than the estimated background rate of speciation (0.05 to 0.2 new species per million species per year), which explains the accumulation of biodiversity over time (De Vos et al. 2015; Rosenzweig 1995). Interestingly, the Devonian (375 Ma) extinction event is thought by some to be more of a "lack of speciation" event (Stigall 2012). Evidence of greater extinctions in narrowly ranging species, invasions by generalist species, and coinciding sea level rises (which removed land barriers to dispersing marine species) during this period support the idea that habitat homogenization and a shift to generalized, wide-ranging species resulted in lower speciation rates and an overall loss of total biodiversity (Stigall 2012).

The Sixth Extinction

Although species have gone extinct since life began, it is irrefutable that during the past 12,000 years, human activity has had dramatic effects on the environment and particularly devastating effects on the rate of species loss. The scientific consensus is that humans are causing Earth's sixth mass extinction (Ceballos et al. 2015). Archaeological and palaeoecological evidence shows distinctive, massive human intervention on nearly three-quarters of Earth's land, including more than 95% of temperate and 90% of tropical woodlands (Ellis et al. 2021). Extinctions can be difficult to attribute directly to humans, but circumstantial evidence abounds. Island surveys for bird skeletal remains document dramatic declines after first contact with humans (Steadman and Martin 2003). For instance, over half of the original terrestrial avifauna in the Hawaiian Islands is estimated to have gone extinct after human contact, and this incredible loss is repeated throughout the Pacific Islands (Pimm et al. 2006). Just as islands, by their geographic barriers, create species, once these barriers are breached, those species become vulnerable to extinction. For example, more than 50 species of honeycreepers once made their homes in the Hawaiian Islands, evolving from a single ancestor population that underwent adaptive radiation to suit varying island conditions. The honeycreepers evolved in isolation, so when humans showed up and brought along new animals and diseases, the birds had no prior co-evolutionary experience and, therefore, no evolved defenses. The consequences were devastating. Now, just 17 honeycreeper species survive in Hawaii, and 15 of them are listed as federally endangered under the U.S. Endangered Species Act

(McGlashen 2017). Similar dire situations have been documented for native plants in Hawaii, where 9% of its 1,159 native flora taxa are extinct, and over 50% of the remainder are at risk of extinction (Sakai et al. 2002).

Islands may be particularly sensitive to extinction due to their endemism, and there is strong evidence of human arrival and subsequent faunal demise not only in the Hawaiian Islands but also in many other islands around the globe, including Mauritius, the Galapagos Islands, New Zealand, Madagascar, and the continent-sized island of Australia (Burney and Flannery 2005). In a comparison of recent island versus continental mainland bird extinctions, Steadman (1997) estimated that 97 island bird species went extinct after the year 1600, whereas the estimate of human-caused continental bird extinctions since 1600 is 17 to 20.

Mammals have been equally impacted. Of 60 recent mammal extinctions, 19 occurred on Caribbean islands and 18 in Australia (Pimm et al. 2008). Ironically, islands can sometimes serve as refuges from extinction. After predators, such as the Pacific rat, Norway rat, and ship rat, were introduced to New Zealand, some species of weta, New Zealand's giant grasshoppers, avoided extinction by living on predator-free island refuges off the main islands (Gibbs 1998). Extinctions of some species can also cause problems for other species. Anderson et al. (2011) compared the pollination, seed production, and density of the shrub *Rhabdothamnus solandri* on the "mainland" North Island of New Zealand, where three pollinating bird species are **functionally extinct,** to those growing on three nearby offshore islands, which serve as sanctuaries for the birds. They found that the loss of these bird species on the mainland reduced seed production per flower by 84% and the number of juvenile plants produced per adult by 55%. This is an example of a **trophic cascade** and demonstrates how species loss may contribute to further species loss, even during a modest extinction event.

The key to classifying these modern extinctions as part of a "sixth mass extinction" is determining whether the current rate of extinction is greater than the background rate of extinction. Pimm et al. (2006) estimated that the rate of modern bird extinctions over the last 2,000 years is about 100 times the background rate. Extinction rates for other groups of animals and plants, such as amphibians and primates, tend to be even higher, reaching up to 1,000 times the background rate (Pimm et al. 2014). More than 400 vertebrate species have become extinct in the last 100 years (Ceballos et al. 2020). Some charismatic examples include the thylacine (declared extinct in 1982), a carnivorous marsupial native to the Australian mainland and the islands of Tasmania and New Guinea (Paddle 2000); the ivory-billed woodpecker, which inhabited old-growth forests in the southeastern United States and Cuba, and was last sighted in 1944 in Louisiana and 1986 in Cuba; and the Round Island burrowing boa, a snake endemic to Mauritius, last seen in 1975 (Cole 2021). Many other examples exist.

Those are just examples where we know extinction has occurred. There are many more species on the brink of extinction. The International Union for Conservation of Nature (IUCN) defines populations comprising fewer than 1,000 individuals as being on the brink of extinction (Ceballos et al. 2021). Accordingly, there are 515 terrestrial vertebrate species, including 74 species of mammals, 335 birds, 41 reptiles, and 65 amphibians, currently on the "Brink of Extinction" list. This, according to Ceballos et al. (2020), represents 1.4% of the total mammal species evaluated by IUCN, 3.2% of birds, 0.6% of reptiles, and 1% of amphibians. These terrestrial vertebrates on the brink include such majestic species as the Sumatran rhino (*Dicerorhinus sumatrensis*), Clarion Island wren (*Troglodytes tanneri*), Española giant tortoise (*Chelonoidis hoodensis*), and harlequin frog (*Atelopus varius*).

There may even be many on the brink that have yet to be evaluated. Cowie et al. (2022) highlight a systematic bias in the IUCN "Red List", which focuses more on vertebrates than on invertebrates. It can be difficult, according to Cowie and colleagues (2022), to evaluate whether an invertebrate has gone extinct or even whether it is on the brink of extinction, because these species are often known only from a single specimen. This lack of information prevents them from being included on the list. Such bias can lead to a serious underestimate of the impact of "the sixth mass extinction."

Moreover, our knowledge of the current (or recent) diversity of invertebrate species is much less comprehensive than our knowledge of vertebrates and plants.

The Value of Biodiversity

According to Wilson (1999), it has been a long, evolutionary journey from "primordial soup" to microbes, eukaryotes, and multicellularity, to an explosion of animals, then finally, to the **fourth great step** in evolutionary innovation that resulted in the origin of the human mind about a million to 100,000 years ago (see Chapter 14 *Human evolutionary ecology*). This innovation allows us to be self-reflective about biodiversity and to place value on it. Why should we care if Earth loses biodiversity? How is biodiversity valuable to us?

The United Nations Environmental Programme (Brown et al. 2014) defines ecosystem services as those parts of the biosphere that play an important role in supporting economic activity, development, and general human well-being. Ecosystem services include "supporting services" such as soil formation, primary production, and nutrient and water cycling; "provisioning services" such as food, fiber, fuel, biochemicals, and natural medicines; "regulating services" such as air quality, climate regulation, erosion control, purification, disease and pest regulation, and pollination; and finally, "cultural services" including spiritual enrichment, cognitive development, recreation, and aesthetic experiences. May (2011) and others assert that biodiversity underpins these ecosystem services upon which humanity depends. For example, in an experimental test that directly controlled the number of prairie species, Tilman et al. (2006) found that higher numbers of plant species led to greater temporal stability in aboveground plant production. This portfolio effect, Tilman et al. (2006) argue, indicates that ecosystem services can be enhanced by greater biodiversity.

Given the value we place on biodiversity, it becomes imperative that we take inventory of what we have in order to prevent its loss.

Inventorying Biodiversity

So far, about 1.9 million species have been discovered and classified (Chapman 2009). But this is only the tip of the biodiversity iceberg, with many millions of species awaiting discovery and description. The current widely accepted estimate of the total number of plant and animal species on Earth is about 8.7 million (±1.3 million SE, range: 7.4–10 million), according to Mora et al. (2011). If that is true, it means over 80% of species have yet to be discovered and documented.

Whatever the absolute numbers, tropical forests are incredibly biodiverse, supporting at least two-thirds of the world's biodiversity (Raven 1988), despite covering less than 10% of Earth's land surface (Bradshaw et al. 2009). These forests receive over 200 centimeters of rain annually, and their three-dimensional habitats feature luxuriant trees that form three or more horizontal layers, the canopy of the tallest reaching 30 meters or more. Most of the world's rainforests are in the Amazon, Congo Basin, and Indonesia. The Amazon is the world's largest rainforest, larger than the next two largest rainforests—in the Congo Basin and Indonesia—combined. The extent of diversity in the rainforests is breathtaking. For instance, from a single leguminous tree in Peru, Wilson (1989) recalled retrieving 43 ant species belonging to 26 genera, which is approximately equivalent to the ant diversity of all of the British Isles. Biologists working in the tropics generally agree that one hectare may hold more than 650 tree species—more than in all of Canada and the continental United States combined (Coley and Kursar 2014).

Difficult as it is to inventory species on land, we know even less about marine diversity, largely because most of the oceans are unexplored. The oceans, covering 71% of the planet, are virtually a black box of biodiversity. Given that the average depth of the world's oceans is 3,800 meters (over 2 miles), the sheer volume of unexplored water and bottom sediments means there are huge uncertainties about the magnitude of marine biodiversity. Citing one example, more than 800 species were found in 21 square meters of ocean bottom sampled off the Eastern Coast of the United States, almost 60 percent of which were previously unknown (Grassle and Maciolek 1992). Among the most explored marine habitats, coral reefs—often called "the rainforest of the seas"—account for nearly one-quarter of the total known marine biodiversity, despite covering only 0.2% of the seafloor (Reaka-Kudla 1997; Knowlton et al. 2010).

Mora et al.'s (2011) estimate was based on species that can be observed with the naked eye; however, the majority of undescribed organisms are bacteria, archaea, and microeukaryotes such as fungi, nematodes, various algae, and others. If we include these microscopic organisms in our estimates of biodiversity, the number of species inhabiting the planet is truly astronomical. Together, these microbes, protists, and fungi are not only the most diverse but also the most abundant, widespread, and longest-evolving forms of life on Earth. There may be a billion species of bacteria, but only about 30,000 have been formally named species (Dykhuizen 2005).

A more recent study of microorganisms suggests that Earth might be home to a staggering 1 trillion species (Loceya and Lennona 2016). Much of this diversity comes from the soil upon which plants—including crops—depend. Almost every phylum known above ground is also represented in soil habitat, and each phylum has a wealth of below-ground species diversity. Less than 10 percent of soil biodiversity has been identified and described. Life in soils includes burrowing vertebrates (naked mole rats, *Heterocephalus glaber*, are perhaps the most highly adapted to this niche), invertebrate macrofauna such as ants, millipedes, and earthworms; mesofauna such as roundworms, mites, and microflora; as well as algae, lichens, fungi, bacteria, archaea, and viruses. To put soil biodiversity into perspective, a cubic meter of grassland soil can harbor 10 million nematodes, 45,000 earthworms and enchytraeids (pale worms), 48,000 mites and Collembola (insect-like springtails), hundreds of thousands of protozoa, algae, and fungi, and billions of bacteria, the majority of which have not yet been identified (Chivian and Bernstein 2008).

Categorizing Threats and Finding Solutions

An inventory is a critical first step toward preserving Earth's biodiversity. An important second step is categorizing threats to biodiversity so that solutions to these threats can be found. In 2005, Edward Wilson coined the acronym *HIPPO* to summarize the five major causes of population decline, leading to local extinctions: "*H*" stands for Habitat loss, "*I*" for Invasive Species, "*P*" for Pollution, "*P*" for human Population, and "*O*" for Overharvesting.

Habitat Loss

Habitat loss can be defined as the conversion of a natural habitat for human use, such as farming or urban development. Habitat fragmentation is the breaking up of once continuous habitat into habitat patches amid a human-dominated landscape. Habitat degradation refers to changes in habitat quality that reduce the fitness of individuals in a species. All three are related.

The United Nations Environmental Programme (2002) reports that habitat loss is the number one contributor to the decline of species worldwide. The impact of habitat loss is most severely felt in tropical rainforests, which, hectare for hectare, contain the most biodiversity compared to other habitats. Examples of particularly destructive human activities include logging in Peninsular Malaysia; oil palm monoculture in Kalimantan, Indonesia; cattle grazing on pastures converted from forests in Mato Grosso, Brazil; and harvesting of wood for charcoal production in Benin, Africa.

Over half of modern Earth's 14 to 18 million km^2 of tropical humid forests have been lost or degraded. Between 2005 and 2013, the tropics lost an average of 5.5 million hectares of forest per year to agricultural land (Pendrill et al. 2019). In 2019, Global Forest Watch estimated a similar amount of deforestation: 5.4 million hectares, 95% of which occurred in the tropics (Curtis et al. 2018), with Brazil and Indonesia accounting for 52%. Most of this loss was to the expansion of pasture for beef production, croplands for soy and palm oil, and, increasingly, the conversion of primary forest to tree plantations for paper and pulp.

Over three decades ago, Edward Wilson warned that it was not only tropical forests that were in peril, but temperate and boreal forests as well. One attempt (Fitzsimmons 2002) to assess deforestation in Canada estimated mean annual rates of loss between 0.19% and 0.43% per year between 1963 and 1990 in two landscapes initially about 4,500 km^2.

Even prairies and savannahs suffer from habitat loss and fragmentation. The native fauna of North America's Great Plains evolved large-scale movement as an adaptation to the region's extreme spatiotemporal variability in resources. However, human land use has fragmented this landscape, restricting these movements and altering natural disturbance regimes, causing strong effects on ecosystem links and function (Augustine et al. 2021). Another disheartening example comes from the African savannah, which once encompassed about 13.5 million km^2 of lion habitat. Now, due to increasing human population density and fragmentation of the habitat, the current extent of the free-ranging lion population has shrunk to 3.4 million km^2, with over 6,000 lions in areas where their long-term viability is in doubt (Riggio et al. 2013).

Freshwater aquatic habitats have also shrunk and become degraded across the world due to widespread creation of dams, altered floodplains, excessive use, and threats such as pollution and invasive species (see below). Although rivers, lakes, and wetlands cover less than 1 percent of Earth's surface, they harbor about 10% of all known species (Strayer and Dudgeon 2010). Habitat loss has affected even large rivers. The natural flow of the planet's great rivers—such as the Colorado in North America, the Nile in Africa, the Ganges and Indus in India, and the Yellow in China—has been so reduced that they no longer flow to the sea during the dry season. These great rivers reflect the fate of many lesser rivers as well, making riverine fishes some of the most threatened taxa on the planet (Cooke et al. 2012). Limitations hinder effective conservation of knowledge about these fish (Cooke et al. 2012), and underscore how much less we know about the enormous diversity of freshwater invertebrates and microbes that could also be impacted by this aquatic habitat loss as well.

As on land, more marine biodiversity is found in the tropics than elsewhere. The planet's diverse corals are themselves an amazing habitat, with coral reefs providing home to almost 100,000 marine species, including an estimated 4,000 to 5,000 species of fish, which comprise almost 40 percent of the world's known marine fishes (Roberts et al. 2002). The global center of marine biodiversity lies in the Southeast Asian archipelago, encompassing the Philippine and Indonesian Islands, while in the Atlantic Ocean, the Caribbean Sea holds the greatest biodiversity. The decline in reef areas across the world is now raging unabated, like an underwater forest fire. An estimated 20 percent of the planet's reefs have already been destroyed and 50 percent are at risk of collapse (Wilkinson 2004). Evidence indicates that this is largely due to the combined human-driven impacts of ocean warming and acidification (Hoegh-Guldberg et al. 2019), accompanied by increasing levels of pollution, unsustainable coastal development, overfishing, and outbreaks of coral predators such as the crown of thorns seastar, *Acanthaster planci* (Kayal et al. 2012).

In sum, habitat loss is utterly dismaying. Are there any strategies in place to stop or mitigate it? Yes, but they are few and often haphazard. One of the most common strategies for addressing habitat loss is to set aside land. In the United States, public lands are set aside, with varying degrees of human use. These lands provide refuge for many species, including endangered species, but past policies have often led to species losses. In response to this, The National Park Service has actively worked to restore species and habitat that occurred in national parks historically but have been degraded or lost, such as reintroducing wolves to Yellowstone and restoring water to the Everglades (Lowry 2009). Private entities also work to prevent habitat loss. The Nature Conservancy and World Wildlife Fund actively purchase and help to set aside protected areas. The Nature Conservancy, which began opportunistically buying and protecting land well over half a century ago, now strategically prioritizes land conservation in areas of biodiversity significance all across the world (Valutis and Mullen 2000).

The modern protected-area movement started in North America, Australia, and Europe in the 19th century, where designated areas were set up to protect spectacular natural features and wildlife

(Phillips 2004). By the middle of the twentieth century, many protected areas were established in developing countries in Asia, Africa, and South America, with dual mandates of landscape and species protection, as well as public use (such as tourism within protected areas). In some developing countries, the income associated with protected areas is significant (for instance, tourism revenue from visits to see mountain gorillas inside Volcanoes National Park in Rwanda). Some tourism revenue is used innovatively to conserve the region's landscape and species. The role protected areas have played in conserving species diversity and landscapes is undeniable. Recently, on 19 December 2022, more than 190 countries agreed during the COP15 International Biodiversity Summit in Montreal, Canada. The key target agreed upon was to protect and restore 30% of the world's land and seas by 2030 (Gilbert 2022).

Instead of reserving or restoring natural habitat, another approach to habitat loss is to find innovative ways to create human habitats that also provide habitat for other species; this is called **reconciliation ecology** (Rosenzweig 2003). Rosenzweig (2003) provides examples of reconciliation ecology, such as the Red Sea Star Restaurant, located underwater on a waterfront in Israel, which grows corals outside its submerged windows to attract both fish and diners. Reconciliation ecology can also occur in our backyard. For instance, across the Midwest, groups like the Iowa Natural Heritage Foundation provide guides to removing lawns and starting backyard prairies, and the Conservation Reserve Program has been encouraging farmers to plant prairie strips among farmland crops to the benefit of both biodiversity and soil health (Dutter et al. 2025). Xeriscaping with native plants is gaining popularity in many parts of the drought-stricken western United States. Additionally, many people place nest boxes and roosts around their property for birds and bats. In most cases, it merely takes good information and word-of-mouth to inspire people to practice reconciliation ecology.

Invasive Species

Invasive species are nonnative species that disperse widely, rapidly, and at the expense of native species when they arrive in a new habitat. When humans migrate, they often intentionally take their domesticated animals and plants with them. For example, dogs accompanied humans migrating to the Americas, the Pacific, and Australia thousands of years ago (Jackson et al. 2017). Humans also inadvertently carry along their pests and pathogens, including rats and diseases, which spread from Asia with the movement of humans in prehistoric times (Aplin et al. 2011). With increasing globalization—beginning with the Silk Road between Asia and Europe in the second century BCE, continuing with the "Columbian Exchange" between Europe and the Americas about 500 years ago, and extending into today's widespread and rapid movement of goods around the world—species introductions, whether intentional or accidental, have increased exponentially around the globe.

Virtually all ecosystems worldwide, including freshwater and marine, desert, grassland, forest, wetland, and even arctic ecosystems, have experienced the human-assisted introduction of nonnative species, including plants, fungi, vertebrates, arthropods, mollusks, worms, bacteria, and others. Despite all this movement, not all introduced species become established. In general, about half of the freshwater fish, mammal, and bird species introduced from Europe to North America, or vice versa, fail to establish and spread (Jeschke and Strayer 2005). Furthermore, not all nonnative species become invasive—that is, harmful to native species or ecosystem services. Some thrive particularly well in environments already modified by humans, as exemplified by the German cockroach (*Blattella germanica*), which has exclusively adapted to human dwellings and even our methods of pest control (Wada-Katsumata et al. 2023). However, even in natural environments, not all introduced species are harmful. Rejmánek and Randall (2004) estimate that only 15–30% of naturalized plant species in the United States end up becoming "noxious weeds," with most naturalized species being "incorporated into native floras without any serious consequences."

Yet some introduced species are extremely disruptive to native ecosystems. For example, invasion by zebra and quagga mussels (*Dreissena*) can cause significant declines in phytoplankton and zooplankton, which in turn causes declines in the abundance and growth rate of pelagic fish species (Higgins and Vander Zanden 2010). Another example is the kudzu plant (*Pueraria montana*), native to Japan and China, which was introduced to North America as an ornamental plant in 1876 and later used between the 1930s and 1950s to battle soil erosion. Millions of seedlings were grown in nurseries and distributed. Unfortunately, it is now considered America's most infamous weed, overtaking as much as three million hectares (more than seven million acres) of native plant habitat (Forseth and Innis 2004).

Invasive species disrupt not only ecosystems but are also implicated in extinctions. In an analysis of 170 animal (largely bird) extinctions across the globe for which causes could be compiled, 91 (54%) included the effects of invasive species (Clavero and Garcia-Berthou 2005). Outstanding examples include the brown tree snake (*Boiga irregularis*), whose accidental introduction caused the extirpation of most bird, bat, and reptile species on the island of Guam (Fritts and Rodda 1998; Figure 5), and the Nile perch (*Lates niloticus*), whose population explosion in Lake Victoria in East Africa caused the extinction of more than 200 cichlid fish species (Pringle 2005).

Between 2012 and 2022, the U. S. Congress appropriated $11 million annually for projects to alleviate invasive carp problems in the Great Lakes as a way to stem their adverse effects on the lakes' fishing industry (Chapman et al. 2023). Likewise, the State of Alaska spends state money to remove invasive northern pike from lakes and waterways on the Kenai Peninsula and Susitna Valley to restore native fish habitat (Dunker et al. 2018). In general, it can be very expensive for local governments to manage, contain, or eradicate invasive species, and as such, prevention is the most efficient and least costly method of dealing with this issue. Efforts by the USDA National Invasive Species Information Center, National Park Service, U.S. Department of Agriculture, U.S. Fish and Wildlife Service, and partnerships by state and local entities all work to educate and inform the public about invasive species. They also encourage public groups, such as "weed warriors," to engage in the manual removal of invasive species. When these efforts fail, costly eradication methods must be employed.

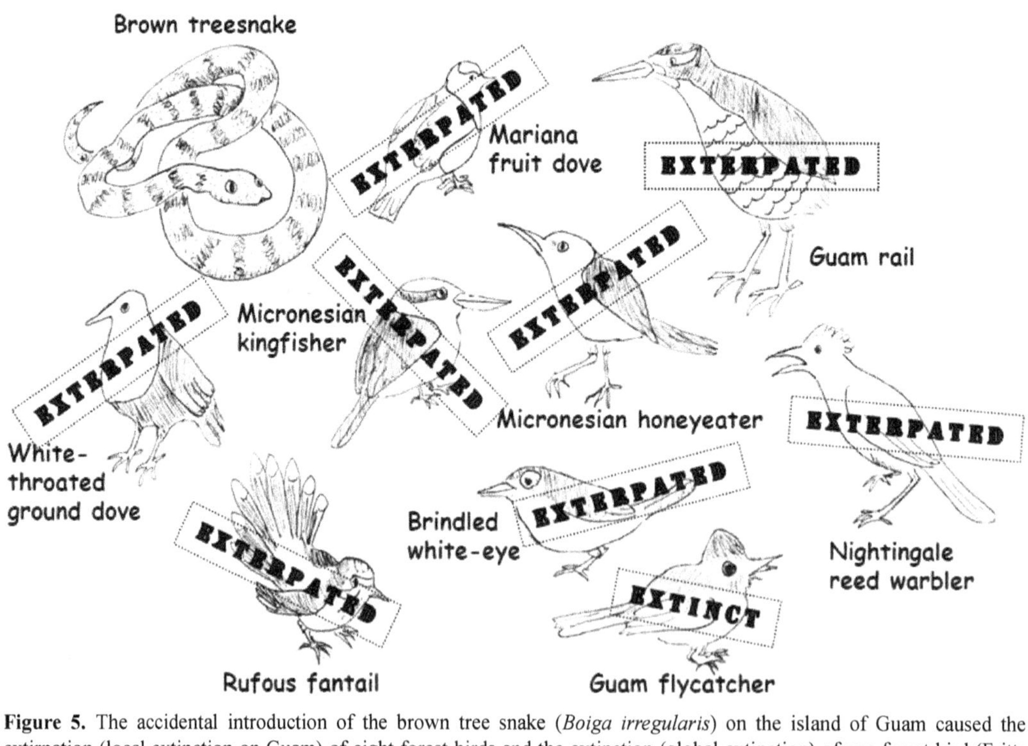

Figure 5. The accidental introduction of the brown tree snake (*Boiga irregularis*) on the island of Guam caused the extirpation (local extinction on Guam) of eight forest birds and the extinction (global extinction) of one forest bird (Fritts and Rodda 1998).

Pollution

Rachel Carson's *Silent Spring* (1962) was a landmark book for the environmental movement and a wake-up call that pollution can lead to extinction. For example, the near collapse of some raptor populations, such as the American bald eagle and the peregrine falcon, is attributed to the disruptive effects of the pesticide DDT on eggshell formation. Although fiercely opposed by chemical companies, *Silent Spring* ultimately swayed public opinion, leading to a nationwide ban on DDT in agriculture.

A pollutant is any chemical in the wrong place or at the wrong concentration with the potential to cause harm. Transportation, general industry, construction, extraction, power generation, and agroforestry all contribute pollutants to the air, land, and water. Tens of thousands of human-made chemicals have been introduced and distributed around the globe. Although many have benefited from human industry, some have been very hazardous. A group of chemicals collectively called "persistent organic pollutants" (POPs) is of particular concern because they travel long distances in air or water, accumulate in fatty tissue, and become progressively more concentrated as they move up the food chain. These pollutants affect reproduction, cause cancer, suppress the immune system, and interfere with the development and function of the nervous system in many animal species, including humans.

Top predators often become serious victims of POPs. For example, POPs are considered the largest threat to polar bears (*Ursus maritimus*), second only to loss of sea ice habitat (Routti et al. 2019). Documented exposure to these contaminants in polar bears has shown adverse effects on thyroid hormones, lipid metabolism, neurochemistry, and immune function (Routti et al. 2019). Toxic health effects have also been documented for cetaceans, which, due to their trophic position and proximity to estuarine and coastal ecosystems, are particularly vulnerable to population declines caused by POPs (Hall et al. 2018).

Heavy metals from mining are a prominent water pollutant, but fertilizers are a much more wide-ranging pollutant affecting ponds, lakes, and coastal ecosystems. Nutrient loading of aquatic systems from agriculture, fossil fuel burning, and wastewater causes eutrophication, of which hypoxia and harmful algal blooms are symptomatic. Eutrophication results in seagrass loss, coral bleaching, fish kills, and marine mammal mortalities, ultimately causing a decline in biodiversity (Dai et al. 2023).

Fortunately, there are very effective and proven means to decrease and prevent pollution through environmental regulations. In the early 1970s, due to heightened concern spurred by Rachel Carson and others, President Richard Nixon created the Environmental Protection Agency, which can conduct research on environmental pollutants, monitor pollutants, and enforce air and water quality standards. Other laws, such as the Clean Air Act and the Clean Water Act, can be used to address and prevent pollution. Similar environmental regulations in other countries and global environmental agreements have been used to address pollution issues worldwide.[4]

Human Population

When human populations are small, what might seem like minor habitat destruction, pollution, or exploitation can be magnified as population density grows. For example, cutting down trees for a modest farm may be harmless in the scheme of things, but when hundreds of farmers cut down trees, little forest will remain. *HIPPO* threats that decrease a species' population size are amplified by increases in this "P" accordingly.

[4] It is worth researching the story of Love Canal to see what happens when toxic substances are not regulated.

Overexploitation

Overexploitation occurs when the harvesting of natural resources by hunting, fishing, or collecting exceeds the rate of population replacement. Overharvesting for food, fashion, or profit has contributed to many local population extinctions of fish, birds, whales, and large African mammals over the past several hundred years. In the United States, the snowy egret, passenger pigeon, and heath hen are all examples of species that fell victim to overexploitation. The passenger pigeon and heath hen went extinct because they were tasty (Schorger 2011; Kamm 2015), but the snowy egret, which has now been brought back from the brink of extinction, was one of many strikingly beautiful birds worldwide whose populations suffered during the "murderous millinery" period in the late 1800s and early 1900s, when bird feathers for women's hats were all the rage (Frohring et al. 1988). It is sobering to note that populations of American bison, once numbering in the millions and dominating North America from the Atlantic coast west to the great plains—extending from Alberta, Canada to northeastern Mexico and across the Rocky Mountains—were reduced to a few thousand after Europeans settled in America and made them the center of a burgeoning hide trade (Lueck 2002). Overharvesting has also led to the decline of many local populations of mammals in other parts of the world as well. In Asia, South America, and Africa, forest animals are taken annually for food. According to one estimate, in western and central Africa, this amounts to a million tons annually (Wilkie and Carpenter 1999). Such exploitation can lead to extinction, as was the case for Miss Waldron's red colobus monkey (*Procolobus badius waldroni*), native to West Africa and last sighted in the Ivory Coast in 1978 (McGraw 2005).

The importation (both legal and illegal) of vast numbers of animals captured from tropical forests, coral reefs, and other ecosystems for food, pets, zoos, and biomedical research can also negatively impact wild populations. In 2002 alone, the United States legally imported 47,000 mammals, 379,000 birds, 2 million reptiles, 49 million amphibians, and 223 million fish (Chivian and Bernstein 2008). Indonesia is considered the "epicenter of the Asian Songbird Crisis" because the cage bird trade there has imperiled many songbird species (Nijman et al. 2021). A market favorite, the Javan pied starlings (*Gracupica jalla*) were declared extinct in the wild in 2021, and the entire remaining population is now captive-bred. However, these starlings are not protected under Indonesian law, and no attempts have been made to reestablish wild populations.

Overharvesting is not limited to animal species and has nearly wiped out some plants. The demise of the African cherry (*Prunus africana*) ultimately came about because extracts of its bark were discovered to work as a therapy for an enlarged prostate, common in older men. International market demand was high, and after a mere 35 years of harvest, it became a CITES endangered species (Stewart 2003). Like animals such as rhinos, overexploitation can also threaten plant species, such as *Citropsis articulata*, when considered a male aphrodisiac (Olupot 2021).

In the oceans, rapid and steep declines have taken place in both the abundance and sizes of predatory fish (Myers and Worm 2003). In the Gulf of Maine and the Grand Banks off Newfoundland, the continental shelf once abounded with Atlantic cod (*Gadus morhua*), a giant predatory fish that hunted in enormous schools. For four centuries, American and Canadian fishermen exploited cod in what was one of the world's richest fisheries. Still, cod numbers began declining with the introduction of trawl fishing methods that replaced traditional hooks, nets, and traps. The stock collapsed in the early 1990s (Olsen et al. 2004).

In general, the slower an organism grows to reach maturity, the more susceptible its population is to overexploitation. Although commercial fishing for orange roughy (*Hoplostethus atlanticus*), a species that reaches maturity after 20–30 years, only started in 1979, the stock has been rapidly fished down below what the population can sustainably yield (Branch 2001). Similar examples of overexploitation of slow-growing species come from whaling in the 18th and 19th centuries for whale oil, which brought many whale species to the brink of extinction. Another marine mammal, the Steller's sea cow (*Hydrodamalis gigas*), which was a giant relative of manatees and dugongs, was discovered and eaten by the naturalist Georg Wilhelm Steller and his fellow shipwrecked crew on Bering Island in the Bering Sea in 1741. After rescuing themselves by building a new boat from the shipwreck, word of this island and the sea cows spread. With the burgeoning growth of the North Pacific maritime fur trade, this remnant population of Steller's sea cow was driven to extinction by 1768, just 27 years after its discovery. Although hunting of sea cows by hungry fur traders almost certainly put the nail in their coffin, Estes et al. (2016) argue that it was the overexploitation by the fur trade of sea otters (*Enhydra lutris*), a keystone species that maintained the kelp forest food of the sea cows, that set the stage for the extinction of the sea cow. This extinction is seen as the harbinger of extinction cascades, a series of extinctions triggered by the disappearance of a keystone species (in this case, the sea otter) in an ecosystem (Estes et al. 2016).

Overexploitation, though often driven by economics, is not an outcome desired by exploiters. Some of the best conservation measures have been implemented by communities of fishers and hunters wishing to sustain their economies and ways of life. Fishing and hunting in the United States are regulated by both national (National Oceanic and Atmospheric Administration, Fish and Wildlife Service) and state (fish and game departments) entities. Some states, like Alaska, even have sustainable harvest written into the state constitution. Other similar regulatory bodies that generate and enforce fishing and hunting regulations can be found across the globe. In some cases, such as with whale oil, bison hides, and fancy hats, simply "devaluing" a species as a commodity can help bring it back from the brink of extinction. This is demonstrated by pioneering legislation such as the Migratory Bird Act of 1918 and the Marine Mammal Protection Act of 1972, which effectively took those species "off the market," allowing for their recovery.

Why are Rare and Endangered Species Prone to Extinction?

If the species is the fundamental unit of biodiversity (Wilson 1999), then avoiding species extinction is the fundamental goal of biodiversity conservation. Unlike many other human impacts on Earth, species extinction is irreversible—pithy dreams of "de-extinction" aside. Not surprisingly, the most widely used conservation approach has been at the species level, and the most widely recognized conservation approach is the International Union for Conservation of Nature's Red List of Threatened Species (www.iucnredlist.org).

The recent 2017–2020 Red List of Threatened Species includes 38,543 species from a broad range of taxonomic groups, including vertebrates (10,119), invertebrates (5,653), plants (22,503), and fungi and protists (268) (IUCN 2021). As of 2021, almost 30% of the world's amphibians and 20% of mammals were threatened, followed by 13% of birds, 14% of reptiles, and 10% of fish. A species is considered "threatened," according to the IUCN Red List, if it is "Critically Endangered," "Endangered," or "Vulnerable." These terms have specific scientific meanings. A species is considered critically endangered when it has lost 80% of its population either over the past decade or three generations (whichever is longer), when its population is less than 250 mature individuals, or when its habitat has shrunk to less than 10 km^2. Critically endangered species have a 50% risk of extinction in the wild within ten years or three generations. Endangered species have a 20 percent chance of extinction in 20 years or five generations, and vulnerable species have a 10 percent extinction risk in 100 years.

An examination of Red List entries for which threats are classified reveals that habitat destruction is the overwhelming driver of species endangerment. This suggests that addressing habitat loss should be the cornerstone of conservation action, ideally by establishing protected areas. However, resources to set aside habitat are limited, so an important question arises: "How do we prioritize conserving biodiverse areas across the world?" At the same time, Edward Wilson was popularizing the term "biodiversity" through his influential book *Biodiversity* in 1988, Norman Myers was also coining a new term that would soon become a conservation buzzword: "hotspots." These are defined as areas that contain high species diversity, many endemic species, and have lost at least 70% of their original native vegetation (Myers et al. 2000). Many conservation hotspots are in the tropics. Examples include the Atlantic forests of Brazil and the rainforests of West Africa, both of which have been severely reduced and degraded. Despite encompassing a small fraction (< 2%) of Earth's land surface, these hotspots contain over half the world's terrestrial species (Myers et al. 2000).

Whether in "hot spots" or not, rare and endangered species are vulnerable to extinction because their populations are already small. Most large carnivores, for instance, have relatively small populations. Small populations are prone to extinction from both "deterministic" and "stochastic" causes. The deterministic factors (i.e., *HIPPO*, see previous sections) make a species' per capita growth rate, r, negative (see Chapter 2 *Population Dynamics*). In other words, the population declines and eventually becomes locally extinct. However, unlike large populations, the dynamics of small populations are also governed by high levels of stochasticity (Caughley 1994). Three types of stochasticity have been recognized: genetic, demographic, and environmental. In genetic stochasticity, alleles are lost by chance, thus eroding genetic diversity. Demographic stochasticity is random variation in births and deaths—variation in r—associated with individuals due to pregnancy failure, injury, etc. In a small population of 10 individuals, for instance, there is a chance that no offspring are produced, or all offspring produced are males. Environmental stochasticity is random variation in births and deaths associated with a population—variation in mean r—caused by unpredictable natural disasters such as hurricanes, floods, fire, or droughts. Thus, a small population is more likely to be reduced to zero or such a low number that makes recovery difficult.

The heath hen in North America, once common from Massachusetts south to Virginia, was prized by hunters. The last mainland Massachusetts heath hen was shot in the 1830s, and the last from Connecticut was shot in the 1840s (Kamm 2015). By 1870, no heath hens could be found on the mainland, and the population was confined to the island of Martha's Vineyard. In 1907, the Commonwealth of Massachusetts created a heath hen reservation on Martha's Vineyard when 77 individuals were counted (Kamm 2015). By 1916, the population had increased to nearly 2,000 birds—a spectacular success. However, in May of that year, an environmentally disastrous fire broke out. Because heath hens tend to remain on their nests rather than fleeing danger, large numbers of birds were killed. A harsh winter, followed by an influx of goshawks, further reduced the population to an estimated 100 birds (Kamm 2015). At this stage, demographic uncertainty played its role, decimating the population. By 1928, no broods had been seen for several years, and only a single living heath hen was known to exist, booming for females at a lekking site. By 1932, this heath hen and all his kin were never seen again.

In restoration and conservation, the goal is to make the population growth of the endangered species positive (births – deaths = $r > 0$), while for management, the goal is to make the population growth of the managed species stable ($r = 0$). In conservation and restoration, r becomes critical for the following reason: it can be used to evaluate success ($r > 0$) or failure ($r < 0$). If r continues to be less than zero, we know that our restoration or conservation plan is not ideal. As we have discovered, a key to understanding the population dynamics of any species is having a good grasp of its interspecific interactions, foraging dynamics, life history tradeoffs, and role in the food web. For those in the field of biological conservation, finding the unique conditions to achieve positive r is the central challenge as stewards of Earth's biodiversity.

References

Anderson, S. H., Kelly, D., Ladley, J. J., Molloy, S. and Terry, J. 2011. Cascading effects of bird functional extinction reduce pollination and plant density. Science 331: 1068–71. doi: 10.1126/science.1199092.

Aplin, K. P., Suzuki, H., Chinen, A. A., Chesser, R. T., Ten Have, J., Donnellan, S. C. et al. 2011. Multiple geographic origins of commensalism and complex dispersal history of black rats. PLoS ONE 6: e26357.

Augustine, D., Davidson, A., Dickinson, K. and Van Pelt, B. 2021. Thinking like a grassland: challenges and opportunities for biodiversity conservation in the Great Plains of North America. Rangeland Ecology and Management 78: 281–295.

Baldwin, B. G., Kyhos, D. W., Dvorak, J. and Carr, G. D. 1991. Chloroplast DNA evidence for a North American origin of the *Hawaiian silversword* alliance (Asteraceae). PNAS 88: 1840–1843.

Barclay, A. W. and Habicht, C. 2015. Genetic baseline for Upper Cook Inlet Chinook salmon 42 SNPs and 7,917 Fish. Alaska Department of Fish and Game, Fishery Manuscript Series No. 15-01, Anchorage.

Barnosky, A. D., Matzke, N., Tomiya, S., Wogan, G. O., Swartz, B., Quental, T. B. et al. 2011. Has the Earth's sixth mass extinction already arrived? Nature 471: 51–57.

Barraclough, T. G. 2019. The Evolutionary Biology of Species. Oxford University Press.

Bell, E. A. Boehnke, P. and Harrison, T. M. and Mao, W. L. 2015. Potentially biogenic carbon preserved in a 4.1 billion-year-old zircon. PNAS 112: 14518–14521.

Berlin, B. 1992. Ethnobiological classification: principles of categorization of plants and animals in traditional societies. Princeton University Press, Princeton, NJ.

Bowles, A. M. C., Bechtold, U. and Paps, J. 2020. The origin of land plants is rooted in two bursts of genomic novelty. Current Biology 30: 530–536.

Bradler, S., Cliquennois, N. and Buckley, T. R. 2015. Single origin of the Mascarene stick insects: ancient radiation on sunken islands? Evolutionary Biology 15: 196.

Bradshaw, C. J. A., Sodhi, N. S. and Brook, B. W. 2009. Tropical turmoil: A biodiversity tragedy in progress. Front. Ecol. Environ. 7: 79–87.

Branch, T. A. 2001. A review of orang roughy *Hoplostethus atlanticus* fisheries, estimation methods, biology, and stock structure. S. Afr. J. Mar. Sci. 23: 181–203.

Brown, C., Reyers, B., Ingwall-King, L., Mapendembe, A., Nel, J., O'Farrell, P. et al. 2014. Measuring ecosystem services: Guidance on developing ecosystem service indicators. UNEP-WCMC, Cambridge, UK.

Burney, D. A. and Flannery, T. F. 2005. Fifty millennia of catastrophic extinctions after human contact. Trends in Ecology & Evolution 20: 395–401.

Carson, R. 1962. Silent Spring. Fawcett Crest, New York.

Caughley, G. 1994. Directions in conservation biology. Journal of Animal Ecology 63: 215–244.

CBD. 1992. Convention on biological diversity. (http://www.cbd.int/doc/legal/cbd-en.pdf).

Ceballos, G., Ehrlich, P. R. and Raven, P. H. 2020. Vertebrates on the brink as indicators of biological annihilation and the sixth mass extinction. PNAS 117: 13596–13602.

Ceballos, G., Ehrlich, P. R., Barnosky, A. D. and García A. 2015. Accelerated modern human-induced species losses: Entering the sixth mass extinction. Science Advances 1: [e1400253]. (https://doi.org/10.1126/sciadv.1400253).

Ceballos, G., Zarza, H., González-Maya, J. F., de la Torre, J. A., Arias-Alzate, A., Alcerreca, C. et al. 2021. Beyond words: from jaguar population trends to conservation and public policy in Mexico. PLoS ONE 16: e0255555. (https://doi.org/10.1371/journal.pone.0255555).

Chapman, A. D. 2009. Numbers of living species in Australia and the world. Report for the Australian Biological Resources Study, Canberra, Australia.

Chapman, D., Amberg, J., Calfee, R. D., Hlavacek, E., Hortness, J., Jackson, P. R., Kazyak, D. C., Knights, B. and Roberts, J. 2023. US Geological Survey invasive carp strategic framework, 2023–27 (No. 1504). US Geological Survey.

Chivian, E. and Bernstein, A. 2008. Sustaining Life: How Human Health Depends on Biodiversity. Oxford University Press, New York.

Clavero, M. and Garcia-Berthou, E. 2005. Invasive species are a leading cause of animal extinctions. Trends in Ecology & Evolution 20: 110. doi: 10.1016/j.tree.2005.01.003.

Cole, D. B., Mills, D. B., Erwin, D. H., Sperling, E. A., Porter, S. M., Reinhard, C. T. et al. 2020. On the co-evolution of surface oxygen levels and animals. Geobiology 18: 260–281.

Cole, N. 2021. "Bolyeria multocarinata". IUCN Red List of Threatened Species: e.T2864A13483086. doi:10.2305/IUCN.UK.2021-2.RLTS.T2864A13483086.en.

Coley, P. D. and Kursar, T. A. 2014. On Tropical Forests and Their Pests. Science 343: 35–36.

Collins, L. and Jahnert, R. 2014. Stromatolite research in the Shark Bay World Heritage Area. Royal Society of Western Australia Journal 97: 189–219.

Cooke, S. J., Paukert, C. and Hogan, Z. 2012. Endangered river fish: factors hindering conservation and restoration. Endangered Species Research 17: 179–191.

Cowie, R. H., Bouchet, P. and Fontaine, B. 2022. The sixth mass extinction: fact, fiction, or speculation? Biological Reviews 97: 640–663.

Curtis, P. G., Slay, C. M., Harris, N. L., Tyukavina, A. and Hansen, M. C. 2018. Classifying drivers of global forest loss. Science 361: 1108–1111.

Dai, M., Zhao, Y., Chai, F., Chen, M., Chen, N., Chen, Y. et al. 2023. Persistent eutrophication and hypoxia in the coastal ocean. Cambridge Prisms: Coastal Futures 1: e19.

Darwin, C. 1859. On The Origin of Species by Means of Natural Selection. John Murray, London, UK.

de Queiroz, K. 1998. The general lineage concept of species, species criteria, and the process of speciation. pp. 57–74. *In*: Howard, D. J. and Berlocher, S. H. [eds.]. Endless forms: species and speciation. Oxford University Press, Oxford.

De Vos, J. M., Joppa, L. N., Gittleman, J. L., Stephens, P. R. and Pimm, S. L. 2015. Estimating the normal background rate of species extinction. Conserv. Biol. 29: 452–462. doi: 10.1111/cobi.12380. Epub 2014 Aug 26.
Del Corso, J., Song, H., Callegaro, S., Chu, D., Sun, Y., Hilton, J. et al. 2022. Environmental crises at the Permian—Triassic mass extinction. Nature Reviews Earth and Environment 3: 197–2014.
Donoghue, P. C. J. 2020. Fossil cells. Current Biology 30: R485–R490.
Dunker, K., Sepulveda, A., Massengill, R. and Rutz, D., 2018. The northern pike, a prized native but disastrous invasive. pp. 356–398. *In*: Skov, C. and Nilsson, P. A. [eds.]. Biology and Ecology of Pike. CRC Press.
Dutter, C. R., Rieke, E. L., Pottebaum, S. and McDaniel, M. D. 2025. Prairie strips improve many measures of soil health in nearly a decade. Journal of Soil and Water Conservation, 1–17.
Dykhuizen, D. 2005. Species numbers in bacteria. Proc Calif Acad Sci. 56: 62–71.
Edwards, C. T., Saltzman, M. R., Royer, D. L. and Fike, D. A. 2017. Oxygenation as a driver of the Great Ordovician Biodiversification Event. Nature Geoscience 10: 925-929.
Ellis, E. C., Gauthier, N., Goldewijk, K. K., Bliege Bird, R., Boivin, N., Díaz, S. et al. 2021. People have shaped most of terrestrial nature for at least 12,000 years. PNAS 118: e2023483118. doi: 10.1073/PNAS2023483118.
Erwin, D. H. 1990. The end-Permian mass extinction. Annual Review of Ecology and Systematics 21: 69–91.
Eskelin, A. and Barclay, A. W. 2018. Eastside set gillnet Chinook salmon harvest composition in Upper Cook Inlet, Alaska, 2017. Alaska Department of Fish and Game, Fishery Data Series No. 18–30, Anchorage.
Estes, J. A., Burdin, A. and Doak, D. F. 2016. Sea otters, kelp forests, and the extinction of Steller's sea cow. PNAS 113: 880–885.
Fitzsimmons, M. 2002. Estimated rates of deforestation in two boreal landscapes in central Saskatchewan, Canada. Canadian Journal of Forest Research 32: 843–851.
Forseth, I. N. and A. F. Innis. 2004. Kudzu (*Pueraria montana*): history, physiology, and ecology combine to make a major ecosystem threat. Critical Reviews In Plant Sciences 23: 401–413.
Fritts, T. H. and G. H. Rodda. 1998. The role of introduced species in the degredation of island ecosystems: a case history of Guam. Annual Review of Ecology and Systematics 29:113–140.
Frohring, P. C., Voorhees, D. P. and Kushlan, J. A. 1988. History of wading bird populations in the Florida Everglades: a lesson in the use of historical information. Colonial Waterbirds 11: 328–335.
Gibbs, G. W. 1998. Why are some weta (Orthoptera:Stenopelmatidae) vulnerable yet others are common? Journal of Insect Conservation 2: 161–166.
Gilbert, N. 2022. Nations forge historic deal to save species: what's in it and what's missing. Nature. 19 Dec 2022. doi: 10.1038/d41586-022-04503-9.
Gogarten, J. P. and Townsend, J. P. 2005. Horizontal gene transfer, genome innovation and evolution. Nature Reviews Microbiology 3: 679–687.
Gomory, D., Yakovlev, I., Zhelev, P., Jedináková, J. and Paule, L. 2001. Genetic differentiation of oak populations within the *Quercus robur/Quercus petraea* complex in Central and Eastern Europe. Heredity 86: 557–563.
Grant, P. R. and Grant, B. R. 2008. How and Why Species Multiply: The Radiation of Darwin's Finches. Princeton University Press, Princeton, NJ.
Grassle, J. F. and Maciolek, N. J. 1992. Deep-sea species richness—regional and local diversity estimates from quantitative bottom samples. The American Naturalist 139: 313–341.
Haeckel, E. 1866. Generelle morphologie der organismen. G. Reimer, Berlin.
Hall, A. J., McConnel, B. J., Schwacke, L. H., Ylitalo, G. M., Williams, R. and Rowles, T. K. 2018. Predicting the effects of polychlorinated bipheyls on cetacean populations through impacts on immunity and calf survival. Environmental Pollution 233: 407–418.
Heckman, D. S., Geiser, D. M., Eidell, B. R., Stauffer, R. L., Kardos, N. L. and Hedges, S. B. 2001. Molecular evidence for the early colonization of land by fungi and plants. Science 293: 1129–1133.
Higgins, S. N. and Vander Zanden, M. J. 2010. What difference a species makes: a meta-analysis of dreissenid mussel impacts on freshwater ecosystems. Ecological Monographs 80: 179–196.
Hoegh-Guldberg, O., Pendleton, L. and Kaup, A. 2019. People and the changing nature of coral reefs. Regional Studies In Marine Science 30: 100699 (20p.). (https://archimer.ifremer.fr/doc/00500/61196/).
Huxley, J. 1942. Evolution: The Modern Synthesis. Allen and Unwin, London, UK.
International Union for Conservation of Nature (IUCN). 2021. IUCN Red List version 2021-2: Table 1a.
Jackman, T. R. and Wake, D. B. 1994. Evolutionary and historical analysis of protein variation in the blotched forms of salamanders of the *Ensatina complex* (Amphibia: Plethodontidae). Evolution 48: 876–897.
Jackson, S. M., Groves, C. P., Fleming, P. J. S., Aplin, K. P., Eldridge, M. D. B., Gonzalez, A. et al. 2017. The wayward dog: is the Australian native dog or dingo a distinct species? Zootaxa 4317: 201–224.
Jeschke, J. M. and Strayer, D. L. 2005. Invasion success of vertebrates in Europe and North America. PNAS 102: 7198–7202.

Jönsson, A. 2011. Linnaeus's international correspondence. The spread of a revolution. pp. 171–191. *In*: Gunnarsson, B. [ed.]. Languages of science in the eighteenth century. De Gruyter Mouton, Berlin/Boston.

Kamm, M. 2015. The life and death of the heath hen. Bird Observer. 43: 9–14.

Kayal, M., Vercelloni, J., Lison de Loma, T., Bosserelle, P., Chancerelle, Y., Geoffroy, S. et al. 2012. Predator Crown-of-Thorns Starfish (*Acanthaster planci*) Outbreak, Mass Mortality of Corals, and Cascading Effects on Reef Fish and Benthic Communities. PLoS ONE 7: e47363. (https://doi.org/10.1371/journal.pone.0047363).

Knowlton, N., Brainard, R. E., Fisher, R., Moews, M., Plaisance, L. and Caley, M. J. 2010. Coral reef biodiversity. pp. 65–77. *In*: McIntyre, A. D. [ed.]. Life in the World's Oceans: Diversity, Distribution, and Abundance. Wiley-Blackwell, Hoboken, N. J.

Kring, D. A. 2007. The Chicxulub impact event and its environmental consequences at the Cretaceous—Tertiary boundary. Paleogeography, Paleoclimatology, Paleoecology 255: 4–21.

Kutschera, U. 2016. Haeckel's 1866 tree of life and the origin of eukaryotes. Nature Microbiology 1: 16114. https://doi.org/10.1038/nmicrobiol.2016.114.

Landing, E. and M. E. Johnson. 2024. Stromatolites and their "kin" as living microbilites in contemporary settings linked to a long fossil record. J. Mar. Sci. Eng. 12: 2127. https://doi.org/10.3390/jmse12122127.

Lerner, H. R. L., Meyer, M., James, H. F., Hofreiter, M. and Fleischer, R. C. 2011. Multilocus resolution of phylogeny and timescale in the extant adaptive radiation of Hawaiian honeycreepers. Current Biology 21: 1838–1844.

Linnaeus, C. 1758. Systema Naturae per regna tria naturae, secundum classis, ordines, genera, species cum characteribus, differentiis, synonymis, locis. Tenth Edition, Vol. 1, Laurentii Salvii, Stockholm.

Loceya, K. J. and Lennona J. T. 2016. Scaling laws predict global microbial diversity. Proc. Natl. Acad. Sci. 113: 5970–5975. doi: 10.1073/pnas.

Lowry, W. R. 2009. Repairing Paradise: The Restoration of Nature in America's National Parks. Rowman and Littlefield.

Lueck, D. 2002. The extermination and conservation of the American bison. The Journal of Legal Studies 31: S609–S652.

Ma, M., Zhang, W., Zhao, M., Qiu, Y., Cai, H., Chen, J., et al. 2022. Deccan traps volcanism implicated in the extinction of non-avian dinosaurs in southeastern China. Geophysical Research Letters 49: e2022GL100342. https://doi.org/10.1029/2022GL100342.

Mallet, J. 1995. A species definition for the Modern Synthesis. Trends in Ecology & Evolution 10: 294–299.

Margulis, L. 1970. Origin of eukaryotic cells. Yale University Press, New Haven, CT.

Marsicano, C. A., Irmis, R. B., Mancuso, A. C., Mundil, R., and Chemale, F. 2016. The precise temporal calibration of dinosaur origins. PNAS 113: 509–513.

May, R. M. 2011. Why worry about how many species and their loss? PLoS Biology 9: e1001130.

Mayr, E. 1948. The bearing of the New Systematics on genetical problems; the nature of species. Advances in Genetics 2: 205–237.

Mayr. E. 2000. The biological species concept. pp. 17–29. *In*: Wheelerand, Q. D. and Meier, R. [eds.]. Species concepts and phylogenetic theory: a debate. Columbia University Press, New York.

McGlashen, A. 2017. The 'I'iwi, a besieged Hawaiian forest bird, is now listed as threatened. Audubon Magazine. https://www.audubon.org/news/the-iiwi-besieged-hawaiian-forest-bird-now-listed-threatened (accessed Dec 7, 2021).

McGraw, W. S. 2005. Update on the search for Miss Waldron's red colobus monkey. International Journal of Primatology 26: 605–619.

Mills, D. B., Boyle, R. A., Daines, S. J., Sperling, E. A., Pisani, D., Donoghue, P. C. J. et al. 2022. Eukaryogenesis and oxygen in Earth history. Nature Ecology & Evolution 6: 520–532.

Mora, C., Tittensor, D. P., Adl, S., Simpson, A. G. and Worm, B. 2011. How many species are there on Earth and in the ocean? PLoS Biology 9: e1001127. doi: 10.1371/journal.pbio.1001127.

Myers, N., Mittermeier, R., Mittermeier, C. G., da Fonseca, G. A. and Kent, J. 2000. Biodiversity hotspots for conservation priorities. Nature 403: 853–8.

Myers, R. A. and Worm, B. 2003. Rapid worldwide depletion of predatory fish communities. Nature 423: 280-283. doi: 10.1038/nature01610.

Naranjo-Ortiz, M. A. and Gabaldón, T. 2019. Fungal evolution: major ecological adaptations and evolutionary transitions. Biological Reviews 94: 1443–1476.

Naumenko, S. A., Logacheva, M. D., Popova, N. V., Klepikova, A. V., Penin, A. A., Bazykin, G. A. et al. 2017. Transcriptome-based phylogeny of endemic Lake Baikal amphipod species flock: fast speciation accompanied by frequent episodes of positive selection. Molecular Ecology 26: 536–553.

Newman, M. E. J. 1997. A model of mass extinction. Journal of Theoretical Biology 189: 235–252.

Nijman, V., Campera, M., Ardiansyah, A., Balestri M., El Bizri, H. R., Budiadi, B. et al. 2021. Large-scale trade in a songbird that is extinct in the wild. Diversity 13: 238 https://doi.org/10.3390/d13060238

Olsen, E., Heino, M., Lilly, G. R., Morgan, M. J., Brattey, J., Ernande, B. et al. 2004. Maturation trends indicative of rapid evolution preceded the collapse of northern cod. Nature 428: 932–935.
Olupot, W. 2021. Restoration of *Citropsis articulata*, a species at risk from medicinal overharvesting in a Ugandan rainforest reserve. International Journal of Forestry Research. 2021: 7264632. https://doi.org/10.1155/2021/7264632.
Pace, N. R. 1997. A molecular view of microbial diversity and the biosphere. Science 276: 734–740.
Paddle, R. 2000. The Last Tasmanian Tiger: The History and Extinction of the Thylacine. Cambridge University Press, Cambridge, UK.
Pendrill, F., Persson, U. M., Godar, J., Kastner, T., Moran, D., Pendrill, F. et al. 2019. Agricultural and forestry trade drives large share of tropical deforestation emissions. Global Environmental Change 56: 1–10.
Phillips, A. 2004. The history of the international system of protected area management categories. Parks 14: 4–14.
Pimm, S. L., Alves, M. A. S., Chivian, E. and Bernstein, A. 2008. What Is biodiversity? pp 3–27. *In*: Chivian, E. and A. Bernstein [eds.]. Sustaining Life: How Human Health Depends on Biodiversity. Oxford University Press, New York.
Pimm, S. L., Jenkins, C. N., Abell, R., Brooks, T. M., Gittleman, J. L. and Joppa, L. N. et al. 2014. The biodiversity of species and their rates of extinction, distribution, and protection. Science 344: 1246752. doi: 10.1126/science.1246752.
Pimm, S., Raven, P., Peterson, A., Sekercioglu, C. H. and Ehrlich, P. R. 2006. Human impacts on the rates of recent, present, and future bird extinctions. PNAS 103: 10941–10946.
Pinto, G., Mahler, D. M., Harmon, L. J. and Losos, J. B. 2008. Testing the island effect in adaptive radiation: rates and patterns of morphological diversification in Caribbean and mainland Anolis lizards. Proc. R. Soc. B 275: 2749–2757.
Pringle, R. M. 2005. The origins of the Nile perch in Lake Victoria. BioScience 55: 780–787.
Puttick, M. N., Morris, J. L., Williams, T. A., Cox, C. J., Edwards, D., Kenrick, P. et al. 2018. The interrelationships of land plants and the nature of the ancestral embryophyte. Current Biology 28: 733–745.
Raven, P. H. 1988. Our diminishing tropical forests. pp. 119–122. *In*: Wilson, E. O. and Peter, F. M. [eds.]. Biodiversity. National Academy Press, Washington, DC.
Rea, A. M. 1997. At the Desert's Green Edge: An Ethnobotany of the Gila River Pima. University of Arizona Press, Tucson, AZ.
Reaka-Kudla, M. L. 1997. The global biodiversity of coral reefs: a comparison with rain forests. pp. 83–108. *In*: Reaka-Kudla, M. L., Wilson, D. E. and Wilson, E. O. [eds.]. Biodiversity II: Understanding and Protecting Our Biological Resources. Joseph Henry Press, Washington, DC.
Rejmánek, M. and Randall, J. M. 2004. Total number of naturalized species can be a reliable predictor of the number of alien pest species. Diversity and Distributions 10: 367–369.
Riggio J., Jacobson, A., Dollar, L., Bauer, H., Becker, M., Dickman, A. et al. 2013. The size of savannah Africa: A lion's (Panthera leo) view. Biodiversity and Conservation 22: 17–35.
Roberts, C. M., McClean, C. J., Veron, J. E. N., Hawkins, J. P., Allen, G. R., McAllister, D. E. et al. 2002. Marine biodiversity hotspots and conservation priorities for tropical reefs. Science 2002: 295: 1280–1284.
Rosenzweig, M. L. 1995. Species diversity in space and time. Cambridge University Press, Cambridge, UK.
Rosenzweig, M. L. 2003. Win-Win Ecology: How the Earth's Species can Survive in the Midst of Human Enterprise. Oxford University Press, New York.
Ros-Rocher, N., Perez-Posada, A., Leger, M. M. and Ruiz-Trillo, I. 2021. The origin of animals: an ancestral reconstruction of the unicellular-to-multicellular transition. Open Biology 11: 200359.
Routti, H., Atwood, T. C., Bechshoft, T., Boltunov, A., Ciesielski, T. M., Desforges, J-P. et al. 2019. State of knowledge on current exposure, fate and potential health effects of contaminants in polar bears from the circumpolar Arctic. Science of the Total Environment 664: 1063–1083.
Sakai, A. K., Wagner, W. L. and Mehrhoff, L. A. 2002. Patterns of endangerment in the Hawaiian flora. Syst. Biol. 51: 276–302.
Schluter, D. and Grant, P. R. 1984. Determinants of morphological patterns in communities of Darwin's finches. The American Naturalist 123: 175–196.
Schorger, A. W. 2011. The passenger pigeon: Its natural history and extinction. University of Wisconsin Press, Madison, WI.
Seehausen, O. 2006. African cichlid fish: a model system in adaptive radiation research. Proc. R. Soc. B 273: 1987–1988.
Selden, P. A. 2016. Land Animals, Origins of. pp. 288–295. *In*: Kliman, R. M. [ed.]. Encyclopedia of evolutionary biology. volume 2. Academic Press, Oxford. doi:10.1016/B978-0-12-800049-6.00273-0.
Sheehan, P. M. 2001. The late Ordovician mass extinction. Annu. Rev. Earth Planet. Sci. 29: 331–364.

Steadman, D. W. 1997. Human-caused extinction of birds. pp. 139–162. *In*: Reaka-Kudla, M. L., Wilson, D. E. and Wilson, E. O. [eds.]. Biodiversity II: Understanding and protecting our biological resources. Joseph Henry Press. Washington, D.C.

Steadman, D. W. and Martin, P. S. 2003. The late Quaternary extinction and future resurrection of birds on Pacific islands. Earth-Science Reviews 61: 133–147.

Stewart, K. M. 2003. The African cherry (*Prunus africana*): Can lessons be learned from an over-exploited medicinal tree? Journal of Ethnopharmacology 89: 3–13.

Stigall, A. L. 2012. Speciation collapse and invasive species dynamics during the Late Devonian "mass extinction". GSA Today 22: 4–9.

Strayer, D. L. and Dudgeon, D. 2010. Freshwater biodiversity conservation: recent progress and future challenges. Journal of the North American Bethological Society 29: 344–358.

Tilman, D., Reich, P. B., Knops and J. M. H. 2006. Biodiversity and ecosystem stability in a decade-long grassland experiment. Nature Letters 441: 629–632.

United Nations Environment Programme. 2002. Global Environmental Outlook—3. Earthscan, London, UK.

Valutis, L. and Mullen, R. 2000. The Nature Conservancy's approach to prioritizing conservation action. Environmental Science and Policy 3: 341–346.

Wada-Katsumata, A., Hatano, E. and Schal, C. 2023. Gustatory polymorphism mediates a new adaptive courtship strategy. Proc. R. Soc. B. 290: 20222337. https://doi.org/10.1098/rspb.2022.2337.

Whittaker, R. H. 1969. New concepts of kingdoms of organisms. Science 1963: 150–159.

Wignall, P. B. and Atkinson, J. W. 2020. A two-phase end-Triassic mass extinction. Earth-Science Reviews 208: 103282.

Wilkie, D. S. and Carpenter, J. F. 1999. Bushmeat hunting in the Congo Basin: an assessment of impacts and options for mitigation. Biodiversity and Conservation 8: 927–955.

Wilkinson, C. [ed]. 2004. Status of Coral Reefs of the World: Vol. 1. Australian Institute of Marine Science, Townsville, Queensland, Australia.

Wilson, E. O. [ed]. 1988. Biodiversity. National Academy Press. Washington, DC.

Wilson, E. O. 1989. Threats to Biodiversity. Scientific American 261: 108–117.

Wilson, E. O. 1999. The Diversity of Life. New York. Norton.

Woese, C. R., Kandler, O. and Wheelis, M. L. 1990. Towards a natural system of organisms: proposal for the domains Archaea, Bacteria, and Eucarya. PNAS 87: 4576–4579. doi: 10.1073/PNAS87.12.4576.

Yoon, H. S., Hackett, J. D., Ciniglia, C., Pinto, G. and Bhattacharya, D. 2004. A molecular timeline for the origin of photosynthetic eukaryotes. Molecular Biology and Evolution 21: 809–818.

Zachos, F. E. 2016. Species concepts in biology: historical development, theoretical foundations, and practical relevance. Springer, Switzerland.

CHAPTER 12

Climate Change

"Men argue. Nature acts." – Voltaire

Farming can be seen as a necessary precursor to civilization as seen in the cultivation of rice and millet for China, maize and potatoes for the Americas, and wheat and barley for South Asia and the Near East (Tauger 2020). A corollary to this argument is that a viable civilization must mitigate weather-related agricultural catastrophes. Prior to modern agronomics, sacrifice and fertility rituals might have been one way to do this, but nearly all ancient civilizations attempted to predict the weather and plan accordingly. The pre-Columbian Mayan people relied on astronomy and mathematics to derive agricultural calendars that dictated dates of cultivation and fertility rituals (Camacho-Villa et al. 2021). During the rise of the Mediterranean and European civilizations, a method called astrometeorology put the Sun, Moon, and planets at the center of weather modulation on Earth (Scofield 2010). In China around 200 BCE, weather was understood in terms of yin (the earthbound elements of clouds and rain) and yang (the heat of the sun and celestial elements), and the Chinese developed almanacs that marked weather events typically occurring during specific 15-day parts of the year (Teague and Gallicchio 2017). Such farming-related, weather-predicting almanacs have been used not only historically but also well into the modern age. In the United States, the oldest and perhaps most widely read is the *Old Farmer's Almanac*, whose "secret weather forecasting formula" was derived by the founder, Robert B. Thomas, in 1792 (Huff 2014).

Although farming communities, both past and present, have invested heavily in weather prediction to optimize planting and production and avoid catastrophic losses, it is the regional climate that determines what type of agricultural production can succeed. For instance, coffee plantations cannot exist where barley is grown, and vice versa. Likewise, if we are to understand why organisms thrive where they do (**biogeography**) or how they thrive where they do (evolution), it is not the weather we should focus on, but the climate.

Weather and Climate

Mark Twain, an American writer and humorist, once said, "Weather is what you get and climate what you expect." Weather—which includes temperature, humidity, precipitation, wind (speed and direction), and cloud cover—is multidimensional. It can vary daily, and even hourly, and across short distances, making its understanding and prediction challenging. Climate, in contrast, is a location's prevailing weather, making it far more predictable, and, therefore, a driver of adaptations for the organisms that live there.

When ecologists study the reasons why, for example, particular butterfly species have declined in number in the mountains of northern California over a 42-year period, a change in the prevailing weather (in this case, a lack of rain for a decade) may be implicated (e.g., Halsch et al. 2021). Climate, as the prevailing weather, can be described in terms of monthly, yearly, or even semi- or multi-decadal averages of weather characteristics such as temperature, humidity, precipitation, etc.

DOI: 10.1201/9781003008538-12

Climate can also be analyzed at a local, regional, or global scale, depending on the spatial and temporal scales important to the organisms of interest. In some places, changeable weather is characteristic of the climate, and so the concepts of minimum and maximum temperatures and precipitation become important as well. In summer, it is not unusual for Chicagoans to enjoy beautiful sunny days interrupted by sudden thunderstorms, bringing hail and a drop in temperature before returning to warm sunshine. By contrast, visitors to Bangkok may encounter monotonously clear skies for days, while residents of Hawaii experience the same daily temperature swing of 5°C (10°F) year-round. Thus, a location's climate can only be fully characterized when variations and extremes are also included; it is the aggregate of the prevailing weather conditions that defines climate.

Most organisms are adapted to the prevailing conditions in their environment, whether it is an ephemeral desert plant that germinates only when enough rain has fallen, or a polar bear that survives months of subzero temperatures. When climatic variation—within its maximum and minimum bounds—is survivable or predictable over several generations, populations of organisms can become adapted to the local climate, and their numbers and distributions can stabilize (though not necessarily be invariable). In other words, climate can become a predictive factor in explaining a region's ecology.

For terrestrial ecologists, temperature and precipitation are among the most important elements of climate, as they strongly correlate with the distribution of the world's major vegetation types. In a classic study of vegetation in the Santa Catalina Mountains, Arizona, USA, Whittaker and Niering (1975) linked the changes in types of vegetation, from desert to fir forest, along an elevation gradient to variations in temperature and precipitation. On a global scale, Whittaker (1975) quantified the relationship between climate and vegetation in terms of mean annual temperature and mean annual precipitation. He separated Arctic, cold temperate, warm temperate, and tropical vegetation along a temperature axis, and categorized desert, grassland, and forest types along a precipitation axis, thereby pinpointing the type of vegetation expected to occur at particular temperatures and precipitation levels (Figure 1).

Why might temperature and precipitation influence differences in vegetation types? A single word—**photosynthesis**—provides much of the answer. Photosynthesis requires carbon dioxide (CO_2), sunlight, and liquid water for plant production. Areas with different climates provide plants with varying levels of access to these critical photosynthetic resources, leading to the evolution of specialized adaptations. Deserts are water-limited due to a lack of precipitation, while forests are not water-limited but may experience light limitation, and so forth. Photosynthesis also requires a particular enzyme, the efficiency of which at CO_2 assimilation peaks at different temperatures across different taxa, reflecting an environmental adaptation of the photosynthetic enzyme to particular temperatures (Galmés et al. 2014).

Adaptations that maximize photosynthetic production in a given climate tend to converge on similar solutions. For example, completely unrelated succulents—the cacti (family Cactaceae) of North America and the euphorbias (family Euphorbiaceae) of Africa—both have adaptations for storing water (Felger and Henrickson 1997) and use a special form of photosynthesis (crassulacean acid metabolism, or CAM) that enables them to conserve water by opening their stomata at night when evaporation is less of an issue, storing the CO_2, and using it for photosynthesis during the day when light is available (Kluge and Ting 1978). In particularly wet climates, "drip-tips" that shed water off the ends of leaves are a widespread feature across multiple plant taxa. This specialized leaf morphology has been shown to provide a functional benefit in areas with periods of very high precipitation (Malhado et al. 2012). In climates with long periods of freezing temperatures, plants must adapt to a short growing season. Whether in the Arctic or on the alpine slopes of tropical mountaintops, solutions to this situation are limited to the tiny perennial plants that characterize this biome (Billings 1974).

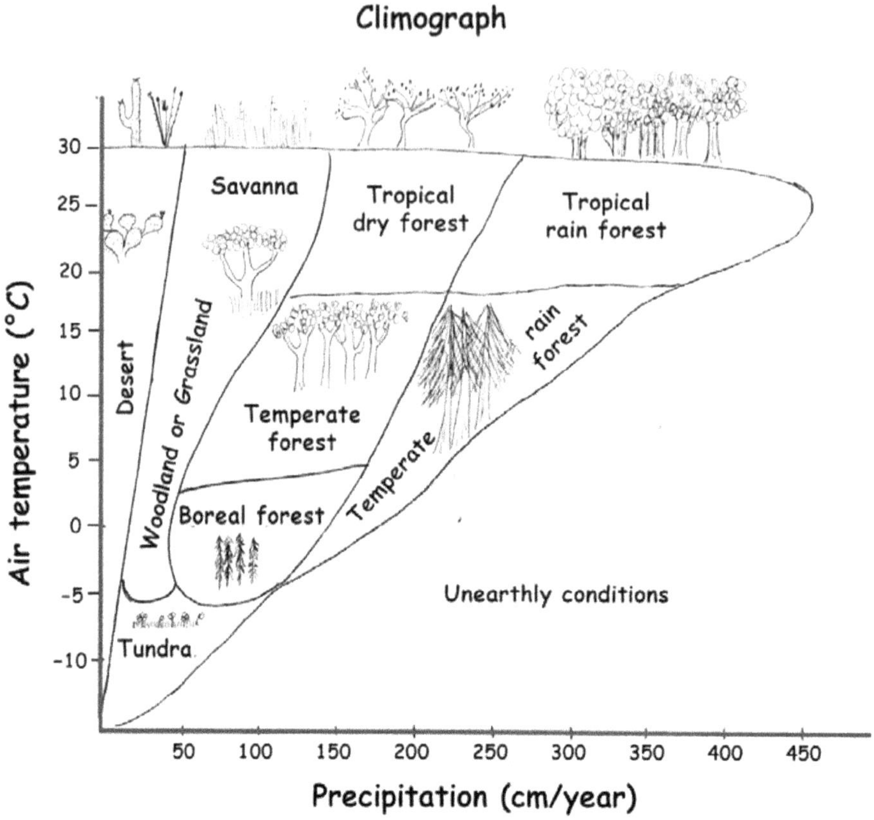

Figure 1. Climograph depicting the relationships between the climatic variables mean annual temperature and mean annual precipitation versus vegetation types (interpreted from Whittaker's [1975] Figure 4.10).

Global Drivers of Temperature and Precipitation

Given that, at a broad scale, temperature and precipitation are predictive of vegetation types—which are essentially the architectural frameworks of terrestrial **biomes**—understanding the factors that drive these climate variables on a global scale can provide a better understanding of biogeography, evolution, and biological responses to a changing climate.

The word "climate" can be traced back to the ancient Greek term "*klima*," which described the angle of the Sun's rays with respect to the Earth's surface (Caccamo and Maguzu 2021). It is one of the keys to understanding climate because the intensity of the sun's rays varies with latitude due to the inclination of the Earth's axis with respect to the Sun. That is why it is hotter in tropical forests than in boreal forests. But this idea barely scratches the surface of what drives climate. One of the earliest people to study both climate and biogeography on global terms was Alexander von Humboldt (1769–1859), who invented isotherms (lines of similar temperature on a map) in his essay *On the Isothermal Lines and the Distribution of Heat on the Earth* (1817) to visualize global climate patterns (Wulf 2015). In doing this analysis, Humboldt demonstrated that temperatures were not the same along the same latitudes, as one might expect. Instead, altitude, landmass, proximity to oceans, and winds appeared to influence the distribution of heat on Earth (Wulf 2015). Humboldt was not wrong, and two hundred years later, with the blossoming of climate science, we've filled in many of the details on the important drivers of Earth's complex, interactive global climate system.

To understand climate, we first need to understand Earth's energy budget—those factors that input energy into the Earth, distribute energy around the planet, and release energy back into the void. These factors include the sun, ice sheets and snow, planetary shape, tilt, and rotation, oceans, mountains and plains, as well as the atmosphere and its composition.

Most of the energy in Earth's climate system comes from the Sun, primarily composed of hydrogen and helium gases, which release energy during the nuclear fusion process occurring within its core. This energy travels in the form of shortwave, visible, and near-infrared radiation away from the Sun. Planetary bodies orbiting the Sun are bombarded by this energy, with **shortwave radiation** being the strongest and most penetrating. Incoming solar energy, which on Earth currently totals about 340 W m^{-2}, is either reflected (about 102 W m^{-2}) or absorbed (about 240 W m^{-2}), depending on the reflective properties (referred to as **albedo**) of the surface and atmosphere at a particular location (Graedel and Crutzen 1995; Trenberth et al. 2009). Some solar radiation never reaches Earth's surface (about 79 W m^{-2}) and is reflected and scattered by clouds or aerosols (very fine particulates) in the atmosphere. Some of the transmitted energy that makes it through this barrier is reflected as well (about 23 W m^{-2}) by surface features with high albedo (Trenberth et al. 2009; Stephens et al. 2015). Ice sheets and especially snow cover increase albedo, as do deforestation and desertification, whereas oceans have very low albedo (Barron et al. 1980; Stephens et al. 2015). Incoming solar radiation that escapes reflection is absorbed by Earth's atmosphere and surface, totaling about 78 W m^{-2} and 161 W m^{-2}, respectively.

Planetary shape, tilt, and rotation affect the distribution and movement of radiant energy arriving from the Sun in both space and time. Because Earth is a sphere, sunlight arrives at higher latitudes at an angle, spreading over a greater area than at the equator, where the sun's rays hit directly. This causes tropical regions to receive a larger amount of energy per unit area than polar regions. At higher latitudes, the same incoming solar radiation is distributed over a larger surface area of the Earth, causing more of the energy to be scattered and reflected into space. Thus, less energy reaches higher latitudes, which translates to lower temperatures, creating the polar zones and contrasting with the higher-energy temperate zones. Furthermore, Earth's axial tilt of about 23° causes predictable seasonal changes in climate in the northern and southern hemispheres, with each hemisphere receiving more or less energy from the Sun as it tilts toward or away from the Sun during Earth's orbit.

Although equatorial and polar locations receive incoming solar radiation unequally, the higher equatorial energy is redistributed across Earth's surface through various physical processes. This redistribution not only creates weather patterns but also contributes to the formation of tropical and desert biomes, as well as the trade winds and ocean currents. Because more solar energy arrives per unit area at the equator, the air masses near the surface, heated from below by radiant energy, expand and, being less dense, rise. This expansion lowers equatorial surface air pressure, causing air masses with higher pressure on either side of the equator to flow in, replacing the rising air masses. This flow of air toward the tropics generates the trade winds. As the rising equatorial air ascends, it cools, reducing its ability to hold moisture, which results in plenty of tropical rains. Once the air reaches the upper atmosphere, which acts like a ceiling, it separates and begins moving north and south toward the poles. When it reaches about 30° north and south, the now dry air cools, becomes denser, and sinks to the ground, creating areas of low rainfall—forming desert biomes at these latitudes. This subtropical high-pressure zone is then set to move back toward the lower-pressure tropics via the trade winds. This continuous equatorial warming and movement of surface air establishes a global atmospheric circulation pattern known as the Hadley circulation, or a **Hadley cell**.

The basic idea was first proposed by Edmond Halley in 1686 and modified later by George Hadley in 1735 to explain the direction of the trade winds. Although a pressure gradient drives surface air from the mid-latitudes to the equator, Earth's rotation causes these air masses to deviate from a direct path. Due to the Coriolis effect, air masses curve to the right (west) north of the equator and the left (west) south of the equator (Figure 2).

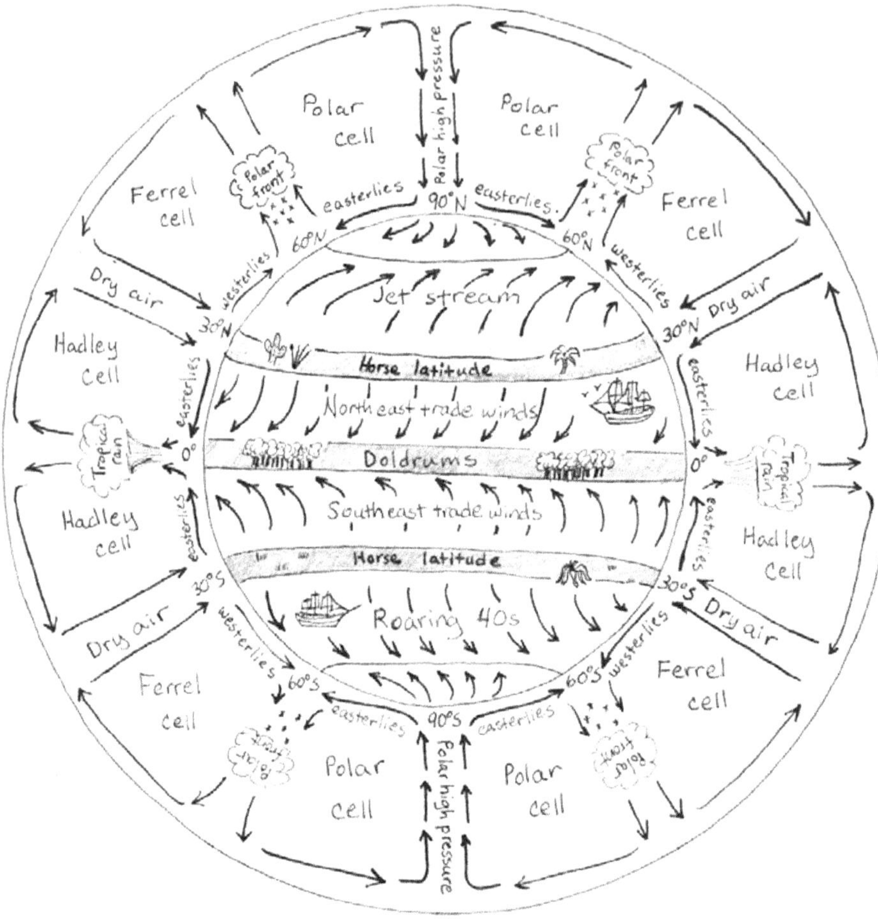

Figure 2. Idealized depiction of global atmospheric circulation (not to scale).

In 1865, meteorologist William Ferrel proposed that a single pair of circulation cells does not govern Earth's atmosphere but by three pairs of cells—one of each pair in each hemisphere—to explain how Earth's atmosphere moves (Liu 2021). These include two Hadley cells between the equator and 30° north and south, two (now called) **Ferrel cells** between 30° and 60° north and south, and two **polar cells** between 60° and the poles. Each of these cells is composed of a band of sinking air with surface high pressure, a band of rising air with surface low pressure, and airflow between them, following pressure gradients at both the surface and in the upper atmosphere (Liu 2021).

Air movement across Earth's surface creates wind; however, rising and falling air do not. Thus, sailors historically encountered the windless "doldrums" at the equator and the "horse latitudes," at approximately 25–35° north and south. However, the movement of air from 25° latitude toward the equator in the Hadley cells creates the **northeast and southeast trade winds** (also called *easterlies*, named for the direction from which they blow) in the north and south latitudes, respectively. In contrast, Ferrel cells drive the *westerlies*, which blow toward the poles from about 35° latitude in both hemispheres, opposite to the direction of the easterlies. In North America, these westerlies are commonly called "the jet stream." In the Southern hemisphere, because there is more ocean than land and nothing to break the flow, the westerlies blow with greater force, earning the name the "roaring forties" for their intensity at those latitudes. After their discovery by Hendrik Brouwer in 1611, sailing the westerlies could cut months off a Dutch spice trader's trip, However, at the time, the difficulty in measuring longitude—and therefore knowing when to turn north to the "East Indies"—often caused ships to wreck off the coast of Western Australia (Van Duivenvoorde 2015).

These global winds—the trades and the westerlies—also influence surface ocean currents, which form systems of gyres across the Earth's surface (Munk 1950). Ocean currents and gyres play a vital role in regulating weather and climate (Yang et al. 2020). The Atlantic Meridional Overturning Circulation (AMOC), which transports warm water north and strongly influences the climate of western Europe, is driven by the energy of winds and tides (Cunningham and Marsh 2010). In addition to affecting climate, ocean currents support marine life with upwelling nutrients (e.g., Ngo and Hsin 2021) and affect marine communities and dispersal (Kuhn et al. 2019).

Unlike the north-south Hadley, Ferrel, and polar cells, there is a fourth major atmospheric cell called the **Walker cell**, which is oriented east-west in the tropical Pacific. This cell is closely tied to ocean conditions and the El Niño phenomenon (Liu 2021). The Walker cell moves air upward over the western Pacific and downward over the eastern Pacific. Its intensity significantly influences global climate, affecting ecosystems (e.g., changing drylands to woodlands), severe weather (e.g., increasing tropical cyclones in eastern Australia), and agricultural production on interannual time scales (Liu 2021). While many factors influence the intensity of the Walker cell, it is particularly influenced by sea surface temperatures through coupled atmosphere–ocean processes. The overall effect of the Walker cell is often referred to as the El Niño–Southern Oscillation or ENSO.

Besides the easterlies (trade winds) and the westerlies (e.g., the jet stream), globally, well over 60 winds occur frequently enough to earn a name (Streever 2016). For instance, the *sirocco* blows from the Sahara toward North Africa and southern Europe; the Mediterranean has the *ostro*; India has the *elephanta*; Cuba has the *bayamo*; Chile has the *puelche*; and the *Santa Ana* winds, dry and hot, sweep down the seaward slopes of California's mountains, sometimes fanning wildfires and driving deadly flames downhill. Together, these winds and the world's ocean currents form a global weather system that redistributes heat energy from the tropics to the polar regions, shaping the climate in which we live.

At the broadest scale, climate varies with latitude, interacting with the atmosphere and ocean to influence temperature and redistribute moisture. However, latitudinal climate effects can be magnified, modified, or mitigated by Earth's varied topography, leading to finer-scale climate heterogeneity to which the world's biota are adapted. As Humboldt discovered from his mapping of isotherms, mountains are also significant climate drivers, though he understood this long before he wrote his essay on isotherms. In June 1802, he and his companions scaled 6,263 m (20,548 ft) Mt. Chimborazo in the Andes, where he observed the distribution of plants in vegetational zones—starting with tropical palms at sea level, progressing upward to humid forests, then upward to European-like conifers, oaks and alders, followed by plants resembling those in the Arctic circle, and finally, the last bit of life—a lichen—at 5,500 meters (18,000 ft; Wulf 2015).

We now know that **adiabatic cooling**—which occurs in an adiabatic system when pressure is decreased—is responsible for the similar distribution of plant life from low to high altitude, as well as from high to low latitude. Temperature decreases as atmospheric pressure drops. At sea level, atmospheric pressure is about 100 kPa; at 5,500 meters, it is about half that. Mountains are also important climate drivers because they collect rain on their windward side (**orographic precipitation**) and create rain shadows on their leeward side. This is explained by prevailing winds and adiabatic cooling, which saturates air as it rises in elevation, leading to rain on the windward side. After crossing the topographic barrier, the air descends and warms on the leeward side, where rainfall ceases (Van den Hende et al. 2021). As a result, depending on the direction of prevailing winds and orientation of mountains, areas just kilometers apart can have drastically different climates and vegetation—such as the wet, windward "Hilo" side of Hawaii versus the dry, leeward "Kona" side.

In addition to its clouds, aerosols, atmospheric cells, and varying pressure, the gas composition of the atmosphere is also a major climate driver. We have primarily focused on how incoming shortwave radiant energy from the sun gets reflected or absorbed and redistributed, but some of that incoming energy—along with radiant energy from all the substances above absolute zero on Earth—radiates back toward space (about 396 W m^{-2}; Trenberth et al. 2009). The atmospheric composition of what we call **greenhouse gases** is the key to determining how much energy returns to space and how much stays on Earth. In March 2000, it was estimated that about 333 W m^{-2} of this back radiation was reabsorbed by Earth's surface, while only 40 W m^{-2} escaped as longwave radiation (Trenberth et al. 2009). This effect is largely responsible for Earth's average surface temperature of approximately 15°C, to which humans have comfortably adapted (Houghton 2004).

The **greenhouse effect**, which allows shortwave energy to enter but prevents longwave energy from escaping, depends strongly on the type and concentration of gases in the atmosphere. Earth's atmosphere is currently made up largely of N_2 at 78% and O_2 at 21%, but these two-atom molecules are too tightly bound to vibrate in a way that absorbs heat (Kweku et al. 2017). In contrast, the greenhouse gases—such as carbon dioxide (CO_2), methane (CH_4), nitrogen oxides (NO_x), chlorofluorocarbons (CFCs), ozone (O_3), and water vapor—consist of more loosely bound molecules capable of vibrating, effectively absorbing longwave radiation. This absorption results in a greenhouse effect, warming the atmosphere and preventing heat from escaping Earth (Kweku et al. 2017). Although carbon dioxide comprises only 0.04% of the atmosphere, it is the most significant greenhouse gas, playing a major role in determining Earth's average surface temperature.

The ocean, while it plays a role in the distribution of moisture, winds, and currents, is also a major climate driver by exchanging heat and carbon dioxide with the atmosphere. For example, although the Southern Ocean covers only 30% of the global ocean surface, it transfers up to 75% of global heat and 40% of anthropogenic carbon from the atmosphere to the ocean (Frölicher et al. 2015). Additionally, it releases natural carbon—stored over eons—in upwelled waters to the atmosphere. Sutton et al. (2021) show that net CO_2 uptake by the Southern Ocean is primarily driven by wind-caused physical mixing. Stronger winds promote greater upwelling and outgassing, reducing net CO_2 uptake (leaving more CO_2 in the atmosphere), while calmer conditions promote ocean stratification, reducing upwelling and outgassing, and increasing net CO_2 uptake (leaving less CO_2 in the atmosphere).

Finally, the biomes of the world—the biosphere—are also closely tied to the concentration of atmospheric CO_2. The iconic dataset of monthly average CO_2 levels at Mauna Loa Observatory in Hawaii (Figure 3) not only shows the year-to-year change in atmospheric CO_2 concentration but also provides evidence of the magnitude of Northern Hemisphere biospheric effects, driven by seasonality in plant growing seasons. The lows depict summer consumption of CO_2 by plants, while the highs reflect winter senescence (Buermann et al. 2007). This "breath" of the biosphere demonstrates that factors affecting the biosphere–atmosphere exchange (e.g., deforestation) can contribute to climate change.

Climate Variability

Because so many factors interact to affect climate, it is unsurprising that climate at a particular location can vary from year to year. However, this variability is sometimes extreme and cyclic, with shifts from one climatic regime to another over the span of years or decades. For example, interactions between the atmosphere and the tropical Pacific Ocean result in the phenomenon known as El Niño, which affects weather patterns worldwide. **El Niño** is a warming of the eastern equatorial Pacific that occurs every two to six years and lasts for several seasons. It brings heavy rainfall along the western South American coast (Peru and Ecuador) and southern and western parts of the United States, while causing drought in Australia, Indonesia, southeastern Asia, northeast Brazil, and Colombia (Philandar 1998).

In Peru and Ecuador, El Niño ("the son," referring to the Child Jesus in Spanish) was once considered benign, as it brought heavy rains to the otherwise arid coastal zones, turning deserts lush with greenery and mountains into fertile pastures, which, in turn, boosted livestock and cotton production. However, these heavy rains can also wash away infrastructure such as houses and roads. Furthermore, the El Niño phenomenon affects the abundance of anchovies off the coast of Peru, changing the once-lucrative cold-water, monospecific anchovy fishery to a less productive, warm-

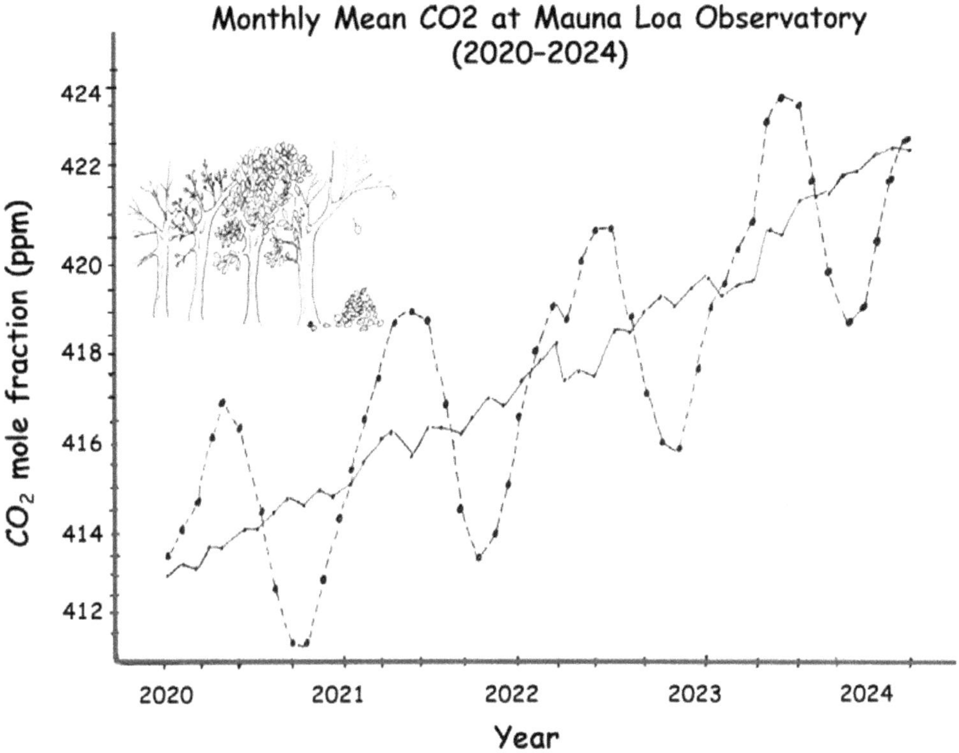

Figure 3. Five complete years (2020–2024) of the monthly Mauna Loa CO_2 records (interpreted from Betts et al.'s [2024] Figure 1). Dashes link monthly means, and year-to-year change is depicted with a solid line (the moving average of seven adjacent seasonal cycles). It is worth pointing out that the seasonal cycle is driven by summertime photosynthesis and wintertime respiration in the Northern Hemisphere (Buermann et al. 2007).

water, multi-species fishery with lower economic value (Niquen and Bouchon 2004). Rapid human population growth has increased vulnerability to these negative effects, so, unlike in earlier times, El Niño is now viewed with disapproval (Rojas et al. 2014).

During "normal" (moderate) conditions, trade winds blow west along the equator, carrying warm water from South America toward Asia. This causes cold-water upwelling along the South American coast, providing an ideal habitat for anchovies. During extreme El Niño conditions, the trade winds weaken, pushing warm water back toward the American coast, resulting in less productive warm-water fisheries. However, there is also an opposite extreme that disrupts this "normal" pattern, called **La Niña**. Under a La Niña climatic regime, the trade winds are stronger than usual, pushing more warm water toward Asia. This typically leads to drought in the southern and western U.S.

In 2012 and 2013, drought gripped the United States during a record-breaking heat wave, wreaking havoc on crops and water supplies, with an economic impact approaching $30 billion (Rippey 2015). This drought can be attributed to back-to-back La Niña episodes during 2010–2011 and 2011–2012 (Rippey 2015). A two-year drought tied to a La Niña event in 1988 and 1989 (Hoell et al. 2014) was even worse, inflicting approximately $40 billion in economic damage. Crops failed so disastrously in the midwestern United States that cattle had to be slaughtered for lack of grass to feed them. The problem was global, as the Soviet Union and China were similarly drought-stricken, but torrential rains plagued parts of Africa, India, and Bangladesh as the La Niña conditions pushed warm water to the west in the Southern Pacific (Philander 1998).

These droughts, while impactful, do not compare to the legendary American Dust Bowl drought of the 1930s, when temperatures exceeded 100 degrees Fahrenheit in the Central High Plains, and more than 12 million pounds of soil from the Great Plains blew into Chicago. Other windblown soil headed for New York City, Washington, D.C., and Boston, destroying towns and farmlands. Over 7,000 people died, and over two million left their homes. The economic and personal hardships the Dust Bowl inflicted on the land and people were immortalized by John Steinbeck in his epic novel "The Grapes of Wrath," published in 1939: "…All day the dust sifted down from the sky, and the next day it sifted down. An even blanket covered the earth, it settled on the corn, piled up on the tops of the fence posts, piled up on the wires; it settled on roofs, blanketed the weeds and trees." What made the drought of the 1930s so unusual and devastating? Its duration. It lasted not just a few years, but the entire decade, plunging the United States, Europe, and the whole world into the Great Depression.

Data from various sources, such as tree rings, lake sediments, and river deposits, indicate that multi-year droughts like the Dust Bowl occur at a rate of one to two per century. However, the searing drought of the 1930s is not even categorized as a megadrought. Megadroughts last not just several years, but several decades, occurring at a rate of two to three every thousand years. The most recent North American megadrought occurred from the 9th to the 16th century and may have contributed to the disappearance of the lost English colony on Roanoke Island in North Carolina (Overpeck 2000). Based on an analysis of climate model forcings (Steiger et al. 2019) and gridded drought reconstructions from tree-ring chronologies (Herweijer et al. 2007), the medieval megadroughts in the American Southwest can be explained by anomalously frequent La Niña conditions, characterized by cold Pacific Sea surface temperatures, warm Atlantic temperatures, and local temperature increases.

In 1899, a few decades prior to the Great American Dust Bowl, the other corner of the world—central India—witnessed a disastrous failure of the monsoon rains. Food crops and livestock died as the drought stretched into a second year, and soon people began to starve. The death toll may have exceeded several million (Amrith 2020). In 1904, Gilbert Walker, a British mathematician and then Director-General of Observatories in India, witnessed the impact of this tragic event and was so appalled that he decided to search for a way to predict the rainfall associated with the Asian monsoons. Walker (1924) found a correlation between barometric (air) pressure across the Pacific Ocean and the amount of rain falling during the monsoons. Reduced barometric pressure in the eastern Pacific was accompanied by increased barometric pressure in the western Pacific, and vice versa. Walker called this oscillation in barometric pressure the **Southern Oscillation**. Low values of the Southern Oscillation Index, he noticed, were associated with drought in Australia, Indonesia, India, and parts of Africa. He even suggested that winter temperatures in Canada were somehow connected to the Southern Oscillation. Unfortunately, despite his brilliant observations and deductions, Walker's notion of a Southern Oscillation and its far-flung effects—although ahead of his time—was criticized and ridiculed by his contemporaries (Philander 1998).

The connection between the Southern Oscillation and ocean temperature patterns during El Niño was eventually described by Bjerknes (1966), who honored Walker by calling this atmospheric system the Walker Circulation or the **Walker cell** (see our discussion of atmospheric cells above). Today, climate scientists recognize the **El Niño–Southern Oscillation** or ENSO, a highly dynamic, large-scale weather system, as a major driver of our modern climatic variability around the globe, affecting the climates of North America, South America, Australia, southern Asia, Africa, and parts of southern Europe.

El Niño, which recurs on average every four to five years, is one of the best-known of Earth's short-term climatic cycles. These cycles are characterized by shifting weather regimes on the order of a few years to decades. Other ocean-related climatic cycles include the **Pacific Decadal Oscillation** (PDO) and the **Atlantic Multidecadal Oscillation** (AMO). Because oceans store more than 30 times as much energy as the atmosphere, the Pacific Ocean—the world's largest and the epicenter of El Niño and the PDO—is one of the most active and important climate drivers. Operating in concert with the PDO (a 30–40-year cycle), El Niño causes ocean temperatures to change by as much as two degrees Celsius. Although this may seem like a small change, its impacts are significant.

For example, fisheries scientists can describe a relationship between Alaskan salmon harvests and changes in climate in the northern Pacific (Hare and Mantua 2001). While the impact of the PDO is primarily located in the Northern Pacific, El Niño primarily affects the tropics. The AMO, which operates on a 60-year cycle, has less impact on global temperatures than El Niño or the PDO, but is still believed to have caused a temperature rise of 0.25°C from 1970 to 2000 (Hare and Mantua 2001).

Droughts and monsoons, warming and cooling sea surface temperatures—and their associated effects—derive from the "short-term" climatic variation caused by ENSO, PDO, and AMO. Periodic occurrences of these effects attest to their somewhat predictable nature; we know a switch in climate regime will occur at some point in the next five years or so, after which it will swing back again. Organisms can respond to these changes in ecological time through population dynamics, migration, life history, and tolerance. However, if change is unidirectional and persists over prolonged periods, then this is no longer considered climatic variability, but rather climate change, which can cause sweeping ecosystem changes. For instance, in parts of the Sahara Desert, we now expect hot and dry conditions, which are unsuitable for human habitation, whereas we know that civilizations once flourished there thousands of years ago. Radiocarbon dating of a chronology of climate proxies shows that abundant rains, along with associated flora, fauna, and human populations, from 9,000 to 6,000 BP, switched to complete aridification by 2,000 BP (Petite-Maire et al. 2018). Climate can change not only over human history but also throughout Earth's history. What drives these monumental changes in climate?

Natural Drivers of Climate Change

Ecologically relevant climate change occurs as a unidirectional shift in climate over a long enough period that it significantly changes a locally adapted ecological community, either by exclusion, extinction, invasion, or evolution of its members. This type of climate change can be caused by well-documented natural drivers, including plate tectonics, volcanic activity, variation in Earth's orbit, and solar variability. We will explore each of these in turn.

Plate Tectonics

The theory[1] of plate tectonics describes Earth's surface as consisting of several vast, incredibly slow-moving rigid slabs (plates). This theory, which provides a sound basis for Wegener's (1912) concept of continental drift, has a climatic dimension. The latitude and relative positions of the continents influence not only the regional climate over lands and oceans but can also affect global climate. For example, when landmasses become concentrated at high latitudes, they tend to become covered with ice, which reflects more solar radiation, thereby lowering global temperature. This phenomenon appears to have been the case for the supercontinent Gondwana (a single continent that eventually broke apart to form modern Australia, Africa, Antarctica, and South America). Gondwana's position during the time of tree-sized lycophytes and horsetails is implicated as a key factor in triggering the Paleozoic ice age (360–255 mya), though other factors are also implicated (Blakey et al. 2008).

As the continental plates play bumper cars across the face of the Earth, they can open and close oceanic gateways between continents, influencing global oceanic circulation and climate. Models of climate and paleogeography for the Cretaceous period (145–66 mya), which was marked by global warming during the early and mid-Cretaceous, followed by cooling toward the end of the Cretaceous, implicate the role of changing marine gateways and the widening of the Atlantic in driving global ocean circulation and ocean basin temperatures, regardless of changes in atmospheric CO_2 levels (Ladant et al. 2020).

---

[1] Here we use the term "theory" in the same sense as for the theory of evolution by natural selection: a set of principles, repeatedly tested by alternative sources of facts and experiments, widely accepted, and that is used to make predictions about natural phenomena.

The movements of these continental plates also result in mountainous upheavals, which in our modern world have created the Himalayas in Asia, the Rockies in North America, the Andes in South America, the Alps and Urals in Europe, and other mountain ranges around the world, profoundly affecting regional climate through their orographic effects. Geologic evidence from the Cenozoic (the last 65 million years) indicates that uplifted plateaus in Southern Asia and the American West created significant climate changes, some of which were felt far from the uplifted regions. Such changes include winter cooling in North America, northern Europe, and northern Asia; summer drying of the North American west coast, the Eurasian interior, and the Mediterranean; and winter drying of the North American plains and Asian interior (Ruddiman and Kutzbach 1989).

There is no doubt that tectonic plates and the changes in landmass positions they create have affected the climate long ago. However, because continental movement is slow (a few centimeters per year) and alters climate on a scale of several million years, the theory of plate tectonics does not adequately explain climate change that has occurred on shorter time scales, such as tens, hundreds, or thousands of years. There must be other explanations to account for these changes.

Variations in Earth's Orbit

Along with feedback mechanisms that work in concert with these fluctuations, fluctuations in the amount of solar radiation received at Earth's surface (called **insolation**) may represent the single greatest cause of global temperature change. Most of us, both north and south of the equator, are familiar with the short-term orbital variations that cause seasonality in temperatures. However, during the first half of the 20th century, Serbian civil engineer and mathematician Milutin Milankovitch (Milankovitch 1930) recognized that long-term variation in Earth's orbit could cause large fluctuations in insolation, affecting global temperatures as well. His theory of orbital forcing, now known as the **Milankovitch cycles**, postulates three kinds of orbital variations affecting Earth's journey around the sun (Cvijanovic et al. 2020).

First, the Earth travels in an elliptical, not a circular, orbit. In 100,000-year cycles, the orbit becomes less elliptical and more circular, then moves back to being elliptical. This is called *eccentricity*. Greater eccentricity increases the magnitude of seasonal changes and lengthens the season occurring at the **perihelion** (the closest point to the Sun). Second, the Earth itself is tilted, and the *angle of the tilt* changes in 41,000-year cycles, from a maximum of 24.5 degrees away from the vertical to a minimum of 21.5 degrees. Greater tilt increases the total annual insolation at higher latitudes and decreases the total insolation closer to the equator. Third, the Earth wobbles like a top when it spins around its axis. This wobble, called *precession*, follows a 23,000-year cycle. It is caused by the tidal forces of the Sun and the Moon, affecting which hemisphere is tilted toward the Sun at perihelion, either enhancing or canceling out the seasonal variation in irradiation, depending on which hemisphere is tilted toward the Sun.

Milankovitch (1941) recognized that these modest periodic changes in orbit, affecting summer insolation at high northern latitudes, can explain the ice ages that impacted the Earth at regular intervals during the most recent 2–3 million years, causing swings between glacial and interglacial periods spanning tens of thousands of years. Support for Milankovitch's provocative 100,000-year temperature cycle is provided by empirical evidence of water temperature changes in past millennia from sediments at the seafloor (Hays et al. 1976). Through oxygen-isotope analysis and statistical analysis of climatically sensitive microorganisms, deep-sea sediment studies established a chronology of temperature change going back nearly 500,000 years. This timescale of climate change was then compared with astronomical calculations of eccentricity, obliquity, and precession to determine whether a correlation did indeed exist. Hays et al. (1976) concluded that changes in the Earth's orbital geometry are fundamental to the succession of **Quaternary ice ages**, that is, the climate changes of the last 2.6 million years (Imbrie et al. 1992).

But if orbital variations explain alternating glacial and interglacial periods, then why have the predictably periodic ice ages only occurred during the most recent 2–3 million years of Earth's history?

First, it is important to understand that these cycles have always been present. Deep-sea sediment cores recording climate-influenced lead isotopes dating back 65 million years show oscillations in the climatic effects of precession and eccentricity (Josso et al. 2021). Even older evidence of Milankovitch cycles is found in banded iron formations dating to 2.46 billion years ago, which indicates that Earth's precession cycle had a higher frequency than present, resulting from shorter day lengths and shorter Earth-Moon distance (Lantink et al. 2022).

Second, it is important to understand that Milankovitch cycles are not the sole drivers of the 100,000-year Northern Hemisphere glacial cycles; the changes in insolation caused by the Milankovitch cycles are the trigger, but the underlying cause of the repeating pattern has been attributed to feedback mechanisms from ice sheets and their effect on albedo, as well as delayed bedrock rebound (Abe-Ouchi et al. 2013), and feedbacks involving CO_2, dust, Southern Westerly Winds, and Southern Ocean upwelling (Ai et al. 2020). The theory of plate tectonics may also provide some insight into the ice-age cycles. Abe-Ouchi et al. (2013) found that North American ice sheets were the dominant force driving their model of the 100,000-year ice-age cycles, whereas the Eurasian ice sheets, which were smaller in volume, followed a 40,000-year cycle. They concluded that this 100,000-year cycling only existed because of the unique geographic and climatological setting of the North American ice sheets in relation to the received isolation. Looking into the past 65 million years using global deep-sea oxygen, carbon isotope records, and sedimentary rocks, Zachos et al. (2001) found that Milankovitch orbital rhythms oscillate around a climatic mean that is constantly shifting due to plate tectonics and its effects on continental geography, topography, oceanic gateway locations, bathymetry, and the resulting concentrations of atmospheric greenhouse gases and humidity (it takes water to make ice). In other words, climatic response to changing insolation due to Milankovitch cycles is dictated by the relative positions and sizes of the land-sea distribution around the globe, which in turn influences ocean-cryosphere (ice)-atmospheric dust and CO_2 feedback mechanisms.

While Milankovitch's theory attributes changes in isolation due to the Earth's position relative to the Sun as the trigger to climatic change, there is also evidence that the Sun's radiation output changes over time.

Solar Variability

Solar irradiance is the primary source of energy into the global climate system. Recent direct observational records of solar irradiance (2003–2020) from NASA's Solar Radiation and Climate Experiment (SORCE) show that the Sun's irradiance varies across all timescales: from seconds to minutes during solar flares, from hours to days during sunspot activity, over the Sun's 27-day rotation period, and throughout its approximate 11-year solar activity cycle (Woods et al. 2022). Total solar irradiance measures all wavelengths of energy hitting the Earth's upper atmosphere, and space-based measurements of total solar irradiance show that it can change by 0.1% through the Sun's 11-year solar cycle (Lean 2010). This translates into a change in radiative forcing of 0.17 W m^{-2} at the Earth's surface, which, while measurable, is small compared to the 3.7 W m^{-2} change estimated to result from doubling atmospheric CO_2 levels (IPCC 2007).

Although changes in the amplitude of solar irradiance may appear small, there is evidence that responses and feedbacks within the ocean–atmosphere system can amplify these variations in ways that affect climate (Engels and van Geel 2012), especially in regions where monsoons and drought drive climate. For example, increased solar activity has been linked to higher temperatures and increased moisture in Alaska, drought in western USA, the Mayan Peninsula, and Equatorial East Africa, warming in the North Atlantic, increased rainfall in Oman, stronger monsoons in China, and warming in Tasmania (Lean 2010).

Space-based measurements have established that solar irradiance follows a regular 11-year cycle, but can solar activity be directly linked to our understanding of climate change? To answer this, we need evidence of changing solar irradiance detected over a period longer than the space-based measurements. Solar activity and sunspots are strongly correlated with irradiance, and telescope observations of sunspots date back to the early 1600s (e.g., by Galileo), and systematic observations beginning in the mid-1800s and continuing to the present (Lean 2018). Records of low solar activity during 1645–1715 AD and 1795–1820 AD coincide with colder climate conditions in the Northern Hemisphere (Engels and van Geel 2012). Longer records of solar activity are derived from isotope records stored in ice cores and tree rings, which also contain climate information. These paleoclimatic records suggest that a decline in solar activity was the most likely forcing agent behind changes in global atmospheric circulation patterns and the cooling in Europe during 2700–2800 BP (years before 1950 AD; Engels and van Geel 2012).

Changes in solar irradiance, which have been detectably increasing in recent years, are estimated to have contributed only about 10% of the global temperature increase over the past century (Lean 2010). However, Rypdal (2012) warns that attempts to tie irradiance to climate on a global scale, at least during the 1959–2004 period, are confounded by four major volcanic eruptions that coincidentally took place just before four dips in both the irradiance and mean surface temperature records. This is because volcanic eruptions also affect the climate.

Volcanic Activity

Volcanic eruptions have the potential to cool the planet by releasing ash and aerosol particles into the atmosphere. These particles can block a portion of the incoming solar radiation, lowering temperatures in the **troposphere** (the layer of atmosphere between 6 and 20 km from the Earth's surface), thereby quickly affecting global temperature. Over 200 years ago, Benjamin Franklin, perhaps the first American weather forecaster, argued that material from the eruption of Iceland's Laki volcano in 1783 could have reflected sunlight into space, possibly contributing to the unusually cold winter of 1783–1784, which led to widespread crop failures in Europe. When Mount Tambora in Indonesia erupted in 1815, Europe and North America plunged into the "year without a summer," characterized by heavy snow in June and frost in July and August. This extreme weather devastated crops, either by frost or a lack of sunshine (Oppenheimer 2003). In the 20th century, three major volcanic events—Mt. Agung in Indonesia in 1963, El Chichon in Mexico in 1982, and Mount Pinatubo in the Philippines in 1991—provided ample opportunities to study the impact of volcanoes on global temperature. For a year after the eruption of Mount Pinatubo, the haze of tiny aerosols increased albedo and lowered global temperature by 0.5°C (McCormick et al. 1995). However, quantifying the effects of volcanic eruptions can be challenging due to complex interactions between various atmospheric processes (Dhomse et al. 2020), meaning not all volcanic eruptions have the same impact. For example, major eruptions in the tropics can inject SO_2 directly into the **stratosphere** (the layer of atmosphere between 20 and 50 km from the Earth's surface), where it becomes trapped by a subtropical atmospheric barrier. Because the intensity of solar radiation is highest at low latitudes, reflection by these subtropical aerosols can cause strong and prolonged surface cooling (Dhomse et al. 2020). However, while particulates from volcanic eruptions can cool the planet, causing unusual hardship during one to several growing seasons, volcanic particles in the atmosphere do not remain there for long (gravity prevails). Therefore, unless eruptions happen in frequent intervals over long periods, this global cooling is typically transitory and may not persist long enough to induce significant ecosystem changes.

On the flip side, volcanoes also emit greenhouse gases, and when extensive volcanism occurs, there is evidence that it can contribute to global climate change. Based on mercury anomalies found in sediments, it appears that extensive volcanism was the likely cause of the Paleocene-Eocene Thermal Maximum (PETM) event, which occurred about 56 million years ago. This event was marked by a burst of greenhouse gases over a few thousand years, leading to global warming (4–8°C), ocean acidification, sea-level rise, and significant changes in the biosphere, including species migrations and extinctions (Tremblin et al. 2022). At the time of the PETM, the Earth was much warmer than it is today, with Arctic temperatures warm enough to support alligators and turtles (Eberle et al. 2010). There were significant impacts on marine ecosystems, including extinctions of benthic foraminifera and substantial increases in the abundance and geographic range of dinoflagellates (McInerney and Wing 2011). Terrestrial systems also saw large changes, with an explosion of mammalian "first appearances" in new habitats, including Artiodactyla, Perissodactyla, and primates, suggesting mammalian dispersal across high-latitude land bridges while the climate was exceptionally warm (McInerney and Wing 2011). The prolonged and extensive volcanism associated with the PETM event is estimated to have emitted CO_2 at a rate eighteen times slower (0.58 PgC per year vs. about 10 PgC per year) than the rate at which CO_2 is currently entering Earth's atmosphere (Gutjahr et al. 2017). Today, volcanoes account for only one to 2% of total CO_2 emissions (Cartier 2019).

Pleistocene Climate Change

For the past million years, Earth's climate has been dominated by glacial–interglacial cycles with an average period of 100,000 years. These cycles are primarily driven by slight variations in Earth's orbit affecting solar insolation on Earth's surface, but these changes in insolation are actually minor and act merely as a trigger. Instead, compelling chemical, physical, and biological evidence from polar ice cores implicates the central role of climate system feedbacks on atmospheric CO_2, which amplify these insolation changes into significant global climate shifts (Ai et al. 2020). Central to this feedback is the Southern Ocean, which vents or sequesters atmospheric CO_2 in complex ways (Ai et al. 2020):

1. CO_2 vents into the atmosphere via deep ocean upwelling when Southern Ocean waters are transported north by the Southern Westerly (SW) Winds.
2. CO_2 is sequestered or removed from the atmosphere:
    a. when ice cover prevents venting;
    b. when dust-carried iron fertilizes greater phytoplankton productivity;
    c. when SW Winds are reduced.

Upwelling, and thus increases in atmospheric CO_2, occur when there are strong SW Winds, which can happen during periods of high summer insolation or a weakening of the North Atlantic overturning circulation (Ai et al. 2020; Williams et al. 2021). This influx of CO_2 to the atmosphere typically occurs during interglacial periods, amplifying this warming. However, during glacial maxima, the SW Winds weaken, causing upwelling to decline, and atmospheric dust to increase. This allows for greater CO_2 drawdown by phytoplankton because they are no longer iron-limited (Shaffer and Lambert 2018), and increased ice cover stratifies the Southern Ocean, preventing upwelling. All of these factors amplify global cooling via decreased CO_2 and increased dust in the atmosphere (Ai et al. 2020).

In an orbital twist, lower obliquity (when the Earth is more upright in its rotation) steepens the tropic-to-polar temperature gradient and strengthens the SW Winds, even during periods of global cooling when the winds typically weaken. Higher obliquity, in contrast, weakens the SW Winds and depresses upwelling, counteracting the effects of weak North Atlantic overturning and high summer insolation. Ai et al. (2020) found that obliquity explains the lag in CO_2 behind temperature during the start of both glaciation and deglaciation during the latter part of the Pleistocene. However, this obliquity-induced lag did not prevent CO_2 from acting as a climate amplifier; it only delayed its action.

The last ice age ended 11,700 years ago at the start of the Holocene, as the Earth's orbit around the Sun became conducive to warming. Our modern climate, along with its attendant distribution of biomes, is a result of this change in climate from "glacial" to "interglacial." However, the interglacial period we are experiencing now differs markedly from the last one, which occurred 128,000 to 123,000 years ago, and had warmer and more stable global temperatures (Bova et al. 2021). In contrast, strong evidence from multiple different climate proxies (terrestrial, marine, and glacial sources) indicates that global temperatures in the Holocene reached a thermal maximum between 8,000 and 6,000 years ago before cooling again until recently (Kaufman and Broadman 2023). This cooling in the latter half of the Holocene coincided with small decreases in solar irradiance, along with decadal-frequency volcanic eruptions, and is consistent with Arctic amplification of sea-ice feedbacks driven by orbital forcings and declining summer insolation at northern high latitudes (Park et al. 2019; Kaufman and Broadman 2023). However, this cooler climate has not persisted past 1850. The ten years between 2014 and 2023 were the warmest decade in global temperature measurements since 1850 (174 years) up to that point, and the five years between 2019 and 2023 were the five warmest years in terms of global ocean heat content, measured from 1960 to 2023 (64 years).[2] This global temperature rise of about 1.1°C since 1850 (Pörtner et al. 2022) already exceeds the 0.8°C rise estimated for the Holocene thermal maximum (Kaufman and Broadman 2023). Are we witnessing the start of yet another climate change? Can natural forcing factors alone explain this rise in global temperature?

We can eliminate plate tectonics as the cause because the continents have not drifted more than a few meters since 1850, Earth's orbital position has not changed substantially over the last 150 years, and recent volcanic activity has cooled rather than warmed the planet due to the ejection of aerosols. This leaves solar variability as the remaining plausible natural cause for this change. In one study that modeled the role of solar radiation as a cause for the recent temperature anomalies, two-thirds of the warming could be attributed to this natural phenomenon (Harde 2022; but see Andrews 2023, for a critique of their general assumptions). In another study that sought to correlate the 11-year solar cycle with recent global temperatures, Chatzistergos (2023) concluded that solar cycle data cannot be used to suggest that the warming on Earth over the last 50 years has a solar origin. Although these perspectives differ, both analyses leave room for a climate forcing that needs to explain anywhere between one-third and all of the recent global climate change we are experiencing.

Human-Caused Climate Change

The Earth has undergone significant changes in two key areas since around 1850: the Industrial Revolution led to the conversion of lithified stored carbon into atmospheric CO_2 as a means of production, and the human population began to grow exponentially. Since the start of the industrial revolution, billions of industrious humans have been unwittingly, unwillingly, or intentionally contributing to a massive climate experiment: What happens if we remove stored carbon from the lithosphere and biosphere—according to United Nations Department of Economic and Social Affairs (2024) approximately 3.6 to 3.9 billion metric tons of oil per year, 6.8 to 7.1 billion metric tons of hard coal per year, 4.5 to 4.9 million metric tons of natural gas per year, among other sources of stored carbon, including wood biomass—convert it to CO_2 (31.2 to 33.4 billion metric tons per year), and pump it into the atmosphere at a very high rate?

Our awareness that changes in atmospheric CO_2 can potentially drive climate change is not new. The role of CO_2 as a greenhouse gas in the atmosphere has been known since 1896, when the Swedish chemist Svante Arrhenius, a creator of ionic theory, suggested that our industrial activities could increase atmospheric levels of CO_2, causing global warming. A geometric increase of CO_2, according to Arrhenius, would produce a nearly arithmetic rise in surface temperature. In 1938, George Callendar, a British engineer, recognized the importance of carbon emissions from deforestation and other land use practices and predicted that a 1.5°C (2.7°F) rise in temperature would result from a doubling of CO_2 levels. At the time, however, the prevailing assumption was that the oceans, which have vastly more carbon dioxide than the atmosphere, would prevent atmospheric accumulation by absorbing the excess gas. In 1957, Roger Revelle and Hans Suess, Americans based in California, argued that there are limits to the amount of carbon dioxide that the oceans can absorb and that the atmospheric levels of CO_2 must be rising. The impact of the paper

[2] NOAA National Centers for Environmental Information, Monthly Global Climate Report for Annual 2023, published online January 2024, accessed February 17, 2024 from https://www.ncei.noaa.gov/access/monitoring/monthly-report/global/202313.

by Revelle and Suess (1957) was so significant that scientists rushed to set up the first station for measuring background atmospheric CO_2 concentration on the northern slope of Mauna Loa on the island of Hawai'i, 3,400 m above sea level and 800 m below the massive Mokuaweoweo Crater. In 1958, daily CO_2 measurements began from the top of the four towers, taken four times every hour, and have continued to this day (see Figure 3).

In 1958, the mean CO_2 concentration at Mauna Loa was 315.5 parts per million (ppm), meaning 315.5 molecules of CO_2 for every 1 million air molecules. In December 2024, it surpassed 425 ppm, according to the Global Monitoring Laboratory of the Earth System Research Laboratories.[3] This represents an increase of more than 100 ppm in just 67 years. To put this in perspective, ice-core data show the concentration of atmospheric CO_2 has only varied by about 50 ppm, around a median of about 250 ppm, over the past 800,000 years (Figure 4; Lüthi et al. 2008). Comparing the 800,000-year ice core data with recent measurements at Mauna Loa reveals the astonishing fact that CO_2 is now at its highest level in the past 800,000 years—approximately 40% higher than the highest previous concentration, which occurred more than 300,000 years ago (Figure 4).

Given the magnitude of CO_2 emissions and what we have learned about the role of CO_2 in the atmosphere as a greenhouse gas and as an amplifier of global change, it would not be surprising if we started to detect the effects of this folly experiment. There are multiple lines of evidence indicating the global climate has changed over the last half-century. These include not only changes in the extremes of precipitation and temperature but also biological changes in phenology, population recruitment, and species ranges (Walther 2002). Change is especially noticeable in the Arctic and sub-Arctic, including the loss of permafrost (James et al. 2013), the loss of Bering Sea ice (Stabeno and Bell 2019), increases in woody plants (Mekonen et al. 2021), and range expansion by beavers (Tape et al. 2022). All of these changes point to significant shifts in the global climate over the last 150 years.

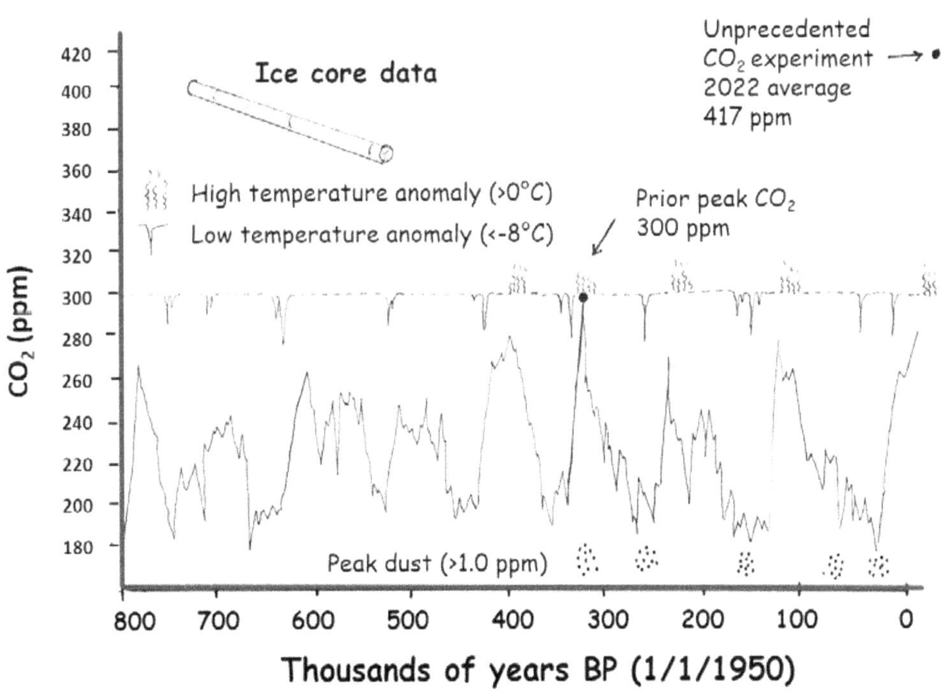

Figure 4. Atmospheric CO_2, in parts per million by volume (ppm; based on trapped air) for the past 800,000 years and high and low temperature anomalies (based on deuterium data) interpreted from Antarctic ice core data (ice core graph interpreted from Figure 2 in Lüthi et al. 2008; peak dust interpreted from Figure 2 in Petit et al. 1999).

[3] https://gml.noaa.gov/ccgg/trends/, accessed on Jan 24, 2025.

It is currently widely accepted, based on multiple sources of evidence, that present-day climate change is an unequivocal fact. However, whether it is caused by humans and the extent of human responsibility remains controversial. Oreskes and Conway (2010) argue that, for the oil industry, sowing doubt under the guise of science is a strategy to stay in business. As Oreskes and Conway say in their book *Merchants of Doubt*, "Doubt is crucial to science…but it also makes science vulnerable to misrepresentation, because it is easy to take uncertainties out of context and create the impression that *everything* is unresolved." Throughout Earth's 4.5-billion-year history, the planet has certainly warmed and cooled many times for various reasons. However, very reliable measurements now exist for both total carbon atmospheric levels and total human carbon emissions, leaving no doubt about the source of the increase in atmospheric carbon (Andrews 2023). Paleo-data support CO_2's role as a climate change amplifier, so even if we found contrary evidence showing that human-caused CO_2 emissions were not the direct cause of current warming, even the most conservative climate models predict that continued emissions of CO_2 will lead to ecosystem-rocking levels of warming in the near future (Hausfather et al. 2022).

Increasing the amount of CO_2 in the atmosphere is not the only way humans can change the global climate. Methane, a more "potent" greenhouse gas (due to its chemical structure), is released by human activities, largely from agriculture, waste disposal, and oil and gas extractions. Its concentration in the atmosphere has more than doubled from pre-industrial levels (Abernathy et al. 2021). Fortunately, it is short-lived in the atmosphere compared to CO_2 (lasting a decade rather than a century; Lelieveld et al. 1998), so reductions in emissions can have an immediate impact. Human-produced black carbon deposition (soot from burning diesel and coal) can also contribute significantly to regional and global warming by reducing snow albedo and increasing radiative forcing (Hadley and Kirchstetter 2012).

On the other hand, humans can also contribute to planetary cooling via aerosol emissions, which may have contributed to some of the observed slowdown in global warming since the mid-1990s. However, more potent contributions to this slowdown include reductions in emissions of ozone-depleting chlorofluorocarbons via the worldwide Montreal Protocol, as well as reductions in global methane emissions (Estrada et al. 2013). Humans can also increase albedo by cutting down forests, which can cause cooling effects. However, such actions also increase the amount of CO_2 in the atmosphere (since forests consume CO_2), thereby offsetting the cooling. Lawrence et al. (2022) found that tropical deforestation leads to strong net global warming, partly as a result of increasing CO_2, but at latitudes greater than 50°N, large-scale deforestation can lead to a net global cooling due to the increased albedo.

In the face of global climate change and a warming Earth, humans have the choice and know-how to either ameliorate or exacerbate climate change. Other species do not. If warming is not advantageous, their option is limited—move, adapt, or die.

Ecological and Evolutionary Responses to Climate Change

Because thermal and precipitation gradients—such as those created across latitudes, altitudes, and water depths—are powerful drivers of ecology and evolution, we can safely assume that species will respond to the current global warming and ongoing climate change in three ways (*cf.* Holt 1990). First, species may shift their geographic ranges by dispersing to suitable habitats elsewhere. Second, organisms can adapt to altered conditions either by individuals adjusting via phenotypic, phenological, or behavioral plasticity, or by populations evolving via exploiting genetic variation (e.g., color patterns, body shape, and size) that may reduce the impact of climate change. Third, failing either to shift range or adapt, populations may decline, and species may go extinct locally.

Range Shifts

Species are capable of shifting geographical distribution and range boundaries over time, and they do not need legs to do it. For instance, multiple lines of evidence (such as pollen, macrofossils, plant fragments preserved in packrat middens) indicate that tree species have moved poleward or to higher elevations over the past 15,000–20,000 years, following the retreat of the North American glacial ice sheets (which covered most of what is now Canada down past the Great Lakes as far as Iowa), opening areas for colonization (Davis and Shaw 2001). However, this was not a movement *en masse*, but rather, responses were individual to each tree taxon. For example, some, like the spruce (*Picea* spp.), moved their entire range, while others, like the oak (*Quercus* spp.), expanded from refuges (Davis and Shaw 2001; Figure 5).

We consider biological communities to be groups of species living and interacting together in the same place and time, generally experiencing the same climate on a broad scale. However, while we may call something a "beech-maple" forest, it is made up of individual species that may respond differently to changes in climate. Therefore, the organisms found together in today's "beech-maple" forest may not have been present together in the past. Various combinations of temperature, seasonality, and precipitation that occurred 12,000–17,000 years ago resulted in "no-analog communities" that were compositionally unlike any found today (Williams and Jackson 2007). Records of plants, mammals, beetles, and diatoms all show repeated shuffling of community members throughout the cycling Quaternary climates (Bennett 1990). In climate simulations by the Intergovernmental Panel on Climate Change, novel climates are predicted to arise by 2100 AD, which are expected to create no-analog communities in the future (Williams and Jackson 2007). Walker et al. (2002) use these predicted climate conditions to model the responses of ten important tree species in the Great Lakes region, suggesting that sugar maple (*Acer saccharum*) and beech (*Fagus grandifolia*) would largely remain, while black walnut (*Juglans nigra*) would shift into the area from the south, and yellow birch (*Betula alleghaniensis*) would shift out of the area to the north.

We do not have to look to the past or the future for range shifts; they are already occurring. In two surveys of 53 bird species in the Sierra Nevada Mountains of California, taken nearly a century apart, results showed that about 90% of the species tracked shifts in temperature and precipitation climate variables (Tingley et al. 2009). In Scandinavia, during the last century, the red fox

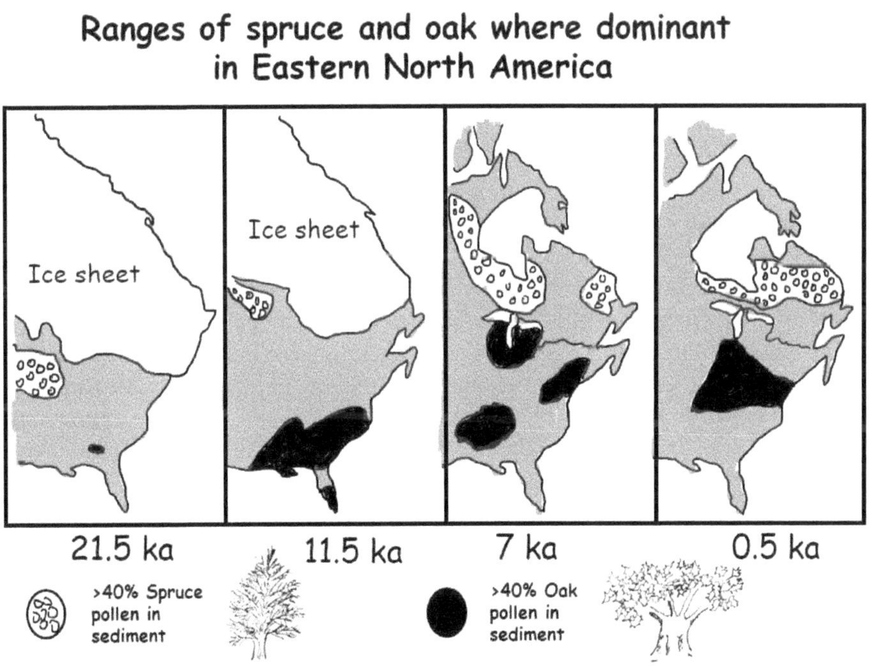

Figure 5. Individualistic patterns of migration in response to climate change shown for spruce and oak based on ranges where they are the dominant component of the pollen (> 40%) in the sediment from four different periods: 21,500 years ago (last glacial maximum), 11,500 years ago (end of last glacial period), 7,000 years ago (preindustrial Holocene thermal maximum), and 500 years ago (preindustrial) (interpreted from Figure 1 in Davis and Shaw 2001).

(*Vulpes vulpes*) has extended its range northward into the tundra zone at the expense of the Arctic fox (*Alopex lagopus*), which has retracted its southern distribution limits. The range expansion of the red fox, according to Hersteinsson and Macdonald (1992), is related to climate warming and the concomitant increased primary productivity of the tundra. Beaver (*Castor canadensis*), a decidedly non-Arctic species, has expanded its range poleward over the past 70 years, constructing over 10,000 new beaver ponds in Arctic tundra regions (Tape et al. 2022).

In general, warm-adapted species have increased in abundance on average since the 1980s, while cold-adapted species have declined (Gregory et al. 2009). Tropical, warm-adapted species are expanding their ranges into environments previously dominated by temperate cold-tolerant species—a trend known as **tropicalization** (Vergés et al. 2014). An analogous phenomenon, referred to as **borealization**, has occurred in the Arctic, where, for example, boreal fish communities have responded to warming in the Barents Sea by shifting northward. This borealization has led to a high turnover in Arctic fish communities (Fossheim et al. 2015).

More than half of a couple of hundred terrestrial species examined between 1970 and 2000 have shifted their ranges into adjacent habitats in the direction expected by a warming global climate (Thomas 2010). Pecl et al. (2017) estimated that terrestrial and marine species are moving poleward at an average of 17 and 72 km per decade, respectively. A global meta-analysis of 434 range-shifting species, from such diverse taxa as trees, fish, birds, and invertebrates, shows that 80% have shifted in accordance with climate predictions (Parmesan and Yohe 2003). Surprisingly, this analysis included only two mammal species. While this may not be indicative of a lack of range shifts in mammals, it does suggest that some species, whether due to an inability to move or other factors, may respond to climate change in ways that do not involve shifting ranges.

Behavioral, Phenological, and Phenotypic Plasticity

Large-bodied species with long generation times require large habitats, making it difficult for them to shift their distributional range to track climate (Cardillo et al. 2005). In other cases, species may be pushed into climate dead ends, such as the tops of mountains or coastal peninsulas, where they lack the ability to escape a shrinking habitat (Moore and Schindler 2022). Species that are unable to shift their range to track climate may depend on behavioral flexibility or sufficient phenotypic plasticity, including malleable phenology, developmental plasticity, and physiological adjustments (Fitzpatrick 2012). Unlike genetic adaptation, **plasticity**—the ability to change phenotype, phenology, or behavior in response to a changing environment without changing genotype—allows the organism itself, rather than its future lineage, to respond to environmental change.

For animals, because it occurs quickly, a behavioral change may be the first observed response to a changing climate. For example, in the Apennine chamois (*Rupicapra pyrenaica ornata*), the goat-antelopes of the Italian mountains, Fattorini et al. (2023) found that as dry, warm weather suppressed plant productivity, the rate and intensity of aggression between chamois increased. The authors found that chamois initiating agonistic contests were more likely to displace opponents over food patches, and they suggest that if global warming continues to reduce the quality and quantity of food resources, this plastic response in behavior could shift to actual selection for more aggressive individuals.

An ability to respond variably to changes in **phenology** (timing of annual growth and reproductive events) may be the first response available to both plants and many animals in a climatically changing environment. If phenological adjustments are possible, the populations of such organisms could be minimally impacted by climate-induced changes. Advancing phenologies (e.g., shifts to earlier timing for spring events and later timing for fall events in temperate latitudes) in plants and animals are widely documented. Parmesan and Yohe (2003), in a meta-analysis of phenological shifts in herbs, shrubs, trees, birds, butterflies, and amphibians over the past 132 years, found a mean change toward earlier spring timing of 2.3 days per decade. Recently, Horton et al. (2023) found that, based on six decades of North American bird banding records (over 2.8 million banded birds from 19 species of North American wood warblers) from 1961 to 2018, these species shifted their migration phenology so that the median timing of passage advanced an average 0.65 days for every 1°C increase in temperature.

Although some species may be able to shift their phenology to advantage, this is likely to alter species interactions, leading to trophic mismatches. For example, in Kangerlussuaq (West Greenland), successful reproduction by the caribou population depends on synchronizing offspring production with the time of year when resources are most abundant. With the warming climate, the onset of the plant growing season advanced by 14.8 days, whereas the onset of calving advanced by only 1.28 days (Post and Forchhammer 2008). In fact, the growing mismatch results from the difference in the environmental cues that trigger the beginning of the reproductive season in caribou and the onset of plant growth in the region. Changes in day length, which hardly vary across years, regulate the timing of seasonal summer migration of caribou, ensuring that calves are born in summer grounds. In contrast, the onset of the plant growing season is correlated with mean spring temperatures, which increased by 4.6° Celsius over the study period. This apparent trophic mismatch resulted in declining calf production (Post and Forchhammer 2008).

Behavioral changes and flexible phenology are relatively "easy" responses to a changing climate, and many examples can be documented, though they may come with associated costs, as discussed above. Physiologically flexible responses to climate change, however, may be less widespread, as preexisting physiological "superpowers" need to be present in the organism. The following example demonstrates this. Normally, calcium carbonate ($CaCO_3$) is formed in marine environments through the reaction of calcium and carbonate ions:

$$CO_3^{2-} + Ca^{2+} => CaCO_3$$

Also, normally, seawater absorbs CO_2 and dissociates it into bicarbonate ions (HCO_3^-) and hydrogen (H^+) ions. But increased absorption of anthropogenic CO_2 by the oceans can potentially impact oceanic life by hindering the production of shells and plates made of calcium carbonate ($CaCO_3$), which is important for survival in many marine organisms. As seawater absorbs more CO_2, it produces more H^+, increasing acidity. Under these conditions, carbonate ions (CO_3^{2-}) tend to neutralize H^+ ions (which are acidic) by forming bicarbonate (HCO_3^-), rather than reacting with calcium to form calcium carbonate. Under both laboratory experiments and paired water and biological sampling, as carbonate ion concentrations decrease and acidity increases, calcifying species—including oysters, clams, sea urchins, shallow water corals, deep sea corals, calcareous plankton and pelagic sea snails—exhibit reduced calcification, slower growth rates, and increased shell dissolution (Doney et al. 2009; Bednaršek et al. 2014). However, a few invertebrate species may be able to buffer increasing ocean acidification to some extent by making acid-base adjustments in their haemolymph (analogous to blood in vertebrates), thus tolerating high bicarbonate levels. However, this comes at a high metabolic cost (Whiteley 2011). Strong iono- and osmo-regulating species, such as those that inhabit shallow coastal environments under freshwater influence, are likely to be most tolerant to ocean acidification, whereas relatively slow-moving, inactive species, such as those inhabiting deep sea or polar environments, are likely to be particularly vulnerable to this change (Whiteley 2011).

Evolutionary Adaptation

When considering physiological, phenological, or behavioral changes, one might assume that plasticity (biological flexibility of the individual) is in play. However, climate change can also set the stage for evolution. For instance, blackcaps (birds), which breed in south-central Europe, have been increasingly overwintering in Britain rather than southern Iberia and North Africa. This migratory divide between British and Iberian overwintering has led to assortative mating and is creating a genetically distinct British subpopulation. Due to earlier northern photoperiod cues, "British" birds arrive earlier at the European breeding grounds, enabling the "British" males to obtain superior breeding territories, which are then more attractive to "British" females, resulting in offspring that are more genetically isolated from their southern-overwintering counterparts (Bearhop et al. 2005).

While some phenological responses to climate change may be plastic, phenology is also heritable, allowing for adaptive selection. This is the case for the North American red squirrel (*Tamiasciurus hudsonicus*) in the southwest Yukon (Canada). Changing climate conditions have caused increasingly warmer spring temperatures, which in turn have allowed white spruce to produce more cones in the fall, providing more food for the squirrels. In a study of a red squirrel population near Kluane Lake, Canada, Réale et al. (2003) observed a shift to earlier birthing times—almost 2 days per year—over a 10-year period. Fall food abundance had a significant effect on the parturition date for individual squirrels. The more cones females collected in the fall, the earlier they gave birth in the spring, which is a plastic phenotypic response. However, Réale and the team also found the mean breeding date was heritable and had shifted earlier from one generation to the next in the population, revealing an evolutionary response to strong selection. This selection may occur because competition for territories is intense. Young female squirrels born earlier have an advantage over those born later because they are the first to find territorial vacancies and are often larger and more competitive (Descamps et al. 2008).

In another example of adaptive change in phenology, European great tits (*Parus major*), a bird species, in Holland's Hoge Veluwe Park depend on caterpillars to feed their young. However, the caterpillars are maturing earlier each year as spring arrives sooner. Great tits have a genetic trigger that spurs them to lay eggs when spring begins, and there is some variation in how much an individual bird can adjust its egg-laying date in response to an earlier spring. Selection is expected to act against birds that are unable to synchronize their hatching schedule with the peak abundance of caterpillars. In a study of 833 great tits in Hoge Veluwe over 32 years, Nussey et al. (2005) found stronger genetic selection for birds that could adjust their egg-laying time to match the caterpillars' arrival. If this trend continues, it could save the species from a decline caused by the nesting-food mismatch. However, it remains unclear whether the great tits can evolve fast enough to keep pace with the shifting caterpillar hatch schedule as the temperatures rise (Nussey et al. 2005).

In Auke Creek, Alaska, one pink salmon (*Oncorhynchus gorbuscha*) population is migrating about two weeks earlier than it was 40 years ago. Kovach et al. (2012) report that between 1983 and 2011, the frequency of a genetic marker for late migration dropped by 10% in response to warmer temperatures (during the same period, the local water temperature increased by about one degree Celsius on average). In the Columbia River, sockeye salmon (*Oncorhynchus nerka*) are migrating earlier each spring and early summer to spawn. Over 60 years, data show that evolutionary changes in the population in response to higher water temperatures are the most likely explanation for about two-thirds of the shift, while individual adaptation to river flow changes may account for the rest (Crozier et al. 2011).

Phenology is not the only area we find examples of selective change; there is also evidence of adaptive phenotypic change. Tawny owls come in two shades: brown and gray, depending on the amount of a pigment called pheomelanin in their plumage (which is dictated by a variety of genes). The gray coloration provides camouflage against snow. With milder winters in Finland, the environment has less snow. One population of tawny owls (*Strix aluco*) showed a significant increase in brown-plumed owls over the last 28 years (Karell et al 2011). As snow cover has decreased in Finland, the proportion of brown owls in the population has grown from about 12% in the early 1960s to 40% in 2010 (Karell et al. 2011). Similarly, warmer temperatures in Europe may lead to a greater abundance of lighter coloration in banded snails (*Cepaea nemoralis*). In populations sampled at 16 sites in the Netherlands in 1967 and again in 2010, the proportion of yellow-shelled snails increased compared to brown ones, coinciding with a 1.5°F to 2.0°F increase in annual temperature in the area (Ozgo and Schilthuizen 2013).

One of the best pieces of evidence of genetic change in response to increasing temperatures (i.e., warmer winters) comes from plants. Genotypic varieties of Mediterranean thyme (*Thymus vulgaris*) produce oils with different chemical compositions, and the ones with stronger-smelling chemical compounds like phenols are more effective at deterring herbivores. However, producing phenols typically comes at a cost—plants with greater amounts of phenols become more sensitive to freezing. In southern France's Saint-Martin-de-Londres basin, winters are getting warmer, and since the 1970s, the basin has experienced fewer freezing nights during the cold season. Examining 24 populations across the basin in 1974 versus 2010, Thompson et al. (2013) found an increase in the proportion of plants with phenolic-producing genotypes (Figure 6). These plants have even begun to spread into areas where they did not grow in the 1970s.

In marine systems, physiological responses to both climate warming and changing ocean conditions are widespread (Poloczanska 2013). For instance, higher temperatures cause corals to "bleach," which is when they spit out the colorful symbiotic algae that live inside their tissues. However, there is evidence that the temperature at which corals bleach has a genetic component. On Ofu Island in American Samoa, the table coral *Acropora hyacinthus* lives in both hot and cool pools. In the lab, Palumbi et al. (2014) found that only 20% of corals from the hot pools bleached compared to 55% of corals from the cool pools. Additionally, corals from cooler pools that were transplanted to hot pools for a year had an advantage—only 32.5% of those corals bleached in the lab tests. The results suggest that the species has both the physiological plasticity and the genetic variability necessary to adapt and survive at increasing temperatures.

Adaptive evolution requires genetic variation upon which natural selection can act. Hybridization is one way in which populations can rapidly acquire genetic variation. Due to changing species ranges across high latitudes and altitudes as the climate warms, hybridization is increasing (Chunco 2014). The southern flying squirrel, *Glaucomys volans*, has expanded its range northward in eastern North America and is now hybridizing with the northern flying squirrel, *G. sabrinus* (Garroway et al. 2010). In North American rivers and streams, hybridization between invasive rainbow trout and native cutthroat trout has become more frequent as the former expand into warming waters (Muhlfeld et al. 2014). Hybridization events like these can create populations with increased genetic variation, providing fodder for an evolutionary response to a changing climate. For instance, on a Galapagos Island, hybridization with a smaller species (*Geospiza fuliginosa*) drove the rapid evolution in the beak size in another species of Darwin's finch (*G. fortis*) in response to a severe

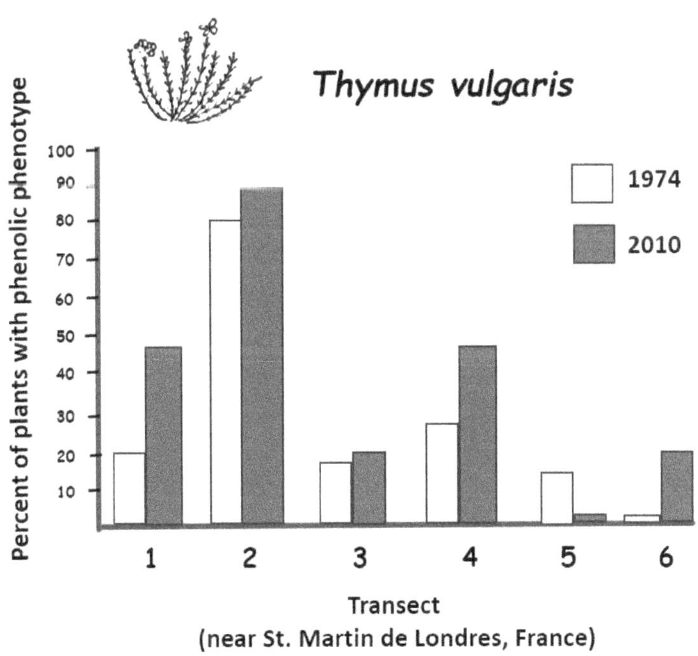

Figure 6. Percentage of plants with a phenolic (freezing sensitive) phenotype in initially mixed phenotype populations along six transects in southern France (interpreted from Table 1 in Thompson et al. 2013).

El Niño event, which changed the seed composition of the island, making smaller beak sizes more adaptive (Grant and Grant 1993).

Increase, Decline, or Extinction

As we recall from the last chapter (Chapter 11 *Biodiversity*), mass extinctions were often associated with large scale climate change. If individuals in a population fail to adjust or the population fails to evolve in response to climate change, the population may decline, and extinction, one way or another, can often result from small population size. Will it be due to limited physiological tolerance to changing environmental conditions, changing biotic interactions, or something else?

In some cases, species physiologically sensitive to rising temperatures will experience population declines. In the Persian/Arabian Gulf, *Acropora downingi* (table coral) populations have declined by over 90% in many areas, primarily due to bleaching and diseases caused by rising temperatures (Riegl et al. 2018). These researchers suggest that the increasing frequency of temperature-induced mortality events, clearly linked to climate change, will lead to functional, if not total, extinction of this species. Another example of decline comes from a comparison of recent and historical surveys of 48 Mexican lizard species at 200 sites. Sinervo et al. (2010) found that since 1975, 12% of local populations have gone extinct due to the rapid pace of temperature change, which has outstripped the species' ability to adapt.

In some cases, competitive interactions may change, and species that were once dominant may face increasing competition from rarer community members that happen to possess traits that are more advantageous under shifting climatic conditions. The best example of this comes from the Quaternary ice ages, when CO_2 closely tracked temperature. Bragg et al. (2013) found evidence, based on stable isotopes from sediment cores, of the key role of CO_2 concentration driving the last glacial-interglacial vegetation shift in southern Africa. Higher CO_2 favors trees (C_3 photosynthesizers, which can respond to increased CO_2 with enhanced growth), whereas lower CO_2 favors tropical grasses (C_4 photosynthesizers, which are better competitors under low CO_2 conditions). During the peak of the glaciers, Southern Africa was dominated by tropical grasses. However, as CO_2 increased after the last glacial maximum, there was a large increase in trees relative to tropical grasses. Bragg et al. (2013) speculate that the current rise in CO_2 concentration is likely influencing tree-grass competition similarly, contributing to the "woody thickening" observed in savannas, prairies, and tundra worldwide.

Endemic organisms are less widespread (by definition) and are often more likely to be locally adapted. In a meta-analysis of predicted extinction risks from climate change, Urban (2015) found that extinction risk was higher for taxa in South America, Australia, and New Zealand, which face predicted no-analog climates in the future and have high rates of endemism. It is clear from broad patterns like this, as well as from the individual cases above, that a changing climate means Earth's biosphere faces significant ecological and evolutionary dynamics and challenges in the immediate future. Equilibrium is a concept of the past.

References

Abe-Ouchi, A., Saito, F., Kawamura, K., Raymo, M. E., Okuno J., Takahashi, K. et al. 2003. Insolation-driven 100,000-year glacial cycles and hysteresis of ice-sheet volume. Nature 500: 190–193.

Abernathey, S., O'Connor, F. M., Jones, C.D., and Jackson, R. B. 2021. Methane removal and the proportional reductions in surface temperature and ozone. Phil. Trans. R. Soc. A 379: 20210104. https://doi.org/10.1098/rsta.2021.0104

Ai, X. E., Studer, A. S., Sigman, D. M., Martínez-García, A., Fripiat, F., Thöle, L. M. et al. 2020. Southern Ocena upwelling, Earth's obliquity, and glacial-interglacial atmospheric CO_2 exchange. Science 370: 1348–1352.

Amrith, S. 2020. The life and possible death of the Great Asian Monsoon. American Scientist 108: 114. doi: 10.1511/2020.108.2.114.

Andrews, D. E. 2023. Clear thinking about atmospheric CO_2. Science of Climate Change 3.1: 33–45.

Barron, E. J., Sloan II, J. L., and Harrison, G. A. 1980. Potential significance of land-sea distribution and surface albedo variations as a climatic forcing factor; 180 M. Y. to the present. Paleogeography, Palaeoclimatology, Palaeoecology 30: 17–40.

Bearhop, S., Fiedler, W., Furness, R. W., Votier S. C., Waldron, S., Newton, J. et al. 2005. Assortative mating as a mechanism for rapid evolution of a migratory divide. Science 310: 502–504.

Bednaršek, N., Feely, R. A., Reum, J. C. P., Peterson, B., Menkel, J., Alin, S. R. et al. 2014. *Limacina helicina* shell dissolution as an indicator of declining habitat suitability owing to ocean acidification in the California Current Ecosystem. Proc. R. Soc. B. 281: 20140123.

Bennett, K. D. 1990. Milankovitch cycles and their effects on species in ecological and evolutionary time. Paleobiology 16: 11-21.

Betts, R. A., Jones, C. D., Knight, J. R., Pope, J. O., Sandford, C. 2024. Mauna Loa carbon dioxide forecast for 2024: Atmospheric CO_2 rise predicted to exceed IPCC 1.5°C scenarios. Met Office Hadley Centre.

Billings, W. D. 1974. Adaptations and origins of alpine plants. Arctic and Alpine Research 6: 129–142.

Bjerknes, J. 1966. A possible response of the atmospheric Hadley Circulation to equatorial anomalies of ocean temperature. Tellus 18: 820–829.

Blakey, R. C., Fielding, C. R., Frank, T. D., and Isbell, J. L. 2008. Gondwana paleogeography from assembly to breakup-A 500 my odyssey. Geological Society of America Special Papers 441: 1–28.

Bova, S., Rosenthal, Y., Liu, Z., Godad, S. P., and Yan, M. 2021. Seasonal origin of the thermal maxima at the Holocene and the last interglacial. Nature 589: 548–553. https://doi.org/10.1038/s41586-020-03155-x.

Bragg, F. J., Prentice, I. C., Harrison, S. P., Eglinton, G., Foster, P. N., Rommerskirchen, F. et al. 2013. Stable isotope and modelling evidence for CO_2 as a driver of glacial-interglacial vegetation shifts in southern Africa. Biogeosciences 10: 2001–2010. https://doi.org/10.5194/bg-10-2001-2013

Buermann, W., Lintner, B. R., Koven, C. D., Angert, A., Pinzon, J. E., Tucker, C. J. et al. 2007. The changing carbon cycle at Mauna Loa Observatory. PNAS 104: 4249–4254.

Caccamo, M. T. and Maguzu, S. 2021. On the breaking of the Milankovitch cycles triggered by temperature increase: the stochastic resonance response. Climate 9: https://doi.org/10.3390/cli9040067.

Camacho-Villa, T. C., Martinez-Cruz, T. E., Ramírez-López, A., Hoil-Tzuc, M., and Terán-Contreras, S. 2021. Mayan traditional knowledge on weather forecasting: who contributes to whom in coping with climate change? Frontiers in Sustainable Food Systems 5: 618453.

Cardillo, M., Mace, G. M., Jones, K. E., Bielby, J., Bininda-Emonds, O. R., Sechrest, W. et al. 2005. Multiple causes of high extinction risk in large mammal species. Science 309: 1239–41.

Cartier, K. M. S. 2019. Human activity outpaces volcanoes, asteroids in releasing deep carbon. Eos. 100. https://doi.org/10.1029/2019EO134727.

Chatzistergos, T. 2023. Is there a link between the length of the solar cycle and Earth's temperature? Rendiconti Lincei. Scienze Fisiche e Naturali 34: 11–21.

Chunco, A. J. 2014. Hybridization in a warmer world. Ecol. Evol. 4: 2019–2031.

Crozier, L. G., Scheuerell. M. D., and Zabel, R. W. 2011. Using time series analysis to characterize evolutionary and plastic responses to environmental change: A case study of a shift toward earlier migration date in sockeye salmon. The American Naturalist 178: 755–773.

Cunningham, S. A. and Marsh, R. 2010. Observing and modeling changes in the Atlantic MOC. Wiley Interdisciplinary Reviews: Climate Change 1: 180–191.

Cvijanovic, I., Lukovic, J. and Begg, J. D. 2020. One hundred years of Milanković cycles. Nat. Geosci. 13: 524–525.

Davis, M. B. and Shaw, R. G. 2001. Range shifts and adaptive responses to Quaternary climate change. Science 292: 673–9.

Descamps, S., Boutin, S., Berteaux, D., McAdam, A. G. and Gaillard, J.-M. 2008. Cohort effects in red squirrels: the influence of density, food abundance and temperature on future survival and reproductive success. Journal of Animal Ecology 77: 305–314.

Dhomse, S. S., Mann, G. W., Marrero, J. C. A., Shallcross, S. E., Chipperfield, M. P., Carslaw, K. S. et al. 2020. Evaluating the simulated radiative forcings, aerosol properties, and stratospheric warmings from the 1963 Mt Agung, 1982 El Chichón, and 1991 Mt Pinatubo volcanic aerosol clouds. Atmospheric Chemistry and Physics 21: 13627–13654.

Doney, S. C., Fabry, V. J., Feely, R. A., and Kleypas, J. A. 2009. Ocean acidification: the other CO2 problem. Annual Review of Marine Science 1: 169–192.

Eberle, J. J., Fricke, H. C., Humphrey, J. D., Hackett, L., Michael G. Newbrey, M. G. et al. 2010. Seasonal variability in Arctic temperatures during early Eocene time. Earth and Planetary Science Letters 296: 481–486.

Engels, S. and van Geel, B. 2012. The effects of changing solar activity on climate: contributions from paleoclimatological studies. J. Space Weather Space Clim. 2: A09. https://doi.org/10.1051/swsc/2012009.

Estrada, F., Perron, P., and Martínez-López, B. 2013. Statistically derived contributions of diverse human influences to twentieth-century temperature changes. Nature Geoscience 6: 1050–1055. doi: 10.1038/NGEO1999.

Fattorini, N., Lovari, S., Franceschi, S., Chiatante, G., Brunetti, C., Baruzzi, C. et al. 2023. Animal conflicts escalate in a warmer world. Science of Total Environment 871: 161789. http://dx.doi.org/10.1016/j.scitotenv.2023.161789.

Felger, R. and Henrickson, J. 1997. Convergent adaptive morphology of a Sonoran desert cactus (*Peniocereus striatus*) and an African spurge (*Euphorbia cryptospinosa*). Haseltonia 5: 77–85.

Fitzpatrick, B. M. 2012. Underappreciated consequences of phenotypic plasticity for ecological speciation. Int. J. Ecol. 2012: 256017.

Fossheim, M., Primicerio, R., Johannesen, E. Ingvaldsen, R. B., Aschan, M. M., and D. A. V. 2015. Recent warming leads to a rapid borealization of fish communities in the Arctic. Nature Climate Change 5: 673–677.

Frölicher, T. L., Sarmiento, J. L., Paynter, D. J., Dunne, J. P., Krasting, J. P. and Winton, M. 2015. Dominance of the Southern Ocean in anthropogenic carbon and heat uptake in CMIP5 models. Journal of Climate 28: 862–886.

Galmés, J., Kapralov, M. V., Copolovici, L. O., Hermida-Carrera, C., and Niinemets, U. 2014. Temperature responses of the Rubisco maximum carboxylase activity across domains of life: phylogenetic signals, trade-offs, and importance for carbon gain. Photosynthesis Research 123: 183–201.

Garroway, C. J., Bowman, J., Cascaden, T. J., Holloway, G. L., Mahan, C. G., Malcolm, J. R. et al. 2010. Climate change induced hybridization in flying squirrels. Glob. Change Biol. 16: 113–121.

Graedel, T. E. and Crutzen, P. J. 1995. Atmosphere, Climate, and Change. Scientific American Library. New York.

Grant, B. R. and Grant, P. R. 1993. Evolution of Darwin's finches caused by a rare climatic event. Proc. R. Soc. Lond. B 251: 111–117.

Gregory, R. D., Willis, S. G., Jiguet, F., Vorísek, P., Klvanová, A., van Strien, A. et al. 2009. An indicator of the impact of climatic change on European bird populations. PLOS ONE 4: e4678. doi: 10.1371/journal.pone.0004678.

Gutjahr, M., Ridgwell, A., Sexton, P. F., Anagnostou, E., Pearson, P. N., Pälike, H. et al. 2017. Vary large release of mostly volcanic carbon during the Palaeocene-Eocene Thermal Maximum. Nature 548: 573–577.

Hadley, G. 1735. Concerning the cause of the general trade-winds. Philosophical Transactions 39: 58–62.

Hadley, O. L. and Kirchstetter, T. W. 2012. Black-carbon reduction of snow albedo. Nature Climate Change 2: 437–440.

Halsch, C. A., Sharpiro, A. M., Fordyce, J. A., Nice, C. C., Thorne, J. H., Waetjen, D. P. et al. 2021. Insects and recent climate change. PNAS 118: e2002543117. https://doi.org/10.1073/PNAS2002543117.

Harde, H. 2022. How much CO_2 and the Sun contribute to global warming: comparison of simulated temperature trends with last century observations. Science of Climate Change 2.2: 105–133.

Hare, S. R. and Mantua, N. J. 2001. An historical narrative on the Pacific Decadal Oscillation, interdecadal climate variability and ecosystem impacts. Report of a talk presented at the 20th NE Pacific Pink and Chum workshop, Seattle, WA, 22 March 2001.

Hausfather, Z., Marvel, K., Schmidt, G. A., Nielsen-Gammon, J. W. and Zelinka, M. 2022. Climate simulations: recognize the 'hot model' problem. Nature 605: 26–29.

Hays, J. D., Imbrie, J. and Shackleton, N. J. 1976. Variations in the Earth's Orbit: Pacemaker of the Ice Ages. Science 194: 1121–1132.

Hersteinsson, P. and Macdonald, D. W. 1992. Interspecific competition and the geographical distribution of red and arctic foxes *Vulpes vulpes* and *Alopex lagopus*. Oikos 64: 505–515.

Herweijer, C., Seager, R., Cook, E. R. and Emile-Geay, J. 2007. North American droughts of the last millennium from a gridded network of tree-ring data. Journal of Climate 20: 1353–1376.

Hoell, A., Funk, C. and Barlow, M. 2014. The regional forcing of Northern hemisphere drought during recent warm tropical west Pacific Ocean La Niña events. Clim. Dyn. 42: 3289–3311.

Holt R. D. 1990. The microevolutionary consequences of climate change. Trends in Ecology & Evolution 5: 311–315.

Horton, K. G., Morris, S. R., Van Doren, B. M. and Covino, K. M. 2023. Six decades of North American bird banding records reveal plasticity in migration phenology. Journal of Animal Ecology 92: 738–750. https://doi.org/10.1111/1365-2656.13887.

Houghton, J. 2004. Global Warming: The Complete Briefing, 3rd edition. Intergovernmental Panel on Climate Change. Cambridge University Press, Cambridge, UK.

Huff, K. 2014. A look at the American Almanac. Iowa Heritage Illustrated Summer 2014: 90–93.

Imbrie, J., Boyle, E. A., Clemens, S. C., Duffy, A., Howard, W. R., Kukla, G. et al. 1992. On the structure and origin of major glaciation cycles 1. Linear responses to Milankovitch forcing. Paleoceanography 7: 701–738.

IPCC. 2007. Climate change 2007: The physical science basis. Contribution of Working Group I to the Fourth Assessment Report of the Intergovernmental Panel on Climate Change. Cambridge University Press, Cambridge, UK.

James, M., Lewkowicz, A. G., Smith, S. L., and Miceli, C. M. 2013. Multi-decadal degradation and persistence of permafrost in the Alaska Highway corridor, northwest Canada. Environmental Research Letters 8: 045013.

Josso, P., van Peer, T., Horstwood, M. S. A., Lusty, P. and Murton, B. 2021. Geochemical evidence of Milankovitch cycles in Atlantic Ocean ferromanganese crusts. Earth and Planetary Science Letters 553: 116651. https://doi.org/10.1016/j.epsl.2020.116651.

Karell, P., Ahola, K., Karstinen, T., Valkama J. and Brommer, J. E. 2011. Climate change drives microevolution in a wild bird. Nature Communications 2: 208. https://doi.org/10.1038/ncomms1213.

Kaufman, D. S. and Broadman, E. 2023. Revisiting the Holocene global temperature conundrum. Nature 614: 425–435.

Kluge, M. and Ting, I. P. 1978. Taxonomy and geographical distribution of CAM plants. pp. 5–28. *In*: Kluge, M. and Ting, I. P. [eds.]. Crassulacean Acid Metabolism. Ecological Studies, vol 30. Springer, Berlin, Heidelberg. https://doi.org/10.1007/978-3-642-67038-1_3.

Kovach, R. P., Gharrett, A. J. and Tallmon, D. A. 2012. Genetic change for earlier migration timing in a pink salmon population. Proc. R. Soc. B. 279: 3870–3878.

Kuhn, A. M., Dutkiewicz, S., Jahn, O., Clayton, S., Rynearson, T. A., Mazloff, M. R. et al. 2019. Temporal and spatial scales of correlation in marine phytoplankton communities. Journal of Geophysical Research: Oceans 124: 9417–9438. https://doi.org/10.1029/2019JC015331.

Kweku, D. W., Bismark, O., Maxwell, A., Desmond, K. A., Danso, K. B., Oti-Mensah, E. A. et al. 2017. Greenhouse effect: greenhouse gases and their impact on global warming. Journal of Scientific Research and Reports 17: 1–9.

Ladant, J.-B., Poulsen, C. J., Fluteau, F., Tabor, C. R., MacLeod, K. G., Martin, E. E. et al. 2020. Paleogeographic controls on the evolution of Late Cretaceous ocean circulation. Clim. Past. 16: 973–1006, https://doi.org/10.5194/cp-16-973-2020.

Lantink, M. L., Davies, J. H. F. L., Ovtcharova, M. and Hilgen, F. J. 2022. Milankovich cycles in banded iron formations constrain the Earth-Moon system 2.46 billion years ago. PNAS 119: e2117146119.

Lawrence, D., Coe, M., Walker, W., Verchot, L. and Vandecar, K. 2022. The unseen effects of deforestation: Biophysical effects on climate. Frontiers in Forests and Global Change 5: 49.

Lean, J. 2010. Cycles and trends in solar irradiance and climate. Wiley Interdisciplinary Reviews: Climate Change 1: 111–122.

Lean, J. L. 2018. Estimating solar irradiance since 850 CE. Earth and Space Science 5: 133–149. https://doi.org/10.1002/2017EA000357.

Lelieveld, J., Crutzen, P. J. and Dentener, F. J. 1998. Changing concentration, lifetime and climate forcing of atmospheric methane. Tellus 50B: 128–150.

Liu, R. 2021. Latest trends of atmospheric cells under global warming. IOP Conf. Ser.: Earth Environ. Sci. 658: 012003.

Lüthi, D., Le Floch, M., Bereiter, B., Blunier, T., Barnola, J. M., Siegenthaler, U. et al. 2008. High-resolution carbon dioxide concentration record 650,000–800,000 years before present. Nature 453: 379–382.

Malhado, A. C. M., Malhi, Y., Whittaker, R. J., Ladle, R. J., ter Steege, H., Fabre´ et al. 2012. Drip-tips are associated with intensity of precipitation in the Amazon Rain Forest. Biotropica 44: 728–737.

McCormick, M., Thomason, L. and Trepte, C. 1995. Atmospheric effects of the Mt Pinatubo eruption. Nature 373: 399–404.

McInerney, F. A. and Wing, S. L. 2011. The Paleocene-Eocene thermal maximum: a perturbation of carbon cycle, climate, and biosphere with implications for the future. Annu. Rev. Earth Planet. Sci. 39: 489–516.

Mekonen, Z. A., Riley, W. J., Berner, L. T., Bouskill, N. J., Torn, M. S., Iwahana, G. et al. 2021. Arctic tundra shrubification: a review of mechanisms and impacts on ecosystem carbon balance. Environmental Research Letters 16: 053001. https://doi.org/10.1088/1748-9326/abf28b.

Milankovitch, M. (Ed.). 1930. Mathematische Klimalehre und Astronomische Theorie der Klimaschwankungen. Borntraeger, Berlin.

Milankovitch, M. 1941, Kanon der Erdbestrahlung und seine Anwendung auf das Eiszeitenproblem (Belgrade: Mihaila Curcica).

Moore, J. W. and Schindler, D. E. 2022. Getting ahead of climate change for ecological adaptation and resilience. Science 376: 1421–1426.

Muhlfeld, C. C., Kovach, R. P., Jones, L. A., Al-Chokhachy, R., Boyer, M. C., Leary, R. F. et al. 2014. Invasive hybridization in a threatened species is accelerated by climate change. Nat. Clim. Chang. 4: 620–624.

Munk, W. H. 1950. On the wind-driven ocean circulation. Journal of Meteorology 7: 80–93.

Ngo, M.-H. and Hsin, Y.-C. 2021. Impacts of wind and current on the interannual variation of the summertime upwelling off southern Vietnam in the South China Sea. Journal of Geophysical Research: Oceans 126: e2020JC016892. https://doi.org/10.1029/2020JC016892.

Niquen, M. and Bouchon, M. 2004. Impact of El Niño events on pelagic fisheries in Peruvian waters. Deep Sea Research Part II: Tropical Studies in Oceanography 51: 563–574.

Nussey, D. H., Postma, E., Gienapp, P. and Visser, M. E. 2005. Selection on heritable phenotypic plasticity in a wild bird population. Science 310: 304–306.
Oppenheimer, C. 2003. Climatic, environmental and human consequences of the largest known historic eruption: Tambora volcano (Indonesia) 1815. Progress in Physical Geography: Earth and Environment 27: 230–259.
Oreskes, N. and Conway, E. M. 2010. Merchants of Doubt: How a Handful of Scientists Obscured the Truth on Issues from Tobacco Smoke to Global Warming. Bloomsbury Press, New York.
Overpeck, J. 2000. Forces of Change. National Geographic Society. Washington D.C.
Ozgo, M. and Schilthuizen, M. 2013. Evolutionary change in *Cepaea nemoralis* shell colour over 43 years. Global Change Biology 18: 74–81.
Palumbi, S. R., Barshis, D. J., Traylor-Knowles, N. and Bay, R. A. 2014. Mechanisms of reef coral resistance to future climate change. Science 344: 895–8.
Park, H.-S., Kim, S.-J., Steward, A. L., Son, S.-W. and Seo, K.-H. 2019. Mid-Holocene Northern Hemisphere warming driven by Arctic amplification. Science Advances 5: eaax8203.
Parmesan, C. and Yohe, G. 2003. A globally coherent fingerprint of climate change impacts across natural systems. Nature 421: 37–42.
Pecl, G. T., Araújo, M. B., Bell, J. D., Blanchard, J., Bonebrake, T. C., Chen, I. C. et al. 2017. Biodiversity redistribution under climate change. Science 355: 1–9.
Petit, J. R., Jouzel, J., Raynaud, D., Barkov, N. I., Barnola, J.-M., Basile, I. et al. 1999. Climate and atmospheric history of the past 420,000 years from the Vostok ice core, Antarctica. Nature 399: 429–436.
Petite-Maire, N., Beufort, L. and Page, N. 2018. Holocene climate change and man in the present day Sahara Desert *In*: Dalfes, H. N., Kukla, G. and Weiss, H. [eds.]. Third millennium BC Climate Change and Old World Collapse. NATO ASI Series: 297–308.
Philander, S. G. 1998. Is the Temperature Rising? Princeton University Press, Princeton, NJ.
Poloczanska, E. S., Brown, C. J. and Sydeman, W. J. 2013. Global imprint of climate change on marine life. Nat. Clim. Chang. 3: 919–925.
Pörtner, H. O., Roberts, D. C., Poloczanska, E. S., Mintenbeck, K., Tignor, M. et al. 2022. Summary for policymakers. pp. 3–33. *In*: Pörtner, H.-O., Roberts, D. C., Tignor, M., Poloczanska, E. S. et al. [eds.]. IPCC 2022: Climate Change 2022: Impacts, Adaptation and Vulnerability. Contribution of Working Group II to the Sixth Assessment Report of the Intergovernmental Panel on Climate Change. Cambridge, UK and New York, NY, USA: Cambridge University Press.
Post, E. and Forchhammer, M. C. 2008. Climate change reduces reproductive success of an Arctic herbivore through trophic mismatch. Phil. Trans. R. Soc. B 363: 2367–2373.
Réale D., McAdam, A. G., Boutin, S. and Berteaux, D. 2003. Genetic and plastic responses of a northern mammal to climate change. Proceedings of the Royal Society B 270: 591–6.
Revelle, R. and Suess, H. E. 1957. Carbon dioxide exchange between atmosphere and ocean and the question of an increase of atmospheric CO2 during the past decades. Tellus 9: 18–27.
Riegl, B., Johnston, M., Purkis, S., Howells, E., Burt, J., Steiner, S. C. C. et al. 2018. Population collapse dynamics in *Aropora downingi*, an Arabian/Persian Gulf ecosystem-engineering coral, linked to rising temperature. Global Change Biology 24: 2447–2462.
Rippey, B. R. 2015. The U.S. drought of 2012. Weather and Climate Extremes 10: 57–64.
Rojas, O., Li, Y. and Cumani, R. 2014. Understanding the drought impact of El Niño on the global agricultural areas: An assessment using FAO's Agricultural Stress Index (ASI). FAO, Rome, Italy.
Ruddiman, W. F. and Kutzbach, J. E. 1989. Forcing of late Cenozoic northern hemisphere climate by plateau uplift in Southern Asia and the American West. Journal of Geophysical Research 94: 18,409–18,427.
Rypdal, K. 2012. Global temperature response to radiative forcing: Solar cycle versus volcanic eruptions. Journal of Geophysical Research 117: D06115. doi:10.1029/2011JD017283.
Scofield, B. 2010. A history and test of planetary weather forecasting. Open Access Dissertations. 221. https://doi.org/10.7275/1562218 https://scholarworks.umass.edu/open_access_dissertations/221.
Shaffer, G. and Lambert, F. 2018. In and out of glacial extremes by way of dust–climate feedbacks. PNAS 115: 2026-2031.
Sinervo, B., Méndez-de-la-Cruz, F., Miles, D. B., Heulin, B., Bastiaans, E., Villagrán-Santa Cruz, M. et al. 2010. Erosion of lizard diversity by climate change and altered thermal niches. Science 328: 894–899.
Stabeno, P. J. and Bell, S. W. 2019. Extreme conditions in the Bering Sea (2017-2018): Record-breaking low sea-ice extent. Geophysical Research Letters 46: 8952-8959. https://doi.org/10.1029/2019GL083816.
Steiger, N. J., Smerdon, J. E., Cook, B. I., Seager, R., Williams, A. P. and Cook, E. R. 2019. Oceanic and radiative forcing of medieval megadroughts in the American Southwest. Science Advances 5: 2375–2548.
Stephens, G. L., O'Brien, D., Webster, P. J., Pilewski, P., Kato, S., and Li, J. L. 2015. The albedo of Earth, Rev. Geophys. 53: 141–163. https://doi.org/10.1002/2014RG000449.

Streever, B. 2016. And soon I heard a roaring wind: A natural history of moving air. Little, Brown and Company, Hachette Book Group, NY.

Sutton, A. J., Williams, N. L. and Tilbrook, B. 2021. Constraining Southern Ocean CO_2 flux uncertainty using uncrewed surface vehicle observations. Geophysical Research Letters 48: e2020GL091748. https://doi.org/10.1029/2020GL091748.

Tape, K. D., Clark, J. A., Jones, B. M., Kantner, S., Gaglioti, B. V., Grosse, G. et al. 2022. Expanding beaver pond distribution in Arctic Alaska, 1949 to 2019. Scientific Reports 12: 7123.

Tauger, M. B. 2020. Agriculture in World History. Routledge. https://doi.org/10.4324/9780367821739.

Teague, K. A. and Gallicchio, N. 2017. The evolution of meteorology: a look into the past, present, and future of weather forecasting. Wiley Blackwell, NJ.

Thomas, C. D. 2010. Climate, climate change and range boundaries. Divers Distrib. 16: 488–95.

Thompson, J., Charpentier, A., Bouguet, G., Charmasson, F., Roset, S., Buatois, B. et al. 2013. Evolution of a genetic polymorphism with climate change in a Mediterranean landscape. Proc Natl. Acad. Sci. USA. 110: 2893-7.

Tingley, M. W., Monahan, W. B., Beissinger, S. R. and Moritz, C. 2009. Birds track their Grinnellian niche through a century of climate change. PNAS 106: 19637–19643.

Tremblin, M., Khozyem, H., Adatte, T., Spangenberg, J. E., Fillon, C., Grauls, A. et al. 2022. Mercury enrichments of the Pyrenean foreland basins sediments support enhanced volcanism during the Paleocene-Eocene thermal maximum (PETM). Global and Planetary Change 212: 103794, https://doi.org/10.1016/j.gloplacha.2022.103794.

Trenberth, K. E., Fasullo, J. T. and Kiehl, J. 2009. Earth's global energy budget. Bulletin of the American Meteorological Society 90: 311–323.

United Nations Department of Economic and Social Affairs. 2024. 2021 Energy Statistics Yearbook. United Nations.

Urban, M. C. 2015. Accelerating extinction risk from climate change. Science 348: 571–573.

Van den Hende, C., Van Schaeybroeck, B. Nyssen, J., Van Vooren, S., Ginderachter, M. V. and Termonia, P. 2021. Analysis of rain-shadows in the Ethiopian Mountains using climatological model data. Climate Dynamics 56: 1663–1679.

Van Duivenvoorde, W. 2015. Dutch East India Company shipbuilding: The archeological study of Batavia and other seventeenth-century VOC ships. Texas A&M University Press, Texas.

Vergés, A., Steinberg, P. D., Hay, M. E., Poore, A. G., Campbell, A. H., Ballesteros, E. et al. 2014. The tropicalization of temperate marine ecosystems: climate-mediated changes in herbivory and community phase shifts. Proc. R. Soc. B Biol. Sci. 281: 20140846.

Walker, G. T. 1924. Correlation in seasonal variations of weather. IX. A further study of world weather. Memoirs of the Indian Meteorological Department 24: 275–332.

Walker, K. V., Davis, M. B. and Sugita, S. 2002. Climate change and shifts in potential tree species range limits in the Great Lakes Region. J. Great Lakes Res. 28: 555–567.

Walther, G. R., Post, E., Convey, P., Menzel, A., Parmesan, C., Beebee, T. J. et al. 2002. Ecological responses to recent climate change. Nature 416: 389–395. doi: 10.1038/416389a.

Wegener, A. L. 1912. Die Entstehung der Kontinente. Geologische Rundschau 3: 276–292.

Whiteley, N. M. 2011. Physiological and ecological responses of crustaceans to ocean acidification. Marine Ecology Progress Series 430: 257–271.

Whittaker, R. H. 1975. Communities and Ecosystems, 2nd edition. Macmillan, New York.

Whittaker, R. H. and Niering, W. A. 1975. Vegetation of the Santa Catalina Mountains, Arizona. V. Biomass, production, and diversity along the elevation gradient. Ecology 56: 771–790.

Williams, J. W. and Jackson, S. T. 2007. Novel climates, no-analog communities, and ecological surprises. Front. Ecol. Environ. 5: 475–482.

Williams, J. W., Martin, E. E., Sikes, E., Starr, A., Umling, N. E. and Glaubke, R. 2021. Neodymium isotope evidence for coupled Southern Ocean circulation and Antarctic climate throughout the last 118,000 years. Quaternary Science Reviews 260: 106915.

Woods, T. N., Harder, J. W., Kopp, G. and Snow, M. 2022. Solar-cycle variability results from the *Solar Radiation and Climate Experiment* (SOURCE) Mission. Solar Physics 297: 43 https://doi.org/10.1007/s11207-022-01980-z.

Wulf, A. 2015. The Invention of Nature: Alexander von Humboldt's New World. Alfred A. Knopf, New York.

Yang, H., Lohmann, G., Krebs-Kanzow, U., Ionita, M., Shi, X., Sidorenko, D. et al. 2020. Poleward shift of the major ocean gyres detected in a warming climate. Geophysical Research Letters 47: e2019GL085868. https://doi.org/10.1029/2019GL085868.

Zachos, J., Pagani, M., Sloan, L., Thomas, E., and Billups, K. 2001. Trends, rhythms, and aberrations in global climate 65 Ma to Present. Science 292: 686–693.

Chapter 13

Evolutionary Ecology

"I feel truly thankful that Darwin had been studying the subject so many years before me..."
(Alfred Russell Wallace 1869)

Legend has it that Daedalus, a mythical craftsman of ancient Greece—credited with the invention of carpentry tools and creator of sculptures so animated, they had to be tied down or they would run away—joined the court of King Minos, the ruler of Crete, as the king's engineer. In due course, after egregiously offending the king, Daedalus and his son were held captive. While imprisoned, the ingenious Daedalus, observing the birds flying outside their enclosure, fashioned two pairs of wings from feathers held together by beeswax and thread, one for himself and the other for his son, Icarus. As Daedalus attached the wings to his son, he warned him to fly carefully, lest the wings be damaged by moisture or heat. They flew off with their wings and escaped Crete, but the freedom of flight made Icarus arrogant, and despite his father's warnings, Icarus sought to fly as close to the sun as he could. Higher and higher he flew until the hot rays of the sun melted the wax that held the wings to his body, and now wingless, the boy fell into the Mediterranean Sea and drowned.

Unbelievable parts of this story aside, the folklore of Daedalus and Icarus reflects our deep yearnings to fly like the birds despite our intrinsic inability. Yet the ways we have since learned to fly, using inventions such as airplanes, hot-air balloons, gliders, rockets, and the like, have never come close to the way birds can fly with the help of their wings. Why then is it impossible that Daedalus and Icarus could have flown the coop, even with wings as finely constructed as a bird's? It turns out birds have not only wings and feathers, but there is hardly an organ in their bodies that is not adapted for flight: massively developed breast muscles attached to the breast bones, necessary to power flapping flight (Tobalske 2016); large air sacs, extensions of the lungs, enhancing balance and agility in flight (Farmer 2006); connected to this respiratory system by membranous extensions are airy (pneumatic) bones (instead of mammal-like bones with heavy marrow) that are modified to be extraordinarily light and particularly strong (Dumont 2010). The list goes on. Overall, the morphology of most birds has made them fit for aerial life. Icarus, even if flying with the best pair of wings, would have crashed to the ground as he flapped his inadequate arm muscles against the weight of his marrow-filled bones. Unlike most birds, humans are not adapted to life in the air.

Organisms—from bacteria to barnacles, fish, lizards, mushrooms, pine trees, orchids, elephants, stick insects, and bonobos—are wonderfully adapted to their environments. This "fit" of form, function, and behavior appears artfully designed. Paley (1802) wrote in his *Natural Theology*,

> "In crossing a heath, suppose I pitched my foot against a *stone*, and were asked how the stone came to be there, I might possibly answer that, for any thing I knew to the contrary, it had lain there forever...But suppose I had found a *watch* upon the ground..."

This was the essence of Paley's argument that adaptations are incontrovertible evidence of the existence of a Supreme Designer (God). A stone has no sentient designer and has no functional intent; a watch, however, implies a maker with an intentional design.

DOI: 10.1201/9781003008538-13

Darwin (1859), the father of evolutionary ecology, did not deny that organisms are "designed" for certain ways of life, with parts tailored for certain functions. The orchid *Angraecum sesquipedale* is designed for pollination, with nectaries the exact length of the proboscis of the moth that feeds on it. But this "design," Darwin argued, comes not from a "designer," but from natural law—specifically, the law of natural selection.

> "The old argument of design in nature, as given by Paley, which formerly seemed to me so conclusive, fails, now that the law of natural selection has been discovered. We can no longer argue that, for instance, the beautiful hinge of a bivalve shell must have been made by an intelligent being, like the hinge of a door by man…" (Darwin, C. in Barlow 1958)

Natural selection—framed in evolutionary ecology—explains ecological relationships through the adaptations of organisms to their environments. Evolutionary biologists, following Darwin, use the term **adaptation** in two distinct but related ways. The term adaptation is used to describe both the *product* (result) of natural selection as well as the evolutionary *process* of natural selection that produced it. As a product of natural selection, an adaptation is a phenotypic trait (feature) of an organism that can reasonably be said to serve a purpose (Williams 1966). The adaptive trait can be morphological, physiological, behavioral, developmental, or reproductive. For a trait to *become* adaptive, the trait must be heritable, and it must maximize fitness, given the circumstances, where **fitness** is defined as the per capita growth rate of a (hypothetical) population entirely composed of individuals with that trait (Vincent and Brown 2005; see Chapter 7 *Light Competition in Forests and Evolutionary Game Theory*). The product of natural selection applies to the phenotypic traits of organisms, whereas the process of natural selection applies to individuals with that feature (Bock 1980).

As we shall see, adaptive traits are not generated *de novo* from natural selection, but they are traits that are shaped, co-opted, or maintained by the evolutionary processes of natural selection within an ecological context. In our introductory example above, we talked about bird adaptations to flight in terms of a historical narrative—essentially a "just so" story. "Look at how these pneumatic bones are adaptive for flight," we exclaim! But was this adaptation the result of natural selection for flight? In the case of pneumatic bird bones, we can make the tautological mistake of thinking pneumatic bones, which are clearly adaptive for flight, evolved because flying birds with aerier bones had greater fitness than flying birds without. However, fossil evidence shows us that birds inherited their pneumatic bones from their theropod ancestors (Benson et al. 2012; Figure 1), who did not fly! As Bock (1980) observes, "the state of being adapted is independent of the process of becoming adapted as adaptations need not evolve under the control of selection forces to which they are now adapted." Sound confusing? It is, and that is why we must be careful when we talk about adaptations with respect to the process that created them. The adaptation we see today resulted in part from natural selection in the past. In the case of birds, greater pneumaticity in their small endothermic non-flying theropod ancestors may have provided the energetic savings (Benson et al. 2012) that gave greater fitness in the past. But the adaptation we see today also resulted from natural selection in progress. By analyzing the fitness consequences of having pneumatic bones in their current ecological role, we see why pneumatic bones have been maintained in flying birds but are reduced in diving birds (O'Connor 2004) and even thickened (by sexual selection!) in the wings of the manakin *Machaeropterus deliciosus* (Bostwick et al. 2012). Therefore, if we say an adaptive trait evolved because it is adaptive in its current situation, it is quite possible, without knowing its phylogenetic history or testing the idea with counterexamples, that we could be completely wrong.

Before we delve deeper into the process of natural selection, we must keep in mind another thought trap. We must remember that selection acts on the individual, which is a whole package of traits. While flight is a major adaptation in birds that enabled them to live in the air, birds have other adaptations that help them survive, grow, and reproduce in particular environments. Some of those traits may be more adaptive (provide greater fitness) in some contexts but not in others, and some may be neutral in all present contexts but were adaptive in the past. Some may even be maladaptive

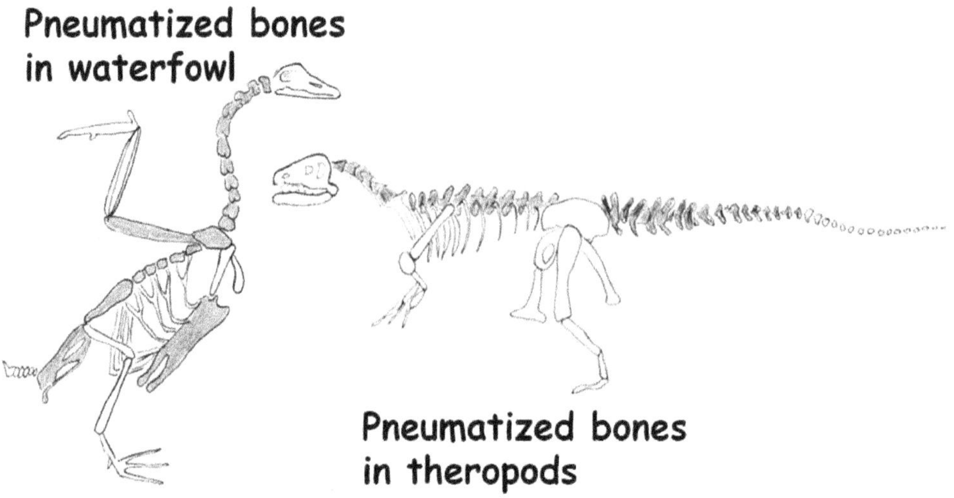

Figure 1. Pneumatized bones (shaded) in waterfowl (after O'Connor 2004) and theropods (after Benson et al. 2012). Not to scale; illustrative, not anatomically exact.

when only a weaker selective context is considered rather than the strongest one. For instance, beak shapes and sizes are an adaptation for the different types of foods birds eat (Lack 1947). Scarlet ibises (*Eudocimus ruber*) forage for food in shallow water with a long downcurved bill, which looks helpful in sweeping back and forth across the bottom substrate while keeping their nostrils out of the water. But males have bills that are 22% longer on average than females (Babbitt 2009). Why? This is because males and females experience different selective contexts. Babbitt (2009) found that in sparring contests with other males over nest sites, winning males had significantly longer bills than losing males, illuminating that the selective context differs for bills in males (reproduction and foraging) from bills in females (foraging). These complexities should be kept in mind as we focus on how adaptive traits evolve.

The Process of Natural Selection

> "Evolution is the physical, genetic, or behavioral change in populations of biological organisms over time. Evolution's more interesting and significant manifestations result from **natural selection,** a process that engineers biological systems. Natural selection works within genetic, developmental, and environmental constraints to shape biological organisms in ways that make them appear adapted to their environments" (Vincent and Brown 2005).

Drawing from Lewontin (1974), Darwin's arguments of natural selection can be distilled into three postulates (Vincent and Brown 2005). According to the first postulate, there is **heritable variation** in traits associated with each type of organism, resulting in different individuals within a species differing slightly from one another. This variation is heritable, so on average, offspring resemble their parents more than they resemble other individuals. Roughly speaking, "like tends to beget like." The second postulate—among organisms, there is a **struggle for existence**—simply means all organisms, interacting with their biotic and abiotic environments in varied ways, compete for limited resources for survival and reproduction under environmental stress. The third postulate states that **heritable traits matter in the struggle for existence.** This postulate is the key to understanding the consequences of natural selection: on average, individuals within a population who are phenotypically fitter (as defined above) will leave more offspring either immediately or in remote generations than phenotypically less fit individuals. These three postulates account for evolutionary change by natural selection.

It is essential to clarify what is meant by "heritable variation" in the first postulate, as it comes in two parts. On the one hand, it is necessary that populations are made up of individuals with similar but varying traits; natural selection cannot act as an "engineer" on a population of identical individuals. The existence of variation within a population can easily be demonstrated. Select around 100 or so specimens and measure the variants of a single characteristic; for instance, the tail lengths of mice, the number of scales on snakes, the rows of kernels on ears of corn, etc. Individuals vary not only in physiological, morphological, and behavioral traits but also in their life history traits, such as how long they grow before reproducing, how long they live, how many and how often they produce offspring, and how fast their offspring can grow (see Chapter 6 *Life Histories and Life History Tradeoffs*). But "variation" is not a sufficient condition for natural selection to act as an "engineer." This takes us to the second part: the "variations" must be transmitted from parent to offspring such that individuals resemble their parents more than they do unrelated individuals.

Many physical, behavioral, or physiological variations in characteristics or traits are transmitted from parents to offspring. Through observation and breeding experiments, we know that many phenotypic traits—such as size, shape, eye color, seed weight, and disease resistance—are passed down to descendants via a genetic code. These traits are therefore heritable. However, some trait variations can be clearly distinguished as not arising from genetic transmission but rather from environmental circumstances. Perhaps most obvious is the effect of growing up in a food-rich vs a food-poor habitat. Taborsky (2006) demonstrated that high-food or low-food environments experienced early in life had a strong influence on reproductive rate and offspring size in cichlid fish. Other types of non-genetic phenotypic variation include sex determination due to temperature (e.g., reptiles; Janzen and Phillips 2006), immune system parameters due to pathogen exposure (e.g., in humans; Brodin et al. 2015), and maternal effects on offspring phenotype (e.g., a female beetle's resource allocation to eggs depends on the host plant she chooses; Mousseau and Fox 1998). Are environmentally derived traits heritable? Often, the answer is no—such is the case of a tree weathered like a bonsai by a prevailing wind, or a bird with a missing foot due to predation or accident. But as we are now discovering, surprisingly often, the answer turns out to be yes.

Many types of phenotypic variation derive from a phenomenon called environmentally conditioned genetics (***epigenetics***). Epigenetics is the process by which environmental conditions turn genes "on" or "off" via DNA methylation under specific environmental circumstances. Epigenetics can not only set the phenotypic fate of an individual, but also, in some cases, this epigenetic reprogramming can be transmitted to the next generation and even beyond (examples can be found in plants and vertebrates; Zhang and Sirard 2021). Epigenetics, therefore, makes "the line between 'inherited' and 'acquired'...fuzzy," while also providing fodder for natural selection (Petronis 2010). Especially confounding is the fact that parents not only pass on genes (strict genetic inheritance), but also influence the imprinting of genes via transmission of environmentally influenced information in sperm or the maternally derived environment of the egg (Crean and Bonduriansky 2014). Even the behavior of a mother in choosing a mate based on his genetically determined traits affects offspring phenotype (as in female sexual selection for rare male eye color; Frost 2006). Thus, traits acquired by offspring from their parents either through the transmission of genetic alleles or by certain non-genetic factors can be considered heritable and subject to adaptive evolution via natural selection.[1]

Darwin's second observation—there is "struggle for existence" among organisms—comes from his reading of Malthus's (1798) book, "Essay on Population," in which Malthus posits that human populations have the potential to grow geometrically, but their resources can increase only arithmetically (Figure 2). Malthus wittily states, "A slight acquaintance with the numbers will shew

[1] Stoltenberg (1997) would strongly disagree, saying "it is inappropriate to speak of traits or characteristics being inherited; only the constituents of the gametes are passed from one generation to the next." But as we argue above, this does not matter to natural selection. "Traits" can evolve adaptively without genetics or even bodies as long as the "like begets like" rule applies; ask any computer scientist.

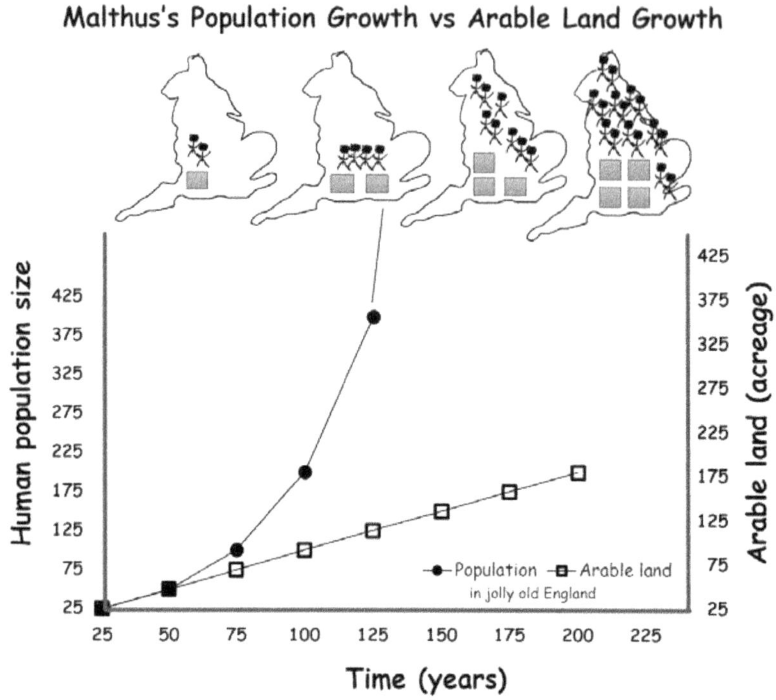

Figure 2. Malthus's (1798) postulate of geometric (increasing by a factor, in this case a factor of 2 every 25 years) human population growth potential versus an arithmetic (increasing by a number, in this case 25 acres every 25 years) growth potential in arable land. He assumes the "difficulty of subsistence" (arising from, e.g., "misery," "vice," etc.) prevents human population growth from actually obtaining a geometric increase.

the immensity of the first power in comparison to the second." He further reasons that the laws of nature must keep these two unequal powers equal, implying "a strong and constantly operating check on [the human] population from the difficulty of subsistence." Darwin (1859), applying this doctrine to "the whole animal and vegetable kingdoms," argued that "...many more individuals of each species are born than can possibly survive..., consequently, there is a frequently recurring struggle for existence".

This massive discrepancy between the number of offspring produced and the number that can be sustained by available resources creates a "struggle for existence" among organisms from which a fraction of individuals will succeed. This "struggle" need not be restricted to competition for food. Consider, for instance, a female Atlantic cod that matures between four and six years and can live up to 25 years. This cod can produce two million eggs in a breeding season. However, as soon as the eggs are released, invertebrates and other fish devour 99% of them, and over 90% of the survivors die within a year. Because of this, the cod population does not explode geometrically; if the population neither grows nor declines, in her lifetime, an average female cod will produce, logically, just two successful offspring (Ridley 2004; see Chapter 2 *Population Dynamics* and Chapter 6 *Life Histories and Life History Tradeoffs*, for reasoning). However, Darwin (1859) argues that the identity of the two surviving offspring is not random. Because of this "struggle for existence," he reasons that "any being, if it varies however slightly in any manner profitable to itself, under the complex and sometimes varying conditions of life, will have a better chance of surviving, and thus be *naturally selected* [his emphasis]."

Because some individuals die before reproducing or fail to reproduce, each generation represents the offspring of just a subset of individuals from the previous generation. Different ancestors leave different numbers of descendants. It is only the individuals who were capable of surviving and reproducing that influence subsequent generations. As stated above, it is Darwin's third observation—heritable variation in traits influences the struggle for existence—that is the key to understanding the consequences of this differential generational contribution. If heritable traits influence the struggle for existence, this means that, in any given environment, individuals with heritable traits that make them better able to survive and reproduce than their brethren will cause those traits to be more represented in the next generation.

For centuries, our experience with animal and plant breeding demonstrates the existence of heritable variation and the power of selective breeding to create functional or desirable properties in the breeds. Darwin applied observations of selective breeding by, for example, pigeon fanciers, to the selection of variation by nature. As Lewontin (2010) succinctly wrote, "Natural selection is human selection writ large." Favorable variations are preserved and multiplied from generation to generation at the expense of less advantageous ones. Darwin's ground finches (*Geospiza*) offer no finer example. Peter and Rosemary Grant, along with Peter Boag (Boag and Grant 1981; Grant and Grant 1993), carefully documented evolutionary change in beak size in response to changes in the food environment over time by following the survival and reproductive success (fitness) of *G. fortis* after a major drought. Birds with large beaks survived the best because they were able to open the large, hard seeds, which were the only food available during the drought; birds with smaller beaks died at a higher rate. As a consequence, the average beak size increased in the population, and offspring of the survivors had larger beaks on average than the population before natural selection occurred.

Although the central argument of natural selection begins with the existence of heritable variation, Darwin struggled with the mechanism of inheritance. He was not aware of Mendel's work on particulate inheritance (1866), but he was on the right path when he stated, "I am strongly inclined to suspect that the most frequent cause of [heritable] variability may be attributed to the male and female reproductive elements having been affected prior to the act of conception." If only he had access to any modern biology textbook's exposition of meiosis and fertilization! Since the rediscovery of Mendel's particulate inheritance (1866), we often think of evolution in terms of genetics—the study of the inheritance of genes from one generation to the next. Yet, Darwin presented his theory of evolution by natural selection in the absence of any understanding of Mendelian genes. Furthermore, Darwin did not always distinguish between heritable and non-heritable variation or between traits that affect fitness and traits that do not. It was not until Darwin published his book *The Descent of Man, and Selection in Relation to Sex* (1871) that Darwin admits, "I had not formerly sufficiently considered the existence of many structures which appear to be, as far as we can judge, neither beneficial nor injurious; and this I believe to be one of the greatest oversights as yet detected in my work." But these weaknesses point to the power of Darwin's three postulates in explaining how natural selection engineers adaptive traits: first, that the actual mechanism of inheritance itself is not important; what matters is that a mechanism exists; and second, that the heritable trait must influence the struggle for existence; otherwise, natural selection is blind to it.

In the 1930s, nearly seventy years after Darwin published the *Origin*, Fisher (1930), Wright (1932), Haldane (1932), Dobzhansky (1937), Mayr (1942), and others combined evolution, mathematics, and genetics into the *Modern Synthesis* (Mayr and Provine 1980). In creating this foundation, heritable traits were considered the result of genes that could be shuffled to produce offspring phenotypes in predictable ways. Novel characteristics were thought to arise from mutations or lasting alterations that arise randomly or in an "undirected" way in individual genes. The Modern Synthesis instigated the primacy of genes over heritable traits as the objects of evolution, as indicated by Dobzhansky's (1937) definition of evolution: "a change in gene frequency over time." This idea became firmly cemented by the discovery of the structure of DNA (Watson and Crick 1953), which made it possible to identify how phenotype arises from genotype (the "Central Dogma"; Crick 1958). The primacy of genes was taken to the extreme in the 1970s in Richard Dawkins' (1976) book *The Selfish Gene*,[2] which used a gene-centric (rather than trait-based) argument to explain the evolution of cooperation.

[2] For those of you who love memes, it is in this book that Dawkins coins the term. It is worth reading his final chapter to see how what he calls a meme differs from our modern usage.

We must remember, however, while selection may act at many levels, including that of the gene (Burt and Trivers 2006) or the group (Simon et al. 2013), in actual operation, natural selection does not act on any one trait per se, but works whole cloth on the individual's ability to survive and leave viable offspring in particular environments, whether that individual is as simple as a virus or as complex as a human being. It is therefore only for convenience that we discuss adaptive characteristics separately; after all, survival can occur only if the whole organism is adapted to the environment (Allen 1929). The receiving end of selection is then the phenotype *in toto*, which includes traits that can appear and disappear throughout an organism's life, and which can be exposed to the gamut of environmental experiences in that lifetime, including abiotic rigors such as minimum and maximum temperatures in winter and summer, droughts and floods, rapid changes in exposure to light, salinity and hydrogen-ion concentration, as well as biotic rigors such as competition and predation. Although these rigors might act periodically or only occasionally, they can do so with a strong evolutionary impact, as we saw for mean beak size in the *G. fortis* population. This brings up another point. Natural selection acting on individuals causes a population to evolve. Consequently, genes do not cause adaptive traits and do not cause populations to evolve; instead, selection on individuals causes adaptive traits to evolve in populations (Lewontin 1970).

Evolution by natural selection—from the evolutionary ecology perspective—describes the change in frequency of phenotypes (rather than genes) within a population over generations (compare to Dobzhansky's (1937) definition above). It recognizes that evolution consists chiefly of adaptive changes in a population in response to the environment. These changes require alterations of both the quality and quantity of population genetic (and other heritable) variation. Because evolution cannot act on identical individuals, continued evolution is possible only if reductions in population variation are balanced by increases. Natural selection is a non-random process (which is why it acts like an engineer); it sieves out the less fit and allows the more fit to pass on their heritable traits; it reduces population variation. What, then, causes heritable variation to increase in populations?

Sources of Variation

Genetic variation entails two mechanisms: the mutation of one variant to another and, in eukaryotes,[3] the sexual process that recombines those variants to form a multitude of variations. The recombination of genes—through sexual reproduction, by crossing over, or by translocations and inversions—produces various gene combinations within a population, resulting in genetic diversity among individuals, even if the gene pool does not change (Dobzhansky 1937). But a deck has only 52 cards and only 2,598,960 possible 5-card poker hands, most of which are non-winning. Over time, cards may be destroyed or lost by chance, resulting in fewer possible poker hands. Even without natural selection, this same type of process can lead to a loss of genetic variation within a population, so the sexual process cannot lead to an increase in variation in the gene pool. Mutations, however, can.

Hugo de Vries (1901), one of the co-discoverers of Mendel's paper on particulate inheritance, introduced the term **mutation** into biology. Mutation—which is the fountainhead of all genetic variation within populations—is a random, permanent biochemical change in an organism's DNA (Cairns et al. 1988), resulting from events such as copying and process errors during cell division, mechanical damage when molecules and cell structures collide with DNA, and exposure to certain chemicals or high energy radiation such as ultraviolet light or X-rays. Mutations can occur as single nucleotide variants (e.g., adenine substituted for thymine), copy number variants (e.g., a gene doubled), transposable element insertions (a sequence of genes that change position), and other structural variants (large regions of the genome rearranged; Dou et al. 2018), including doubling

[3] Without breaking the narrative too much, we will mention here that although they do not reproduce sexually, prokaryotes also have various mechanisms of gene exchange including conjugation, transduction, and transformation that provide genetic variation to cell-lines.

or loss of whole chromosomes. Mutations, which can occur at all gene loci (locations of genes on chromosomes), might thus produce alterations in every possible morphological and physiological character.

For multicellular organisms, there are two types of mutations: **somatic mutations** and **germline mutations**. Somatic mutations occur in the cells that make up an organism's body and have been studied extensively in the context of cancer, but they also appear in normal cells (Dou et al. 2018) and, in fact, accumulate as organisms age (Ren et al. 2024). Germline mutations occur in male and female gametes. Can you guess which is fodder for evolutionary change in populations? As you can probably tell, changes to gametes are heritable, but alterations to somatic cells are not. Interestingly, the mutation rate is higher in somatic cells than in gametic cells, and a paternal bias exists in germline mutations in animals (that is, fathers contribute more to germline mutations; Ren et al. 2024).

Germline mutations can be **beneficial**, **deleterious,** or **neutral** when it comes to their impact on fitness. Beneficial mutations are rare (and context dependent), and deleterious mutations in the germline often never come to fruition—a result of natural selection (Tenaillon and Matic 2020). But neutral mutations, defined as having a null effect on fitness, may occur in such a way as to change one gene (allele) into another gene (different allele), as in the case of "purple A" and "white a" alleles in Mendel's peas, or they may occur in genes that affect traits additively (as opposed to the dominant-recessive manner of Mendel's peas), or in genes that affect traits polygenically (when traits result from multiple genes). Mutations in these latter types of genes may not manifest in morphological, behavioral, or physiological ways that are noticeable, and they may be "invisible" to natural selection as well (by definition, as they do not affect fitness). Because this variation is not pruned by natural selection, it accumulates over time, often at predictable rates, making it helpful in reconstructing genealogies and phylogenies. However, it turns out neutral mutations may be the most significant source of variation involved in the evolution of natural populations. This is because it tends to accumulate in a population over several generations, creating a sort of latent variation. Tenaillon and Matic (2020) give an example at the molecular level: a change in the codon CGG (which codes for the amino acid arginine) to AGG would still code for arginine and is therefore a neutral mutation by any definition; however, if in a subsequent generation another mutation were to occur in this codon, the sets of possible amino acids coded by this mutation would dramatically differ from the original. Thus, a single neutral mutation can set up the possibility for even greater genetic variation in subsequent generations of the population. One can imagine that once enough neutral genetic variation has accumulated, natural selection might start to "see" this variation and then drive adaptive evolution.

Most populations also exhibit what's called **standing variation**, defined as the presence of more than one allele at a locus in a population (Barrett and Schluter 2008). Standing variation may be effectively neutral in terms of the fitness of any individual in the population until the population experiences a novel selective pressure or colonizes a new environment that makes this variation relevant. A century of quantitative genetics has established the ubiquity of standing variation in natural populations and its role in the response of those populations to selection (Hill 1982). Instead of waiting around for the appropriate mutation to occur, standing variation is immediately available under a new selective regime, resulting in faster evolution, the fixation of more alleles of small effect, and the spread of more recessive alleles (Barrett and Schluter 2008). A noteworthy breeding experiment with corn showed the oil content of kernels changed rapidly after the very first generation in which selection was applied (Dudley and Lambert 2004), exemplifying the role of such standing variation in rapid evolution. The adaptation in the Atlantic killifish *Fundulus heteroclitus* to toxic pollutants in urban estuaries is another example of the importance of standing variation in evolution in a "natural" environment (Reid et al. 2016).

Hybridization between two closely related species may be especially effective at introducing novel alleles, genes, and traits into a population. After the major drought event we discussed earlier (Figure 3), 40 years of research followed the fate of the medium ground finch *Geospiza fortis* on the

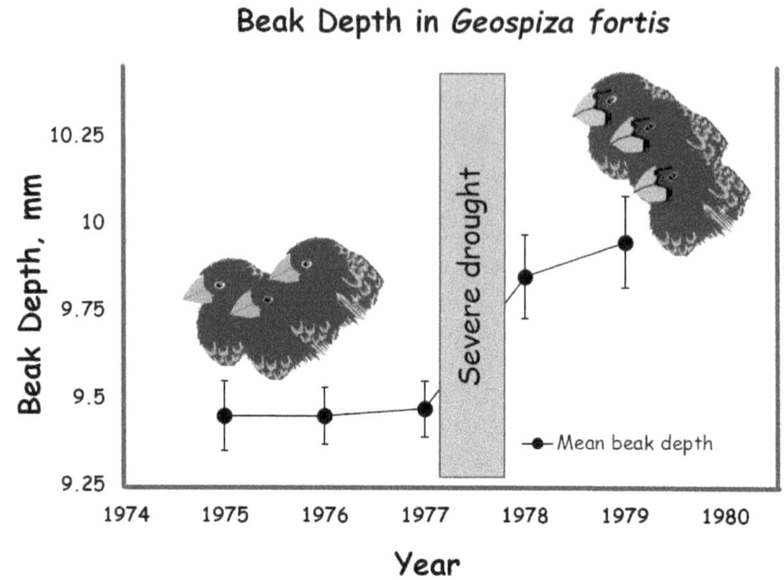

Figure 3. Example of natural selection in action after a severe drought on Daphne Major (Galapagos Islands), showing an increase in the mean beak depth in response to food availability during the drought. Larger beaks have the ability to crack the hard woody tissues protecting the seeds of *Tribulus cistoides* that occurred in relatively high abundance during the drought (interpreted from Grant and Grant's [2008] Figure 5.6, which was adapted from Boag 1981).

island of Daphne Major. The population shrank from more than 1,400 individuals to a mere 200 due to the intense competition and selection following the drought, and moreover, the effective population size (the number of individuals that could potentially father or mother the next generation) was even smaller, resulting from a 6:1 (male: female) sex ratio. This was because selection favored larger individuals (see Figure 3), and males are larger than females (Boag and Grant 1981). This episode of selection decreased genetic variation in the population, yet the population still evolved over the 40-year study period, largely due to the injection of variation from interspecific hybridizations with *G. scandens* and *G. fulinginosa*. Hybridization increased the heritable variation in beak traits by 8% to 11% and in size-related traits by 19% to 40% (Grant and Grant 2014), providing fodder for natural selection.

Natural selection uses population variation to drive evolutionary change in populations over time. At the end of the introduction to the first edition of the *Origin of Species*, Darwin explained: "I am convinced that selection has been the main but not exclusive means of modification." What other processes, besides natural selection, may cause phenotypes to change over time?

Other Evolutionary Processes

When populations are small, chance alone may determine which individuals (and the alleles they carry and the traits they bear) prevail and survive. Small populations can arise at the extremes of a species' range, through colonization of new habitats, or as a result of catastrophic environmental events, as well as from demographic dynamics generated by biotic interactions (such as predation, disease, invasive species, etc.). Chance may dictate that a rare trait, rather than the most common one, is the only one to prevail in the reduced population. These random fluctuations in a trait (and its corresponding allele(s)), resulting from chance alone, are referred to as **genetic drift** (Wright 1931).

Genetic drift is the evolutionary equivalent of sampling error, with each successive generation representing only a subset or sample of the gene pool from the previous generation. In a large population, genetic drift has minimal effect on each generation because the random selective effects will tend to average out. But, in a small population, the chance loss of a single individual or a few individuals with an allele of a certain form could cause the allele to disappear from the population in a single generation. Character variations formed in this manner often appear to be nonadaptive (Wright 1931), and Wright (1932) envisioned this as a way in which a population might "wander" off an adaptive peak in a fitness landscape. For example, Arctic fox populations repeatedly undergo boom-and-bust dynamics as they cycle with prey abundance. Individual Arctic foxes may have either white or "blue" pelage. Over most of the Arctic fox's range, the blue trait is rare compared to the white, but on certain islands, only the blue trait occurs. Because the fox's blue pelts are easy to see against the snow-blanketed islands by both potential predators and potential prey, Elton (1930) originally thought of this mismatch between organisms and their environments—this lack of adaptation—as "ecology explaining what natural selection could not explain." In other words, demography, not natural selection, was the cause of the evolution of this trait in the population. At some point in the past, during a low point in the population cycle, by chance, only homozygous blue foxes must have survived on those islands, rendering their offspring and the entire population blue. Elton's explanation would soon be integrated within population genetics and formalized as random genetic drift (Huneman 2019).

A special case of genetic drift occurs when a small group of individuals colonizes a place previously uninhabited by their species. The first human settlement on Tristan da Cunha, an island in the middle of the South Atlantic Ocean, was documented in 1816. The approximately 270 people living on the island (according to the 2001 census) are all descended from five female and seven male founders, with a few additional male contributions (Soodyall et al. 2003). Historical records indicate that at least two of the founders may have been asthmatic, and in the living descendants, over half show at least partial evidence of asthma, with almost a quarter having a definitive diagnosis of the condition (Zamel et al. 1996). This is considered a typical example of the **founder effect**. It occurs when a small group of individuals—sometimes as few as a single mating pair (or a single inseminated female insect, for instance)—breaks off from a larger population, and establishes a colony in a new area. When the new colony is isolated from the original population and the founding individuals do not represent the full genetic diversity of the original population, some alleles may be missing entirely or be overrepresented in the new population (Mayr 1942).

Another special case of genetic drift can occur when a population passes through a "**bottleneck**." Bottlenecks occur when only a few individuals survive, but the population later expands again. As in the case of the founder effect, the allele frequencies in a small population of survivors may differ significantly from those of the original population, and some alleles may be completely missing. Bottlenecks can occur from many types of events. The New Zealand fur seal (*Arctocephalus forsteri*) passed through a population bottleneck due to commercial hunting during the 18th and 19th centuries (Emami-Khoyi et al. 2017). Darwin's finches, as we saw, suffered from drought in the late 1970s. An outbreak of biting flies decimated the lions of the Ngorongoro Crater in the early 1960s (Packer et al. 1991). All of these various events left indelible marks on the genetic composition of these populations. The cheetah (*Acinonyx jubatus*) has possibly experienced three bottlenecks over the last 12,000 years, the most recent of which was caused by humans: a loss of 75% of their original habitat, resulting in inbreeding, high rates of reproductive defects, elevated juvenile mortality, and increased vulnerability to disease (Dobrynin et al. 2015). Like cheetahs, most other endangered and vulnerable mammal species may be experiencing a genetic bottleneck effect due to small population size.

The frequency of alleles or traits in a population can also change when individuals or propagules move in or out of a population. Such movement is known as **gene flow**. *Hybridization* is an extreme case of gene movement, which involves two species, and it generally introduces variation into the receiving population (as we discussed above). *Gene flow* is the movement of genes between semi-isolated populations within a species, and in the absence of natural selection and genetic drift, gene flow leads to the erosion of genetic differences between these populations, making them more homogenized. On the one hand, it might be argued that if these various populations were locally adapted, gene flow would tend to introduce "maladaptive" genes into any particular population. But this is not necessarily the case, especially if the population is small and inbred. For instance, in a narrow strip of grassy meadow along the coast of Sweden bordering the Baltic Sea and surrounded by heavily human-altered habitat, an isolated population of adders (*Vipera berus*) went through a genetic bottleneck that resulted in a population with a high proportion of deformed or stillborn offspring, and very low genetic variability. After discovering the population was nearly extinct, with falling numbers and little recruitment, Madsen et al. (1999) introduced 20 adult male adders from large, genetically diverse populations in other parts of Sweden, and the results were dramatic. Within three years, both the population size and the genetic variability increased rapidly. After the fourth year, the eight surviving introduced males were returned to their original populations. Eleven years after the single translocation event, this isolated population was still growing (Madsen et al. 2004). Gene flow into an inbred population is called **genetic rescue**, and it has been used in conservation efforts involving species such as bighorn sheep, Florida panther, Mexican wolf, and Prairie chickens, resulting in increased population growth, reproduction, and survival (Whiteley et al. 2014).

Recall how Dobzhansky (1937) defined evolution as a change in gene frequency over time. Technically, all of these "other evolutionary processes" we have discussed—gene flow and the kinds of genetic drift—cause changes in gene frequencies in a population. Like natural selection, these processes can winnow variation, but unlike natural selection, the changes in gene frequency may be "hidden" or not lead to observable changes in phenotype (e.g., the population bottleneck in New Zealand fur seals was only evident when mitochondrial DNA sequences were examined; Emami-Khoyi et al. 2017). Additionally, unlike natural selection, some changes in phenotype can be maladaptive (as we saw with the humans of Tristan da Cunha). In fact, if mutation, migration, and drift were suddenly the only processes driving evolutionary change, it would probably bring about the entire demise of the Earth's ecosystems. None of these mechanisms can produce adaptive traits, so by random walk, populations would gradually become unfit for their environment—especially if the environment changes. Only selection (natural or otherwise) can engineer adaptive traits.

Adaptation From Natural Selection

Natural selection explains an organism's **fit of form and function** to its environment—that is, the fit of an organism to where it lives, what it eats, and who eats it, playing a major role in determining a species' distribution and abundance. Thus, adaptations to variation in abiotic and biotic environmental conditions in time and space lead to a diversity of lifeforms (**biodiversity**). A close look at these adaptations can explain how organisms are connected by common ancestry and common design rules in the vaunted **procession of life** (see Chapter 11 *Biodiversity*).

Although life is ubiquitous on Earth, no single species can tolerate the full range of Earth's environments and inhabitants. Recall our discussion of tradeoffs in Chapter 6, *Life Histories and Life History Tradeoffs*. Certain species find some environments either too warm, too cold, or too saline. This is due to adaptations; you would not find polar bears living in the boreal forest or brown bears living on polar ice. Characteristics of the physical environment can thus limit geographical distributions. Polar bears (*Ursus maritimus*) have dense bones, which reduce buoyancy and may benefit swimming, and they have wide feet with rough pads, which enhance traction on sea ice (Whiteman 2021). These are adaptations to their abiotic Arctic marine environment, but this only partially explains the distribution of this species. Polar bears also interact with other organisms in their biotic environment, and they are highly specialized for hunting their seal prey. When compared to their close relative, the brown bear (*U. arctos*), polar bears have highly flattened skulls with a larger olfactory area in their turbinates (delicate bones in the nasal cavity), giving them an advantage in detecting seal prey on far-off ice floes (Whiteman 2021). However, their skull morphology leaves them poorly able to chew vegetation. Some brown bears, in contrast, can have a predominantly vegetarian diet (Bojarska 2012), and although they are carnivorous as well, taking everything from salmon to moose calves, they are ill-equipped to hunt seals on pack ice. Thus, not only are organisms constrained by the abiotic and biotic components that make up their environment, but they also become adapted to that environment as natural selection acts across generations. The ranges of the brown bear and polar bear lie adjacent to one another but hardly overlap. This is because brown bears are better suited to a boreal environment, while polar bears are better suited to hunting marine mammals on sea ice. In fact, polar bears diverged from the generalist, widespread brown bear population into their specialized niche about 500,000 years ago (Wang et al. 2022).

The one constant in evolutionary ecology is that environments change over time. Changes in the abiotic environment can cause sweeping changes in distribution (recall the spruce and oak in Eastern North America since the last ice age: Figure 5 in Chapter 12 *Climate Change*), but this can also bring about changes in the biotic environment, and changes in both the abiotic and biotic environment are capable of bringing about evolutionary changes in populations. We can find a good example of this by returning to the polar and brown bears. Genetic analysis by Wang et al. (2022) of an ancient polar bear genome compared to modern brown and polar bear genomes revealed that when temperatures warmed 5–8°C above present and sea levels rose 6–10 m above present during the warm interglacial that occurred between 110,000 and 70,000 years ago, the ranges and interactions of these two species differed drastically from today. Based on their results, Wang et al. (2022) imagine a period when plant communities capable of supporting brown bears spread northward to the Beaufort Sea coastline, and sea ice shrank. These changes created a contact zone between terrestrial foraging brown bears and marine-foraging polar bears. Polar bears would have been restricted to the shoreline, where they would have hunted walrus and other pinnipeds (which must now haul out on shore) and scavenged whale carcasses that drifted ashore. In this newly created contact zone, the two bear species exchanged genes. When the warm period ended and glaciers advanced again, brown bear genes became maladaptive for the polar bear, and natural selection acted strongly; the population shed all of its detectable brown bear genes. Polar bear genes had a minimal impact on the brown bear, and as they retreated southward, they retained genetic evidence of this mixing.

Species that have broad diets and unrestricted ranges are typically called **generalists**, whereas those with very narrow diets and habitats are considered **specialists**. Generalists typically have high local abundance as well as broad geographic distributions because they are able to exploit a broad range of resources, thus giving them a demographic advantage over specialists (Brown 1984). This is the case for the modern brown bear. There is, however, likely a trade-off between generalization and the ability to successfully exploit any given resource. Remember the adage: the jack-of-all-trades is a master of none (MacArthur 1972). The issue of trade-offs is fundamental throughout evolutionary ecology, as trade-offs can influence the evolution of life-history parameters (Chapter 6 *Life Histories and Life History Tradeoffs*), dietary specialization (Chapter 5 *Foraging Dynamics*), interspecific interactions (Chapters 3 and 4), and, quite generally, all aspects of ecological specialization versus generalization (Rosenzweig 1995). In the context of species' distribution and abundance, trade-offs become important when environments are spatially varying, and selection favors different phenotypes in different places. When encroaching glaciation made hunting on Arctic Sea ice advantageous, the specialized adaptations of the polar bear were selected for, and the extraneous brown bear genes were purged. Although the polar bear population is smaller compared to the brown bear population, they are highly successful in their Arctic environment, and brown bears cannot use it. Thus, the fit of form and function to the environment governs the distribution of species and the stability of local ecological communities and ecosystems (MacArthur 1972).

We may think of the evolution of polar bears as a response to a changing environment, but sometimes it is simply the interaction between organisms that can cause evolution to occur. Populations can evolve in response to predators, herbivores, parasites, and competitors, and examples abound. Lemanis and Zlotnikov (2023) found evidence that the driving force behind the increasing complexity in shell structure in ammonites (over their 351-million-year history) was the evolution of resistance to *predation*-induced buckling. Nettles (*Urtica thunbergiana*), a plant characterized by stinging hairs on its stems and leaves, have higher densities of these hairs in Nara Park (Japan), where a population of sika deer (*Cervus nippon*) has been maintained for more than

1,200 years, compared to other areas without deer browsing (Kato et al. 2008). Stinging hairs are a defensive structure against *herbivory* under genetic control, thus subject to natural selection. In an experiment with three-spined sticklebacks (*Gasterosteus aculeatus*), Brunner et al. (2017) exposed the fish to the *parasite* (*Gyrodactylus*), which evolved changes in metabolic, immune, and stress response genes that limited parasite infection in subsequent generations. Gonçalves et al. (2011) believe they found evidence that pairs of co-occurring yeasts in the *Saccharomyces* genus evolved specialized temperature niches, and speciated in response to *competition*; the list goes on. Sometimes the evolutionary response to an interaction causes a reciprocal evolutionary response; this is called coevolution. Examples of coevolution include mutualisms (e.g., orchids and bees; Baguette et al. 2020) and predator-prey interactions (e.g., poisonous newts and neurotoxin-resistant garter snakes; Brodie III and Brodie, Jr. 1999).

In nature, change is perpetual (nature is not an art gallery and a species is not art): conditions change, individuals adapt, populations evolve. "Natural selection," Darwin (1859) writes in *On the Origin of Species*, "is daily and hourly scrutinizing, throughout the world, every variation, even the slightest; rejecting that which is bad, preserving and adding up all that is good..."

Stabilizing Selection—A Paradox?

After the publication of *On the Origin of Species*, one of Darwin's many critics asked him something like this, "If evolution is true and if change characterizes the living world, then why have modern ostriches not changed since they were depicted in ancient Egyptian documents four thousand years ago?" (Hardin 1966: p. 171). Darwin (1859) acknowledged his era's limitations, writing, "Our ignorance of the laws of variation is profound." But he recognized that natural selection was the antithesis of variation, and he devoted considerable thought to how variation is maintained despite the forces of natural selection. Darwin would have seen the traits that make up an ostrich as having become "fixed" by natural selection over time. He would have also maintained that the ostrich's environment had not changed enough in the past 4,000 years for those "fixed" characters to become disadvantageous.

Today, we might frame that question in a more nuanced way. We know that organisms accumulate mutations across generations. Ren et al. (2024) estimate that humans might have a mutation rate of 60 new single-nucleotide (e.g., A to T) variations per generation. Conservatively, then, we might expect that each newly born ostrich would contain at least one new mutation in a million vulnerable loci. Since most mutations breed true, as the generations pass, we would expect them to accumulate and be assorted into new combinations through sexual reproduction, ultimately leading to phenotypic change—maybe ostriches of unusual size or color. To explain the constancy of ostriches over many millennia, some process must counteract the dispersive force of mutation. Darwin, in fact, knew that selection could stabilize variation. Modern evolutionary ecologists call this concept **stabilizing selection.** Stabilizing selection acts to impede changes in populations, which is why we see many species remain relatively unchanged over generations.

Although Darwin (1859) considered stasis, he provided no convincing evidence why organisms maintain stability in their features. Herman Bumpus (1899) did it instead.[4] Bumpus collected 136 English sparrows, who were rendered helpless by a severe snowstorm. He was able to revive 72 of them. Then he measured size, wing length, and other characteristics and compared the measurements of the birds that died with those that survived. He found the dead birds had the greatest and least overall lengths, the greatest and least wingspreads, the longest and shortest craniums, and the heaviest and lightest weights. To Bumpus' surprise, the deviants suffered a higher mortality, whereas the measurements of the survivors were nearest to the means of the group as a whole. He postulated that this selective event kept the mean population traits constant. Stabilizing selection, which is also

[4] We note that Bumpus's data have been reanalyzed multiple times using updated statistics to remove biases in the data. While we use Bumpus's analysis illustratively here, it appears that, from his data, both larger size and shorter wings were advantageous (Pugesek and Tomer 1996).

referred to as **normalizing** or **optimizing selection**, is expected to be common in nature because phenotypic traits, such as height, weight, number of offspring, and lifespan, typically show greater numbers of individuals with average characteristics and fewer towards the extremes (Figure 4). If a population remains in the same environment for hundreds or thousands of generations, it will acquire the "best" adaptations to that environment available in the gene pool. Stabilizing selection thus favors the phenotypic norm over all deviations from the norm (Schmalhausen 1949; Stebbins 1966).

Ample evidence of phenotypic stability comes from paleontology. In fact, many plant and animal assemblages from terrestrial and marine fossil records are characterized by long periods of stasis, sometimes for millions of years (DiMichele et al. 2004). Estes and Arnold (2007) used historical and paleontological data on phenotypic traits in organisms lasting one to 10 million generations and

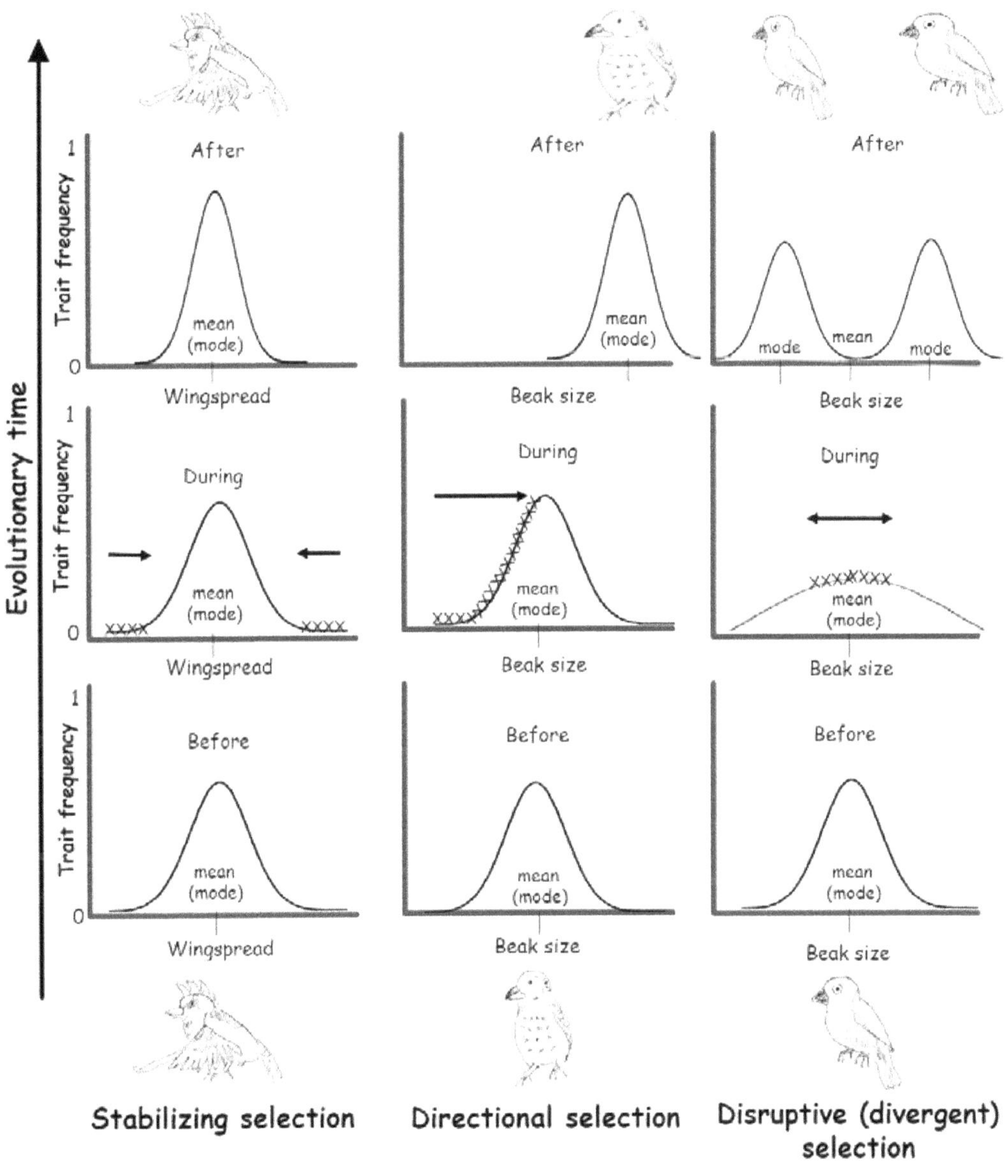

Figure 4. The three basic types of selection (interpreted from Mather 1953). During selection, X phenotypes are selected against. Arrows indicate direction of selection. Note that divergent selection is the only type that produces bimodal phenotypes.

found, after considering different models of evolution, that phenotypic stasis is best explained by adaptation to an optimum—in other words, stabilizing selection. Estes and Arnold (2007) suggest that optimal traits for a given environment remain relatively consistent over time, and, as a result, a species' phenotypic mean will mirror the optimal traits for that environment.

An early 20th century example of stabilizing selection comes from studies of birth weight in humans: very small or very large infants suffered higher mortality (Karn and Penrose 1951). By the 1990s, evidence of stabilizing selection on human birth weight had disappeared in developed countries, but not in developing countries (Ulizzi et al. 1998). This resulted from improved care for premature deliveries and the increasing trend of Cesarian deliveries in developed countries. More evidence of stabilizing selection comes from Lack's (1947) clutch size study in birds. Lack observed that if too many eggs are laid in a clutch, there will be insufficient resources available to parents to fledge all of the resulting nestlings. However, too few eggs laid reduces the number of chicks hatched, and, therefore, reduces the fitness of the parents. Overall total fitness to the parent (measured as young fledged per nest), then, is maximized at a clutch size that is neither too small nor too large.

Directional Selection—The Road to Adaptation

Directional selection, as implied by its name, shifts the mean population phenotype in a new direction by selecting against traits on one side of the trait distribution (Figure 4). Adaptive evolution results from directional selection, and when directional selection occurs, the rate of evolution can be rapid (Otto 2018), especially in species introduced into new regions or subjected to human alterations to their environment. The reason we know directional selection can act rapidly is that we have seen and documented it. Human actions, both intentional and unintentional, are effective causes of directional selection. If we imagine a normal distribution of traits within a population (where the mean trait is the most common trait and the extreme tails are rarer), then rather than favoring the mean (stabilizing selection), directional selection favors traits at one of the extremes. If directional selection continues for many generations, favored traits can become more and more dominated by the extreme type, leading to distinct changes in traits (and allele frequencies) in a population.

The earliest empirical evidence for rapid directional evolution comes from studies of plant and animal responses to anthropogenic effects. Weldon (1899), in his presidential address to the Linnaean Society, reported on changes in the shell shapes of the crab *Carsinus maenas* over a few years in response to the polluting of Plymouth Sound from industrial use of mineral clay, effluent from towns, and shipping activity. The evolution of heavy metal tolerance in the grass *Agrostis tenuis* in just one generation (Walley et al. 1974) revealed how rapid directional selection can be. Another example of human-driven rapid adaptation is the evolution of more petite body and horn sizes in bighorn sheep (*Ovis canadensis*) in Ram Mountain, Alberta, Canada, in response to intense hunting that annually removed 10% of males with large horns from 1971 to 2002 (Coltman et al. 2003). Under this human exploitation environment, smaller rams left more offspring (had higher fitness). Classic examples of unintentional human-caused directional selection are the rapid spread of antibiotic resistance in bacteria and pesticide resistance in insects.

Humans are not the only cause of rapid evolution. Urban et al. (2020) scanned the literature to find 500 case studies of evolution in action over human-observed timescales—everything from guppies evolving less sensitivity to intraspecific competition in dense populations (Bassar et al. 2013) to evolution of lower dispersal in wind-dispersed plants on islands (Cody and Overton 1996), to big bluestem grass evolving a greater interaction with mycorrhizae to extract nutrients under low-nutrient conditions (Schultz et al. 2001).

Another classic example of rapid directional selection, and one we include here because we would be remiss not to, is industrial melanism—the gradual darkening of the wings of many species of moths and butterflies living in woodlands darkened by industrial pollution. Kettlewell (1955a,b) was the first to provide evidence of the mechanism involved in this directional selection: selective bird predation on moths. Before the Industrial Revolution, the light (peppered) form of the peppered moth *Biston betularia* was well camouflaged among the light-colored lichens growing on trees around London. A dark (melanic) form of the peppered moth existed in the population (as a dominant allele), but it was rarely observed because melanic individuals were easily spotted and eaten by birds, their main predator. Following the Industrial Revolution, soot killed the lichens, exposing the dark tree bark below; consequently, the dark form of the moth—which had existed in low frequency among the more abundant lighter-colored morphs—was better camouflaged from birds than the peppered form; selection against the peppered form allowed the dark form to increase in number. One hundred years after the first dark moth was discovered in 1848, 90% of the peppered moths were dark-colored. Dozens of other species of moths have changed in the same way as the peppered moth in industrialized areas throughout Eurasia and North America, with dark forms becoming more common from the mid-19th century onward as industrialization spread. After clean air laws were enacted in Britain in 1956, the frequency of the melanic form dropped from 94% in 1960 to 19% in 1994, at Caldy Common outside Liverpool (Grant et al. 1996).

When conditions make directional selection rapid, changes in the direction of selection are not unusual. As we illustrated earlier (Figure 3), the Grants were able to document the rapid evolution of beak size in the medium ground finch on the Galápagos Island of Daphne Major. Interestingly, this is not the end of the story. Because the Grants studied the finches on this island for so long (Grant and Grant 2008; Grant and Grant 2014), they observed and documented that the pattern of directional selection could change from year to year depending upon the weather. After the drought-caused selective event in 1977, a very wet El Niño year (see Chapter 12 *Climate change*) in 1983 produced lots of small, soft seeds, which the large-billed individuals were not well adapted to handle. Bill characteristics selected for in the dry years were all selected against during the wet year. In the long term, **temporally fluctuating selection** is likely to maintain some genetic variability in traits because different character values are favored at different times and will therefore persist in a population (Ellner and Hairston 1994).

In the short term, fluctuating environments can cause fluctuating directional selection, which in turn leads to greater variability around a mean phenotype. But when the environment shifts "permanently" (as on a geological time scale), the population's mean phenotype shifts toward a new phenotype, which is better adapted to the altered environment. Eventually, an equilibrium is reached, in which the trait stops evolving. At this point, we would expect to observe stabilizing selection in action, where "average" individuals may be the fittest members of the population. In short, directional selection shapes organisms into a particular form to fit an ecological niche, and then stabilizing selection maintains that form. Over geological time, we can expect directional selection to lead to major changes in the morphology and behavior of a population or a species. Thus, from this, we now understand how adaptation can arise and be maintained. However, we have yet to discuss a third type of selection, which holds the key to the origin of species: disruptive or divergent selection (Figure 4).

Disruptive Selection, Speciation, and Diversity of Life

Recall that Darwin (1859), with his theory of evolution by natural selection, aimed to explain two different aspects of the living world: Adaptation (the fit of form and function) and something he famously called "that mystery of mysteries," the origin of species (the diversity of life forms). Darwin had a simple answer to the question of how both occur: by means of natural selection. However, in Chapter 4 of *Origin*, Darwin laid out two conditions he believed most likely to lead to diversity: (1) isolation and the prevention of intercrossing, and (2) divergence of characters brought about by a "diversification of structure" (due to a heterogeneous environment). Although he does not call it disruptive selection, Darwin describes this mechanism in explaining how divergence of characters might come about, and he speculates that disruptive selection, leading to greater and greater divergence, is what generates the procession of life as a population evolves over a long period of time—into varieties, into species, into sub-genera, into genera, into sub-families, into families, and so on. However, it took biologists nearly 150 years to fully agree with Darwin on this latter point, because, as always, the devil lies in the details.

The Fitness Landscape

Before we begin our discussion on disruptive selection, it is important to firmly understand a few concepts. The first is something called an **adaptive** or **fitness landscape** (Wright 1932). In its simplest form, it resembles Figure 4, with a continuously varying trait value on the *x*-axis, but it differs from Figure 4 because "fitness" is on the *y*-axis (Figure 5) instead of "frequency." Fitness landscapes can be multidimensional, but easier to understand when we focus on a single heritable trait, like the size of a beak. For example, imagine a habitat where plants produce seeds of different sizes and a seed-eating bird that uses its beak to open these seeds. This bird is part of a population with a normal distribution of beak sizes—where the mean beak size is the mode—and there is variation around the mean (e.g., the "before" scenario in Figure 4). In this population, birds with larger beaks are better at opening larger seeds, and birds with smaller beaks are better at opening smaller seeds. Now, suppose this population has a relatively small mean beak size, but there is a greater abundance of larger seeds. The birds with larger beaks will have more resources, allowing them to produce more young and, therefore, have greater fitness than those with smaller beaks.

Natural selection, acting on this variance, directionally selects larger beaks. Over time, the population mean moves toward the beak size that has the greatest fitness, which, in this case, lies to the right (the larger size) of the original population mean in Figure 5a. As natural selection continues, the population evolves "up the fitness peak" to a larger beak (Figure 5b). Evolution is driven by the fitness gradient of this peak, with one direction (smaller beak size) associated with lower fitness, and the other direction (larger beak size) associated with higher fitness. Rule 1 of the fitness landscape is that populations always evolve in the direction of higher fitness. When the population finally reaches the peak, it stops evolving because it has reached a maximum on the fitness landscape. The only way off this peak is to suffer a decrease in fitness, but we cannot violate Rule 1. Therefore, stabilizing selection keeps the population from evolving further. The population also cannot continue to grow and therefore reaches an equilibrium density ($dN/dt = 0$; see Chapter 2 *Population Dynamics*).

When populations grow at low densities but stop growing or decline at high densities, ecologists call this **negative density dependence,** which is exemplified by the logistic growth curve (see Chapter 2 *Population Dynamics*). Now, let us add an evolutionary twist. Let us look back at the stabilizing selection example in Figure 4. "After" stabilizing selection, the population reaches an equilibrium density; at this point, some individuals are doing worse fitness-wise than others and are no longer contributing to population growth. When natural selection is involved, these individuals are not a random subset of the population. Under stabilizing selection, individuals with traits at the extremes of the distribution—such as those with very small or very large beaks (or wings in Figure 4)—are selected against. On the trait frequency scale (*y*-axis), these individuals represent the *rare* phenotypes.

Because nature is selective at particular phenotype frequencies, this process is called **frequency-dependent selection**. When natural selection favors the most common phenotypes and selects *against* the rarest (i.e., fitness increases as the phenotype becomes more common, as seen "during" stabilizing selection in Figure 4), evolutionary biologists call this positive frequency-dependent selection. Conversely, when natural selection favors the rarest phenotypes (i.e., fitness is highest when rare) and selects against the most common phenotypes (i.e., fitness decreases as the phenotype becomes more common, as seen "during" disruptive selection in Figure 4), this is called negative frequency-dependent selection. The key difference between density dependence and frequency dependence is that the former concerns the absolute number of individuals in a population, while the latter focuses on the relative frequency of different types in a population. Frequency is expressed

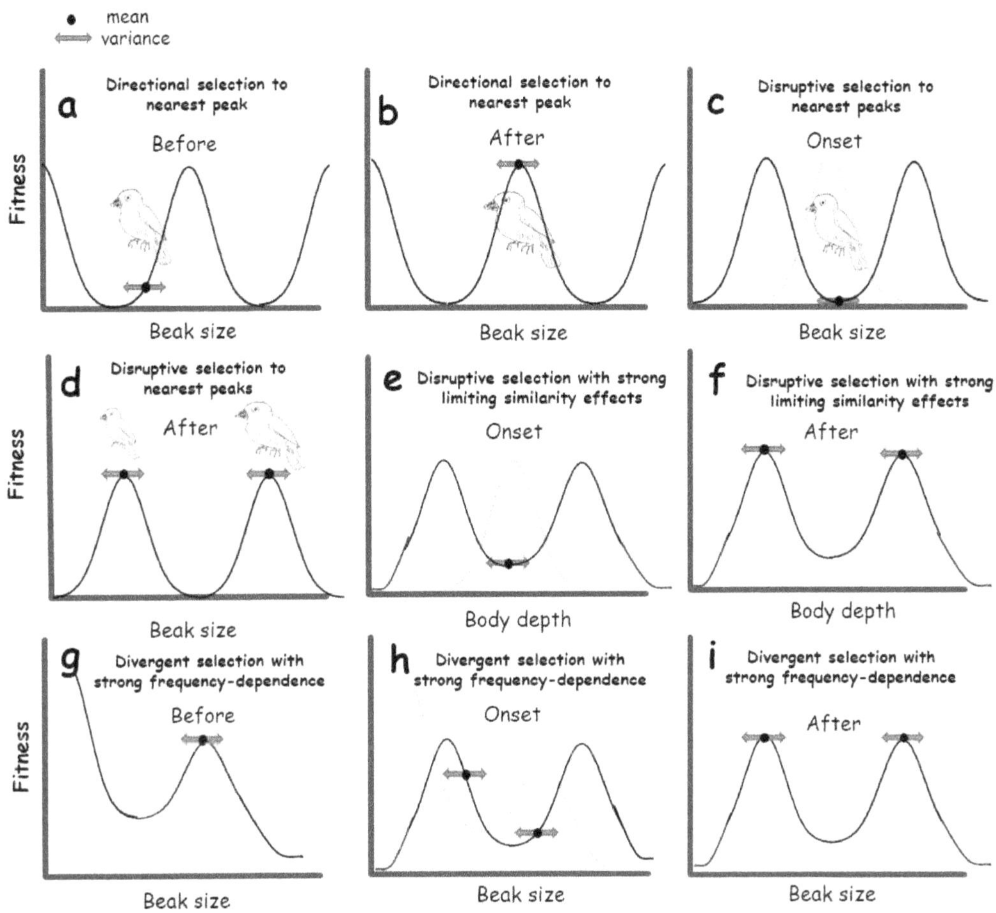

Figure 5. Directional and divergent selection on adaptive landscapes. (a-b) Directional selection in a reproductively isolated ecotype. (c-d) Disruptive selection after an environmental change from a fitness peak (grey lines indicate old environment) to a fitness valley (black lines indicate new environment), where only the tails of a trait distribution can survive. (e-f) Disruptive selection after an environmental change from a fitness peak (grey lines indicate old environment) to a fitness valley (black lines indicate new environment), where competition and limiting similarity select against the intermediate traits (Vincent and Brown 2005). (g-i) The addition of a similar competing species changes the adaptive landscape (grey lines are the fitness landscape before introduction; black lines are the fitness landscape after adding a new, smaller species), causing divergent selection via competitive effects reinforced by assortative mating (Vincent and Brown 2005).

in terms of the **relative abundance** of forms (whether phenotype, genotype, species, etc.) in the system, and it always scales between zero and one. Furthermore, rarity or commonness (low or high frequency) can occur at any population density, although changes in density, if not concurrent between forms, can affect frequency. This distinction is important because selection can act differentially on members of a population depending on whether their traits are common or rare (Doebeli 2011).

Now, returning to the "devil" that prevented early and mid-20th century biologists from agreeing with Darwin on the origin of species, consider this: Once a population evolves to a peak in the fitness landscape, it stays there because any trait that deviates from the mean trait value has lower fitness (Figure 5b); this is stabilizing selection. So, how can a population "split" and cross "valleys" (fitness minima) to reach another peak in the adaptive landscape, creating "diversity"? Wright (1932), who invented adaptive landscapes, did not believe natural selection could do this and invoked increased mutation, inbreeding, and "division into local races" (via the genetics of small populations). However, nature is diverse, and there is evidence that natural selection is responsible for much of that diversity.

Ecotypes

Ecologists in the early twentieth century realized that variation within species reflected adaptations to specific environmental conditions across geographical landscapes. Instead of existing as a single **panmictic** population, many widespread plant and animal species consist of several geographically separate populations. Geographic isolation of groups within a population, where mating is more likely to occur within groups than between groups, is one way natural selection can efficiently operate in different directions on a widespread population at the same time (multi-directional selection). The overall mean population trait is divided into two or more modes due to differences in environmental conditions, such as temperature, rainfall, and soil types, which can lead to widely varying phenotypes and genotypes between localized groups in a widespread population. The more isolated the populations (or subpopulations) are, the faster genetic differences are expected to accumulate as each subpopulation adapts to its locality (recall Darwin's first condition that isolation and prevention of intercrossing help generate diversity).

Turesson (1922) was foundational in demonstrating that locally adapted geographic variation can exist within widespread species. Turesson coined the term **ecotype** (a twist on the term genotype) to distinguish his concept of "hereditary habitat types" from phenotypic plasticity or a chance product of genetic isolation. Ecotypes evolve through an adaptive response to specific environmental conditions (recall Darwin's second condition that diversification of structure helps to generate diversity). Turesson (1925) meticulously documented multiple ecotypes in 16 species of plants gathered from across Europe and cultivated in experimental gardens to test for adaptations. While reproductive isolation via geographical separation and directional selection up the nearest adaptive peak (e.g., Figure 5a-b) might explain ecotypes, Turesson (1925) did not establish that his ecotypes were reproductively isolated. Can ecotypes arise even with substantial gene flow?

A similar meticulous study in animal populations was carried out on the western terrestrial garter snake (*Thamnophis elegans*) in Eagle Lake, California (Gangloff et al. 2015). The snake occupies both lakeshore and mountain meadow habitats and exhibits different life histories for each habitat. The lakeshore populations grow fast, are short-lived, and reproduce annually or biannually, while the meadow populations grow slowly, are long-lived, and reproduce intermittently. In this case, significant gene flow was detected, so Gangloff et al. (2015) tested whether hatchlings taken from the two habitats exhibited differences in resting metabolic rate when grown under common conditions. Their findings revealed a significantly higher metabolic rate in the lakeshore population, indicating that genetic divergence had taken place between the lakeshore and meadow populations. Thus, despite gene flow, these snake populations are ecotypes, shaped by the differing temperature and resource availabilities in each habitat.

When there is gene flow between ecotypes, we must invoke **disruptive selection** as the mechanism of divergence. Disruptive selection occurs when two or more modal phenotypes have higher fitness than the intermediate phenotypes between them (Mather 1955; Figure 4). But if there is gene flow, how can disruptive selection occur? What is the mechanism by which selection, rather than driving the population to one fitness peak or another, winnows the mean phenotype and favors the tails of the distribution? Just because habitat varies does not mean individuals in a generalist population cannot exist there, nor does it mean they need to specialize on that variation in different ways. What, exactly, is driving them to do so?

Disruptive Selection

Let us go back to our adaptive landscapes. Wright (1932) viewed adaptive peaks as fixed, which is easy to do when you don't consider environmental changes or ecological interactions. When you do consider these factors, the adaptive landscape can be flexible (Calsbeek et al. 2012; Christie and McNickle 2023). Adaptive landscapes can flex and shift in a number of ways. The first occurs when the environment changes, either due to external forces (climate, habitat modification, etc.)

or because the population may colonize a new habitat. The second occurs when biotic interactions change in the system, but we will get to this after we discuss the first.

What if the fitness peak in Figure 5b shifted to two fitness peaks, as shown in Figure 5c (with the old fitness peak in grey and the new fitness peaks in black)? We can imagine that, for our seed-eating birds, a midsized seed to which they are adapted suddenly disappears—perhaps due to climate change, habitat destruction, or they colonize an offshore island with differing plants. What remains are plants with very small seeds and plants with very large seeds. Disruptive selection would then favor the extremes of the beak size distribution in the bird population, resulting in a bimodal distribution of this trait (Figure 5d). It is possible, in this scenario, that only the individuals at the extremes of the distribution—those adapted to either the small or the large seeds—would survive. Any "mean" trait birds would have zero fitness because they cannot handle or consume either the small or the large seeds. This would result in a bimodal population in which each form does not compete with the other, and the relative abundance of each form depends solely on density-dependent factors, such as the carrying capacities set by the abundances of small or large seeds. The two phenotypic forms can coexist at their respective environmental carrying capacities because the intermediate phenotypes are eliminated. The large-beaked phenotype cannot eat the small seeds, while the small-beaked phenotypes cannot eat the large seeds (resulting in no competition between the phenotypes). At precisely this point, Darwin's isolating mechanisms become helpful: it becomes adaptive for individuals to become picky about their mates, as producing intermediate-sized offspring would lower their fitness, since none would survive. This may be one reason why mate choice evolves under disruptive selection (Rueffler et al. 2006).

We have made the former scenario (Figure 5c-d) a special case. However, a more likely case is that the mean phenotype can survive perfectly well where it is located in the new adaptive landscape; that is, the new environment puts the population in a fitness valley rather than a fitness peak, but where fitness is not zero. However, when the mean phenotype competes against phenotypes closer to the adaptive peaks (such as the smallest and largest body depths in Figure 5e), the mean phenotype has a lower relative fitness. Selection then acts against the mean and for the extremes, resulting in disruptive selection for smaller and larger body depths (Figure 5f).

Interactions such as those stemming from intraspecific competition, predation, mutualism, and parasitism, can play a significant role in disruptive selection (Rosenzweig 1978; Doebeli and Dieckmann 2000; Martin and Pfennig 2012). For example, when a population contains differing phenotypes that compete, this can affect the density and, consequently, the frequency of each phenotype, changing the **relative abundances** of phenotypes in the overall population (or system of interest). Note also that as a phenotype evolves, its competitive effect (alpha; see Chapter 3 *Species Interaction: Competition*) with respect to other phenotypes in the population can change, depending on how close those phenotypes are on the trait spectrum: the closer they are, the stronger the competition; the farther they are, the less strongly they will compete. Although Darwin (1859) did not perform the mathematical analysis, this was a key point he invoked during his discussion of "Divergence of Character." MacArthur and Levins (1967), building on Darwin's principle of divergence in characters, did the math, which suggested that closely related organisms could inhabit the same environment only if they differed in morphology, habitat use, or other characteristics that reduced their ecological similarity. They called this **limiting similarity.** The fitness gradient caused by limiting similarity selects for divergent traits because it reduces the negative competitive effects on carrying capacity (see Chapter 3 *Species Interaction: Competition*) for both parties. The fitness of the extreme forms is therefore greater than that of an intermediate form (MacArthur and Levins 1967). The corollary: it follows that an intermediate form may be "fit" if and only if the competing forms (those closer to the adaptive peak(s)) are extremely rare in frequency (Rosenzweig 1978).

One final note: for disruptive selection to create the coexisting phenotypes (Figure 5f), the phenotypes must have greater competitive effects on individuals of their own type rather than on individuals of the other types; otherwise, selection will not be disruptive (Christie and McNickle 2023). This creates the negative frequency-dependent component of selection. For example, as a large beak phenotype increases and eats more of the large seeds, it becomes more and more advantageous to have a small beak to take advantage of the relatively more abundant small seeds; the reverse is also true. This means that either form can invade at low frequency when the other is present at high frequency (negative frequency-dependent selection for the low frequency phenotype). However, if the large-beak phenotype can consume both seeds equally well, but the small-beak phenotype can only consume small seeds, then birds with larger beaks will outcompete those with smaller beaks, and disruptive selection will not occur. Instead, selection will favor the high-frequency form.

Examples of intraspecific (inter-phenotype) competition causing disruptive selection include an African finch, the black-bellied seedcracker (*Pyrenestes ostrinus*). This bird is common in the rainforests of Equatorial Africa and displays a non-sex-related polymorphism in bill size (Smith 1993). Although breeding is random with respect to bill size, the polymorphism causes differences in diet and feeding performance relative to seed hardness. Birds with smaller mandible sizes feed on one species of soft seeds, while birds with larger mandibles feed on another species of hard seeds. Differences between the small and large-billed morphs, which vary in bill size but not body size, involve differential regulation of a single gene (vonHoldt et al. 2018). Birds with mandible sizes that are either relatively small or relatively large have been shown to occupy distinct adaptive peaks (Smith 1993), indicating that bill size is under disruptive selection.

Other examples include bimodal gill-raker length in three-spine sticklebacks (Bolnick and Lau 2008) and bimodal body shape in Eurasian perch (Svanbäck and Persson 2009), which allow the different morphs in each case to specialize on distinct diets in different habitats. The case of the Eurasian perch is especially instructive. Svanbäck and Persson (2009) demonstrated that disruptive selection occurred only at high overall population density, indicating that disruptive selection was strongest when the negative effects of intraspecific competition were greatest. This is a clear demonstration of a flexible adaptive landscape as shown in Figure 5e, where the grey mode reflects the adaptive landscape at low overall population density, and the black modes reflect the adaptive landscape at high population density.

In the previous competition examples, negative frequency-dependence—where intra-phenotype competition is stronger than inter-phenotype competition—acts to cause disruptive selection. Yet, due to limits in environmental carrying capacity, the disruptive selection disappears once the traits have evolved toward their respective fitness peaks (i.e., when they reach the adaptive trait values), and stabilizing selection takes over, maintaining the two forms at their adaptive trait values and equilibrium relative abundances (Christie and McNickle 2023).[5] There are, however, cases where disruptive selection never ceases, and the relative abundance of phenotypes fluctuates. This is also confusingly called negative frequency-dependent selection in the evolutionary ecology literature. To avoid confusion, we will call this second type simply "negative frequency selection," omitting the modifying "dependent."

Like frequency-dependent selection, frequency selection occurs when organisms interact (e.g., through competition, predation, mutualism, etc.) in such a way that the fitness of one phenotype is limited by the relative frequency of other phenotypes, but it happens without regard to trait *values*. Instead, selection acts directly on a phenotype when it attains a certain frequency. **Positive frequency selection**, when the fitness of a phenotype increases as it becomes more common with respect to other phenotypes, can cause *directional selection* and is thought to be important in the evolution of signaling and communication. For example, Chouteau et al. (2016) experimentally determined that as the relative frequency of a warningly colored prey increases, the efficacy of the warning increases, up to a threshold where protection stabilizes. Thus, positive frequency selection helps explain the evolution of warning coloration complexes in chemically defended butterflies in the Amazon. However, positive frequency selection is not a source of disruptive selection. **Negative**

[5] Side note: Equilibrium relative abundances do not have to be 50:50 as depicted "after" disruptive selection in Figure 4; they can also result in one being rare and one being common "after," but at that point, it is stabilizing selection and equilibrium population dynamics that keeps them at those relative abundances at the two modes.

frequency selection, where the fitness of a phenotype decreases as it becomes more common with respect to other forms, can cause disruptive selection by allowing rare phenotypes to increase in frequency, while preventing abundant phenotypes from excluding others (Christie and McNickle 2023).

Examples of negative frequency selection include female mate choice where rare males have the advantage (Hughes et al. 2013), host-parasite systems where abundant genotypes suffer higher infection loads (Koskella and Lively 2009), and predators forming search images for cryptic prey (Allen 1988). Colonies of bluegill sunfish (*Lepomis macrochirus*) often maintain two male reproductive strategies: parental (the purported father) and cuckolder (who secretly fertilizes eggs). When cuckolders are rare, they can be very successful and achieve high fitness, but as this strategy becomes more common in the population, their mating success declines, and the parental phenotype regains success. This reflects negative frequency dependence between the competing strategies and results in the persistence of the two strategies in the population (Gross 1991).

We have seen how disruptive selection can occur from either a minimum in the adaptive landscape or from negative frequency selection. Disruptive selection may be at least as common as stabilizing selection (Kingsolver and Diamond 2011). The difference between stabilizing selection and disruptive selection is illustrated in Figure 4: stabilizing selection reduces population variability, whereas disruptive selection maintains or even enhances variation within populations. Disruptive selection favors the evolution of alternative phenotypes (Thoday 1958), sexual dimorphism (Bolnick and Doebeli 2003), and, as we expect, speciation (Udovic 1980). Darwin (1859) was the first to realize that disruptive selection could cause the variation needed for the successively greater divergence seen in the tree of life. He firmly believed that divergence in characters is what allows a greater diversity of species to coexist. But Darwin also believed that divergent selection is behind the origin of species. This idea was not widely embraced by early and mid-20th century evolutionary ecologists, as there was another "devilish detail" to be worked out—gene flow. Although disruptive selection may be quite common and can explain ecotypes and morphological polymorphisms despite gene flow, the question remained: how can a polymorphic population ultimately evolve into separate species? Once again, Darwin saw the solution, even though he did not know the underlying mechanism: for a diversity of life forms to evolve, there must be isolation and/or the prevention of intercrossing between morphs. Perhaps he would not be surprised to find that isolating mechanisms may evolve because they are adaptive.

Flexible Landscapes and Isolating Mechanisms

There is increasing evidence that, given enough time, reproductive isolation can evolve *as a consequence* of adaptation to different environments. Hendry (2009) studied two distinct ecotypes of the threespine stickleback (*Gasterosteus aculeatus*) occurring in a few small lakes in southern British Columbia, Canada. One ecotype is adapted to a limnetic feeding mode and has a shallow body and numerous long gill rakers, and the other is a benthic feeder with a deep body and a small number of short gill rakers. The two types exchange very few genes and can be considered effectively separate species that have evolved since the last glaciation. Divergence onto their limnetic and benthic adaptive peaks has led to reproductive isolation in three ways. First, because each habitat creates its own fitness peak at a particular trait value, there is ecological selection against "migrants" from the other habitat. This is because individuals adapted to one environment show higher survival and growth in that environment than individuals adapted to the other environment. Second, competitive exclusion of "hybrids" occurs because hybrids have lower growth and survival relative to the locally adapted ecotypes within each environment. Third, "assortative mating" is present because individuals of each ecotype prefer to mate with their own type even across the different lakes (recall how we found mate selection to be adaptive in the face of disruptive selection). Thus, speciation can follow from adaptive divergence between populations.

Nonrandom patterns in mate choice may play a large role in reinforcing adaptive divergence of Galapagos Islands ground finches. Based on their extensive research, Grant and Grant (2008) found that changes in finch beak shape also affect the production of song, so diverging beak sizes can lead to diverging song quality. Songs also tend to drift due to "copy errors." As the ancestral population of ground finches colonized various islands, they would have formed ecotypes in response to the differing habitats, and at some point, as the beaks diverged, so would the songs. Ground finches prefer mates based on similarity in beak shape and song, so this premating barrier to interbreeding between the ecotypes could result in a lack of gene flow if the two ecotypes ended up on the same island. This scenario could further enhance disruptive selection.

Consider the medium ground finch *(Geospiza fortis)* and the small ground finch (*G. fuliginosa*), which feed on an overlapping array of seed sizes. The two species, when occurring alone on different islands, show overlap in their beak size distributions and, consequently, their seed size preferences. However, when they live on the same island (e.g., Santa Cruz), they show no overlap in beak size and little overlap in resource use, exhibiting a divergence in characters. Figure 5g-i illustrates how this might have happened. Suppose the medium ground finch was the first to arrive on Santa Cruz. Its moderate beak size allows it to take many of the seed types on the island. By itself, the medium ground finch population may become adapted to the most abundant seed in that moderate size range, and the population reaches an eco-evolutionary equilibrium at a particular trait value with a specific population size at carrying capacity (Figure 5g). But suppose some very small seeds are left unutilized. This creates a negative frequency-dependent scenario, where a species that can utilize some of the smaller seeds would be at a fitness advantage and could invade the island (left-hand uptick in the adaptive landscape in Figure 5g).

But remember how the adaptive landscape is flexible? Once the small ground finch invades, the adaptive landscape changes instantly. Competition with the small ground finch shifts the medium ground finch's adaptive peak to the left, suddenly placing it in an adaptive valley, while an adaptive peak appears for the small ground finch because the population will eventually reach carrying capacity (discounted by competitive effects). Now, each species is forced to climb the nearest adaptive peaks, causing both species' mean beak sizes to shift apart. This same process could happen with two different ecotypes of the same species. When disruptive selection combined with assortative mating causes a divergence of characters, disruptive selection is also known as **divergent selection**, and the resulting scenario is called **character displacement** (Brown and Wilson 1956). Because disruptive selection, coupled with barriers to interbreeding, can create distinct species, this process is called **ecological speciation** (Schluter 2001, 2009).

Up to this point, we have focused on competitive mechanisms that drive a single population to split into two species. But it is important to recognize that frequency-dependent selection on species-specific traits can occur whenever species interact—with predators, competitors, mutualists, parasites, etc. This, in turn, can drive coevolution and increase the diversity of entire communities (McPeek 2017).

Procession of Life and Macroevolution

Darwin (1859) concluded that something like divergent selection could result in ecological speciation, reasoning that the diversity and procession of life follows from this principle. As we have seen, divergent selection can drive the evolution of different characters (or the fixation of different alleles), and once reproductive isolation occurs—considered an adaptation to divergent selection—speciation has taken its course (Schluter 2009).

The process of ecological speciation lends itself well to explaining adaptive radiations, such as the evolution of 13 species of Darwin's finches within the Galapagos archipelago, the hundreds of species of cichlid fish in the Great Lakes of Africa, and the dozens of honeycreepers (forest birds) endemic to the Hawaiian Islands. **Adaptive radiation**, defined by Schluter (2000) as the evolution of ecological and phenotypic diversity within a rapidly multiplying lineage, illustrates how adaptive tweaks in the basic body plan, reinforced by genetic isolation or assortative mating, can arise based on the theories and experiments presented above. Furthermore, it suggests the commonality and connectedness of design across life forms. But what we have discussed so far—the evolution of adaptation and the maintenance of divergent forms—falls under the auspices of **microevolution.** We have not yet addressed the mysteries of the origin of taxonomic hierarchy above the species level or the development of complex organs, such as the vertebrate eye or a bat's wing, which we cannot experimentally test or witness because they evolve over vast spans of time. These phenomena are referred to as **macroevolution**, a term coined by Philiptschenko (1927) simultaneously with microevolution.

Taken at face value, Darwin's theory of the gradual descent of species, with new species slowly emerging from older ones, would seem to predict smooth, gradual transitions by means of small-scale adaptive changes in populations (microevolution). As these changes accumulate, higher taxonomic groupings form, with no missing links. But macroevolutionary puzzles pose a problem to this idea. We do not often see the expected transitions. For instance, the closest living relative to the whales and dolphins is the strikingly different hippopotamus (Gingerich 2012). The fossil record is rife with gaps or discontinuities, especially among high-level taxa such as orders and classes. Soft organs such as eyes or delicate bones such as bat wings, which appear fully formed in present-day organisms, leave no trace as to how they evolved.

Many evolutionary biologists (e.g., Dobzhansky 1937; Mayr 1942; Simpson 1953; Rensch 1959; Stebbins 1966; Charlesworth et al. 1982; Reznick and Ricklefs 2009; Hancock et al. 2021) treat macroevolution as reducible to microevolution—a view closely associated with the Modern Synthesis. Is macroevolution simply microevolution writ large? Not everyone agrees. To address these macroevolutionary puzzles, Gould and Eldredge (1977) proposed the hypothesis of punctuated equilibrium—long periods of little change (**stasis**) interrupted by brief episodes of much more rapid change—to explain the origin of new body plans (**bauplans**) in the fossil record. Just how does one get from a fish to a tetrapod? Or a small terrestrial mammal to a bat?

The proponents of punctuated equilibrium reason that the fossil record is more complete than it seems, taking the rapid appearance of new species at face value. As Gould and Eldredge (1993) write, "…to validate our profession's primary data as signal rather than void." Natural selection (microevolution), they reasoned, could fine-tune organisms during periods of stasis, but another mechanism had to account for punctuated change. They applied Mayr's (1954) model of founder-effect speciation (**peripatric speciation**) to macroevolutionary change, calling their synthesis "**punctuated equilibrium**." In their view, stasis occurs because it is difficult for a large population to overcome the stabilizing selection of an adaptive peak. To evolve to a new form, the population would have to cross a valley in the adaptive landscape. Small populations, on the other hand, find it easier to cross this valley because of genetic drift. Small populations are also most likely to be at the edge of a large population's distribution, subject to geographic separation, or to be a founding population. Under these conditions, rapid genetic change is now possible via natural selection toward a new adaptive peak, and they speculated that this is how speciation occurs.

Gould's view of punctuated equilibrium treats populations as evolutionary individuals, wandering the peaks and valleys of fitness landscapes. One of the main arguments for punctuated equilibrium has been that the fossil record often reveals species sorting, meaning that some lineages rapidly diversify into new species, whereas others decline. To Gould (2002), this implied a higher level of selection—species selection. The idea of species selection treats "species" as the unit of selection in the same way that natural selection treats "individuals" as the unit of selection, to explain both equilibrium and stasis (Gould 2002).

Darwin's notes from 1837 to 1838 (over 20 years before the publication of *Origin*) contain references to stasis as well as rapid evolutionary change (Rhodes 1987), so he was clearly aware that the tempo of evolution could vary. Nonetheless, his "one long argument" repeatedly emphasized that natural selection worked incrementally over long periods of time. We may speculate that he was trying to make a strong case against "catastrophism," a familiar view in geology at the time (see discussion of Cuvier in Chapter 1 *Relationships Between Organisms and Their Environment*), which was used to explain both stasis and the rapid appearance of new species in the fossil record. This would also serve to support his theory of natural selection within the broader framework of geological "gradualism" (see Lyell in Chapter 1). However, he was certainly aware that directed artificial selection by "man" could bring about new breeds in the span of just a few generations, a mere micrometer of geological time.

Although punctuated equilibrium inspired new research on the relative importance of stasis and rapid evolution vs. gradual directional change (e.g., Uyeda et al 2011), one enduring criticism of punctuated equilibrium and species sorting is that the theory gives no mechanism for maintaining stasis over long periods of time in the face of varying environmental, population size, and genetic conditions. That lack of explanation prompted many evolutionary biologists (e.g., Estes and Arnold 2007) to attribute stasis at larger scales to such things as habitat tracking, which is the shifting of species' geographic distributions to match their typical habitat (recall Figure 5 in Chapter 12 *Climate Change*). Others have pointed out the mismatch between fossil evidence (where stasis is common) and studies of extant populations (or species) showing rapid directional evolutionary change (Endler 1986, Otto 2018, Reznick et al. 2018). Certainly, since punctuated evolution was first proposed, many new fossils have filled in mysterious gaps—such as the transitional forms of whales, from the four-legged aquatic *Pakicetus*, to the intermediate, whale-like *Basilosaurus* with leg, ankle, and toe bones, to *Dorudon* with front fins and vestigial hind limbs (Gingerich 2012). Still, there are some who think there is paleontological evidence that macroevolution does not necessarily come from "repeated rounds of microevolution" (Grantham 2007).

Reznick and Ricklefs (2009) remind us that it is often forgotten that Darwin also proposed a mechanism for macroevolution, and ironically, it shares some of the same elements of punctuated equilibria—specifically the idea of species as units of selection. Darwin's link between micro- and macroevolution is depicted in the only figure in *Origin*. It shows both microevolution as divergence in character (like Darwin's tiny sprays of dotted lines emerging from a nodal species) and macroevolution as the extinction of certain lines across the course of time (Figure 6). This art depicts the three critical components of Darwin's theory of the procession of life. The first is that, after diversification, some species go extinct. This we know, and in fact, good examples can be found in the fossil record for horses and humans; it is striking how many branches arose but were subsequently snuffed out, leaving only a few phylogenetic lines remaining. Darwin's second component is that extinctions are nonrandom. Specifically, he proposed that there is species-level selection that favors species that diverge further from the ecological requirements of other species. And finally, extinctions make room for further diversification. These last two points are controversial and harder to prove, although cases of diversification and extinction during faunal exchange in the fossil record are suggestive and supported by phylogenetic studies. Pires et al. (2015) describe the strongly asymmetrical pattern of North American and European carnivore exchanges. Unlike the constrained evolution following the appearance of European carnivores in North America, North American lineages invading Eurasia underwent explosive radiation (like Darwin's tiny sprays of branches), but soon after, many of the diversified forms went extinct, decelerating the rate of radiation over time. Pires et al. (2015) suggest there is evidence that in many cases, the taxa that replaced those from the declining lineages were morphologically similar to those that went extinct, providing evidence of the role of ecology in shaping radiations, just as Darwin predicted.

Vincent and Brown (2005) provide a mechanistic explanation of Darwin's macroevolutionary process. Using evolutionary game theory and treating species as strategies, they show how adaptive radiation can arise from a single parameterized model, which they call a "G-function." This model illustrates how the burgeoning of species arises out of different strategies that coexist along a trait spectrum (much like the varying beak depths of Galapagos finches or varying beak curvatures of Hawaiian honeycreepers). The G-function captures the nodes and sprays depicted in Darwin's figure within *Origin* (Figure 6). Vincent and Brown (2005) envision that different nodes and sprays represent different G-functions (for example, sticklebacks have a different G-function from finches

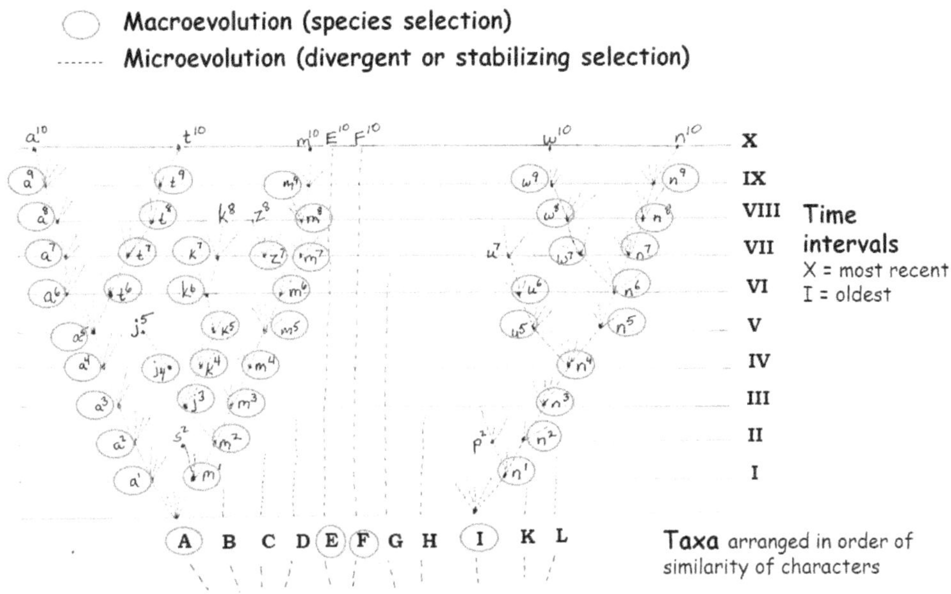

[DIAGRAM OF DIVERGENCE OF TAXA]

Figure 6. Interpretation of a portion of Darwin's (1859) original drawing with added explanation. Originating taxa A–L are arranged in order of resemblance (e.g., A–D are more similar to each other than I–L). Microevolution (represented by dotted lines) causes character displacement, creates variation, and causes taxa to diverge in character (descendants of A, I). At particular intervals in time (indicated by roman numerals), only certain species survive (species selection; surviving species are circled). Of the originating taxa, A and I persist long enough to produce descendant species a^1, m^1, and n^1. B–D, G, H, K, and L eventually go extinct, with only E and F remaining extant through time interval X. Note that these two species exhibit stasis in character (stabilizing selection).

because evolution is working on body depth in the former and beak size in the latter). They also envision that once a G-function is established, it may reduce or increase the possibility of other G-functions. For instance, the appearance of the "grass G-function" (originating from the "forest floor herb" G-function due to climate changes) could pave the way for a new "grazer G-function". Furthermore, they expect new G-functions to arise from old G-functions, as when at least one member of the horse lineage took advantage of the new "grazer G-function" (see figure in MacFadden 2005). These dynamics represent Darwin's connecting lineages between nodes (Figure 6). In the context of evolutionary game theory, Vincent and Brown (2005) hypothesize that the procession of life is driven by the addition and replacement of G-functions over evolutionary time.

Meanwhile, amid evolutionary biologists' quibbles over how evolution proceeds, a notable event occurred in the Galapagos Islands on Daphne Major (home to the medium ground finch, *G. fortis*, and the cactus finch, *G. scandens*). In the early 1980s, a large male *fortis-scandens* hybrid from nearby Santa Cruz Island arrived and mated with *G. fortis* females, producing descendants larger than the resident finches and with distinct vocalizations. After the drought in 2004, the offspring of the survivors produced offspring only with one another, revealing a new reproductively isolated lineage. This was perhaps due to the immigrant male's descendants singing his unique, mixed song, which likely affected female mate choice. This event prompted Grant and Grant (2014) to announce, "a species is in the making," revealing how natural selection leads populations down the path to speciation. Other striking examples of incipient speciation include the European blackcap (Rolshausen et al. 2009), sticklebacks (Lescak et al. 2015), and the mosquitos of the London Underground, *Culex pipiens* (Byrne and Nichols 1999). Clearly, ecologically driven, diversifying evolution is occurring as we speak.

The face of evolutionary biology has changed significantly since the mechanism of inheritance was first discovered in the 1950s. We have sequenced whole genomes, constructed detailed phylogenies of the procession of life, and even extracted precious DNA from extinct organisms—learning how the past is tied to the present. However, until recently, "ecology" had been neglected in a dark corner of the otherwise shining compendium of evolutionary research. Now, however, the flurry of new "evo-eco" discoveries and research resonates with Darwin's much-criticized "principle of divergence" and suggests that the emerging views of today's evolutionary biologists are not so different from those espoused by Darwin over 150 years ago (Schemske 2010).

Reznick and Ricklefs (2009) said it best: "Many people see On the Origin of Species as a beautiful fossil, but we view it as a living document that continues to offer insights and enlighten modern research."

References

Allen, E. J. 1929. The origin of adaptations. Nature 123: 841–843.

Allen, J. A. 1988. Frequency-dependent selection by predators. Phil. Trans. R. Soc. B 319: 485–503.

Babbitt, G. A. 2009. Selection for sexual bill dimorphism in ibises: an evaluation of hypotheses. Waterbirds 30: 199–206.

Baguette, M., Bertrand, J. A., Stevens, V. M. and Schatz, B. 2020. Why are there so many bee-orchid species? Adaptive radiation by intra-specific competition for mnesic pollinators. Biological Reviews 95: 1630–1663.

Barlow, N. (ed.). 1958. The autobiography of Charles Darwin 1809-1882. With the original omissions restored. Edited and with appendix and notes by his grand-daughter Nora Barlow. Collins, London, UK.

Barrett, R. D. and Schluter, D. 2008. Adaptation from standing genetic variation. Trends in Ecology & Evolution 23: 38–44.

Bassar, R. D., Lopez-Sepulcre, A., Reznick, D. N. and Travis, J. 2013. Experimental evidence for density-dependent regulation and selection on Trinidadian guppy life histories. The American Naturalist 181: 25–38.

Benson, R. B. J., Butler, R. J., Carrano, M. T. and O'Connor, P. M. 2012. Air-filled postcranial bones in theropod dinosaurs:physiological implications and the 'reptile'-bird transition. Biol. Rev. Camb. Philos. Soc. 87: 168–93.

Boag, P. T. and Grant, P. R. 1981. Intense natural selection in a population of Darwin's finches (Geospizinae) in the Galapagos. Science 214: 82-85.

Bock, W. J. 1980. The definition and recognition of biological adaptation. American Zoologist 20: 217–227.

Bojarska, K. 2012. Spatial patterns in brown bear *Ursus arctos* diet: the role of geographical and environmental factors. Mammal Rev. 42: 120–143.

Bolnick, D. I. and Doebeli, M. 2003. Sexual dimorphism and adaptive speciation: two sides of the same ecological coin. Evolution 57: 2433–2449.

Bolnick, D. I. and Lau, O. L. 2008. Predictable patterns of disruptive selection in stickleback in postglacial lakes. The American Naturalist 172: 1–11.

Bostwick, K. S., Riccio, M. L. and Humphries, J. M. 2012. Massive, solidified bone in the wing of a volant courting bird. Biological Letters 8: 760–763.

Brodie III, E. D. and Brodie Jr., E. D. 1999. Costs of exploiting poisonous prey: evolutionary trade-offs in a predator-prey arms race. Evolution 53: 626–631.

Brodin, P., Jojic, V., Gao, T., Bhattacharya, S., Lopez angel, C. J., Furman, D. et al. 2015. Variation in the human immune system is largely driven by non-heritable influences. Cell 160: 37–47.

Brown, J. H. 1984. On the relationship between abundance and distribution of species. The American Naturalist 124: 255–279.

Brown, W. L. and Wilson, E. O. 1956. Character displacement. Syst. Zool. 5: 49–65.

Brunner, F. S., Anaya-Rojas, J. M., Matthews, B. and Eizaquirre, C. 2017. Experimental evidence that parasites drive eco-evolutionary feedbacks. PNAS 114: 3678–3683.

Bumpus, H. C. 1899. The elimination of the unfit as illustrated the introduced sparrow, *Passer domesticus*. Biol. Lectures, Marine Biol. Lab., Woods Hole, pp. 209–226.

Burt, A. and Trivers, R. 2006. Genes in Conflict: The Biology of Selfish Genetic Elements. Harvard University Press.

Byrne, K. and Nichols, R. A. 1999. *Culex pipiens* in London Underground tunnels: differentiation between surface and subterranean populations. Heredity 82: 7–15.

Cairns, J., Overbaugh, J. and Miller, S. 1988. The origin of mutants. Nature 335: 142–145.

Calsbeek, R., Gosden, T. P., Kuchta, S. R. and Svensson, E. I. 2012. Fluctuating selection and dynamic adaptive landscapes. pp. 89–109. *In*: Svensson, E. and Calsbeek, R. [eds.]. The adaptive landscape in evolutionary biology. Oxford University Press.

Charlesworth, B., Lande, R. and Slatkin, M. 1982. A Neo-Darwinian commentary on macroevolution. Evolution 36: 474–498.

Chouteau, M., Arias, M. and Joron, M. 2016. Warning signals are under positive frequency-dependent selection in nature. PNAS 113: 2164–2169.
Christie, M. R. and McNickle, G. G. 2023. Negative frequency dependent selection unites ecology and evolution. Ecology and Evolution 13: e10327.
Cody, M. L. and Overton, J. M. 1996. Short-term evolution of reduced dispersal in island plant populations. The Journal of Ecology 84: 53–61.
Coltman, D. W., O'Donoghue, P., Jorgenson, J. T., Hogg, J. T., Strobeck, C. and Festa-Bianchet, M. 2003. Undesirable evolutionary consequences of trophy hunting. Nature 426: 655–658.
Crean, A. J. and Bonduriansky, R. 2014. What is a paternal effect? Trends in Ecology & Evolution 29: 554–559.
Crick, F. H. 1958. On protein synthesis. Symp. Soc. Exp. Biol. 12: 138–163.
Darwin, C. 1859. On The Origin of Species by Means of Natural Selection, or Preservation of Favoured Races in the Struggle for Life. John Murray, London, UK.
Darwin, C. 1871. The Decent of Man, and Selection in Relation to Sex. 2nd ed. John Murray, London, UK.
Dawkins, R. 1976. The Selfish Gene. Oxford University Press, London.
de Vries H. 1901. Die mutationstheorie. Vol. I and II. Leipzig (Germany): Verlag von Veit and Company; 1901–1903.
DiMichele, W. A., Behrensmeyer, A. K., Olszewski, T. D., Labandeira, C. C., Pandolfi, J. M., Wing, S. L. et al. 2004. Long-term stasis in ecological assemblages: evidence from the fossil record. Annu. Rev. Ecol. Evol. Syst. 35: 285–322.
Dobrynin, P., Liu, S., Tamazian, G., Xiong, Z., Yurchenko, A. A., Krasheninnikova, K. et al. 2015. Genomic legacy of the African cheetah, *Acinonyx jubatus*. Genome Biology 16: 1–20.
Dobzhansky, T. G. 1937. Genetics and the Origin of Species. Columbia University Press, New York (3rd Ed., 1951).
Doebeli, M. and Dieckmann, U. 2000. Evolutionary branching and sympatric speciation caused by different types of ecological interactions. The American Naturalist 156(Suppl): 77–101.
Doebeli, M. 2011. Adaptive Diversification. Princeton University Press.
Dou, Y., Gold, H. D., Luquette, L. J. and Park, P. J. 2018. Detecting somatic mutations in normal cells. Trends Genet. 34: 545–557.
Dudley, J. W. and Lambert, R. J. 2004. 100 generations of selection for oil and protein in corn. Plant Breeding Reviews 24 (part 1): 79–110.
Dumont, E. R. 2010. Bone density and the lightweight skeletons of birds. Proceedings of the Royal Society B 277: 2193–2198.
Ellner, S. and N. G. Hairston. 1994. Role of overlapping generations in maintaining genetic variation in a fluctuating environment. The American Naturalist 143: 403–417.
Elton, C. 1930. Animal Ecology and Evolution. Oxford: The Claredon Press, London, UK.
Emami-Khoyi, A., Paterson, A. M., Hartley, D. A., Boren, L. J., Cruickshank, R. H., Ross, J. G. et al. 2017. Mitogenomics data reveal effective population size, historical bottlenecks, and the effects of hunting on New Zealand fur seals (*Arctocephalus forsteri*). Mitochondrial DNA Part A. doi: 10.1080/24701394.2017.1325478
Endler, J. A. 1986. Natural Selection in the Wild. Princeton University Press, Princeton, NJ.
Estes, S. and Arnold, S. J. 2007. Resolving the paradox of stasis: Models with stabilizing selection explain evolutionary divergence on all timescales. The American Naturalist 169: 227–244.
Farmer, C. G. 2006. On the origin of avian air sacs. Respiratory Physiology & Neurobiology 154: 89–106.
Fisher, R. A. 1930. The Genetical Theory of Natural Selection, Clarendon Press, Oxford.
Frost, P. 2006. European hair and eye color: a case of frequency-dependent sexual selection? Evolution and Human Behavior 27: 85–103.
Gangloff, E. J., Vleck, D. and Bronikowski, A. M. 2015. Developmental and immediate thermal environments shape energetic trade-offs, growth efficiency, and metabolic rate in divergent life-history ecotypes of the garter snake *Thamnophis elegans*. Physiological and Biochemical Zoology 88: 550–563.
Gingerich, P. D. 2012. Evolution of whales from land to sea. Proceedings of the American Philosophical Society 156: 309-323.
Gonçalves, P., Valério, E., Correia, C., de Almeida, J. M. and Sampaio, J. P. 2011. Evidence for divergent evolution of growth temperature preference in sympatric *Saccharomyces* species. PloS one 6: e20739.
Gould, S. J. 2002. The Structure of Evolutionary Theory. Harvard University Press. Cambridge, USA.
Gould, S. J. and Eldredge, N. 1977. Punctuated equilibria: the tempo and mode of evolution reconsidered. Paleobiology 3: 115–151.
Gould, S. J. and Eldredge, N. 1993. Punctuated equilibrium comes of age. Nature 366: 223–227.
Grant, B. R. and Grant, P. R. 1993. Evolution of Darwin's finches caused by a rare climatic event. Proc. R. Soc. B 251: 111–117.
Grant, B. S., Owen, D. F. and Clarke, C. A. 1996. Parallel rise and fall of melanic peppered moths in America and Britain. Journal of Heredity 87: 351–357.

Grant, P. R. and Grant, B. R. 2008. How and Why Species Multiply. Princeton University Press, Princeton, NJ.
Grant, P. R. and Grant, B. R. 2014. 40 Years of Evolution: Darwin's Finches on Daphne Major Island. Princeton University Press, Princeton, NJ.
Grantham, T. 2007. Is macroevolution more than successive rounds of microevolution? Palaeontology 50: 75–85.
Gross, M. R. 1991. Evolution of alternative reproductive strategies: frequency-dependent sexual selection in male bluegill sunfish. Phil. Trans. R. Soc. B 332: 59–66.
Haldane, J. B. S. 1932. The Causes of Evolution, Harper and Row, London, UK.
Hancock, Z. B., Lehmberg, E. S. and Bradburd, G. S. 2021. Neo-darwinism still haunts evolutionary theory: A modern perspective on Charlesworth, Lande, and Slatkin (1982). Evolution 75: 1244–1255.
Hardin, G. 1966. Biology. 2nd edition. W. H. Freeman and Company, San Francisco and London.
Hendry, A. P. 2009. Ecological speciation! Or the lack thereof? Can. J. Fish. Aquat. Sci. 66: 1383–1398.
Hill, W. G. 1982. Rates of change in quantitative traits from fixation of new mutations. PNAS 79: 142–145.
Hughes, K. A., Houde, A. E., Price, A. C. and Rodd, F. H. 2013. Mating advantage for rare males in wild guppy populations. Nature 503: 108–110.
Huneman, P. 2019. How the Modern Synthesis came to ecology. Journal of the History of Biology 52: 635-686. https://doi.org/10.1007/s10739-019-09570-9.
Janzen, F. J. and Phillips, P. C. 2006. Exploring the evolution of environmental sex determination, especially in reptiles. Journal of Evolutionary Biology 19: 1775–1784.
Karn, M. N. and Penrose, L. S. 1951. Birth weight and gestation time in relation to maternal age, parity, and infant survival. Annals of Eugenics 16: 147–164.
Kato, T., Ishida, K. and Sato, H. 2008. The evolution of nettle resistance to heavy deer browsing. Ecological Research 23: 339–345.
Kettlewell, H. B. D. 1955a. Recognition of appropriate backgrounds by the pale and black phases of Lepidoptera. Nature 175: 943–944.
Kettlewell, H. B. D. 1955b. Selection experiments on industrial melanism in the Lepidoptera. Heredity 9: 323–342.
Kingsolver, J. G. and Diamond, S. E. 2011. Phenotypic selection in natural populations: what limits directional selection? The American Naturalist 177: 346–357.
Koskella, B. and Lively, C. M. 2009. Evidence for negative frequency-dependent selection during experimental coevolution of a freshwater snail and a sterilizing trematode. Evolution 63: 2213–2221.
Lack D. 1947. Darwin's Finches. Cambridge University Press, Cambridge, UK.
Lemanis, R. and Zlotnikov, I. 2023. Fractal-like geometry as an evolutionary response to predation? Science Advances 9: eadh0480.
Lescak, E. A., Bassham, S. L., Catchen, J., Gelmond, O., Sherbick, M. L., von Hippel, F. A. et al. 2015. Evolution of stickleback in 50 years on earthquake-uplifted islands. Proc. Natl Acad. Sci. USA 112: E7204-12. doi: 10.1073/PNAS1512020112.
Lewontin, R. C. 1970. The units of selection. Annual review of ecology and systematics 1: 1–18.
Lewontin, R. C. 1974. The Genetic Basis of Evolutionary Change, Columbia University Press, New York.
Lewontin, R. C. 2010. Not so natural selection. The New York Review, What Darwin Got Wrong (Jerry Fodor and Massimo Piattelli-Palmarini). Farrar, Straus and Giroux, 264 pp.
MacArthur, R. and Levins, R. 1967. The limiting similarity, convergence, and divergence of coexisting species. The American Naturalist 101: 377–385.
MacArthur, R. H. 1972. Geographical Ecology. Princeton University Press, Princeton, NJ.
MacFadden, B. J. 2005. Fossil horses—evidence for evolution. Science 307: 1728–1730.
Madsen, T., Shine, R., Olsson, M., and Wittzell, H. 1999. Restoration of an inbred adder population. Nature 402: 34–35.
Madsen, T., Ujvari, B., and Olsson, M., 2004. Novel genes continue to enhance population growth in adders (*Vipera berus*). Biological Conservation 120: 145–147.
Malthus, T. 1798. An Essay on the Principle of Population. J. Johnson, London, UK.
Martin, R. A. and Pfennig, D. W. 2012. Widespread disruptive selection in the wild is associated with intense resource competition. Evolutionary Biology 12:136. doi: 10.1186/1471-2148-12-136.
Mather, K. 1953. The genetical structure of populations. Symp. Soc. Exp. Biol. 7: 66–95.
Mather, K. 1955. Polymorphism as an outcome of disruptive selection. Evolution 9: 52–61.
Mayr, E. 1942. Systematics and the Origin of Species. Columbia University Press, New York.
Mayr, E. 1954. Change of genetic environment and evolution. pp. 157–180. *In*: Huxley, J., Hardy, A. C. and Ford, E. B. [eds.]. Evolution as a Process. Allen & Unwin, London, UK.
Mayr, E. and Provine, W. B. 1980. The Evolutionary Syntheses: Perspectives on the Unification of Biology. Harvard University Press, Cambridge, MA.
McPeek, M. A. 2017. Evolutionary Community Ecology. Princeton University Press.

Mendel, G. J. 1866. Experiments in plant hybridization. Proceedings of the Natural History Society in Brunn iv: 3–47.
Mousseau, T. A. and Fox, C. W. 1998. The adaptive significance of maternal effects. Trends in Ecology & Evolution 13: 403–7. doi: 10.1016/s0169-5347(98)01472-4.
O'Connor, P. M. 2004. Pulmonary pneumaticity in the postcranial skeleton of extant aves: a case study examining Anseriformes. Journal of Morphology 261: 141–161.
Olsen, E., Heino, M., Lilly, G. R., Morgan, M. J., Brattey, J., Ernande, B., et al. 2004 Maturation trends indicative of rapid evolution preceded the collapse of northern cod. Nature 428: 932–935.
Otto, S. P. 2018. Adaptation, speciation and extinction in the Anthropocene. Proceedings of the Royal Society B 14: 285: 20182047. doi: 10.1098/rspb.2018.2047.
Packer, C., Pusey, A. E., Rowley, H., Gilbert, D. A., Martenson, J. and O'Brien, S. J. 1991. Case study of a population bottleneck: lions of the Ngorongoro Crater. Conservation Biology 5: 219–230.
Paley, W. 1802. Natural Theology: or, Evidences of the Existence and Attributes of the Deity. 1st edition. J. Faulder, London, UK.
Petronis, A. 2010. Epigenetics as a unifying principle in the aetiology of complex traits and diseases. Nature 465: 721–727.
Philiptschenko, J. 1927. Variabilität und Variation. Gebruder Borntraeger, Berlin.
Pires, M. M., Silvestro, D. and Quental, T. B. 2015. Continental faunal exchange and the asymmetrical radiation of carnivores. Proceedings of the Royal Society B 282: 20151952.
Pugesek, B. H. and Tomer, A. 1996. The Bumpus house sparrow data: a reanalysis using structural equation models. Evolutionary Ecology 10: 387–404.
Reid, N. M., Proestou, D. A., Clark, B. W., Warren, W. C., Colbourne, J. K., Shaw, J. R. et al. 2016. The genomic landscape of rapid repeated evolutionary adaptation to toxic pollution in wild fish. Science 354: 1305–1308.
Ren, P., Zhang, J., and Vijg, J. 2024. Somatic mutations in aging and disease. GeroScience 46: 5171–5189.
Rensch, B. 1959. Evolution Above the Species Level. Columbia University Press, New York.
Reznick, D. N. and Ricklefs, R. E. 2009. Darwin's bridge between microevolution and macroevolution. Nature 457: 837–842.
Reznick, D. N., Losos, J. and Travis, J. 2018. From low to high gear: there has been a paradigm shift in our understanding of evolution. Ecology Letters 22: 233–244.
Rhodes, F.H. 1987. Darwinian gradualism and its limits: The development of Darwin's views on the rate and pattern of evolutionary change. Journal of the History of Biology 20: 139–157.
Ridley, M. 2004. Evolution. 3rd edition. Wiley-Blackwell, NJ.
Rolshausen, G., Segelbacher, G., Hobson, K. A. and Schaefer, H. M. 2009. Contemporary evolution of reproductive divergence in sympatry along a migratory divide. Current Biology. 19: 2097–2101.
Rosenzweig, M. L. 1978. Competitive speciation. Biol. J. Linn. Soc. 10: 274–289.
Rosenzweig, M. L. 1995. Species Diversity in Space and Time. Cambridge University Press, New York.
Rueffler, C., Van Dooren, T. J., Leimar, O. and Abrams, P. A., 2006. Disruptive selection and then what? Trends in Ecology & Evolution 21: 238–245.
Schemske, D. W. 2010. Darwinian thinking: 150 years after the Origin. The American Naturalist 176: S1–S3
Schluter, D. 2000. The Ecology of Adaptive Radiation. Oxford University Press, Oxford.
Schluter, D. 2001. Ecology and the origin of species. Trends Ecology & Evolution 16: 372–380.
Schluter, D. 2009. Evidence for ecological speciation and its alternative. Science 323: 737–741.
Schmalhausen, I. 1949. Factors of evolution; the theory of stabilizing selection. Blakiston, Philadelphia.
Schultz, P. A., Michael Miller, R., Jastrow, J. D., Rivetta, C. V. and Bever, J. D. 2001. Evidence of a mycorrhizal mechanism for the adaptation of *Andropogon gerardii* (Poaceae) to high- and low-nutrient prairies. Am. J. Bot. 88: 1650–1656.
Simon, B., Fletcher, J. A. and Doebeli, M. 2013. Towards a general theory of group selection. Evolution 67: 1561–1572.
Simpson, G. G. 1953. The Major Features of Evolution. Columbia University Pres, New York.
Smith, T. B. 1993. Disruptive selection and the genetic basis of bill size polymorphism in the African finch *Pyrenestes*. Nature 363: 618–620.
Soodyall, H., Nebel, A., Morar, B. and Jenkins, T. 2003. Genealogy and genes: tracing the founding fathers of Tristan da Cunha. European Journal of Human Genetics 11: 705–709.
Stebbins, G. L. 1966. Organic Evolution: Processes of Organic Evolution. Prentice-Hall, Englewood Cliffs, N.J.
Stoltenberg, S. F. 1997. Coming to terms with heritability. Genetica 99: 89–96.
Svanbäck, R. and Persson, L. 2009. Population density fluctuations change the selection gradient in Eurasian perch. The American Naturalist 173: 507–516.
Taborsky, B. 2006. The influence of juvenile and adult environments on life-history trajectories. Proceedings of the Royal Society B 273: 741–750.

Tenaillon, O. and Matic, I. 2020. The impact of neutral mutations on genome evolvability. Current Biology 30: R527–R534.

Thoday, J. M. 1958. Effects of disruptive selection: experimental production of a polymorphic population. Nature 181: 1124–1125.

Tobalske, B. K. 2016. Evolution of avian flight: muscles and constraints on performance. Phil. Trans. R. Soc. B 371: 20150383. http://dx.doi.org/10.1098/rstb.2015.0383.

Turesson, G. 1922. The species and the variety as ecological units. Hereditas 3: 100–113.

Turesson, G. 1925. The plant species in relation to habitat and climate, contributions to the knowledge of genecological units. Hereditas 6: 147–236.

Udovic, D. 1980. Frequency-dependent selection, disruptive selection, and the evolution of reproductive isolation. The American Naturalist 116: 621–641.

Ulizzi, L., Astolfi, P. and Zonta, L. A. 1998. Natural selection in industrialized countries: A study of three generations of Italian newborns. Annals of Human Genetics 62: 47–53.

Urban, M. C., Strauss, S. Y., Pelletier, F., Palkovacse, E. P., Leiboldf, M. A., Hendryg, A. P. et al. 2020. Evolutionary origins for ecological patterns in space. PNAS 117: 17482–17490.

Uyeda, J. C., Hansen, T. F., Arnold, S. J. and Pienaar, J., 2011. The million-year wait for macroevolutionary bursts. PNAS 108: 15908–15913.

Vincent, T. L. and Brown, J. S. 2005. Evolutionary Game Theory, Natural Selection, and Darwinian Dynamics. Cambridge University Press, Cambridge, UK.

vonHoldt, B. M., Kartzinel, R. Y., Huber, C. D., Underwood, V. L., Zhen, Y. Ruegg, K. et al. 2018. Growth factor gene IGF1 is associated with bill size in the black-bellied seedcracker *Pyrenestes ostrinus*. Nature Communications 9: 4855. https://doi.org/10.1038/s41467-018-07374-9.

Walley, K. A., Khan, S. I. and Bradshaw, A. D. 1974. The potential for evolution of heavy metal tolerance in plants. Heredity 32: 309–319.

Wang, M. S., Murray, G. G., Mann, D., Groves, P., Vershinina, A. O., Supple, M. A. et al. 2022. A polar bear paleogenome reveals extensive ancient gene flow from polar bears into brown bears. Nature Ecology & Evolution 6: 936–944.

Watson, J. D. and F. H. Crick. 1953. The structure of DNA. Cold Spring Harb. Symp. Quant Biol. 18: 123–31. doi: 10.1101/sqb.1953.018.01.020.

Weldon, W. F. R. 1899. Presidential address, British Association for the Advancement of Science (Report of Meeting). Transactions, section D: 887–902.

Whiteley, A. R., Fitzpatrick, S. W., Funk, W. C. and Tallmon, D. A. 2015. Genetic rescue to the rescue. Trends in Ecology & Evolution 30: 42–49.

Whiteman, J. P. 2021. Polar bear behavior: Morphologic and physiologic adaptations. pp. 219–246. *In*: Davis, R. W. and Pagano, A. M. [eds.]. Ethology and Behavioral Ecology of Sea Otters and Polar Bears. Springer International Publishing. https://doi.org/10.1007/978-3-030-66796-2_12.

Williams, G. C. 1966. Adaptation and natural selection: A critique of some current evolutionary thought. Princeton University Press, Princeton, NJ.

Wright, S. 1931. Evolution in Mendelian populations. Genetics 16: 97–159.

Wright, S. 1932. The roles of mutation, inbreeding, crossbreeding, and selection in evolution. Proc. 6th Int. Cong. Genetics 1: 356–366.

Zamel, N., McClean, P. A., Sandell, P. R., Siminovitch, K. A. and Slutsky, A. S. 1996. Asthma on Tristan da Cunha: looking for the genetic link. The University of Toronto Genetics of Asthma Research Group. American Journal of Respiratory and Critical Care Medicine 153: 1902–1906.

Zhang, Y. and Sirard, M-A. 2021. Epigenetic inheritance of acquired traits through DNA methylation. Animal Frontiers 11: 19–27.

CHAPTER 14

Human Evolutionary Ecology

So human an animal (René Dubos, 1901–1982)

Each human being is unique and unprecedented. We recognize each distinctive person by their voice, facial expression, the way they walk, and many other characteristics. Each person is even more unique in their creative responses to surroundings and events. Is *Homo sapiens* special or simply one of the ten million species on the planet? Where do humans belong in the evolutionary tree of life? What is our place in this grand design of nature?

The answer was partly given in the 18th century by Carl Linnaeus (1758), who, using phenotypic characters, classified humans with monkeys, apes, and other primates. One hundred years later, on 30 June 1860—seven months after the publication of *On the Origin of Species* by Charles Darwin—the British Association held its annual meeting at the Oxford University Museum of Natural History. Clergy and scientists gathered to discuss evolution and humanity's place as the pinnacle of creation. Samuel Wilberforce, the Lord Bishop of Oxford, renowned for his eloquence and influence, vehemently attacked Darwin's theory of evolution by natural selection.

At the end of the debate, Wilberforce is said to have asked Julian Huxley, Darwin's peer and erstwhile proponent of evolution, "whether it was through his grandfather or grandmother that he claimed descent from a monkey." The audience went wild with glee, immediately followed by a pin-drop silence when they heard Professor Huxley's measured response: "If the question is put to me whether I would rather have a miserable ape for a grandfather or [be] a man highly endowed by nature and possessed of great means of influence and yet employs these faculties and that influence for the mere purpose of introducing ridicule into a grave scientific discussion, I unhesitatingly affirm my preference for the ape" (Thomson 2000). Critics like Wilberforce attacked Darwin's theory of natural selection largely because it implied the same process produced humans (indeed, Darwin ended up writing a whole book on this subject)[1]. *Ipso facto,* that meant humanity was related to monkeys and apes!

No matter which path we follow—be it anatomy, developmental biology, paleontology, physiology, genetics, or phylogenetics, over and over—the evidence concurs that humans not only share a common ancestor with the great apes (chimpanzees, gorillas, etc.), but that our relationships with other animals reach far, far back in time. For example, the cells in our bodies lack a cell wall, and we produce embryos that pass through a blastula stage; if we look back into the Precambrian, our last common ancestor with all animals also had these traits. During the Cambrian period, our last common ancestor with all vertebrates was bilaterally symmetrical, and, like the early developmental stages of our embryos, had visceral pouches, a notochord, and a body divided into trunk and tail regions. Our last common ancestor with all placental mammals was *Juramaia*, a shrew-like creature that lived 160 million years ago in the late Jurassic (Luo et al. 2011). These small mammals,

[1] (1871).

DOI: 10.1201/9781003008538-14

scampering amid the dinosaurs, reproduced in abundance, employing placentas to nourish their young during gestation (Klein 1999)—just like us.

About 66 million years ago, most dinosaurs, except birds (avian dinosaurs), were wiped out when a massive meteor crashed into Earth near the present-day Yucatan Peninsula in Mexico. But mammals did not have to wait for that extinction event to diversify into "open niches," as we used to believe. Major fossil finds, especially from China, now indicate that mammalian diversification began well before that event, filling the Earth with not only insectivores like *Juramaia* but also gliding mammals, river dwellers with webbed feet, diggers, herbivores, and even carnivores that preyed on dinosaurs (Lou 2007; Pickrell 2019).

The first mammals appeared 178 million years ago in the Triassic and diversified into many subgroups, only three of which still occur today: placentals, marsupials, and monotremes (Luo 2007). However, these three groups of modern mammals make up 6,399 extant mammalian species (Burgin et al. 2018), which inhabit almost every environment—including oceans, freshwater, underground, treetops, and even the sky.

What Makes us Fundamentally Mammals?

No matter where they live and how they live, all mammals possess fundamentally similar characteristics, and humans share these attributes.

Humans, like all mammals, have **mammary glands** that produce milk. Mammary glands, which are highly modified sweat glands, are found in both sexes but cease to develop in males well before puberty. Their origin may be correlated with the development of milk teeth and the pattern of tooth replacement seen in most modern mammals. Milk teeth are less efficient than adult teeth at chewing, due to their small size and ever-changing pattern of occlusion as the young mammal grows. The inefficiency is offset by dependence on a food that needs no chewing—milk—a fluid rich in protein, fat, sugars (especially lactose), and some amounts of vitamins, and salts. Besides milk as a source of water and nutrients, lactation also promotes immune competence and endocrine maturation in the newborn, thus meeting both immediate and long-term developmental needs (Lefèvre et al. 2010). There are also behavioral and psychological aspects of suckling and nurturing between mother and offspring that strengthen parental care and ensure newborn survival. No such universal provisioning requirement is found in birds or other vertebrates, such as fish, amphibians, and reptiles.

Suckling in mammals is indeed an amazing biological innovation. Drinking milk is made possible by the ability to suck and swallow, aided by the hyoid bones in the throat and muscles that support it. This apparatus also forms the voice box. According to a recent finding (Zhou et al. 2019), hyoid bones first appeared at least 165 million years ago in vole-sized docodonts *Microdocodon gracilis*, a close relative of the true mammals.

Hair and fur, like mammary glands, are uniquely mammalian. Both birds and mammals evolved endothermy—the ability to regulate body temperature physiologically—and both evolved feathers or fur, which provide insulation critical for maintaining endothermy (Maderson 2003). But which came first, endothermy or insulation? Maderson (2003) suggests that fur first arose in the therapsid ancestors of mammals as an adaptation to prevent water loss and abrasion. Selection for increased hair density would have conferred an insulating function, allowing mammals to occupy nocturnal niches and setting the stage for the evolution of endothermy. The evolution of fur—and hence the ability to regulate body temperature—may have enabled mammals to invade colder climates that ectothermic reptiles could not inhabit. Additionally, it may have ensured the survival of mammals when the non-avian dinosaurs perished.

Fur functions primarily as an insulator, influencing the rate at which heat is transferred between mammals and the environment. Mammalian pelage has readily evolved in response to a variety of thermal selective pressures (Riddell et al. 2022). In cold climates, a longer, denser pelage—often shed seasonally—is adaptive for trapping warm air for insulation. But in hot arid climates, this can be disadvantageous when individuals are active. Instead, a thin pelage that allows convective cooling but with very low potential for heat transfer allows desert rodents to cool when conditions are hot during the day and maintain body heat when the temperature plummets at night (Riddell et al. 2022). Another adaptive function of hair is camouflage. Camouflage is evidently the single most important evolutionary force in explaining overall coloration in mammals, as the coloration and pattern of a mammal's coat usually match its background (Caro 2005). A little brown mouse is practically invisible against the brown leaf litter of a forest floor, while the orange and black stripes of a Royal Bengal tiger help it disappear against the orange-brown color of the tall grass, its hunting ground.

However, not all mammals are covered in hair. We humans are less hairy than our closest cousin, the chimpanzee, but with copious hair on our head, armpits, and nether-regions, we are still hairier than naked mole-rats (*Heterocephalus glaber*). They have sensory whiskers on their faces and tactile hairs distributed in a grid-like pattern throughout their body (Sarko et al. 2011). Even the least hairy mammals—the whales, dolphins, and porpoises—still retain vestiges of their hairy ancestors. Some grow sensitive bristles on their snouts, while even those without grow and lose their whiskers as fetuses (Chen et al. 2013).

Mammalian jaw bones are uniquely identifiable based on their functionally differentiated teeth, a condition called **heterodont dentition**. However, not all mammals have teeth—baleen whales and anteaters, for example. A mammal's set of teeth provides an open book on its diet and feeding behavior because mammalian teeth are adapted not only to capture and hold food, as is typical of other vertebrates, but to process that food as well. Specialized teeth bite, pierce, cut, grind, shear, and crush food in a process called mastication, which requires adaptive modifications to the jaw muscles that allow horizontal jaw movement (Unger and Sues 2019).

Although mammals typically have four types of teeth—incisors and canines for food procurement and manipulation, and premolars and molars for food processing—the varying morphology of each type reveals adaptive evolution in action, reflected in tongue-twisting terms. For example, the upper molars of insectivorous mammals exhibit zalambdodonty and dilambdodonty, while those of herbivorous mammals are bilophodont, selenodont, or loxodont (Figure 1). Carnivorous mammals have bladed teeth called carnassials. Some mammals have very specialized teeth, like homodont pegs, or wildly modified tusks, which are not used for feeding at all, but instead function as displays or defense (Unger and Sues 2019).

Mammals are also **diphyodont**, meaning they have two successive sets of teeth: an initial "deciduous" set and a subsequent "permanent" set. Diphyodonts contrast with polyphyodonts, such as many fish, whose teeth are constantly replaced. They also differ from monophyodonts, which have a single set of teeth that do not change over a long period of growth. Diphyodonty evolved from mammalian ancestors with polyphyodonty, in which the frequency of tooth replacement was reduced (Luo et al. 2009). In mammals, dental eruption is delayed while neonates nurse on maternal milk. This period of nursing allows for dramatic skull growth, followed by a period of deciduous tooth eruption and a slowing of growth. The skull stops growing with the arrival of permanent teeth, which are perfectly occluded for their very specialized work of food acquisition and processing (Luo et al. 2009).

Like most other mammals, we humans go through a period of exclusive nursing—typically lasting between six months to a year—before our first set of milk teeth erupts (20 deciduous teeth). These are subsequently replaced during childhood and adolescence by an entirely new set of 32 adult teeth. Our smiles rarely reveal our heterodonty, but we also have incisors, canines, premolars, and molars. However, these highly reduced and generalized forms reflect a soft, varied diet (Andrews and Johnson 2020).

Our Inner Primate

There is paleontological and molecular evidence that primates evolved from a common ancestor that lived 65 million years ago, just after the demise of the non-avian dinosaurs (Steiper and Seiffert 2012).

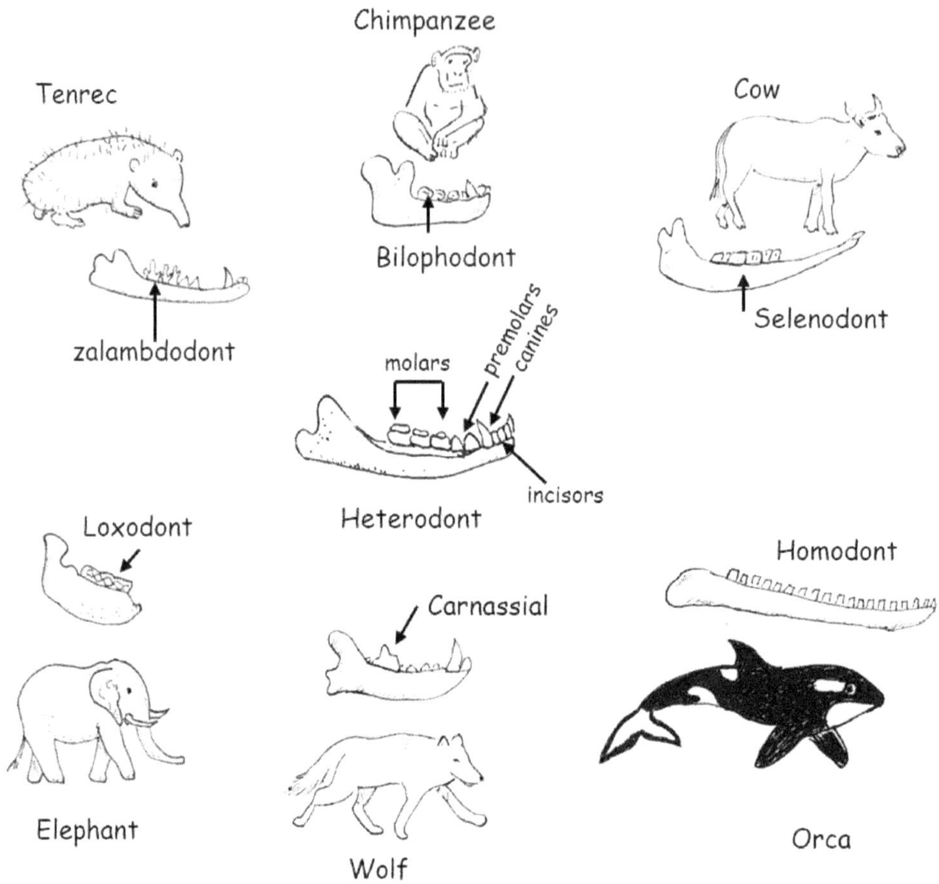

Figure 1. The basic mammalian jaw has molars, premolars, canines, and incisors, but the presence or shape of these teeth can vary based on diet and feeding behavior. Examples of variation in dentition are illustrated here.

Around this time, a small mammal with a somewhat rodent-like appearance—characterized by large incisors and small canines—and with locomotor adaptations consistent with arboreal living, diversified into a group called Plesiadapiforms. Although there is some debate about which, if any, members of this widely diversified group gave rise to true primates, there is no question that by the time of the Paleocene–Eocene Thermal Maximum, a global warming event that occurred about 55.5 million years ago, the first inarguable fossil primates had appeared (Silcox and López-Torres 2023). These early true primates showed all the adaptations that characterize modern primates: eyes in the front of the head, nails on most digits, and brains with some expansion in the neocortex but with proportionately small olfactory bulbs.

Many different types of primates have evolved since then. While many no longer exist, survivors gave rise to the 633 different species of living primates (Perkin 2019). Almost all of today's nonhuman primates live in tropical and subtropical areas of Africa, Madagascar, Asia, Central America, and South America. They include lemurs, lorises, tarsiers, monkeys, and apes. Although humans may seem different from strange-looking aye-ayes, baboons, or lemurs, we share many of the same primate characteristics.

Early primate fossils support the idea that the diversification of tropical angiosperms and the expansion of the arboreal niche went hand in hand with the evolution of primates. As tropical forests developed extensively, they provided excellent habitat for arboreal insects and their predators, including small frogs, lizards, and snakes—as well as our tiny proto-primate, who feasted on all of these in an arboreal buffet. Silcox and López-Torres (2023) argue this ecological scenario—characterized by the visually directed capture of insectivorous and other small prey, along with the grasping and leaping behaviors needed by arboreal predators—catalyzed the evolution of the distinctive primate traits. These include forward-facing primate eyes and opposable digits on palmate hands and feet, adaptations that our bodies still possess today.

Arboreal habitats differ from ground habitats owing to their complex three-dimensional structure, including multiple supports of varying diameters, textures, and aspects. Additionally, access to food and shelter is often intermittent. Thus, adept use of this habitat requires maintaining balance, the ability to bridge gaps, and resistance to slipping and falling (Nyakatura 2019). There is little debate that the first true primates were adapted to these arboreal habitats, but unlike squirrels, which are also arboreal, phylogenetic analysis reveals primates faced different evolutionary pressures (Nyakatura 2019). The common ancestor of squirrels, primates, rabbits, and rodents was a clawed non-grasping mammal. However, once the ancestor of squirrels and primates took to the trees, pedal (foot) grasping evolved, and both groups show this adaptation. This adaptation is essential for exploiting resources on the tiny terminal branches of trees and is also seen in the convergent evolution of pedal grasping in marsupials such as possums (Nyakatura 2019).

What makes squirrels different from primates? The next evolutionary divergence leading to the true primates appears to be the opposable "hallux" (thumb and big toe) with nails rather than claws (Nyakatura 2019). A clawed animal, like a squirrel, is always better at climbing a tree than an animal with nails. For squirrels, claws and paws are adaptive for climbing and running up and down trunks. Since they eat nuts and seeds, plucking and nibbling them at leisure, there is no need for a quick, accurate grasp. Evolving primates, however, must have exploited a different niche—one that required grasping and grabbing, which are best accomplished with opposable digits without a claw. This was characteristic of the Plesiadapiforms, which appeared around 65 million years ago. The evolution of **opposable thumbs and first toes** allowed primates and their immediate forebears to perform a power grip, similar to how we grasp a door handle or a hammer (Figure 2). A power grip in the foot allows hands to be free to do other things, while a power grip in one hand can give

Figure 2. Although both inhabit an arboreal niche, Plesiadapiforms (appearing 65 million years ago) differed from squirrels by having finger nails rather than claws on their opposable hallux (thumb and big toe), which may have been adaptive for terminal branch feeding.

support as the other holds food. Unlike the loss of digits in many different mammalian lineages (for example, horses and deer), primates kept all five digits on all four limbs. A hand with five fingers, one of them opposable, allows dexterity and precision. As a result, primates can not only power grip but also manipulate objects skillfully (Preuschoft 2019).

The final evolutionary divergence, arriving at the true primate lineage, coincided with a period of global warming at the boundary of the Eocene. The common ancestor of all true primates differed from the Plesiadapiforms by possessing leaping adaptations in the hindlimbs, nails on all digits (Nyakatura 2019), and front-facing eye orbits (Silcox and López-Torres 2023). The present-day bush babies or galagos, small arboreal primates native to sub-Saharan Africa, are considered one of the most saltatory (leaping) of primates. They are known to jump fourteen times their body length with faster takeoff velocities than frogs (Burr et al. 1982). These are fast and agile hunters, indicating that leaping is adaptive for arboreal predation. A morphological study of New World monkeys testing adaptive explanations for flattened nails (Hamrick 1998) found that this trait, along with well-developed epidermal ridges (fingerprints) and large, flexible finger pads, was better for foraging in small branch habitat than claws, which are more useful for large-diameter substrates (similar to the way squirrels climb tree trunks).

The grasping of branches and the grabbing of prey is made easier by the capacity to judge distances in height and depth accurately. Without this ability, one can end up empty-handed or, at worst, with a fatal fall. Humans are equipped with precise distance judgement via **binocular vision**, a unique trait to all true primates, both extant and extinct. Our eyes are located frontally, facing forward, which creates large overlapping fields of view and allows stereoscopic or 3-D vision. In contrast, some animals like rabbits, songbirds, or antelopes have two eyes positioned on opposite sides of their heads. While this can give them a wider field of view, sometimes up to 360°, it does not provide them with binocular vision. In fact, primates are the most visually adapted order of mammals, with wide binocular vision, the ability to perceive form, contrast, depth, and motion, and possess multiple variants of color vision (Heesy 2009).

The characteristics of true primates—convergent orbits, grasping hands and feet, reduced claws, and leaping—all suggest ecological selection pressure driven by carnivorous feeding in the small-branch habitat. Molecular phyloecological analysis of digestive system-related genes confirms a switch to increasing carnivory in the diets of ancestral true primates (Wu et al. 2022). Binocular vision has been considered adaptive for both acrobatic arboreal locomotion and predators that target, track, and seize prey. However, comparative data on binocular vision and evolutionary analysis of activity patterns strongly support the idea that nocturnal visual predation drove the evolution of binocular vision in primates (Heesy 2009).

Once true primates appeared, they spread and diversified rapidly. The dawn of the Eocene, about 56 million years ago, with its extreme warmth and spread of tropical climates to high latitudes, allowed these "dawn" primates to migrate across high latitude land bridges between Asia and the New World via Beringia, and between the New World and Europe via Greenland (Beard 2008). These connections allowed for new dietary and habitat opportunities. By the Middle Eocene (about 48 mya), the climate had cooled, and the connections between Asia, Europe, and North America weakened (though Africa was becoming more accessible to Europe). Primate lineages, now isolated in tropical areas of the globe, began to diverge (Fleagle and Gilbert 2006). The earliest anthropoids (monkeys and apes) appeared in North Africa and Asia by the Middle Eocene, and based on morphological similarities, New World monkeys may have originated from immigrants that rafted from Africa to South America in the Early Oligocene (34 mya), when the continents were closer and prevailing currents could have allowed a journey of 8- to 15-day journey (Fleagle and Gilbert 2006).

The traits that characterize primates engendered other evolutionary innovations in response to these new opportunities. Skulls with convergent orbits allowed the enlargement of skull capacity, providing the opportunity to evolve a larger brain. At the same time, many primates became bigger-bodied. While there is a correlation between brain size and body size, an analysis of the effects of body size, group size, and diet on the brain sizes of primates suggests the evolution of large brains in many primates was driven by an increasingly frugivorous diet. This diet supports greater energy turnover, lower energy allocation to digestion, and correspondingly higher energy allocation to the brain (DeCasien et al. 2017). Furthermore, phylogenetic congruence between modern Neotropical primates and Neotropical forest plants suggests that both share an evolutionary history driven by frugivory (Fuzessy et al. 2021).

Brain size is not the only thing that diversified; locomotor patterns did as well. Preference for ground foraging is one of the most significant differences between New World and Old World monkeys, which is reflected in adaptations for terrestriality in Old World monkeys, such as quadrupedal walking in palms in baboons (McCrossin et al. 1998). Apes diverged from Old World monkeys in Africa during the Late Oligocene or Early Miocene, about 30 to 20 million years ago (Fleagle and Gilbert 2006). This divergence was characterized by the evolution of a more upright (**orthograde**) posture, as opposed to the monkey-like horizontal **pronograde** posture (Figure 3). The orthograde body plan includes a lack of external tail, a shorter spine, and a greater range of joint mobility (Almécija et al. 2021). Some apes, especially the gibbons and siamangs from South Asia, are adapted to swinging (brachiation), while other apes forage on the ground and have adaptations for knuckle-walking. Orangutans are as arboreal as gibbons, but they also engage in partial knuckle-walking. Gorillas are true knuckle-walkers, supporting their weight on the backs of their fingers, while monkeys, in contrast, use the palms of their hands. The gorilla is the only ape that is almost entirely terrestrial. Being primarily terrestrial, it manages to avoid predators by having a large, powerful body and fierce demeanor (Johanson and Edey 1981). Chimpanzees are also knuckle-walkers, but they are adept at climbing trees and running or scampering on the ground. Among all apes, chimps are capable of versatile locomotion and are therefore adapted to various ways of living. They mostly live on the ground and, like humans, have a diet that includes a variety of fruits and meat.

The serum precipitin test, which provides a means of establishing degrees of blood relationship, has confirmed that, of the living primates, humans are most closely related to apes and more distantly to monkeys. This close relationship to apes implies that humans and apes shared a common ancestor and that the descendant populations diverged, not that humans evolved from apes. From a genetic perspective, humans share more DNA with other primates—lemurs, monkeys, and apes—than they

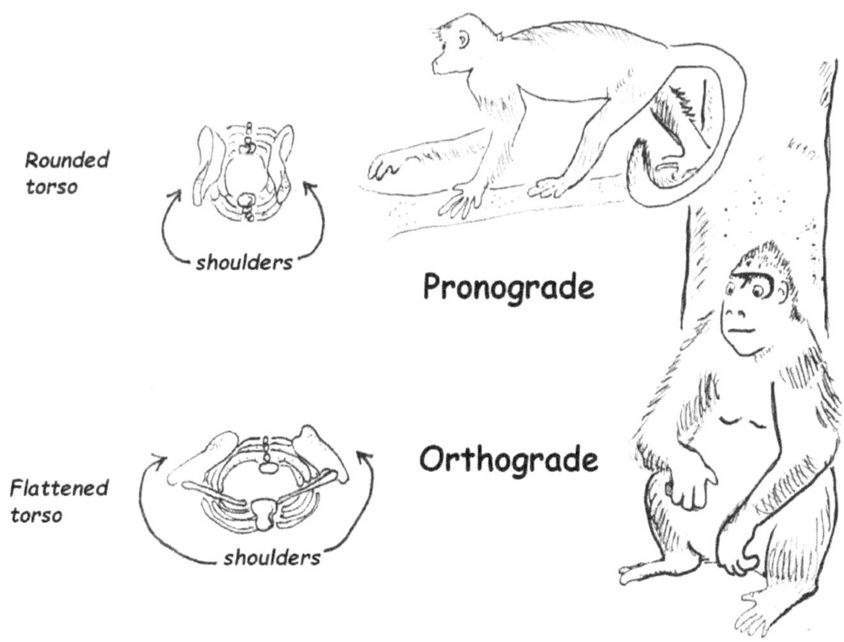

Figure 3. Pronograde vs. orthograde posture.

do with any other mammal, and humans and chimps differ by only one set of chromosomes: Humans have 23 pairs of chromosomes, whereas chimps have 24 (Yunis and Prakash 1982). A whole-genome comparison of DNA nucleotides revealed that humans and chimps, whose common ancestor lived between 6.5 and 7.5 million years ago (Amster and Sella 2016), differ by a mere 1.23% (Mikkelsen et al. 2005). Current genomic evidence indicates that, of all the living mammals on Earth, humans are most closely related to chimpanzees and bonobos (Prado-Martinez et al. 2013).

An Unprecedented Rise of Hominins

Julian Huxley, the British evolutionary biologist (1887–1975), provided the first systematic comparison between humans and apes in the 19th century, concluding that there were enough structural differences to put humans in a family apart from apes (though not outside the primate order). We now know from fossil and molecular discoveries since then that many "human features," such as binocular color vision, grasping hands with nails, and even increased encephalization and slower life history, are all primitive traits inherited from our primate ancestors (Almécija et al. 2021). Thus, the "human" Family (Hominidae or "Hominids") is not singular. Hominids are members of a superfamily known as the Hominoidea that arose in East Africa about 20 million years ago (Larsen 2019). This group (Hominoidea) includes the Hylobatids (gibbons) and the four living genera of Hominids: *Pongo* (orangutans), *Gorilla* (gorillas), *Pan* (chimpanzees), and *Homo* (us). Of the Hominidae, the Tribe Homini (Hominins) encompasses the human lineage, of which we are the last survivors.

The split between Hominins and *Pan* (Tribe Panini) occurred at some point during the Miocene between 9.3 million and 6.5 million years ago (Moorjani et al. 2016). However, due to a paucity of fossils, the exact timing and location of the split remain uncertain. Questions as to who was the last common ancestor (*Pan-Homo* LCA) and whether they walked upright, on their knuckles, or something in between are still up for debate (Almécija et al. 2021). There is general agreement, however, that Hominins can be distinguished by only two traits: the **loss of canine honing** (that is, the premolars no longer sharpen the canines) and features related to **bipedal locomotion**–the ability to walk upright on two feet (Almécija et al. 2021). The ecological-evolutionary drivers of the *Pan–Homo* split, like most of the evolutionary history described in this chapter, remain lost to time; so we are left with fitting circumstantial evidence to adaptive "just-so" stories and adjusting them as new evidence comes to light.

The classic "East Side Story" (Coppens 1994) posits that environmental changes happening ten million years ago triggered the origin of hominins. Following a line that runs from present-day Turkey, through Israel, the Red Sea, Ethiopia, Kenya, and all the way down to Mozambique, two tectonic plates were moving apart, causing the once continuous rainforest to fragment in eastern Africa. During this important geological upheaval, the great Rift Valley was formed, and the east side of the Rift Valley fell into a rain shadow, creating a mosaic of different habitats where an uninterrupted, dense tropical forest had once existed, providing a grand evolutionary opportunity. Chimpanzees adapted to the wetter, more heavily wooded environments prevalent in the west of the Rift Valley, while the hominins remained in the drier east, adapting to savannas and open woodlands (Coppens 1994)—a nice scenario, which now faces serious challenges (Almécija et al. 2021). First, this scenario is based on the living ape geographic distribution, which may not have matched the distribution 10 million years ago, before the Pleistocene ice ages impacted modern distributions. Second, the "story" relies on an outdated fossil record from the 1980s. Subsequent discoveries of early hominin fossils in Chad—2,500 km west of the rift—along with finds in north-central Africa, do not fit the narrative. Furthermore, other subsequent environmental reconstructions of the period during the split suggest a woodland rather than tropical rainforest origin. Despite newer hominin fossil finds, the human origin story remains clouded, largely due to a lack of fossils from between 14 and 10 million years ago, just prior to the *Pan–Homo* split (Almécija et al. 2021).

So, what do we know about human origins? There is general agreement that canine reduction, enhanced manipulative capabilities, and bipedalism were interrelated during human evolution. We also know from molecular biology that chimpanzees and humans share a closer genetic relationship than chimpanzees do with gorillas. This is an important distinction because, as recently as the 1980s, it was still argued, based on similar knuckle-walking anatomy, that chimps belonged with gorillas in the same taxonomic family (Almécija et al. 2021). This raises the question: Are chimpanzees a model for the *Pan-Homo* last common ancestor (LCA), which necessarily assumes chimps have not evolved in the past 10 million years and implies that bipedalism must have evolved from knuckle-walking? Or are chimps and humans two divergent lines from a *Pan-Homo* LCA that was neither a knuckle-walker nor bipedal? Currently, there is not enough fossil evidence to even identify the *Pan-Homo* LCA, let alone determine how they moved. However, we can speculate that walking upright was more energetically efficient than the knuckle-walking gaits of chimpanzees for the activities ancestral hominins engaged in within their habitat, since bipedalism evolved in the hominin line. Standing upright may also have helped hominins to keep cooler in hot sunshine (Lieberman 2015), and bipedal locomotion frees up the hands for other activities, such as carrying tools (Washburn 1960), food acquisition, and general carrying (Richmond et al. 2001). Over time, this would have also selected for greater manual dexterity (Boyd and Silk 2009), plus long-distance walking and running as adaptations for scavenging and hunting (Lieberman 2015).

The earliest hominin fossils date from the Pliocene Period (2.6 to 5.3 million years ago), and the first fossil discovery of a hominin occurred in 1924 in South Africa. The skull, estimated to be 2.8 million years old, belonged to *Australopithecus africanus*, which means "the man-ape from southern Africa" (Dart 1925). *A. africanus*, with a chimp-sized brain (less than 500 cubic centimeters), had a characteristic foramen—the hole that marks the entry of the spinal cord into the head—which suggested an upright stance. Then came the astonishing discovery of Lucy, a remarkably complete *Australopithecus afarensis* skeleton, by anthropologists Donald Johanson and Tom Gray in Ethiopia in November 1974. Discovery of this 3.2-million-year-old australopithecine (Johanson and Edey 1981) revealed a small, chimp-sized brain, but she did have one defining human trait: she walked upright. We now know from hundreds of fossils and even fossilized footprints that australopiths were bipedal, like the humans that came after (Figure 4).

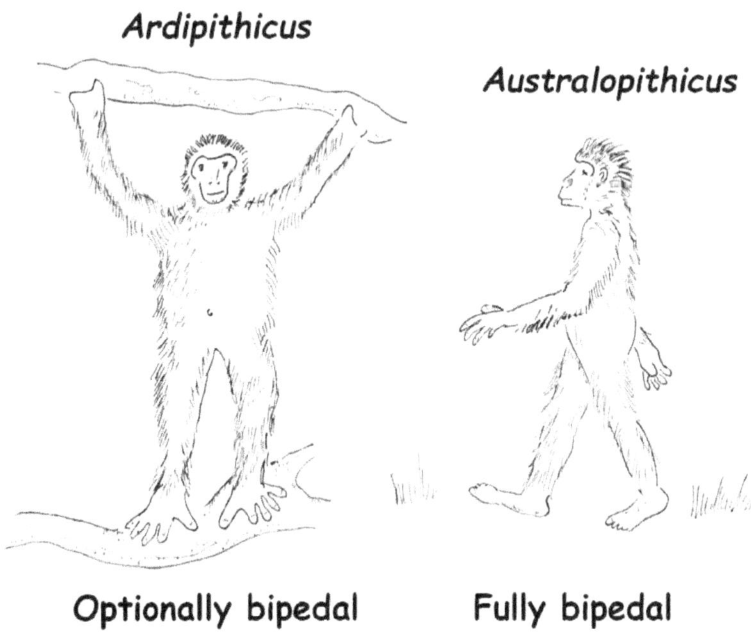

Figure 4. Optionally vs. fully bipedal locomotion.

Australopiths predate and also co-occurred with early *Homo*. However, subsequent fossil finds since 2009 have revealed the earliest hominin yet, living before Australopiths and *Homo*, but after the *Pan-Homo* LCA. The discovery and description of *Ardipithecus ramidus* (White et al. 2009) challenged long-held beliefs that the earliest hominins would resemble chimpanzees and that the burgeoning savannahs drove the evolution of bipedalism. In fact, *Ardipithecus*, which lived about 4.5 million years ago, was a partially arboreal, optionally bipedal, woodland omnivore (White et al. 2009) that looked more like fossil Miocene apes than modern chimpanzees and shared several characters with all later hominins (White et al. 2015).

For example, unlike chimpanzees but like hominins, the hand joints allowed the ability to "cup" the hand, and the lumbar region was longer. Chimpanzees have short lower backs stiffened by the pelvis, which results in a bent-hip, bent-knee waddling bipedal gait; but like humans, earlier *Homo*, and australopiths, *Ardipithecus*, had a longer lower back and a pelvis shaped to allow the center of mass to be placed over the feet, enabling terrestrial bipedality (Figure 4). Interestingly, unlike either chimps or humans, *Ardipithecus's* feet were unique, with a widely divergent and opposable big toe that was not as derived for grasping as a chimpanzee's foot. It was instead able to be used as an effective "toe-off" during bipedal walking (White et al. 2015). However, *Ardipithecus* fossils, all of which point to formative bipedality, have not been found in savannah habitats. Instead, they were a woodland species alongside arboreal colobine monkeys. As White et al. (2015) say, "the hypothesis that opening grasslands led to hominid emergence and bipedality now stands effectively falsified."

The fossil record, being an incomplete, biased, and small sample of the past, has its limitations and must necessarily provide an evolving story of human evolution as new fossil (and, increasingly, genetic and protein) evidence comes to light. The fossil record provides widely varying specimens of both *Australopithecus* and early *Homo*, leading to great debates on which of these specimens represents the direct ancestral line to modern humans. However, there is general agreement that our genus *Homo* can be distinguished from earlier (and contemporary) australopiths by having larger average body and brain sizes (Antón et al. 2014). Antón et al. (2014) argue that early *Homo* evolved "in a context of habitat instability and fragmentation on seasonal, intergenerational, and evolutionary time scales" (recall that the Pleistocene began 2.58 million years ago), and that this context gave rise to selectively advantageous traits such as dietary flexibility mediated by tool use and larger brains.

The first *Homo* that we can agree was much like us appeared about 1.9 million years ago. *Homo erectus* was fully bipedal and created toolkits including scrapers, knives, cleavers, picks, and handaxes (Semaw et al. 2020). Their brains were larger than those of australopiths, with an impressive increase in volume to about 900 cubic centimeters (cc) and 62 billion neurons, compared to 500 cc and 27–35 billion neurons. This imposed an increase in energetic cost of 6 kcal per day per billion neurons (Herculano-Houzel 2012). Although that may not seem like much in our modern world, it is an additional 180 kcal extra every day that *Homo erectus* had to acquire in all seasons, compared to *Australopithecus*, to meet the additional energy demands of those neurons. Where did this energy come from?

Australopithecus skulls indicate powerful masticatory muscles, and in living primates, the MYH16 gene is responsible for their development. However, Stedman et al. (2004) found that in humans, a frameshift mutation in MYH16 inactivated this gene, which is responsible for smaller jaw muscles that, in turn, allow our skulls to expand upward to accommodate our larger brains. They also estimated this mutation arose in early *Homo*. Zink and Lieberman (2016) suggest that morphological changes, such as a decrease in the size of the jaw, molars, and digestive tract (indicated by a smaller rib cage) in *Homo erectus*, all indicate a major shift in diet to softer, easily digested foods—*processed foods*.

There is evidence that as far back as 2.6 million years ago in Ethiopia, early hominins were using simply shaped tools for pounding and butchery, indicating plant and animal tissue processing (Plummer et al. 2023). But evidence from this period is scarce. By the time *Homo erectus* arrived, stone toolkits were abundant and complex, allowing a diet enriched by mechanically processed tubers and rhizomes (which often require crushing and soaking to detoxify), as well as sliced meat.

In experiments testing the chewing force of humans consuming meat and underground storage organs (root vegetables), Zink and Lieberman (2016) found that by pounding tubers and rhizomes, and simply slicing meat, suddenly more calories are available from the former and less chewing energy is required by the latter, making smaller masticatory features and a larger brain possible. Associated with this larger brain, tools for and evidence of butchery of large animals also indicate the development of social hunting, and, by inference, imply the sharing of meat, development of prosocial interactions, and larger group sizes (Linares Matás and Yravedra 2021). Evidence of butchery-marked bones indicates that *Homo erectus* regularly accessed bone marrow (Pobiner et al. 2008), supporting the idea that *Homo erectus* survived on a high-fat diet (Ben-Dor et al. 2011). Thus, tool use by *Homo erectus* is associated with a larger brain, a wider, more flexible diet with more calories to support that brain, and possibly cooperation and group cohesion. These innovations seem to have induced wanderlust, as *Homo erectus* had already dispersed to different parts of Africa, Europe, and Asia by 1.8 million years ago (Antón et al. 2014).

Homo heidelbergensis had a higher, rounder cranium and a larger brain, which most researchers believe evolved from a *Homo erectus* population in Africa between 800,000 and 500,000 years ago. They migrated to Europe and Asia, bringing with them more sophisticated, elaborately flaked tools, some even attached to handles. Those migrants gave rise to Neanderthals (known from Europe) and Denisovans (known from East Asia) about 400,000 years ago (Boyd and Silk 2009). The exodus of *Homo erectus* and then *Homo heidelbergensis* constituted the first and the second waves of migration from Africa (**out-of-Africa hypothesis:** *sensu* Weidenreich 1947). In the third or final wave, modern *Homo sapiens,* presumably evolved from *Homo heidelbergensis* in Africa between 200,000 and 130,000 years ago, migrated into Europe and Asia, replacing the groups from the two previous waves (Jin and Su 2000).

Genetic data from extant humans suggest that all modern humans are descendants of *H. sapiens* that lived and dispersed from Africa at least 55,000 years ago. They reached Australia by at least 45,000 years ago, and by 30,000 years ago, had replaced the Neanderthals across Eurasia (Oppenheimer 2009). Climate reconstructions (Beyer et al. 2021) show that wetter conditions, when hominins could have crossed the Sahara and Arabian Peninsula, occurred 130,000 years ago (as *Homo heidelbergensis* was leaving Africa) and about 67,000–78,000 years ago. However, a southern route to Asia was blocked until 65,000–30,000 years ago (when *Homo sapiens* was traveling to southern Asia and Australia). When modern humans were able to leave Africa, they met the earlier waves of dispersed hominins. These humans sometimes, as DNA data shows, interbred with Neanderthals (Kuhlwilm et al. 2016) and with Denisovans (Browning et al. 2018) before these two earlier groups became extinct. Genomes of Neanderthals and Denisovans have been sequenced and used to detect "introgressed" genetic material in modern humans (Browning et al. 2018). **Genetic introgression** occurs when a hybrid offspring is backcrossed to individuals in only one of the parental lines, whereas genetic "hybridization" refers to extensive intermixing that creates a new population. Sankararaman et al. (2016) estimate present-day humans inherit up to 5% of their ancestry from Denisovans and up to 2% from Neanderthals, although these proportions vary in modern human populations. Their study also supports the hypothesis that reduced male fertility may be a common feature of hybridization between modern humans, Neanderthals, and Denisovans, making the admixtures of these genes into the human genome potentially disadvantageous. However, there is also genomic evidence that some of these genes were retained under positive selective pressure. These include genes related to skin and hair from Neanderthals and genes related to fat metabolism and detecting subtle scents from Denisovans (Sankararaman et al. 2016).

Cranial capacity increased from *Homo erectus* to *Homo heidelbergensis* and their presumed descendants—Neandertals, Denisovans, and modern humans—with an estimated change from 900 cc and 62 billion neurons to 1,200–1,700 cc and 76–90 billion neurons, along with the associated caloric cost (Ehrlich 2000; Herculano-Houzel 2012). Once again, between *Homo erectus* and *Homo heidelbergensis*, selection for a bigger brain appears to have outweighed selection for larger teeth, as *Homo heidelbergensis* has smaller teeth than its predecessor. This suggests yet more changes to a softer, but somehow more energy-rich diet. How could this have happened?

The first evidence that Prometheus gave hominins the gift of fire (metaphorically speaking) comes from Wonderwerk Cave in South Africa. Unambiguous evidence, in the form of burned bones and ashed plant remains, that hominins used fire to cook their food, is dated to about 1 million years ago (Berna et al. 2012). Recall this is the time of *Homo erectus*. However, evidence of controlled use of fire from this time to half a million years later is remarkably sparse and controversial until 400,000 years ago, when the archeological signal of fire use became very visible and widespread across multiple continents (MacDonald et al. 2021). Recall this is the time of *Homo heidelbergensis*, Neanderthals, and Denisovans.

What does the power of fire do? For one, fire can be used for cooking. Cooking dramatically increases calories by making starchy plant organs edible, reducing their toxicity, lowering the calories needed for processing and digestion, and increasing the availability of glucose (Carmody and Wrangham 2009). There is also evidence that starchy plant organs were cooked at least 170,000 years ago in Africa (Wadley et al. 2020). It is tempting to believe that meat also benefits from cooking. However, while Carmody et al. (2011) found experimental evidence in mice that cooked meat has a positive effect on gross energy gain, presumably through protein denaturation, the gelatinization of collagen, and the killing of pathogens, the story is complicated because fats are lost in drippings, and muscle fibers are toughened.

Interestingly, ethnohistoric accounts examined by Speth and Morin (2022) show that hunter-gatherers in both northern and tropical habitats often prefer putrid (decomposing) meat over cooked meat, perhaps indicating predigestion (in this case by microbes), similar to the protein denaturation that comes from cooking, which is the most important factor in caloric gain from meats. Furthermore, Niven et al. (2012) found remains of 18 reindeer that were butchered in one season by Neanderthals, yet despite abundant evidence of filleting and marrow exploitation, there is no evidence of fire hearths, suggesting the Neanderthals were not using fire to process their meat in this case. Regardless, there is ample evidence that fire was used in the Middle Pleistocene as a process to harden stone tools, and research into the effects of such heat treatment suggests that this increases the sharpness and performance of stone cutting edges (Key et al. 2021). Ultimately, better tools make for more efficient food processing (18 reindeer), and presumably, whether or not it is cooked, more meat means more calories to feed a larger brain.

There is a definite bias when it comes to distinguishing early humans from our extinct cousins, the Neanderthals and Denisovans (Peeters and Zwart 2020). We consider ourselves evolutionary "winners," unique in having a "spark," or unprecedented "symbolic consciousness," or dare we say, "superiority" over those "others." Are *Homo sapiens* truly special—and if so, in what way?

Early humans left remains of sophisticated tool-making between 50,000 and 100,000 years ago in Africa, including complex projectile weapons (Shea and Sisk 2010). But are humans the only tool users? Shumaker et al. (2011) document nearly a thousand cases of tool use by non-humans across 221 genera. What about tool makers? Again, no, although most cases are limited to primates and corvids (Colbourne et al. 2021). As we have learned, Neanderthals and Denisovans were accomplished tool makers, making spears and digging sticks as long as 300,000 years ago and using fire to transform materials, including making pitch glue (Villa and Soriano 2010; Breyl 2020).

Evidence of the earliest human symbolic behavior (engraved ochre and ostrich eggshell fragments) dates to about 100,000 years ago in South Africa (Tylén et al. 2020). Are humans the only artists (indicating symbolic consciousness)? The bower birds (*Amblyornis inornatus*) of New Guinea and Australia have a definite eye for beauty, and Diamond (1986) argues that the geographically varying woven towers decorated by males with colorful objects such as fruits, flowers, and butterfly wings to attract their mates manifest artistic styles that may be culturally transmitted, similar to human art styles. Male palm cockatoos (*Probosciger aterrimus*) from Northern Australia make percussive instruments and beat them to attract mates. Heinsohn et al. (2017) documented different beat patterns and drumming styles among individual birds and found that the longest drumming sequences were highly regular and predictable, like human music. Although evidence is patchy and often controversial, and mainly from 50,000 years ago, Neanderthals may have created art (bone engravings; Leder et al. 2021) and music (a flute? Tuniz et al. 2011). There is also evidence they used fibers (Hardy et al. 2020) and buried their dead (Pomeroy et al. 2020), suggesting symbolic thinking.

Is language unique to humans? Many birds and mammals communicate extensively. For instance, vervet monkeys can make different kinds of noises to warn their fellow monkeys about different kinds of predators, and even chickens can make at least 24 distinct vocalizations (Collias and Joos 1953). But unlike chickens and vervet monkeys, humans can invent new words at will to describe different things, actions, and states of mind. Our language is underpinned by rules of syntax, enabling us to manipulate our words to convey just about anything. René Descartes (1596 –1650), a French philosopher and mathematician, said that what makes us different is that we speak. Could Neanderthals produce complex speech as well?

FOXP2 is a gene known to be involved in the development of speech and language, and the inactivation of one *FOXP2* copy can lead to deficits in mouth and facial movement and linguistic processing (Krause et al. 2007). Krause et al. (2007) explain that two amino acid substitutions show up in both Neanderthals and humans but not chimpanzees, and they suggest language abilities brought about by these substitutions were selected for between 300,000 and 400,000 years ago. Breyl (2020) argues that evidence for long-distance transport of materials, local workshops, production of hafting materials, pigment use and production of ornaments, and social care both for individuals and groups, makes it difficult to categorize Neanderthal cognition and technologies as somehow "inferior" to that of human hunter-gatherers.

There is a central theme here: we are not unlike other organisms on this planet, and it is quite possible that early humans were no different in their cognitive and language abilities from Neanderthals and Denisovans. Considering modern examples of demographic replacement (e.g., in early Neolithic Britain; Brace et al. 2019), Breyl (2020) even suggests that cognitive differences were not necessarily to explain the replacement of Neanderthals and Denisovans by humans—ecological and demographic factors may have played a larger role.

Herculano-Houzel (2012) concludes that our brains are "remarkable, yet not extraordinary," and she presents evidence that the human brain is simply a scaled-up primate brain. There has been selection for larger brains in hominins, and the caloric cost seems to have been compensated for by technological innovations generated by the brains themselves. However, there is a tradeoff between bipedal locomotion and increasing brain size (encephalization). Whitman and Wall (2007) observe that evolutionary modifications have occurred in the human female pelvis, as well as in the malleability of the fetal head, rotational birth mechanics, and the earlier timing of delivery, which help compensate for the "obstructed labor injury complex" engendered by a larger brain. They also observe that the accelerated brain growth outside the womb found in modern humans is likely an adaptation that has allowed encephalization despite the size restrictions of the birth canal. Furthermore, the difficulties of childbirth have necessitated social adaptations that require "sympathetic helpers" to accomplish successful deliveries. This suggests brain size may have reached its limit about the time of *Homo heidelbergensis*, as their brain size falls within the range of variation found in *Homo sapiens* (Herculano-Houzel 2012).

Why, then, do we persist with an innate sense of superiority? Why do we think we are special compared to all the other organisms on Earth? Could it be that we are mesmerized by our technological prowess and achievements? After all, we look around and see televisions, microwaves, satellites, nuclear reactors, you name it. Voila, superiority!

Perhaps we can speculate. Having a large enough brain with the capacity for symbolic thought and a couple of mutations that enable complex language, hominins gained the ability to grow communal "brains." We can experiment, learn, teach, tell stories, and pass what we know not only to other adults but also to children, who pass the accumulated knowledge on to their children. Eventually, and this happened a number of times on different continents, we invented writing—

suddenly enabling our thoughts to be shared not just between our close comrades and family but across cultures and time. We can share our thoughts, in principle, with any other human being on the planet.

This also means we are capable of collective brainstorming. We now occupy a state that is unique in the biological world: we can predict the future, imagine the existence of things that do not exist in nature, and those things in nature that are not detectable by our senses, and we can turn imagined things into real things. We can play a symphony written 200 years ago, and we can construct a space capsule or a marvelous cathedral. We have gained control over our environment—even the environments of other species. This capability has made us, in a short period, the commonest, most dominant large mammal on Earth, while our nearest genetic relatives, the chimp and the other great apes—four out of six—are on the verge of extinction, according to the IUCN Red List (IUCN 2021). Our communal brain is still evolving: artificial intelligence can now win games, write its papers, identify individual humans, and create its own art. However, as explored in numerous post-apocalyptic stories, if we took away our cultural legacy, institutions, and communal brain, what would we be?

So far, the weight of evidence suggests that modern humans who lived in and left Africa between 100,000 and 60,000 years ago (*Homo sapiens*) were different from Neandertals and Denisovans. Archeological evidence shows physical differences: we are more gracile, our brows are understated, and our faces are flat (Weaver et al. 2007; Sawyer and Maley, 2005). Although advances in sequencing the Neanderthal genome have shown that only 1.5% to 7% of the human genome is unique, Schaefer et al. (2021) found genetic differences related to brain function and development. But whether there were differences in cognition, language, and the capacity to build communal "brains" between our human ancestors and their cousins, the Neanderthals and Denisovans—or whether, given enough time and population density, they could have developed civilizations to rival our own—remains uncertain. There is not enough evidence to decide. Perhaps the only way we could ever know would be if we could talk to one.

Our human brains compel us to find meaning in our lives and to understand our origin. Evolutionary biology tells us this: You are the culmination of an unbroken line of organisms, stretching back to the dawn of life itself, each of which survived on Earth long enough to reproduce successfully. You are special. But so is everyone else. Or at least, that is what we think. The conundrum is if each of us is "special," then no one is.

References

Almécija, S., Hammond, A. S., Thompson, N. E., Pugh, K. D., Moyà-Solà, S. and Alba, D. M. 2021. Fossil apes and human evolution. Science 372: eabb4363.

Amster, G. and Sella, G. 2016. Life history effects on the molecular clock of autosomes and sex chromosomes. Proceedings of the National Academy of Sciences 113: 1588–1593.

Andrews, P. and Johnson, R. J. 2020. Evolutionary basis for the human diet: consequences for human health. Journal of Internal Medicine 287: 226–237.

Antón, S. C., Potts, R. and Aiello, L. C. 2014. Evolution of early *Homo*: an integrated biological perspective. Science 345: 1236828. doi: 10.1126/science1236828.

Beard, K. C. 2008. The oldest North American primate and mammalian biogeography during the Paleocene-Eocene Thermal Maximum. PNAS 105: 3815–3818.

Ben-Dor, M., Gopher, A., Hershkovitz, I. and Barkai, R. 2011. Man the fat hunter: the demise of *Homo erectus* and the emergence of a new Hominin lineage in the Middle Pleistocene (ca. 400 kyr) Levant. PlosOne 6: e28689.

Berna, F., Goldberg, P., Horwitz, L. K., Brink, J., Holt, S., Bamford, M. et al. 2012. Microstratigraphic evidence of in situ fire in the Acheulean strata of Wonderwerk Cave, Northern Cape province, South Africa. PNAS 109: E1215-1220. doi: 10.1073/PNAS1117620109.

Beyer, R. M., Krapp, M., Eriksson, A. and Manica, A. 2021. Climatic windows for human migration out of Africa in the past 300,000 years. Nature Communications 12: 4889. doi.org/10.1038/s41467-021-24779-1.

Boyd, R. and Silk, J. B. 2009. How Humans Evolved. Fifth edition. W. W. Norton, New York.

Brace, S., Diekmann, Y., Booth, T. J., van Dorp, L., Faltyskova, Z., Rohland, N. et al. 2019. Ancient genomes indicate population replacement in Early Neolithic Britain. Nature Ecology & Evolution 3: 765–771. doi: 10.1038/s41559-019-0871-9.

Breyl, M. 2020. Triangulating Neanderthal cognition: a tale of not seeing the forest for the trees. Cognitive Science 2021: 12e1545. doi.org/10.1002/wcs.1545.

Browning, S. R., Browning, B. L., Zhou, Y., Tucci, S. and Akey, J. M. 2018. Analysis of human sequence data reveals two pulses of archaic Denisovan admixture. Cell 173: 53–61.

Burgin, C. J., Colella, J. P., Kahn, P. L. and Upham, N. S. 2018. How many species of mammals are there? Journal of Mammalogy 99: 1–14.

Burr, D. B., Piotrowski, G., Martin, R. B., and Cook, P. N. 1982. Femoral mechanics in the lesser bushbaby (*Galago senegalensis*): Structural adaptations to leaping in primates. The Anatomical Record 202: 419–429.

Carmody, R. N. and Wrangham, R. W. 2009. The energetic significance of cooking. Evolutionary Anthropology 57: 379–391.

Carmody, R. N., Weintraub, G. S. and Wrangham, R. W. 2011. Energetic consequences of thermal and nonthermal food processing. PNAS 108: 19199–19203.

Caro, T. 2005. The adaptive significance of coloration in mammals. BioScience 55: 125–136.

Chen, Z., Wang, Z., Xu, S., Zhou, K. and Yang, G. 2013. Characterization of hairless (*Hr*) and FGF5 genes provides insights into the molecular basis of hair loss in cetaceans. BMC Evolutionary Biology 13: 34. doi: 10.1186/1471-2148-13-34.

Colbourne, J. A. D., Auersperg, A. M. I., Lambert, M. L, Huber, L. and Völter, C. J. 2021. Extending the reach of tooling theory: a neurocognitive and phylogenetic perspective. Topics in Cognitive Science 13: 548–572.

Collias, N. E. and Joos, M. 1953. The spectrographic analysis of sound signals of the domestic fowl. Behaviour 5: 175–188.

Coppens, Y. 1994. East Side story: The origin of humankind. Scientific American 270: 88–95.

Dart, R. A. 1925. *Australopithecus africanus*: The man-ape of South Africa. Nature 115: 195–199.

Darwin, C. 1871. The Descent of Man, and Selection in Relation to Sex (Vol. I) (John Murray, 1871).

DeCasien, A. R., Williams, S. A. and Higham, J. P. 2017. Primate brain size is predicted by diet but not sociality. Nature Ecology & Evolution 1: 0112. doi: 10.1038/s41559-017-0112.

Diamond, J. 1986. Animal art: variation in bower decorating style among male bowerbirds *Amblyornis inornatus*. PNAS 83: 3042–3046.

Ehrlich, P. R. 2000. Human Natures: Genes, Cultures, and the Human Prospect. Island Press. Washington D.C.

Fleagle, J. G. and Gilbert, C. C. 2006. The biogeography of primate evolution: the role of plate tectonics, climate and chance. Chapter 13. pp. 375–418. *In*: Lehman, S. M. and Fleagle, J. G. [eds.]. Primate Biogeography. Springer, New York.

Fuzessy, L., Silveira, F. A. O., Culot, L., Jordano, P. and Verdú, M. 2021. Phylogenetic congruence between Neotropical primates and plants is driven by frugivory. Ecology Letters 25: 320–329.

Hamrick, M. W. 1998. Functional and adaptive significance of primate pads and claws: evidence from New World anthropoids. American Journal of Biological Anthropology 106: 113-127.

Hardy, B. L., Moncel, M.-H., Kerfant, C., Lebon, M., Bellot-Gurlet, L. and Mélard, N. 2020. Direct evidence of Neanderthal fibre technology and its cognitive and behavioral implications. Scientific Reports 10: 4889. https://doi.org/10.1038/s41598-020-61839-w.

Heesy, C. P. 2009. Seeing in stereo: the ecology and evolution of primate binocular vision and stereopsis. Evolutionary Anthropology 18: 21–39.

Heinsohn, R. Zdenek, C. N., Cunningham, R. B., Endler, J. A. and Langmore, N. E. 2017. Tool-assisted rhythmic drumming in palm cockatoos shares key elements of human instrumental music. Science Advances 3: e1602399.

Herculano-Houzel S. 2012. The remarkable, yet not extraordinary, human brain as a scaled-up primate brain and its associated cost. PNAS 109: 10661–10668.

IUCN 2021. The IUCN Red List of Threatened Species. Version 2021-2. https://www.iucnredlist.org.

Jin, L. and Su, B. 2000. Natives or immigrants: modern human origin in East Asia. Nat Rev Genet. 1: 126–133.

Johanson, D. C. and Edey, M. A. 1981. Lucy: The beginnings of humankind. Simon and Schuster, New York, NY.

Key, A., Pargeter, J. and Schmidt, P. 2021. Heat treatment significantly increases the sharpness of silcrete stone tools. Archaeometry 63: 44–466.

Klein, R. G. 1999. The Human Career: Human Biological and Cultural Origins. 2nd edition. University of Chicago Press, Chicago.

Krause, J., Lalueza-Fox, C., Orlando, L., Enard, W., Green, R. E., Burbano, H. A. et al. 2007. The derived FOXP2 variant of modern humans was shared with Neandertals. Current Biology 17: 1908–1912.

Kuhlwilm, M., Gronau, I., Hubisz, M. J., de Filippo, C., Prado-Martinez, J., Kircher, M. et al. 2016. Ancient gene flow from early modern humans into Eastern Neanderthals. Nature 530: 429–433.

Larsen, C. 2019. Our Origins. Fifth edition., W. W. Norton, New York.
Leder, D., Hermann, R., Hüls, M., Russo, G., Hoelzmann, P., Nielbock, R. et al. 2021. A 51,000-year-old engraved bone reveals Neanderthals' capacity for symbolic behaviour. Nature Ecology & Evolution 5: 1273–1282.
Lefèvre, C. M., Sharp, J. A. and Nicholas, K. R. 2010. Evolution of lactation: ancient origin and extreme adaptations of the lactation system. Annual Review of Genomics and Human Genetics 11: 219–238.
Lieberman, D. E. 2015. Human locomotion and heat loss: an evolutionary perspective. Comprehensive Physiology 5: 99–117.
Linares Matás, G. J. and Yravedra, J. 2021. 'We hunt to share': social dynamics and very large mammal butchery during the *Oldowan-Acheulean* transition. World Archaeology 53: 224–254. doi: 10.1080/00438243.2022.2030793.
Linnaeus, L. 1758. Systema Naturae per Regna Tria Naturae, Secundum Classes, Ordines, Genera, Species, cum Characteribus, Differentiis. Synonymis, Locis (Laurentius Salvius, Holmiae, 1758), vol. Tomus I. Editio Decima, reformata.
Luo, Z.-X, Kielan-Jaworowska, K. and Cifelli, R. L. 2009. Evolution of dental replacement in mammals. Bulletin of the Carnegie Museum of Natural History 36: 159–175.
Luo, Z.-X., Yuan, C.-X., Meng, Q.-J. and Ji, Q. 2011. A Jurassic eutherian mammal and divergence of marsupials and placentals. Nature 476: 442–445.
Luo, Z.-X. 2007. Transformation and diversification in early mammal evolution. Nature 450: 1011–1019.
MacDonald, K., Scherjon, F., van Veen, E., Vaesen, K. and Roebroeks, W. 2021. Middle Pleistocene fire use: the first signal of widespread cultural diffusion in human evolution. PNAS 118: e2101108118.
Maderson, P. F. A. 2003. Mammalian skin evolution: a reevaluation. Experimental Dermatology 12: 233–236.
McCrossin, M. L., Benefit, B. R., Gitau, S. N., Palmer, A. K. and Blue, K. T. 1998. Fossil evidence for the origins of terrestriality among old world higher primates. pp. 353–396. *In*: Strasser, E., Fleagle, J., Rosenberger, A. L. and McHenry, H. [eds.]. Primate locomotion. Plenum Press, New York.
Mikkelsen, T., Hillier, L. W., Eichler, E. E., Zody, M. C., Jaffe, D. B., Yang, S. P. et al. 2005. Chimpanzee Sequencing and Analysis Consortium. Initial sequence of the chimpanzee genome and comparison with the human genome. Nature 437: 69–87.
Moorjani, P. Amorim, C. E. G., Arndt, P. F. and M. Przeworski. 2016. Variation in the molecular clock of primates. PNAS 113: 10607–10612.
Niven, L., Steele, T. E., Rendu, W., Mallye, J-B., McPherron, S. P., Soressi, M. et al. 2012. Neandertal mobility and large-game hunting: the exploitation of reindeer during the Quina Mousterian at Chez-Pinaud Jonzac (Charente-Maritime, France). Journal of Human Evolution 63: 624–635.
Nyakatura, J. A. 2019. Early primate evolution: insights into the functional significance of grasping from motion analyses of extant mammals. Biological Journal of the Linnean Society 20: 1–20.
Oppenheimer, S. 2009. The great arc of dispersal of modern humans: Africa to Australia. Quaternary International 202: 2–13.
Peeters, S. and Zwart, H. 2020. Neanderthals as familiar strangers and the human spark: how the 'golden years' of Neanderthal research reopen the question of human uniqueness. History and Philosophy of the Life Sciences 42: 33. doi.org/10.1007//s40656-020-00327.
Perkin, A. 2019. Rondo Dwarf Galago Paragalago rondoensis (Honess in Kingdon, 1997). pp. 24–27. *In*: Schwitzer, C., Mittermeier, R. A., Rylands, A. B., Chiozza, F., Williamson, E. A., Byler, D. et al. [eds.]. Primates in Peril: The World's 25 Most Endangered Primates 2018–2020. IUCN SSC Primate Specialist Group, International Primatological Society, Global Wildlife Conservation, and Bristol Zoological Society, Washington, DC.
Pickrell, J. 2019. How the earliest mammals thrived alongside dinosaurs. Nature 574: 468–472.
Plummer, T. W., Oliver, J. S., Finestone, E. M., Ditchfield, P. W., Bishop, L. C., Blumenthal, S. A. et al. 2023. Expanded geographic distribution and dietary strategies of the earliest *Oldowan hominins* and *Paranthropus*. Science 379: 561–566.
Pobiner, B. L., Rogers, M. J., Monahan, C. M. and Harris, J. W. K. 2008. New evidence for hominin carcass processing strategies at 1.5 Ma, Koobi Fora, Kenya. Journal of Human Evolution 55: 103–130.
Pomeroy, E., Bennett, P., Hunt, C. O., Reynolds, T., Farr, L., Frouin, M., et al. 2020. New Neanderthal remains associated with the 'flower burial' at Shanidar Cave. Antiquity 94: 11–26.
Prado-Martinez J., Sudmant, P. H., Kidd, J. M., Li, H., Kelley, J. L., Lorente-Galdos, B. et al. 2013. Great ape genetic diversity and population history. Nature 499: 471–475.
Preuschoft, H. 2019. Power grip or precision handling? What determines hand morphology in primates, including Hominidae? Biological Journal of the Linnean Society. 127: 694–706.
Richmond, B. G., Begun, D. R. and Strait, D. S. 2001. Origin of human bipedalism: the knuckle-walking hypothesis revisited. Am. J. Phys. Anthropol. 44: 70–105.
Riddell, E. A., Patton, J. L. and Beissinger, S. R. 2022. Thermal adaptation of pelage in desert rodents balances cooling and insulation. Evolution 76: 3001–3013.

Sankararaman, S., Mallick, S., Patterson, N. and Reich, D. 2016. The combined landscape of Denisovan and Neanderthal ancestry in present-day humans. Current Biology 26: 1241–1247.
Sarko, D. K., Rice, F. L. and Reep, R. L. 2011. Mammalian tactile hair: divergence from a limited distribution. Ann. N. Y. Acad. Scie. 1225: 90–100.
Sawyer, G. J. and Maley, B. 2005. Neanderthal reconstructed. The Anatomical Record (Part B: New Anat.) 283B: 23–31.
Schaefer, N. K., Shapiro, B. and Green, R. E. 2021. An ancestral recombination graph of human, Neanderthal, and Denisovan genomes. Science Advances 7: eabc0776.
Semaw, S., Rogers, M. J., Simpson, S. W., Levin, N. E., Quade, J., Dunbar, N. et al. 2020. Co-occurrence of *Acheulian* and *Oldowan artifacts* with *Homo erectus* cranial fossils from Gona, Afar Ethiopia. Science Advances 6: eaaw4694.
Shea, J. J. and Sisk, M. L. 2010. Complex projectile technology and *Homo sapiens* dispersal into Western Eurasia. PaleoAnthropology 2010: 100–122. doi:10.4207/PA.2010.ART36.
Shumaker, R. W., Walkup, K. R. and Beck B. B. 2011. Animal tool behavior: the use and manufacture of tools by animals. Johns Hopkins University Press. Baltimore, MD.
Silcox, M. T. and S. López-Torres. 2023. The earliest primates and euprimates and their role in the evolution of the Order. pp. 365–380. *In*: Clark Spencer Larsen [ed.]. Primate Origins: A companion to biological anthropology, second edition. John Wiley and Sons Ltd.
Speth, J. D. and Morin, E. 2022. Putrid meat in the tropics: it wasn't just for Inuit. PaleoAnthropology 2022: 327–383.
Stedman, H., Kozyak, B., Nelson, A., Thesier, D. M., Su, L. T., Low, D. W. et al. 2004. Myosin gene mutation correlates with anatomical changes in the human lineage. Nature 428: 415–418.
Steiper, M. E. and Seiffert, E. R. 2012. Evidence for a convergent slowdown in primate molecular rates and its implications for the timing of early primate evolution. PNAS 109: 6006–6011.
Thomson, K. S. 2000. Marginalia: Huxley, Wilberforce and the Oxford Museum. American Scientist 88: 210–213.
Tuniz, C. Bernardini, F., Turk, I., Dimkaroski, L., Mancini, L. and Dreossi, D. 2011. Did Neanderthals play music? X-ray computed microtomography of the Divje Babe 'flute'. Archaeometry 54: 581–590.
Tylén, K., Fusaroli, R., Rojo, S., Heimann, K., Fay, N., Johannsen, N. N. et al. 2020. The evolution of early symbolic behavior in *Homo sapiens*. PNAS 117: 4578–4584.
Ungar, P. S. and Sues, H. D. 2019. Tetrapod teeth: diversity, evolution, and function. pp. 385–429. *In*: Bels, V. and Whishaw, I. [eds.]. Feeding in Vertebrates. Fascinating Life Sciences. Springer, Cham, Switzerland.
Villa, P. and Soriano, S. 2010. Hunting weapons of Neanderthals and early modern humans in South Africa: similarities and differences. Journal of Anthropological Research 66: 5–38.
Wadley, L., Backwell, L., d'Errico, F. and Sievers, C. 2020. Cooked starchy rhizomes in Africa 170 thousand years ago. Science 367: 87–91.
Washburn, S. L. 1960. Tools and human evolution. Scientific American 203: 63–75.
Weaver, T. D., Roseman, C. C. and Stringer, C. B. 2007. Were neandertal and modern human cranial differences produced by natural selection or genetic drift? Journal of Human Evolution 53: 135–145.
Weidenreich, F. 1947. Facts and speculations concerning the origin of *Homo sapiens*. American Anthropologist 49: 187–203.
White, T. D., Asfaw, B., Beyene, Y., Haile-Selassie, Y., Lovejoy, C. O., Suwa, G. et al. 2009. *Ardipithecus ramidus* and the paleobiology of early hominids. Science 326: 75–86.
White, T. D., Lovejoy, C. O., Asfaw, B., Carlson, J. P. and Suwa, G. 2015. Neither chimpanzee nor human, *Ardipithecus* reveals the surprising ancestry of both. PNAS 112: 4877–4884.
Whittman, A. B. and Wall, L. L. 2007. The evolutionary origins of obstructed labor: bipedalism, encephalization, and the human obstetric dilemma. Obstetrical and Gynecological Survey 62: 739–748.
Wu, Y., Fan, L., Bal, L., Li, Q., Gu, H., Sun, C. et al. 2022. Ambush predation and the origin of euprimates. Science Advances 8: eabn6248.
Yunis, J. J. and Prakash, O. 1982. The origin of man: A chromosomal pictorial legacy. Science 215: 1525–1530.
Zhou, C. F., Bhullar, B. S., Neander, A. I., Martin, T. and Luo Z. X. 2019. New Jurassic mammaliaform sheds light on early evolution of mammal-like hyoid bones. Science 365: 276–279.
Zink, K. D. and Lieberman, D E. 2016. Impact of meat and Lower Palaeolithic food processing techniques on chewing in humans. Nature 531: 500–503.

Chapter 15

Cancer as Ecology and Evolution

Consider a population of living organisms which are taken to be cancer cells
(Swan and Vincent 1977)

In the high-stakes world of medicine, where anatomy and physiology, neurosciences, microbiology, and psychology dominate, ecology and evolution might seem like distant theories—until they rewrite the rules of survival itself. Evolutionary ecology looms large in diverse medical disciplines, including infectious disease transmission and pandemics, epidemiology (the ecology of human parasites, including eukaryotic, prokaryotic, and viral ones), antibiotic and drug resistance, and diverse human syndromes with genetic, lifestyle, and environmental bases. An emerging and increasingly important field of health sciences called evolutionary medicine uses principles of ecology, evolution, anthropology, and the historical evolutionary ecology of humans to address diseases, in particular, and human health issues more generally (Stearns and Medzhitov 2024). Perhaps surprising at first, cancer is increasingly being defined, understood, and treated in the context of ecological and evolutionary principles.

Consider a Phase 1 clinical trial for men with advanced metastatic prostate cancer (Zhang et al. 2017). At this stage, the cancer has metastasized to the bone, perhaps forming 20 or more large, separate tumors spread across long bones, the rib cage, and even the skull. Additional metastases may occur in the liver or lungs. The disease has become incurable and fatal. In the control group, all patients died, with a median time to treatment failure of about 14 months and a median survival time of about 30 months. However, patients who received a novel treatment strategy based on ecological and evolutionary principles saw significantly better outcomes: median times to treatment failure extended to 33 months, and median survival increased to 55 months (Zhang et al. 2022). Remarkably, several men in this group remain alive 9 years later, at the time of writing. This success was achieved despite both groups receiving the same drug. The difference lay in the treatment approach: the control group followed the standard of care with continuous drug therapy, while the more successful "adaptive therapy" group received treatment discontinuously, guided by the cancer's cell dynamics within each patient. The adaptive therapy cohort received drugs less than 50% of the time—yet lived nearly twice as long. How is this possible? The concepts and principles of the preceding chapters provide the answer.

For men diagnosed with late-stage prostate cancer, the news is devastating—both for them and their loved ones. If detected early, most cancers can be cured (Hubbell et al. 2024). However, when **metastatic**—meaning it reaches an advanced stage and spreads to other parts of the body—it is often lethal, even after months or even years of surgery, radiation, and various therapies, including chemotherapy, hormonal therapy, and immunotherapies. While these treatments may prolong life, they can also be toxic and debilitating. Even with continued advances in research and patient care, cancer continues to claim far too many lives. At the time of writing, it accounts for 1 in 3.5 deaths in Sweden, 1 in 4 in the Netherlands, and 1 in 5 in the U.S. In Africa, both the incidence and mortality rates of cancer are rising. In some ways, this reflects progress. While some of Africa's

DOI: 10.1201/9781003008538-15

increase in cancer diagnoses is due to lifestyle factors and exposure to carcinogens, much has to do with increases in life expectancies over the past few decades (Hamdi et al. 2021). People are living longer, dying less from other causes, and hence are more likely to experience cancer.

The "War on Cancer" was a popular metaphor in the 1960s to focus attention on finding ways to cure cancer. The emphasis on "war" suggested that curing cancer would require applying the newest and best therapeutics at the maximum dose tolerable to the patient. How did it pan out over the decades? In terms of overall technological advancement, we have seen remarkable progress—space exploration, high-speed trains, all manner of computers and computing capacity, AI, LED screens and lights, and cellphones that function as cameras, storage devices, internet and computer devices, and much more. In terms of cancer care, the decades since the 1960s have seen remarkable developments in robotic surgery, stereotactic radiation therapies, and a stream of new drugs—even as drug development has grown more costly and less likely to succeed. Currently, significant advances are occurring in immuno-therapy. For example, in adoptive therapies such as CAR-T cell therapy, a patient's immune cells (cytotoxic T-cells) are extracted, trained *ex vivo* using cancer cells from live biopsies (Figure 1), multiplied, and then reinjected into the patient. Like tiny bloodhounds, they seek out and destroy cancer cells. Yet, despite these advances, cure rates have improved only modestly for advanced, metastatic cancers, such as those originating in the liver, colon, breast, or prostate and spreading to other organs. Similarly, lethal cancers like those of the lung and brain remain challenges.

Explanations for why curing advanced cancers has been so challenging include the inherent complexities of tumors, the vast variety of cancers across tissue types and individuals, the need to kill the cancer cells without causing fatal or serious harm to the patient, and the genetic and phenotypic heterogeneity within even a single tumor. Perhaps the most critical factor, however, is

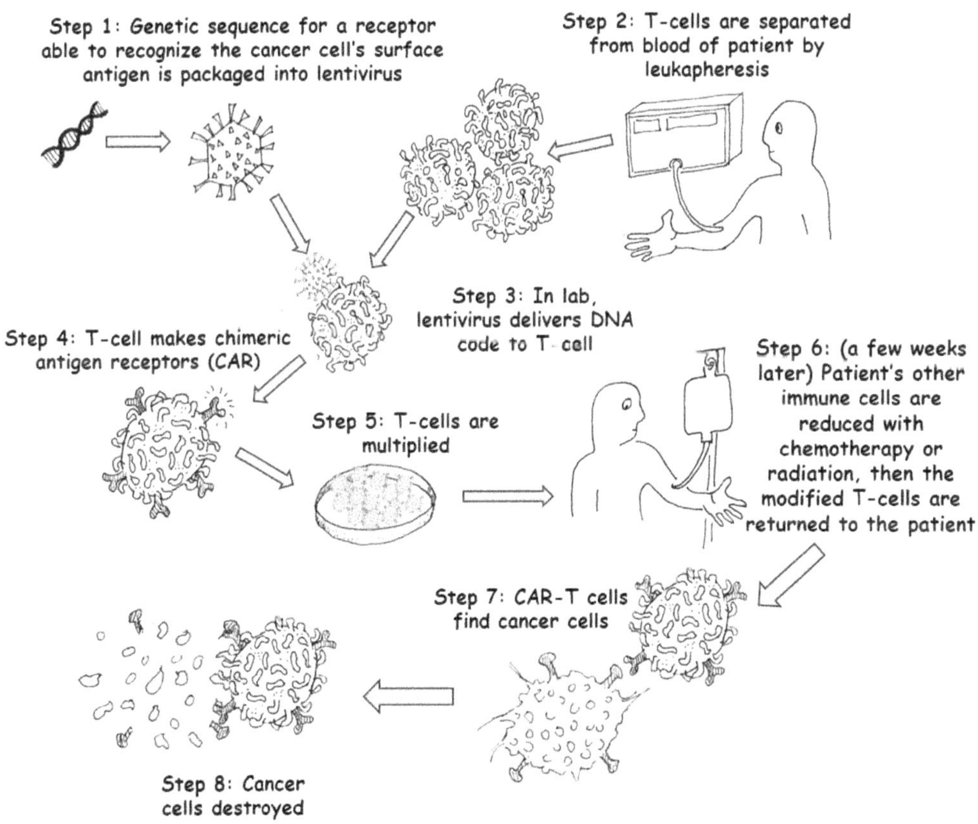

Figure 1. CAR-T cell cancer therapy (conceptual, not to scale).

the evolution of resistance—a key driver of treatment failure. If you do not kill all the cancer cells (often attempted with surgery and radiation in early-stage cancers), the surviving cells will develop resistance, rendering subsequent treatment ineffective.

The challenge in studying and treating cancer begins with how we define the disease itself. Written records reveal that ancient Mesopotamians, as early as 6,000 years ago, recognized cancerous tumors, describing them in cuneiform as "ulcers of a bad kind," "ulcers which spread," and "malignant ulcers" (Fitzgerald 2000). Similarly, in Egypt, the Papyrus Ebers (c. 1538 BCE) categorized tumors with possible reference to cancers described as "eating of the uterus" and breasts with "bleeding, discharge, and eating" (Hajdu and Darvishian 2010).

The ancient Greeks, such as Hippocrates (c. 410–c. 360 BCE), were probably the first to classify cancer as a distinct disease, distinguishing between benign and malignant tumors (Hajdu and Darvishian 2010). They observed that malignant tumors grew unusually fast, exhibited bloody features (akin to modern understanding of **angiogenesis**), and recognized that early-stage cancers could be surgically removed, whereas late-stage interventions often proved fatal (Faguet 2014). In the Middle Ages, Gabriele Fallopius (1523–1562) began describing the physical attributes of malignant cancer—dense (like a breast lump), irregular in shape, with congested blood vessels—and he saw tumors as being at once emergent from a tissue or organ, and quite distinct from the tissue of origin (Faguet 2014). Once the microscope was invented after the 1600s and resolution had improved in the 1800s, it could be seen that tumors included not only normal cells but also abnormal cancer cells, and that cancer cells had derived from normal cells (Faguet 2014). By the 1950s and 1960s, cancer cells were known to proliferate rapidly and seemingly uncontrollably, thus promoting the expansion and growth of the tumor. They also exhibited chromosomal and genetic changes to their morphology and cellular processes. Nowell (1976) later established cancer as an evolutionary disease, wherein sequential and phenotypic changes transform normal cells into malignant ones. Critically, these cancer cells continue evolving, including—and most problematically—the development of therapy resistance.

Currently, the definition of cancer found in national health services dictionaries and medical dictionaries states: "Cancer is a disease of uncontrolled proliferation of cells with the capacity to invade other parts of the body" (Brown et al. 2023). Notably, this definition omits the fact that cancer cells evolve. While it describes what cancer does, it fails to capture what cancer is.

Julian Huxley may have been the first to articulate this concept. As a key architect of the *Modern Synthesis* (Huxley 1942; see Chapter 13 *Evolutionary Ecology*), which harmonized Mendelian genetics with evolution by natural selection, Huxley was also a grandson of Thomas Huxley—famously known as "Darwin's bulldog" for his defense of Charles Darwin's theory of natural selection in the 19th century. In 1956, Julian Huxley proposed that cancer represents a speciation event: a cell (or group of cells) that was once part of a multicellular organism breaks away, effectively becoming an independent single-celled organism subject to evolution by natural selection (Huxley 1956).

Normal cells are integral components of the traits that make up a whole organism. But when seen as a speciation event, cancer cells become individual organisms, the tumor serves as their ecosystem, and the patient acts as their planet.[1] It was not until 2000 that a cancer biologist formally saw cancer as a speciation event (Duesberg and Rasnick 2000). Later, in 2010, the oncologist Mark Vincent made the case for this perspective, describing cancer as "the animal within" (Vincent 2010). More recently, Jason Somarelli and Norman Johnson eloquently framed cancer as "the species within" and argued for an evolutionary ecology approach to understanding and treating human cancers (Somarelli 2021; Somarelli and Johnson 2024). An updated definition of cancer now seeks

1 Intriguingly, we note that there are rare cases of communicable cancers such as the Tasmanian devil (*Sarcophilus harrisii*) facial tumor and the canine transmissible venereal tumor (Woods et al. 2018); a parasite-host metaphor would be more appropriate in these cases.

to define it by what it is rather than only by what it does: "Cancer is a disease of uncontrolled proliferation by transformed cells subject to evolution by natural selection" (Brown et al. 2023).

Thinking back to Chapter 13 *Evolutionary Ecology*, we can ask whether cancer cells meet the criteria for evolution by natural selection: namely, do cancer cells have heritable variation, a struggle for existence, and does the heritable variation influence the struggle for existence?

Heritable Variation

The cancer cells of a patient display diverse sources of heritable variation, including genetic and epigenetic variations, chromosomal rearrangements, gene copy-number variations, polyploidy, and aneuploidy (Laisné et al. 2025). A population of cancer cells generates far more mutations of all kinds compared to the normal cells from which they originate. This property of cancer cells has been termed the "mutator phenotype" (Loeb 2001). Much of this variation shapes cancer cell phenotypes, influencing morphology, motility, cell metabolism, and interactions with neighboring normal and cancerous cells. Crucially, this variation is inherited and expressed in daughter cells generated from the original cell variant. So, yes, cancer cells develop and display considerable heritable variation.

Struggle for Existence

Recall that the struggle for existence recognizes how organisms can grow exponentially under ideal conditions, yet no population can grow exponentially forever; there are limits to growth (Chapter 2 *Population Dynamics*). Empirical evidence from *in vivo* and *in vitro* studies using patient-derived cell lines demonstrates that aggressive cancer cells (particularly in lung and ovarian cancers) can exhibit intrinsic growth rates ranging from 0.35 to 0.70 per day (Sarapata and de Pillis 2014). This corresponds to a doubling time of every 24–48 hours[2]—a growth rate quite similar to that observed in many aquatic, single-celled eukaryotic algae.

When grown in cell culture, an initial inoculum of cancer cells can grow exponentially, but for only 1–2 days. Because of this, after 3–6 days, cells growing in culture are typically **passaged** by taking a fraction of the now substantially larger population of cancer cells and then **plated** onto a fresh culture dish. If left alone for too long, the cancer cells will reach **confluence**—covering the entire surface area—and cease proliferation as they run out of space. Besides space as a limiting resource, growth limitations in cultured systems arise from multiple factors: nutrition depletion (glucose, lipids, amino acids, and growth factors), accumulation of metabolic byproducts (e.g., lactic acid), and acidification of the medium (pH 6–6.5), which both cancer and normal cells find intolerable. Notably, most aquatic single-celled eukaryotic species can tolerate a much wider range of pH (Amaral-Zettler 2013).

To circumvent these limitations on cancer cell growth, researchers have developed alternative culture systems: 3-D spheroids and organoids—where the cells are grown in 3-D droplets of media—or "living biopsies"—where the cancer cells exist in a matrix of inedible gel along with normal cells such as fibroblasts and immune cells typical of actual tumors. In 3-D cultures, rather than passaging the cells, the cancer cells either have their growth media refreshed every few days or specialized microfluidic devices create a constant flow of new media in and old media out of the cell culture. Regardless of the culture method, one often sees exponential growth of cancer cells followed by either a levelling off or decline—clear evidence of growth-limiting mechanisms.

Cancer cells require what at first seems like a very surprising resource: growth hormones. Hormonal dependence varies by cancer type; breast cancers often require estrogen to proliferate, prostate cancers depend on testosterone, and lung cancers need epidermal growth factor.[3] This specific nutritional requirement explains why cancer cells will generally fail to thrive in growth

---

[2] Doubling time for exponential growth $T = \ln(2)/r$.

[3] A substance made by the body that functions to regulate cell division and cell survival.

media that would suit most free-living single-celled microbes. To provide these necessary growth factors, most cancer cell media incorporate fetal bovine serum. Why the need for growth factors? Multicellular organisms bathe their cells in serums rich in nutrients, hormones, steroids, and other types of growth factors, and can regulate the proliferation or reduction of certain cell types within specific tissues and organs. The need for such growth factors is a remnant of the cancer cell's multicellular origin. Within tumors, these same growth factors can become key limiting resources to the cancer cells. This phenomenon was clearly demonstrated in our prostate cancer trial patients, where growth factor availability proved to be a critical determinant of tumor progression.

The need for growth factors in many cancers has led to the successful development of drugs that either suppress the patient's production of hormones (such as estrogen or testosterone) or specifically target and bind to the cancer cell's receptor proteins. Hormonal therapies in breast cancer and prostate cancer provide successful first-line therapies, but resistance becomes inevitable, especially in late stages of the disease. In breast cancer, drugs like tamoxifen bind to estrogen receptors, but the cancer cells may evolve resistance—either by increasing their response to estrogen with additional receptors or by "rewiring" metabolic pathways to function independently of estrogen. A similar pattern occurs in metastatic prostate cancer. First-line therapies can include drugs like Lupron and Relugolix that suppress the body's production of testosterone, and drugs such as Abiraterone that target the cancer cells' androgen receptors, thus starving cancer cells reliant on testosterone as a growth factor. In the clinical trial with the late-stage prostate cancer patients we discussed earlier (Zhang et al. 2017, 2022), the resistant cancer cells adapted by becoming testosterone-independent or producing their testosterone. Stay tuned—more to come on this trial.

In patients, the rate of growth of the cancer cell population is hard to measure precisely. Imaging techniques such as MRI, PET, or other radiographic scans can be used to estimate both the sizes and distributions of cancer cells within the body. Some cancers release molecules into the bloodstream—such as PSA (prostate-specific antigen) for prostate cancer, CEA (carcinoembryonic antigen) for some colorectal, breast, ovarian, lung, liver, and pancreatic cancers. These blood biomarkers are generally low in the absence of cancer and become substantially elevated in its presence, thus providing an estimate of the overall burden of cancer cells. While imperfect, such blood biomarkers can be measured frequently and often inexpensively. They can indicate the cancer's growth rate in the absence of treatment, the effectiveness of treatment in lowering the cancer cell population, and the point at which treatment efficacy ceases due to the evolution of resistance. Newer blood biomarkers now include circulating RNA or DNA, but PSA remains the primary biomarker used in prostate cancer trials.

The rate of cancer cell growth varies considerably among patients, types of cancer, and stages of cancer. Glioblastomas (a type of brain cancer) show growth rates ranging from as slow as 0.21% per day to as fast as 64% per day (mean: 0.73%; James et al. 2007), with larger tumors showing significantly lower growth rates than smaller ones (Stensjøen et al. 2015). Sarcomas generally grow at a rate of 2.5% per day but can reach up to 8.0% per day (Robertson-Tessi et al. 2023). Slower-growing cancers include breast cancer (0.2% to 1.5% per day; Dahan et al. 2021), prostate cancer (0.1%; Schmid et al. 1993), and colon cancer (0.5 to 1.5%), though growth accelerates when metastasized to the liver (Tanaka et al. 2004). Having cancer cells growing at even 1% per day is scary. Compounded over time, such a growth rate will see the cancer cell population double every 70 days.[4] However, relative to their cultured intrinsic growth rates of r_{max} = 0.35–0.70 (doubling every 1–2 days; Sarapata and de Pillis 2014), cancers in patients grow much more slowly, indicating that conditions for cancer cells within their tumor ecosystem are far from ideal.

In patients, cancer cells experience limits to growth. Within a tumor, they likely achieve local equilibria or carrying capacities set by the availability of blood supply. These tumor microenvironments

[4] Doubling time for discrete growth T = ln(2)/ln(1+percent increase).

foster intense competition among cancer cells for space, oxygen, glucose, lipids, amino acids, trace nutrients (e.g., zinc, iron, selenium), and growth factors. Interference competition among cancer cells can involve the release of toxic metabolites, reduction of pH, or secretion of signaling proteins that interfere with the metabolic processes of neighboring cells. Some tumor regions—particularly areas of **necrosis** caused by insufficient blood flow or serum profusion—can become so hypoxic, nutrient-deprived, and toxin-filled that few, if any, cancer cells can survive or thrive. These necrotic regions, including the formation of a "necrotic core," are common features of tumors as they grow. Blood perfusion, which starts from the edges of the tumor, may fail to reach deeper into the tumor. Furthermore, blood flow within the tumor can be erratic temporally and spatially due to the dynamic nature of blood vessel formation, blockages, and the influence of various cancer cell populations. Some cancer cells co-opt the blood flow to other neighborhoods by producing **vascular endothelial growth factors** (VEGF; Gillies et al. 2018), which promote angiogenesis and direct the formation of new blood vasculatures. So, yes, cancer cells experience a struggle for existence.

Heritable Variation Influences the Struggle for Existence

How heritable variation matters in the struggle for existence is most clearly seen when a patient's cancer cells evolve resistance to therapy. Yet, cancer cells also evolve diverse adaptations that promote proliferation and survival. These multiple adaptations allow cancer cells to exploit opportunities such as nutrient availability, evade hazards such as the immune system, engineer their environment by modifying the collagens in the extracellular matrix, and compete with other cells. The **hallmarks of cancer**, as described by Hanahan and Weinberg (2000, 2011), represent broad categories of phenotypes exhibited by cancer cells. Hanahan and Weinberg (2000) recognized that these hallmarks reflect a Darwinian process. We can use their hallmarks framework to more closely examine cancer as ecology and evolution (Bhattacharya et al. 2025).

Becoming an "Individual" Subject to Natural Selection

Three of the hallmarks of a cancer cell—insensitivity to anti-growth signals, evading programmed cell death, and limitless replicative potential—describe the fundamental prerequisites for a cancer cell to function as an organism subject to evolution by natural selection. Under normal conditions, cells are subject to tissue control mechanisms to ensure homeostasis and proper tissue function. If a normal cell divides (proliferates) and produces daughter cells that are unnecessary or even detrimental to tissue function, it receives **anti-growth signals** from surrounding cells or even other parts of the body, instructing it to stop dividing. During cancer initiation, however, the cell evolves the ability to ignore or suppress these anti-growth signals.

Normal cells that become superfluous, damaged, or mislocated either "self-diagnose" or receive signals from other cells to "commit suicide," or programmed cell death, known as **apoptosis**. When the apoptotic pathway is initiated, the cell breaks itself down and dies in a manner that minimizes the release of toxins and facilitates debris removal by fibroblasts and immune cells. With cancer initiation, however, the cancer cell disables this process. It evades programmed cell death and no longer "voluntarily" dies when internally or externally "commanded" to do so.

Finally, except for progenitor and stem cells, normal cells in our body differentiate to perform their whole-organism function. For instance, **keratinocytes** (cells that compose our skin) are produced through the asymmetric division of progenitor cells residing on the skin's basal membrane. As part of the normal skin turnover (which occurs every 28–40 days) or during wound healing, progenitor cells receive signals to divide. Upon division, one daughter cell remains a progenitor, while the other begins a process of differentiation into mature keratinocytes. During differentiation, the cell may continue to divide briefly, but all resulting progeny will migrate from the basal membrane to the skin surface, where they become fully differentiated and incapable of undergoing division. Similar processes of regeneration, growth, and healing occur throughout all the tissues in the body. To be a cancer cell and transition from being part of you to being its single-celled organism, it must evolve to bypass differentiation and remain in a state that permits limitless replicative potential. At this point, the cancer cell essentially behaves as a single-celled eukaryotic organism.

Evolving Growth Promoting Adaptations

The following three hallmarks of a cancer cell—self-sufficiency in growth signals, sustained angiogenesis, and tissue invasion and metastasis—indicate limits to growth. These adaptations free cancer cells from relying on external signals for cell division, increase resource availability through **angiogenesis** (the recruitment of blood vessels to the neighborhood of cancer cells), and allow for range expansion into adjacent tissue spaces or invasion into another organ of the host.

Everything from turtles to humans uses UV radiation to catalyze reactions that promote the production of vitamin D, a vitamin necessary for regulating calcium and phosphate for the maintenance, repair, and growth of bones. For various species of pond turtles, this means crawling out of the water and basking in sunlight. For humans, daily outdoor activity exposes the skin to UV radiation, where moderate exposure promotes healthy vitamin D levels, while excessive exposure can damage skin and increase mutation rates in skin cells, ultimately leading to skin cancers. While vitamin D can be acquired through the vertebrate's diet, the ability to synthesize it internally removes the need to rely solely on dietary sources. Similarly, for cancer cells, self-sufficiency in growth signals frees them from the need to forage for, or be limited by, the growth signals emanating from normal cells as part of whole-tissue homeostasis.

Most ecosystems have autotrophs or primary producers that form the base of food chains. Organic carbon, as an energy storage molecule, enters the food web either through chemosynthesizers around deep-sea vents or photosynthesizers using sunlight (see Chapter 8 *Food Webs*). Not so for cancer cells. All energy and nutrients available to cancer cells within their tumor ecosystem either diffuse directly from the blood or lymph, perfuse through interstitial fluids, or are scavenged from the detritus left by dead intra-tumoral cells, both normal and cancerous. However, scavenging macromolecules or macromolecule aggregates from dead cells (a process often referred to as micropinocytosis) cannot permanently sustain cancer cells. (Recall from Chapter 10 *Nutrient Cycling*, that while nutrients can be recycled in a system, energy cannot due to thermodynamic laws). Even if nutrients are recycled within the tumor, there must be new inputs of oxygen and other nutrients to provide energy to living and dividing cells. These new inputs ultimately come from the blood, which is continually replenished with cellular resources by the lungs (oxygen), liver, bones, and other organs.

Highly regulated cell-cell signaling processes between diverse types of normal cells maintain healthy blood vessels within our tissues. However, unlike normal cells, a tumor does not serve the body it inhabits, so the normal processes of maintaining and growing blood vessels cease. This does not mean that the tumor lacks blood flow. Cancer cells co-opt normal cells within the tumor microenvironment to produce blood vessels by secreting VEGF or by signaling other normal cells to do so. The production of VEGF by cancer cells becomes a form of ecosystem engineering—much like a beaver dam—where cancer cells enhance the quality of their habitat by inducing blood flow and hence increasing nutrient availability. The tumor's sustained angiogenesis results in a highly disorganized network of vessels, where blood flow can be erratic—sometimes slow, sometimes fast—and may vary locally within the tumor in a matter of seconds, minutes, or hours (Gillies et al. 2018).

That angiogenesis provides an ecosystem service to the cancer cells of a tumor is well established. What remains unknown is what kind of evolutionary game drives cancer cells to produce VEGF. Some have suggested it represents a form of cooperation, where a cancer cell pays a cost for the benefit of neighboring cancer cells. However, as Archetti and Pienta (2019) point out, cooperation cannot evolve in games like the Prisoner's Dilemma (see Chapter 7 *Light Competition in Forests and Evolutionary Game Theory*), where the cost of cheating among cooperators is less than the cost of cooperating. More likely, VEGF production by a cancer cell provides a public good. While the cell that produces VEGF gains a net benefit, other nearby cells also reap some reward from this VEGF production (potentially leading to a game with producers and free-loaders; Salimi Sartakhti et al. 2017). Another intriguing possibility is that VEGF production represents a tragedy of the commons or an arms race game—similar to how trees compete for light by growing taller (Chapter 7). In this scenario, a neighborhood of cancer cells may "steal" blood flow from another by producing more VEGF, just as taller trees capture more sunlight at the expense of their shorter neighbors. If a neighborhood of cancer cells reaps rewards (nutrients) proportional to its production of VEGF, then an arms race may ensue where both neighborhoods end up producing significantly more VEGF than would be optimal for either group. Yet, whichever game the cancer cells are playing, in the end, it results in angiogenesis and a sustained flow of blood into the tumor ecosystem.

Range expansion associated with the colonization of new habitats is commonly seen in nature and can result in subsequent adaptation to local conditions (ecotypes; see Chapter 13: *Evolutionary Ecology*). This has been the case for the three-spine stickleback in the Lake Geneva region of central Europe. Since its introduction 140 years ago, the species has expanded its range and evolved habitat-dependent phenotypic divergence over time (Lucek et al. 2014). Tissue invasion and metastasis have much in common with the range expansions of introduced invasive species (Brown et al. 2024). The metastatic cascade comprises a series of recognized steps through which a cancer cell spreads from the primary tumor to establish a new tumor in a distant location—sometimes within the same tissue type, but often in a different tissue type or organ. These steps include entry of cancer cells into a blood vessel (**intravasation**), movement through the circulatory system (**circulation**), exit from the blood vessel in surrounding tissue (**extravasation**), and finally establishment and growth of a new tumor in the colonized environment (organ). It is easy to see a parallel with the typical steps of biological invasion: transport, introduction, establishment, and spread.

Evolving Foraging Strategies and Predator Defense

Another three hallmarks—deregulated cell metabolism, tumor-promoting inflammation, and avoiding immune responses—are adaptations that help cancer cells save and acquire resources, improve resource availability, and evade detection and death by the immune system.

Cancer cells evolve deregulated metabolism in much the same way cave fish evolve through the loss of unnecessary functions. Cave fish have independently evolved numerous times, as caves become colonized by freshwater fish from adjacent rivers. Once inhabiting this new lightless environment, cave fish invariably evolve blindness and lose skin pigments, often becoming translucent (Behrmann-Godel et al. 2024). Eyes and pigments are no longer necessary in the dark, but more importantly, they are costly to maintain and, in the case of eyes, can pose a source of infection or injury.

For cancer cells, adaptive changes to metabolism are drastic, evolving from the metabolic traits of the originating tissue into cells with the hallmarks of cancer. Imagine a normal epithelial cell of the lung: it takes up nutrients such as glucose only as needed and is positioned among its neighboring cells to optimize lung function and architecture. Its positioning is based on signals emitted by other cells that dictate cell-cell adhesion or contact inhibition, ensuring ideal spacing for the organ's overall function. The normal lung cell proliferates, differentiates, and dies in response to tissue-level control processes. Such is the state of a newly initiated cancer cell derived from lung epithelium. In direct violation of orders, this rogue cell produces descendants, which become a population of now single-celled organisms inhabiting an expanding tumor ecosystem within the lung. A predictable set of energy-saving loss-of-function adaptations soon evolves. Lung cancer cells lose the ability to produce surfactants, signaling molecules, and other extracellular products that contribute to lung function. Because none of these products contributes to the fitness of the cancer cell, shutting down their production saves energy. As discussed in Chapter 5 *Foraging Dynamics*, saving energy is adaptive: cancer cells that continue to produce these products will be selected against in favor of those that do not.

Furthermore, due to resource limitations, a population of cancer cells will evolve the ability to increase the activity of transporters, such as GLUT1, to increase the speed of resource acquisition. This is a gain-of-function, analogous to the evolution of new adaptations in cave fish that allow survival in darkness. For instance, cavefish develop enhanced external pressure sensors in their lateral line system, exhibit head asymmetry, and adapt "wall-following" behavior—improving spatial orientation and positioning in areas with higher feeding rates (Chen et al. 2022). Similarly, cancer cells also gain ways to speed metabolic processes near the cell membrane by evolving a form of anaerobic respiration (glycolysis) that bypasses the mitochondria. This process mirrors the glycolytic pathway used by muscle cells when activity demands exceed what can be supported by aerobic respiration (think lactic acid production, muscle pain, and next-day stiffness). However, cancer cells often maintain high levels of glycolysis even in the presence of oxygen, thus contributing to an acidic environment within and adjacent to the tumor. This acid production can be a source of pain experienced by patients.

Tumor-promoting inflammation and avoiding immune responses both involve cancer cells exploiting opportunities and avoiding hazards posed by the patient's immune system, and both have parallels in other ecological systems, often in the form of direct or indirect mutualistic interactions.

Nature has evolved a variety of symbioses or mutualisms in cases that might otherwise be predator–prey systems. One unique association occurs between particular ant species that "ranch" aphids for their honeydew (Nelson and Mooney 2022). Rather than preying upon the aphids, the ants keep them safe from predators and parasitoids while the aphids sit stationary on plants, accessing plant sap with their specialized mouthparts. The aphids excrete a sugar and protein-rich honeydew that rewards the ants for their service with an abundant and rich supply of food. It is almost as if the ants are operating their dairy farm.

Let us see how cancer cells evolve to mimic ants via tumor-promoting inflammation. In the process of wound healing, components of the immune system, such as Type A macrophages, secrete nutrients (a process referred to as co-feeding) for other cells to facilitate the rehabilitation of the damaged tissue. Cancer cells evolve to exploit this wound healing property of macrophages by inducing the production of cytokines—chemical messengers—which fool the host's body into treating the tumor as a wound. Macrophages are thus recruited to the tumor site, where they provide co-feeding—nourishing the cancer cells at the patient's expense. This relationship mirrors the symbiosis between aphids and ants: Aphids secrete honeydew for their ant minders, benefiting the ants but harming the host plant that feeds the aphids. Because the body is continuously supplying these macrophages, tumors have been described as "wounds that never heal."

Our immune systems can be thought of as a security agency dedicated to identifying and attacking pathogens of all sorts, including cancer. This agency is staffed by diverse cell types that respond both locally and systemically to infections. Some cells identify the infection, while others are trained to attack infected cells or the pathogen directly, and still others, like a branch of internal affairs, prevent the overproduction of the "killer" components of the immune system. The immune system is finely balanced to be effective at eliminating infections while minimizing the risk of **autoimmunity** (where our immune cells mistakenly attack perfectly normal cells of our body, as seen in many allergies).

Cancer cells can evade the pathogen-attacking component of the immune system in much the same way that some plants have evolved to avoid herbivores. For example, when subjected to herbivory from caterpillars such as *Manduca quinquemaculata*, wild tobacco (*Nicotiana attenuata*) in Utah's Great Basin desert emits volatile organic compounds that attract predators to the *Manduca* eggs laid on the plant (Kessler and Baldwin 2001). This mutualistic interaction benefits both the plant—by triggering a trophic cascade (see Chapter 8 *Food Webs*)—and the predators, which gain easy access to prey. Like tobacco plants suffering herbivory, cancer cells within a tumor are attacked and destroyed by cytotoxic T-cells, a "killer" component of the immune system. Regulatory T-cells prevent the over-proliferation or hyperactivity of these cytotoxic T-cells, much like insectivorous predators consuming herbivorous moths. As part of avoiding the immune response, cancer cells also evolve mechanisms to signal and recruit regulatory T-cells, thus duping our immune system into shutting down the proliferation and activity of cytotoxic T-cells. This is analogous to how some plants emit volatile chemical signals to attract predators to their herbivores. However, unlike the evolved mutualism between plant and predator, where both parties benefit, in this case, the regulatory T-cells are deceived, ultimately to the detriment of the patient's body. This is just one way cancer cells evolve adaptations for evading the immune system. Many other examples exist that draw fascinating parallels to predator-prey dynamics in nature (Peplinski et al. 2021).

It is easy to envision a kind of predator-prey interaction as cytotoxic T-cells kill and prey upon cancer cells. Models of cancer cell–immune cell interactions share many features with predator-prey models discussed in Chapter 4 *Species Interaction: Predation* (Kareva et al. 2021). Yet cytotoxic T-cells are peculiar predators. While they do proliferate within the tumor, their proliferation rate—unlike most predators—is not determined by how many cancer cells they kill (their functional response). This is partly because these T-cells do not consume the cancer cells they kill—other cells, such as macrophages, may clean up the mess—and also because the signal to proliferate for cytotoxic T-cells comes from the overall presentation of antigens in the environment or from signals emanating from other immune cells. In other words, cytotoxic T-cells do not work on commission like typical predators. In fact, as these T-cells wrestle in combat with cancer cells—rupturing the cancer cell's membrane and injecting toxins—they can sustain damage or enter a state of "T-cell exhaustion," where, after many "kills," they lose effectiveness. To add complexity, cancer cells compete with T-cells for glucose. By foraging efficiently and maintaining a low threshold for giving-up density for glucose (Chapter 5 *Foraging Dynamics*), cancer cells can starve the immune cells, creating regions of the tumor devoid of cytotoxic T-cells.

Cancer cells evolve other "anti-predator" adaptations to evade the immune system in ways that mimic (a pun as we shall see) natural predator-prey systems. A simple approach is to cease the production of the antigen (often a cell membrane surface protein) that allows the cytotoxic T-cells to distinguish cancer cells (intruders) from normal cells ("self"). While masquerading as a normal cell may sound like camouflage, it is not; in fact, it is a form of Batesian mimicry, where a palatable prey (like a Viceroy butterfly) to a bird looks just like an unpalatable prey (like a Monarch butterfly). In this case, normal cells, for reasons of autoimmunity, are unpalatable to a cytotoxic T-cell. So, if a cytotoxic T-cell encounters a cancer cell mimicking normal cells, the cytotoxic T-cell will leave the cancer cell unharmed, just as a predator will avoid preying upon an unpalatable mimic of a toxic model species.

Some normal cells present antigens; however, our body has two signals to distinguish these normal cells and prevent autoimmunity. Both involve increasing the production of proteins known as ligands. One of these ligands, CD44, acts as a "do not eat me" signal. When a T cell comes into physical contact with a CD44-positive cell, it simply releases the putative prey and moves on. A stronger anti-immune response involves the PD-L1 ligand. When this protein attaches to the PD receptor on the T-cell, it acts as an off switch. When activated by PD-L1, the T-cell not only moves on—it ceases to function altogether. These mechanisms allow normal cells to suppress autoimmunity when experiencing an overabundance of cytotoxic T-cells. However, cancer cells across many different cancer types will evolve to permanently increase the production of one or both of these ligands, thereby evading immune detection.

All the hallmarks of cancer converge on a central theme: cancer cells live in an environment in which they interact with other kinds of cells. It should now be clear that when different types of organisms interact within an environment, we are entering the realm of ecology.

The Ecology of Tumors

Tumors are not homogeneous masses of cells but rather complex ecosystems composed of a diverse array of cancer cells, normal cells, and extracellular matrices rich in collagens (Egeblad et al. 2010). As a rule of thumb, a cubic centimeter of tumor has a wet weight of about 1 gram and may contain

100 million to 1 billion cells. In the case of breast cancer or other cancers that can be palpated, a tumor typically reaches 1–2 cm in diameter before it can be felt, though untreated primary breast tumors can exceed 5 cm in diameter (approximately 65 grams). In advanced stages of diseases such as colon cancer, tumors may manifest not only in the colon but also as metastases in the liver and lungs, collectively reaching 500 grams or more in total mass. Among these tumors, the total population size of cancer cells within a single patient—what we might call the cancer's world—can far exceed the global human population. A conservative estimate (100 million cells/gram x 500 grams = 50 billion cells) far exceeds Earth's 8.3 billion people (2025 projection). Given such vast numbers, it is not unexpected that tumors, much like Earth's ecosystems, exhibit striking heterogeneity. This variation occurs across multiple scales: between tumor sites of a patient (think biomes), within separated regions of a single tumor (think macrohabitats), and at the scale of a hundred or fewer cells occurring in what is referred to as the tumor microenvironment (think microhabitat). At all these scales, cancer cells are both influenced by and influencers of heterogeneity.

Like nature at large, environmental heterogeneities and associated tradeoffs among cancer cells lead to mechanisms of coexistence. An emerging field of **cancer community ecology** (Kotler and Brown 2020) examines how different cancer types (analogous to closely related species) diversify and evolve from the original cancer cell lineage, occupying distinct ecological niches. On a microhabitat scale, Strobl et al. (2020) suggest that tumors containing two coexisting cancer "species"—one that produces acid that kills growth-inhibiting stromal cells and the other that degrades the movement-blocking extracellular matrix—are more invasive than tumors where these "species" are spatially separated. Coexistence of cancer cell types is possible at small spatial scales when each species exerts stronger intraspecific competition than interspecific competition (Strobl et al. 2020).

A clear source of heterogeneity at the microhabitat scale is proximity to blood vessels. Near a vessel, resource supply is high, and competition among cancer cells is consequently intense. In contrast, areas farther from the vasculature are resource-poor and consequently less competitive. Using the Lotka-Volterra competition model, Freischel et al. (2021) showed *in vitro* the coexistence conditions for two breast cancer cells—one acting as a "cream skimmer" and the other as a "crumb picker." As expected, these distinct cell types can coexist along the resource gradient in a manner reminiscent of the dynamics observed in Connell's barnacles (Connell 1961; Chapter 3 *Species Interaction: Competition*). The cream skimmer can exist only in the resource-rich habitat, whereas the crumb picker, capable of surviving in both, is competitively restricted to the poorer habitat by the former (Joanne et al. 2001).

Cancer biologists often distinguish between three idealized macro-habitats within a tumor: the edge, the interior, and the necrotic core. Depending on the tumor's shape and its vasculature, the edge may be more or less distinct, appearing distinctly ellipsoid or irregular, with protrusions extending into the surrounding tissue. Necrotic regions can be multiple and may shift over time and space. The tumor's edge interfaces with intact, non-degraded normal (host) tissue, where resources are abundant, reducing competition among cancer cells. However, this region is also highly exposed to the immune system, making it a high-reward, high-risk habitat. In contrast, the tumor's interior is characterized by scarce resources, more intense competition among cancer cells, and lower immune system infiltration. In a study of 15 breast cancer patients, Lloyd et al. (2016) demonstrated the existence of distinct cancer cell types between the tumor edge and interior. The predominant cancer cell type at the edge exhibited traits such as immune evasion, rapid cell turnover, and efficient nutrient uptake. In contrast, the dominant cell type in the interior showed a contrasting set of traits.

What about the regions of necrosis? These are areas generally devoid of cancer cells due to a lack of vasculature. Because of extreme hypoxia and nutrient deprivation, these regions are seen as places where cancer cells die and fail to proliferate, much akin to the Dead Sea. Yet, like the Dead Sea, could there be "extremophiles" here? That is, rare cancer cell types that occur at very low abundance and can tolerate physical harshness while enjoying less intense biotic interactions? It remains an open question whether the necrotic regions have low abundances of extremophile cancer cells and whether these cells are clinically relevant for therapy and the patient.

Metastases further contribute to cancer cell biodiversity. Cancer cells that have transited from the primary tumor to a new site in a different tissue experience different challenges and opportunities. These drive additional adaptations and diversification of cancer cell types. For instance, breast or colon cancer cells metastasizing to the liver may evolve traits that resemble those of primary liver cancer cells (Cunningham et al. 2015). Like nature at large, the ecological and evolutionary dynamics of cancer within a patient exhibit a richness of cancer cell types at many scales—from the local environment around a blood vessel, across the heterogeneous tumor landscape, to the full body-scale involving metastases.

Evolutionary Therapies

There are three key approaches to developing novel therapies and therapeutics for cancer patients: discovery of new therapies, including novel drugs, immunotherapies, radiation therapies or surgical techniques; combination therapies—such as repurposing a drug effective in lung cancer for ovarian cancer, or testing moderately effective drugs in tandem; and evolutionary therapies—strategies that account for the ecological and evolutionary dynamics of cancer to improve outcomes.

For the first two approaches, drugs and procedures are often given at what is termed the **maximum tolerable dose**. The goal is to maximize the kill rate of cancer cells by exposing them to as high a dose as possible, without too much toxicity to the patient. Such an approach has appeal: since the rapid growth of cancer cells leads to declining health, metastases, and often death, eliminating as many cancer cells as possible seems sensible. If a cure is achievable, this strategy is ideal. But for late-stage cancers deemed incurable, the maximum tolerable dose poses a problem as it exerts maximal selection pressure, accelerating the evolution of therapy-resistant cancer cells. In certain cases, cancers evolve resistance to all known therapies. If a therapeutic approach does not kill all the cancer cells, resistance inevitably emerges, often leading to a more aggressive and malignant cancer than before. Evolutionary therapies, in contrast, take a different approach. Instead of solely maximizing cell kill, they use current drugs to achieve better outcomes by anticipating and steering the ecological and evolutionary dynamics of the patient's cancer.

Evolutionary therapies have been used in other fields of resource and pest management where humans induce evolution (Ashley et al. 2003). In commercial fisheries, for example, the use of specific net sizes and harvesting practices can mitigate the evolution of younger and smaller sizes at maturity in commercially valuable species (Heino 1998). Similarly, in agricultural pest management, integrated strategies include resistance management plans to mitigate the evolution of pesticide-resistant pests. The emergence of antibiotic-resistant strains of infectious bacteria (for instance, methicillin-resistant *Staphylococcus aureus* or MRSA) has led to programs to use antibiotics in a more restrained and focused manner. What distinguishes evolutionary therapies from other conventional therapies is their explicit incorporation of resistance management strategies (Cunningham 2021; Seyedi et al. 2024).

There is one striking difference when considering ecological and evolutionary dynamics in cancer treatment compared to other pests and diseases. In the case of antibiotic-resistant strains of bacteria, the primary concern is not that the bacteria will evolve resistance within a given individual or livestock patient. Rather, the concern is that repeated applications across millions—or even billions—of individual patients will result in the emergence of resistant bacteria. These resistant bacteria can then spread into the human or livestock population, rendering the drug ineffective for all. For cancer, the evolution of resistance is deeply personal. When cancer cells in a patient evolve resistance, it only renders therapy ineffective for just that patient, with no effects on other patients.

When anticipating and steering the ecological and evolutionary dynamics of a cancer, the relationship between the cancer and a physician's therapy can be thought of in terms of an evolutionary game. This game is asymmetric because the physician can be rational when applying therapy, whereas the evolving cancer can only adapt to current conditions; it cannot anticipate or evolve adaptations to therapies the physician might use in the future (Staňková et al. 2019).

This type of asymmetry creates "leader-follower" dynamics: the physician acts first, and then cancer cells respond and adapt to therapy. Such dynamics can be strategically exploited to improve cancer care.

A team of scientists led by Robert Gatenby first proposed and then successfully tested this type of evolutionary therapy in mice (Gatenby et al. 2009). Their **adaptive therapy** relies on mathematical models and measures of the past and present state of the individual patient's cancer to anticipate and steer the ecological and evolutionary dynamics of the disease by adjusting therapy timing accordingly. The first adaptive therapy was based on the principles of **competitive release**—a concept from ecology where removing a dominant competitor allows a subordinate to thrive. Similarly, Gatenby imagined a tumor composed of two cell types, one sensitive to the drug and the other resistant. However, this resistance may come at a cost. In the absence of therapy, the resistant cells may be outcompeted or contained by the sensitive cells. However, if one applies therapy at the maximum tolerable dose (MTD), eventually all the sensitive cells will be eliminated, allowing for the competitive release of the resistant cells. Consequently, the MTD therapy becomes ineffective, and the patient's disease burden recovers and progresses. The mouse experiments asked what would happen if therapy were stopped before all the sensitive cells were eliminated. Would the population of sensitive cells recover and competitively suppress the resistant cancer cells as they do so?

In 2014, Jingsong Zhang, a prostate oncologist and physician, initiated the first clinical trial of adaptive therapy (NCT024156210—the description of such trials, filed with the U.S. Federal Drug Administration, can be found using their trial number). This is the trial we have been following throughout this chapter. It is based on the ecological theory of competitive release and modelled using a three-species system of Lotka–Volterra equations.

In this trial, the men had three distinct types of cancer cells: one requiring testosterone (T+), one independent of testosterone (T-), and the third that produces testosterone on its own (TP). These three "species" react differently to prostate cancer drugs that include Lupron, which suppresses testosterone production, and Abiraterone, which targets androgen receptors needed by cells to utilize testosterone. Lupron suppresses the T+ cells but not the TP and T- cells. Abiraterone suppresses TP cells but not the T-cells. The men in the trial were on continuous treatment with Lupron while Abiraterone was toggled on and off based on the patient's PSA level—a blood biomarker of overall disease burden. The rationale was that continuous therapy with both Lupron and Abiraterone would eventually eliminate all T+ and TP cells, leaving the untreatable T- cancer cells, ultimately leading to the patient's death. To avoid this, Abiraterone was withdrawn when a patient's PSA dropped to 50% of his initial value upon entering the trial to ensure that some T+ and TP cells remained, and their subsequent increase during the drug break was intended to suppress the competitively inferior T- cells. Then, when his PSA level climbed back up to the original value, treatment with Abiraterone was resumed to reduce again the populations of T+ and TP cells. By cycling Abiraterone on and off based on PSA values, periods of treatment served to suppress the T+ and TP cells, and periods of no treatment allowed the T+ and TP cells to act as treatment against the T-cells (Figure 2).

Compared to the control group, which received continuous Lupron and Abiraterone, the adaptive therapy regimen was a success. Median overall survival increased from 31 to 53 months, and the men receiving adaptive therapy were only on Abiraterone about 50% of the time. This approach meant less drug use, lower monthly cost, higher quality of life, and more prolonged survival—a clear win. Yet, most patients still ultimately died from their disease. Zhang et al. (2022) provide a complete "after-action report" on this trial, including lessons learned. This trial has amplified interest in defining, understanding, and treating cancer using ecological and evolutionary principles. It has also led to multiple additional clinical trials using evolutionary therapies and the growing sophistication of the treatment strategies. Today, more physicians are incorporating evolutionary therapies into patient care.

The overarching goal of evolutionary therapies is to harness the principles of ecology and evolution, patient data, and mathematical modelling—to anticipate and steer the eco-evolutionary dynamics of cancer. The topics, concepts, and models covered throughout this book are not merely

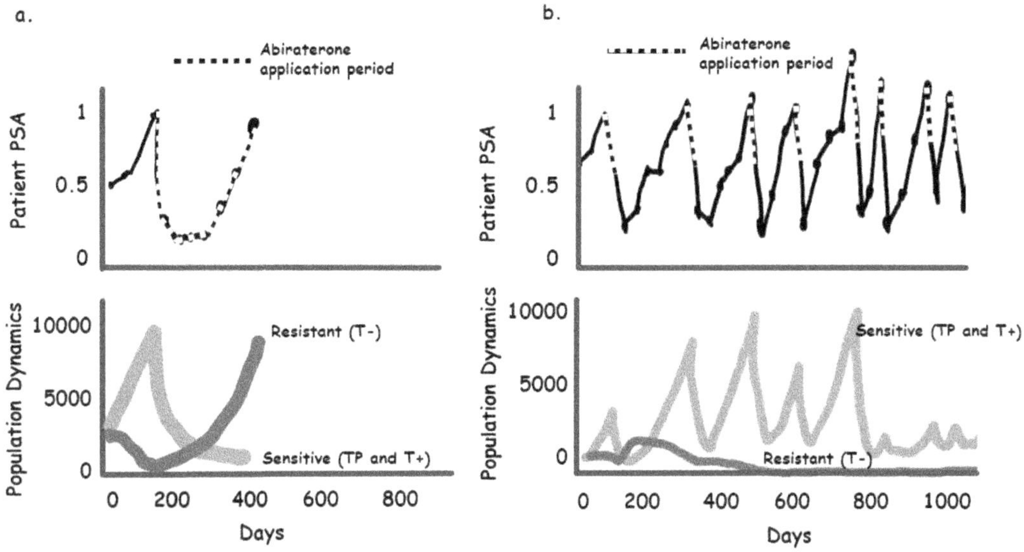

Figure 2. PSA (cancer biomarker) levels (top) and population dynamics of sensitive and resistant cancer cells (bottom) for (a) Abiraterone applied continuously and (b) Abiraterone applied adaptively (conceptualized from Zhang et al. [2022]).

academic; they are vital tools for anyone determined to control one of humanity's greatest challenges. This is more than science—it is a battle for lives, a race against time, and a bold step toward a future where cancer can be not only treated but also ultimately overcome.

References

Amaral-Zettler, L.A. 2013. Eukaryotic diversity at pH extremes. Frontiers in Microbiology 3: 441.

Archetti, M. and Pienta, K. J. 2019. Cooperation among cancer cells: applying game theory to cancer. Nature Reviews Cancer: 19: 110–117.

Ashley, M. V., Willson, M. F., Pergams, O. R., O'Dowd, D. J., Gende, S. M. and Brown, J. S. 2003. Evolutionarily enlightened management. Biological Conservation 111: 115–123.

Behrmann-Godel, J., Roch, S., Böhm, A., Jolles, J. W. and Brinker, A. 2024. Genetic differentiation and phenotypic plasticity drive troglomorphic character development in European cavefish. Evolution 78: 734–745.

Bhattacharya, R., Avdieiev, S., Bukkuri, A., Whalen, C. J., Gatenby, R. A., Tsai, K. Y. and Brown, J. S. 2025. The hallmarks of cancer as eco-evolutionary processes. Cancer Discovery 15: 685–701.

Brown, J. S., Amend, S. R., Austin, R. H., Gatenby, R. A., Hammarlund, E. U. and Pienta, K. J. 2023. Updating the definition of cancer. Molecular Cancer Research 21: 1142–1147.

Brown, J. S., Amend, S. R. and Pienta, K. J. 2024. Cancer and the evolutionary ecology of invasions. pp. 143–160. *In*: Somarelli, J. A. and Johnson, N. A. [eds.]. Cancer through the Lens of Evolution and Ecology. CRC Press.

Chen, B., Mao, T., Liu, Y., Dai, W., Li, X., Rajput, A. P. et al. 2022. Sensory evolution in a cavefish radiation: patterns of neuromast distribution and associated behaviour in *Sinocyclocheilus* (Cypriniformes: Cyprinidae). Proceedings of the Royal Society B 289: 20221641.

Connell, J. H. 1961. The influence of interspecific competition and other factors on the distribution of the barnacle *Chthamalus stellatus*. Ecology: 710–723.

Cunningham, J. J. 2021. Evolutionary game theory and optimal control for integrated metastatic management of prostate cancer. [Doctoral Thesis, Maastricht University]. Maastricht University. https://doi.org/10.26481/dis.20210927jc.

Cunningham, J. J., Brown, J. S., Vincent, T. L. and Gatenby, R. A. 2015. Divergent and convergent evolution in metastases suggest treatment strategies based on specific metastatic sites. Evolution, Medicine, and Public Health 2015: 76–87.

Dahan, M., Hequet, D., Bonneau, C., Paoletti, X. and Rouzier, R. 2021. Has tumor doubling time in breast cancer changed over the past 80 years? A systematic review. Cancer Medicine 10: 5203–5217.

Duesberg, P. and Rasnick, D. 2000. Aneuploidy, the somatic mutation that makes cancer a species of its own. Cell Motility and the Cytoskeleton 47: 81–107.

Egeblad, M., Nakasone, E. S. and Werb, Z. 2010. Tumors as organs: complex tissues that interface with the entire organism. Developmental Cell 18: 884–901.

Faguet, G. B. 2015. A brief history of cancer: age-old milestones underlying our current knowledge database. International Journal of Cancer 136: 2022–2036.

Fitzgerald, P. J. 2000. From Demons and Evil Spirits to Cancer Genes: The Development of Concepts Concerning the Causes of Cancer and Carcinogenesis. American Registry of Pathology, Washington, D.C.

Freischel, A. R., Damaghi, M., Cunningham, J. J., Ibrahim-Hashim, A., Gillies, R. J., Gatenby, R. A. et al. 2021. Frequency-dependent interactions determine outcome of competition between two breast cancer cell lines. Scientific Reports 11: 4908.

Gatenby, R. A., Silva, A. S., Gillies, R. J. and Frieden, B. R. 2009. Adaptive therapy. Cancer Research 69: 4894–4903.

Gillies, R. J., Brown, J. S., Anderson, A. R. and Gatenby, R. A. 2018. Eco-evolutionary causes and consequences of temporal changes in intratumoural blood flow. Nature Reviews Cancer 18: 576–585.

Hajdu, S. I. and Darvishian, F. 2010. Diagnosis and treatment of tumors by physicians in antiquity. Annals of Clinical & Laboratory Science 40: 386–390.

Hamdi, Y., Abdeljaoued-Tej, I., Zatchi, A. A., Abdelhak, S., Boubaker, S., Brown, J. S. et al. 2021. Cancer in Africa: the untold story. Frontiers in Oncology 11: 650117.

Hanahan, D. and Weinberg, R. A. 2000. The hallmarks of cancer. Cell 100: 57–70.

Hanahan D. and Weinberg, R. A. 2011. Hallmarks of cancer: the next generation. Cell 144: 646–74.

Hanson, H. E., Mathews, N. S., Hauber, M. E. and Martin, L. B. 2020. The house sparrow in the service of basic and applied biology. Elife 9: e52803.

Heino, M. 1998. Management of evolving fish stocks. Canadian Journal of Fisheries and Aquatic Sciences 55: 1971–1982.

Hersteinsson, P. and Macdonald, D. W. 1992. Interspecific competition and the geographical distribution of red and arctic foxes *Vulpes vulpes* and *Alopex lagopus*. Oikos 64: 505–515.

Hubbell, E., Clarke, C. A., Smedby, K. E., Adami, H., and Chang-E.T. 2024. Potential for cure by stage across the cancer spectrum in the United States. Cancer Epidemiology, Biomarkers & Prevention 33: 206-214.

Huxley, J. 1942. Evolution: The Modern Synthesis. Allen and Unwin, London, UK.

Huxley, J. 1956. Cancer biology: comparative and genetic. Biol Rev 31: 474–514.

James, B. M., Gillard, J. H., Antoun, N. M., Scott, I. S., Coleman, N., Pinto, E. M. et al. 2007. Glioblastoma doubling time and cellular proliferation markers. Clinical Oncology 19: S33.

Joanne, L.Y., Rak, J. W., Carmeliet, P., Nagy, A., Kerbel, R. S. and Coomber, B. L. 2001. Heterogeneous vascular dependence of tumor cell populations. The American Journal of Pathology 158: 1325–1334.

Kareva, I., Luddy, K. A., O'Farrelly, C., Gatenby, R. A. and Brown, J. S. 2021. Predator-prey in tumor-immune interactions: A wrong model or just an incomplete one? Frontiers in Immunology 12: 668221.

Kessler, A. and Baldwin, I. T. 2001. Defensive function of herbivore-induced plant volatile emissions in nature. Science 291: 2141–2144.

Kotler, B. P. and Brown, J. S. 2020. Cancer community ecology. Cancer Control, 27: 1073274820951776.

Laisné, M., Lupien, M. and Vallot, C. 2025. Epigenomic heterogeneity as a source of tumour evolution. Nature Reviews Cancer 25: 7–26.

Lloyd, M. C., Cunningham, J. J., Bui, M. M., Gillies, R. J., Brown, J. S. and Gatenby, R. A., 2016. Darwinian dynamics of intratumoral heterogeneity: not solely random mutations but also variable environmental selection forces. Cancer Research 76: 3136–3144.

Loeb, L.A. 2001. A mutator phenotype in cancer. Cancer Research 61: 3230–3239.

Lovari, S., Kachel, S., Xueyang, L. and Ferretti, F. 2024. Snow leopard, common leopard, and wolf: are they good neighbors? pp. 137–147. *In*: McCarthy, T., Mallon, D. and Schwartz, K. R. [eds.]. Snow Leopards. Academic Press.

Lucek, K., Lemoine, M. and Seehausen, O. 2014. Contemporary ecotypic divergence during a recent range expansion was facilitated by adaptive introgression. Journal of Evolutionary Biology 27: 2233–2248.

Nelson, A. S. and Mooney, K. A. 2022. The evolution and ecology of interactions between ants and honeydew-producing hemipteran insects. Annual Review of Ecology, Evolution, and Systematics 53: 379–402.

Nowell, P. C. 1976. The clonal evolution of tumor cell populations: acquired genetic lability permits stepwise selection of variant sublines and underlies tumor progression. Science 194: 23–28.

Peplinski, J., Malone, M. A., Fowler, K. J., Potratz, E. J., Pergams, A. G., Charmoy, K. L. et al. 2021. Ecology of fear: spines, armor and noxious chemicals deter predators in cancer and in nature. Frontiers in Ecology and Evolution 9: 682504.

Robertson-Tessi, M., Brown, J. S., Poole, M. I., Johnson, M., Marusyk, A., Gallaher, J. A. et al. 2023. Feasibility of an evolutionary tumor board for generating novel personalized therapeutic strategies. MedRxiv: 2023-01.

Salimi Sartakhti, J., Manshaei, M. H., Basanta, D. and Sadeghi, M. 2017. Evolutionary emergence of angiogenesis in avascular tumors using a spatial public goods game. PloS ONE 12: e0175063.
Sarapata, E. A. and de Pillis, L. G. 2014. A comparison and catalog of intrinsic tumor growth models. Bulletin of Mathematical Biology 76: 2010–2024.
Schmid, H. P., McNeal, J. E. and Stamey, T. A. 1993. Observations on the doubling time of prostate cancer. The use of serial prostate-specific antigen in patients with untreated disease as a measure of increasing cancer volume. Cancer 71: 2031–2040.
Seyedi, S., Harris, V. K., Kapsetaki, S. E., Narayanan, S., Saha, D., Compton, Z. et al. 2024. Resistance management for cancer: lessons from farmers. Cancer Research 84: 3715–3727.
Somarelli, J. A. 2021. The hallmarks of cancer as ecologically driven phenotypes. Frontiers in Ecology and Evolution 9: 661583.
Somarelli, J. A. and Johnson, N. A. 2024. A species within a species. pp. 1–4. *In*: Somarelli, J. A. and Johnson, N. A. [eds.]. Cancer through the Lens of Evolution and Ecology. CRC Press.
Stearns, S. C. and Medzhitov, R. 2024. Evolutionary medicine. Oxford University Press.
Stensjøen, A. L., Solheim, O., Kvistad, K. A., Håberg, A. K., Salvesen, Ø. and Berntsen, E. M. 2015. Growth dynamics of untreated glioblastomas *in vivo*. Neuro-oncology 17: 1402–1411.
Strobl, M. A., Krause, A. L., Damaghi, M., Gillies, R., Anderson, A. R. and Maini, P. K. 2020. Mix and match: phenotypic coexistence as a key facilitator of cancer invasion. Bulletin of Mathematical Biology 82: 1–26.
Swan, G. W. and Vincent, T. L. 1977. Optimal control analysis in the chemotherapy of IgG multiple myeloma. Bulletin of Mathematical Biology 39: 317–337.
Tanaka, K., Shimada, H., Miura, M., Fujii, Y., Yamaguchi, S., Endo, I. et al. 2004. Metastatic tumor doubling time: most important prehepatectomy predictor of survival and nonrecurrence of hepatic colorectal cancer metastasis. World Journal of Surgery 28: 263–270.
Vincent, M. D. 2010. The animal within: carcinogenesis and the clonal evolution of cancer cells are speciation events sensu stricto. Evolution 64: 1173–1183.
Woods, G. M., Fox, S., Flies, A. S., Tovar, C. D., Jones, M., Hamede, R., Pemberton, D. et al. 2018. Two decades of the impact of Tasmanian devil facial tumor disease. Integrative and Comparative Biology 58: 1043–1054.
Zhang, J., Cunningham, J. J., Brown, J. S. and Gatenby, R. A. 2017. Integrating evolutionary dynamics into treatment of metastatic castrate-resistant prostate cancer. Nature Communications 8: 1816.
Zhang, J., Cunningham, J., Brown, J. and Gatenby, R. 2022. Evolution-based mathematical models significantly prolong response to abiraterone in metastatic castrate-resistant prostate cancer and identify strategies to further improve outcomes. Elife 11: e76284.

Index

For Product Safety Concerns and Information please contact our EU
representative GPSR@taylorandfrancis.com
Taylor & Francis Verlag GmbH, Kaufingerstraße 24, 80331 München, Germany

www.ingramcontent.com/pod-product-compliance
Ingram Content Group UK Ltd.
Pitfield, Milton Keynes, MK11 3LW, UK
UKHW052219190926
475702UK00006B/1107
* 9 7 8 0 3 6 7 4 3 7 6 3 3 *